Aquaculture of grey mullets

THE INTERNATIONAL BIOLOGICAL PROGRAMME

The International Biological Programme was established by the International Council of Scientific Unions in 1964 as a counterpart of the International Geophysical Year. The subject of the IBP was defined as 'The Biological Basis of Productivity and Human Welfare', and the reason for its establishment was recognition that the rapidly increasing human population called for a better understanding of the environment as a basis for the rational management of natural resources. This could be achieved only on the basis of scientific knowledge, which in many fields of biology and in many parts of the world was felt to be inadequate. At the same time it was recognised that human activities were creating rapid and comprehensive changes in the environment. Thus, in terms of human welfare, the reason for the IBP lay in its promotion of basic knowledge relevant to the needs of man.

The IBP provided the first occasion on which biologists throughout the world were challenged to work together for a common cause. It involved an integrated and concerted examination of a wide range of problems. The Programme was co-ordinated through a series of seven sections representing the major subject areas of research. Four of these sections were concerned with the study of biological productivity on land, in freshwater, and in the seas, together with the processes of photosynthesis and nitrogen fixation. Three sections were concerned with adaptability of human populations, conservation of ecosystems and the use of biological resources.

After a decade of work, the Programme terminated in June 1974 and this series of volumes brings together, in the form of syntheses, the results of national and international activities.

INTERNATIONAL BIOLOGICAL PROGRAMME 26

Aquaculture of grey mullets

EDITED BY
O. H. Oren
Israel Oceanographic and Limnological Research

CAMBRIDGE UNIVERSITY PRESS
CAMBRIDGE
LONDON NEW YORK NEW ROCHELLE
MELBOURNE SYDNEY

Published by the Press Syndicate of the University of Cambridge
The Pitt Building, Trumpington Street, Cambridge CB2 1RP
32 East 57th Street, New York, NY 10022, USA
296 Beaconsfield Parade, Middle Park, Melbourne 3206, Australia

First published 1981

Printed in Great Britain at the
University Press, Cambridge

British Library Cataloguing in Publication Data

Aquaculture of grey mullets.
1. Gray mullets 2. Fish-culture
I. Oren, O. H. II. International Biological Programme
639′.375′8 SH351.G/ 80-40305

ISBN 0 521 22926 X

Contents

Table des matieres

Содержание

Contenido

Contributors

Editor

O. H. Oren, Israel Oceanographic and Limnological Research, PO Box 8030, Haifa, Israel

Authors

M. Ben-Yami, Fisheries Technology Unit, Haifa, Israel

J. Brusle, Laboratoire de Biologie Marine, Centre Universitaire de Perpignan, Avenue de Villeneuve, 66025 Perpignan, Cedex, France

H. Farrugio, Université des Sciences et Techniques du Languedoc, Laboratoire d'Ichthyologie et de Parasitologie Générale, Place Eugène Bataillon, F-34060, Montpellier, Cedex, France

E. Grofit, Fisheries Technology Unit, PO Box 699, Haifa, Israel

R. M. Koningsberger, Oceanic Foundation, Hawaii

M. N. Kutty, Fisheries College, Tamil Nadu Agricultural University, Tuticorin-628003, Tamil Nadu, India

I.-C. Liao, Tungkang Marine Laboratory, Tungkang, Pingtung, Taiwan 916, Republic of China

C. E. Nash, KCM International Inc., 1917 First Avenue, Seattle, Washington 98101, USA

R. M. Overstreet, Gulf Coast Research Laboratory, Ocean Springs, Mississippi 39564, USA

I. Paperna, The Heinz Steinitz Marine Biology Laboratory, The Hebrew University of Jerusalem, Elat, Israel

J. P. Quinard, Laboratoire de Biologie Marine, Centre Universitaire de Perpignan, Avenue de Villeneuve, 660255 Perpignan, Cedex, France

S. Sarig, Laboratory for Research of Fish Diseases, Nir-David, Israel

J. M. Thomson, University of Queensland, St Lucia, Brisbane, Queensland, Australia

L. Zismann, Israel Oceanographic and Limnological Research, PO Box 8030 Haifa, Israel

Foreword

The present volume represents another milestone on the road towards the endeavours which the International Biological Programme (IBP) tried to achieve when it formulated its objectives: to carry out integrated studies on the biological basis of productivity and human welfare, throughout the world.

Although the IBP has ceased to exist and while the activities under its flag are fading out, the final products of its existence, for less than a decade, are being shaped into their final forms: the 'Synthesis Volumes', of which the present book is one.

These 'Synthesis Volumes' are starting to gather on the bookshelves of many laboratories and offices at a moment when the whole civilized world becomes conscious, concerned and worried about the prospects of adequately feeding the ever-increasing human population inhabiting our planet.

One should look upon the present volume '*Aquaculture of grey mullets*' from at least two points of view: one, restricted to the narrow and direct meaning of its title: aquaculture of mullets, when the term 'mullets' refers mainly to *Mugil cephalus*. The other point of view is a much broader one: it is that of aquaculture of all the fish species, invertebrates, algae and planktonic organisms inhabiting brackish and sea waters of the world, suitable for cultivation under controlled conditions for the ultimate benefit of man as food or raw materials.

This volume can thus be looked upon as a model-review of the knowledge needed, methods used, problems encountered and actions undertaken in order to be able to determine intelligently the suitability of a given species of fish or invertebrate for rational cultivation. The selection of such a species for cultivation would be based on market research, adequate life-history studies, including its suitability for artificial reproduction, fast growing rates, rearing under mono and/or polyculture conditions, resistance to diseases and parasites, etc.

In spite of the vast body of knowledge accumulated over the years regarding the rearing of mullets, as presented in this book, there is still no guarantee that the interested reader will find all the answers to specific problems of a local or even a general nature. This book should rather provide a, hopefully useful, tool to newcomers as well as to oldtimers in this field of work and serve administrators and investors to support research for the improvement of the means of production of these important food organisms.

When dealing with studies of a single species of mullet, *Mugil cephalus*, the direct application of the knowledge acquired in one place can be relatively easy to apply and adapt to another locality, because we deal here with the same cosmopolitan species consisting of separated populations. This is not the case with the other tens of existing mullet species. Of these, the majority are probably not suitable, for various reasons, for aquaculture, while some may be cultivated under certain circumstances. It is for the local scientist and farmer to check and ascribe the local species to one category or the other.

The extensive number of references included in this volume shows the degree of interest and of the real need to 'synthesize' all the knowledge scattered in several thousands of papers in many dozens of international and local scientific journals and reports, some of them not generally available, into one single volume. This is what we try to present in this volume and this is what the founders and supporters of IBP, in fact, wanted among other aims, to achieve.

A natural continuation of the 'Mullet Project' should have been studies on the brackish waters of the extremely large number of lagoons, marshes and other peripheral waters the world over (but largely in areas where food is needed), at present used neither for aquaculture or agriculture nor for industry or other human needs. In the southern Mediterranean alone about 500000 ha of such waters have been listed (UNESCO, 1979). There is no doubt that such a project, together with the one described in this book, could be able to add hundreds of thousands of tons of protein, if not millions of tons to the needing world. Hopefully someone, somewhere, will initiate such a project in the near future and be successful.

Aquaculture has become a very popular theme in the last several years, maybe through the interest raised by the IBP in the subject, and much more has been done, written and said about it than ten years ago. It is true that international organizations dealing with the food problem of the world have been considering aquaculture for a long time as one of the unexploited sources for proteins production, although practical results are still rather poor.

The reasons for this are given in several very recent papers (Naegel, 1979; Schmittou, 1979; Tal, 1979) all of them claiming that all the international funding and supporting agencies and governments supported the administrative and scientific aspects of the problem, and in the practical sense they have supported mainly small scale fish farming enterprises. Such small scale farms, with all the assistance they have obtained and the fish they have produced, were able to produce only so much as to solve their own food problem, and, maybe the food problem of their immediate community. Aquaculture can solve the food problems of the world only if it is done on a very large scale with huge investments and infrastructure (land, energy, installations, water, etc.) supported financially by governments, international funding agencies and banks, as well as by large private investors. If such large-scale investments are made in time and all the other necessary administrative and scientific support given, Naegel (1979) estimates that in the year 2000, aquaculture would be able to increase tenfold and to supply the world with a quantity of cultivated fish similar to the quantity that the sea fisheries of the world supply today. This would mean about 70 million tons of cultivated fish in comparison with about 6 million today.

Again, if this volume will help to contribute toward this goal (as mullets will have a great share in the future tonnage of the aquaculture production), then the IBP should be praised.

At the end of this Foreword, I should like to add a personal note. I greatly enjoyed working for the IBP. This gave me profound satisfaction, inspiration, and generated friends and friendship among knowledgeable people motivated by the same goals. I will always look on my years of association with the International Biological Programme as memorable ones, as I am sure other colleagues similarly involved with IBP activities will do. It was a truly unique experience.

O. H. OREN

References

Naegel, L. (1979). Aquaculture in the Third World. *Animal Res. Devpt.*, **9**, 77–115.
Schmittou, R. (1979). Fish farming in the 1980s. The *Commerc. Fishfarmer* **5**(3), 18–21.
Tal, S. (1979). The concept of aquaculture. *Bamidgeh* **31**(2), 23–5.
UNESCO (1979). Coastal ecosystems of the southern Mediterranean: lagoons, deltas and salt marshes. *UNESCO reports in Marine Science*, no. 7 Paris.

1. The taxonomy of grey mullets

J. M. THOMSON

Introduction

The classification of organisms or the ordering of them into recognizable categories on the basis of their degrees of similarity is a basic activity of biologists. No biochemical, physiological, ethological or ecological study has significance unless the species of organisms involved have been clearly identified, for otherwise the results are not reproducible nor can they be tested by other scientists. For the aquaculturist there is a very practical problem involved in sorting out the young fish which are to be released into his ponds. The need to distinguish the species of mullet when they are small is important for the aquaculturist wishes to select faster growing species rather than slower growing, and to choose species tolerant of farm pond conditions rather than those of limited tolerance.

Classical taxonomy has relied heavily on morphological characters, but in recent years various other characteristics have been used. Biochemical and physiological applications to taxonomy date back for more than a century but they have had a significant degree of acceptance only in recent times. The taxonomy of the majority of groups has been based on the characteristics of adult or advanced juvenile specimens. In particular, the young stages of marine animals have received relatively little taxonomic attention (see Chap. 2).

These general statements are true of Mugilidae. Within the family even the classical morphological concepts have been applied for the most part to regional groupings rather than to the family as a whole and until recently very little attention had been paid to the characteristics of the larval, post-larval and juvenile forms. The use of biochemical and physiological tools in the taxonomy of the family is in its infancy.

Consequently this review of the state of knowledge of the taxonomy of Mugilidae is characterized by the incompleteness of the data, especially with respect to characters that may be used to distinguish the eggs and young stages and with respect to the modern biochemical and physiological approaches. However, the relatively few and regionally restricted studies that have been made may indicate fruitful leads for others to follow.

Suprageneric taxonomy

The Mugilidae are actinopterygian fishes and it has been part of the conventional ichthyological wisdom that the family is closely related to the

Atherinidae (Jordan, 1907; Berg, 1947). Berg separated the Mugilidae, with the Sphyraenidae and the Atherinidae from the Perciformes as the order Mugilformes because they have abdominal pelvic fins, a relatively primitive feature. But doubt has been thrown on the validity of this grouping by the studies of Rosen (1964) who has been followed, more recently, by Greenwood *et al.* (1966) in placing the Atherinidae in a separate order, the Atheriniformes, on the basis of a considerable number of morphological features. Rosen (1964) was not concerned to consider in detail the position of the Mugilidae but he indicated that their nearest relatives were to be found among the Sphyraenidae and Polynemidae and he assigned all three families to the order Perciformes.

Support for the separation of the Mugilidae from the order Perciformes is provided by the studies of blood plasma by Sulya, Box & Gunter (1960) and perhaps more strongly by the study of plasma proteins by Gunter, Sulya & Box (1961) who showed that the plasma proteins of mugilids are less complex than those of any other family generally assigned to the Perciformes and indeed are more similar to those of some species of Cypriniformes and Clupeiformes. On the other hand the plasma proteins of particular species in the three perciform families the Carangidae, the Sciaenidae and the Scombridae are not markedly different from those of the Mugilidae. On this evidence the Mugilidae could be regarded either as amongst the most primitive of Perciformes or as a branch from some ancestral form of Perciformes.

Jordan & Evermann (1896) recognized two subfamilies of mullet, the Mugilinae and the Agonostominae. The latter have sessile teeth which attach directly to the jaws, a flat preorbital and only two anal spines in the adult. The Mugilinae have labial teeth (if any) connected to the jaws by elongated fibres, a ridged and grooved preorbital and three anal spines in the adult, though the early juvenile form has only two anal spines. Aquacultural attention has been paid almost wholly to members of the Mugilinae. While the Mugilinae are almost worldwide except for the polar regions, the Agonostominae are confined to central America, the western Indian Ocean, the tropical west Pacific, and the coastline of Australia.

Generic and specific taxonomy

At the generic and specific levels the taxonomy of the Mugilidae remains confused. Those species with extensive ranges are almost certainly known by different names in various parts of their distribution. A revision of the family, with careful reference to type material, is badly needed. The difficulties of revision in this anatomically uniform family are exacerbated by the fact that some type material has been lost and that no types were kept for some of the earliest species described. A revision in manuscript form by Thomson recognizes fourteen genera and sixty-four species amongst the nominate species of the family which total 282.

It is not the purpose of this review to anticipate a revision of the family but to indicate recent morphological, physiological and biochemical studies which contribute to the distinction of species, and in particular those which refer to the young stages in the life history. The fact that morphological material dominates the discussion reflects the fact that physiological and biochemical studies on mullet are few.

The Maxilla

The main evolutionary trends in the external anatomy of the Mugilidae are to be found in the mouth parts as Schultz (1946) first pointed out. In his generic distinctions, he made much of the form of the maxilla and premaxilla. One consequence of the increasing curvature of these bones had been recognized by earlier taxonomists in the use of their criterion of 'end of maxilla exposed' or 'end of maxilla not exposed'.

Schultz (1946) did not comment upon the functional significance of the evolving form of the maxilla which is the consequence of a shift down the shaft of the maxilla of the point of attachment of the tendon from the adductor muscle (Thomson, 1974). This tendon passes inside the line of the maxilla to attach by a flange on the anterior face of the maxilla. In order to maintain a straight line of action as the point of attachment moves lower on the shaft the bone itself has been forced to adopt the outward curvature which characterizes the more highly evolved genera. This alteration in the level of attachment of the tendon is associated with an increasing protrusibility of the mouth and possibly also with unrecognized changes in feeding habit and diet. The curvature of the pre-maxilla is forced upon it to maintain its terminal connection with the maxilla.

Associated with the outward curve of the upper jaw bones in the more advanced genera there is also a downward curve which brings the end of the maxilla below the mouth corner, whereas in the more primitive genera the end of the maxilla lies on the line of the mouth gape, as it does also in the post-larval specimens of all Mugilidae, the curvature, where it develops, appearing during growth in the early juvenile or querimana stage.

What has been described as 'the end of the maxilla exposed' is not the bone itself but a pad of tissue on the end of the bone which may be sunken well below the surface of the face. The pad of tissue is exposed due to the distortion of tissue forced by the curvature of the shaft of the maxilla. A close examination will show that in some genera the exposed pad is double, the larger posterior section being the pad over the end of the maxilla, the narrower anterior portion lying over the tendon running to the corner of the mouth. This double pad is typical of the genus *Liza* and its derivatives but has led to the inclusion within that genus of species of the genus *Valamugil* whose exposed pad at the corner of the mouth is homologous only with the anterior portion of the double pad of *Liza* .The species of *Valamugil* are distinctive in other ways, having a long pointed pectoral axillary scale which is absent, or small and distorted, in *Liza* and also having distinctive scales with flexible membranous edges (Thomson, 1974).

Besides the varying position relative to the corner of the mouth, the end of the jaws and the corner of the mouth itself also vary in position relative to both vertical and longitudinal axes, reaching back under the eye in some genera but ending well in front in others, and being variously below the eye and at various levels up to about mid-eye level in the vertical.

The preorbital

The preorbital varies in several characteristics which have diagnostic value. The fact that the preorbitals of members of the Agonostominae are flat has been mentioned. They are also massive and have a very broad edge between the eye and the mouth corner. In Mugilinae the surface is grooved with a posterior ridge, and there is also a tendency in the genera for the edge between eye and mouth to become shorter, ultimately being almost pointed. Associated with the curvature of the maxilla is a shortening of the mouth gape, with a subsequent rise of the mouth corner in front of the eye. To accommodate this the preorbital, which is straight-edged in the more primitive genera, becomes notched. The notch is low on the front edge of the preorbital in the genus *Valamugil*, about a third of the way up in *Liza* and at least half-way up in *Oedalechilus*. In this last genus the posterior margin of the notch reaches behind the vertical from the lower corner of the preorbital. In at least some species of *Liza* and *Valamugil* the notch tends to fill in aged fish leading to a curious sequence of events, the anterior edge of the preorbital being straight in the post-larval stages, becoming notched in the querimana and young adult stages and then becoming almost or quite straight again in the large fish. This changeable feature has led to some confusion in taxonomy.

The number and arrangement of the denticulations on the lower and front edges of the preorbital varies both between genera and between species within the one genus. In most species there is only a single row of denticulations along the edge of the preorbital but in species of *Liza* and *Valamugil* there are often a few denticulations situated above the denticulation which is posterior-most on the lower edge. In most genera the denticulations are continuous along the two free edges, but in species of *Valamugil* and *Liza* there may be only a few scattered denticulations between the anterior corner of the lower edge and the notch, though above the notch the denticulations are normally side by side up to a point about two-thirds along the edge.

The Lips

The lips of the primitive Agonostominae are thick. The majority of Mugilinae have thin lips; those that do not seem to be derived from thin-lipped forms and are therefore secondarily thick-lipped. This is confirmed by their dentition, which is typical of Mugilinae, not of Agonostominae. Typically the lower lip of Mugilinae is wafer-thin and ends in a sharp edge. *Mugil thoburni* has unusual lips which were characterized by Schultz (1946) as thick. They are perhaps better spoken of as wide. The original specimens had the lower lip folded down and these two characteristics were used by Schultz (1946) to erect a new genus *Xenomugil*. But only some of the specimens in the British Museum and the Museum d'Histoire Naturelle de Paris have down-turned lips. As others have normally oriented lips this character may well be the result of muscular contraction at death. Apart from the unusual width of the lips, this species is typical of the genus *Mugil* and may well be retained there.

Various species of mullet have ornamentations on the lip. These take the form of either papillae or ridges. All these originate in the juvenile fish as small swellings of the epidermis which only gradually acquire the characteristics which distinguish one ornamented species from another. Trewavas and Ingham (1972) have suggested that these various species may form an evolutionary series within a single genus. But a comparison of other characters, including scales and the maxillae suggests that there are several groups of species with ornamented lips which have developed independently from several genera. The species usually attributed to *Crenimugil* are very similar to *Valamugil* spp. while the papillate *Oedalechilus* and *Chelon* appear to represent two separate lines of evolution from *Liza*.

In *Crenimugil* the papillae, when fully developed, are flask-shaped with a terminal splaying whose pattern varies with the species. In *Chelon* adjacent papillae tend to fuse to form horn-tipped transverse ridges. In *Oedalechilus* the upper lip is ornamented by a row of close-packed horny-tipped protrusions whose form differs in the species. The lower lip is also sheathed in horn but it has the form of a scalloped edging to the lip.

Nostrils

In the post-larval mullet the posterior nostril is high up over the front half of the eye, with the anterior nostril close in front. As the snout lengthens, the nostrils become displaced forwards. The degree of movement and the relative positions of the anterior and posterior nostrils in the post-querimana stage provide taxonomic criteria. In most species the posterior nostril moves to a position where its upper rim barely reaches the level of the upper rim of the eye as viewed laterally, but in some it remains above eye level. The anterior nostril, as viewed laterally, may be entirely within the vertical span of the posterior nostril or reaching partly below it. In only a few species is the anterior nostril entirely below the span of the posterior nostril. In the genus *Rhinomugil*, the nostrils move well down the face, presumably to keep the nostrils immersed while the fish are using their aerial vision. The anterior nostril even reaches below the level of the lower rim of the eye in this genus.

Scales

Species of mullet from fresh water tend to have ctenoid scales whereas scales from mullet inhabiting marine waters tend to be only weakly ctenoid or cycloid. But within a genus both types may occur (e.g., *Liza*). Most cycloid scales have a firm edge but the genera *Crenimugil* and *Valamugil* have pliable membranous edges with digitations ornamenting the posterior margin. These two genera are similar to *Mugil* in having an elongate pointed scale in the axil of the pectoral fin. The other genera lack the pectoral axillary scale or have only a small and distorted scale in its place. A secondary squamation of small cycloid scales is developed in the genus *Mugil*, but is only poorly developed in other genera.

Each species has only a small range of numbers of scales laterally between the axis of the pectoral fin and the base of the tail, and an only slightly variable number of transverse scale rows between the origin of the first dorsal fin and the origin of the ventral (pelvic) fin. But there are many species with overlapping counts which makes this feature difficult to use alone in identifying mugilid species.

A proportion of scales on mugilid fish carry a mucous canal which in some species, particularly the genus *Liza*, tend to bifurcate or to be duplicated. In the species *Liza saliens* and *L. dumerili* each scale may have up to ten mucous canals.

The teeth

Mugilid teeth are usually described as being ciliate or setiform. But at even small magnifications distinct differences at both generic and specific levels become apparent. The pattern of dentition also varies and has been used in systematic studies by Schultz (1946), Thomson (1954) and by Ebeling (1957).

The two subfamilies of grey mullet are distinguished by a major difference in dentition. In the Agonostominae the teeth are attached directly to the jaw bones and are generally canine-like. The Mugilinae have labial teeth, a condition described by Troschel (1865). The sessile teeth of the Agonostominae emerge from the lip well inside the mouth: the labial teeth of Mugilinae line the edge of the lips and attach indirectly to the jaw bones by intervening fibres whose length in some species is several times the length of the shaft of the tooth.

Only a single row of labial teeth is developed in many species, but in others there may be several rows of which the outermost usually consists of teeth larger than those of the inner rows. In most species the inner teeth are similar in form to those of the outer row, but in some species they differ. Several species of *Mugil* for example have simple teeth in the outer row and bifid teeth in the inner rows. There is a tendency for teeth to be lost with age, and aged toothless specimens are known. The species with lip ornamentation either have a reduced dentition or are toothless.

Teeth are not confined to the jaws, but may also be found on the vomer, the palatine, the pterygoids and the tongue. The more primitive genera have teeth on all these bones, but more advanced genera lack one or more sets.

Despite their minute size the teeth of mullet are structurally identical with those of other teleosts. In some species the tips are golden-yellow contrasting with the white shafts. The outer row of teeth (called the primary row by some authorities) are canine-like in the Agonostominae and the teeth of the inner rows may be similar, but in a number of species the teeth of the inner rows (often distinguished as the secondary teeth) are laterally extended with a linear row of multi-cusped crowns. Bicuspid and tricuspid teeth are most common, but up to seven cusps are found.

Multi-cusped teeth are not common in Mugilinae, though bifid and trifid teeth are not unknown, usually in the inner secondary rows. *Liza tricuspidens* is exceptional in having all its teeth tricuspid. Otherwise it is so similar to other species of *Liza* there seems little justification for placing it in a monospecific genus *Heteromugil* as proposed by Schultz (1946).

The mullet from the Galapagos Islands frequently lack unicuspid teeth in the outer row. On this criterion Ebeling (1961) proposed to separate the species *Mugil cephalus* as *M. galapagensis*. However, some specimens from the Galapagos Islands in the British Museum of Natural History have solely unicuspid teeth in the outer row, while others have a mixture. It is quite possible that the harsher feeding conditions at the Galapagos causes early loss of the outer primary unicuspid teeth.

There is great variation in shape in mugiline teeth. In some species the teeth are canine-like, but usually more elongate and slender than in Agonostominae. In the *Liza* species the teeth become finer, almost straight and reduced in size. In other genera they become flattened antero–posteriorly, usually curving inwards towards the tips which are visually demarcated from the shaft by a distinct suture. In some species the tooth is strangulated at this point by marked lateral incisions.

The species of the genus *Chaenomugil* have unusually long gleaming white shafts whose diameter is about that of other mullets. They have an extreme, practically right-angled flexure at the tips which are either bifurcated or trifurcated. To appreciate the form of the teeth in mugilids it is necessary to view them both laterally and anteriorly.

The lingual teeth of Agonostominae are canine-like and are scattered over the whole surface of the tongue. Lingual teeth persist in the more primitive genera of the Mugilinae including Indo-Pacific species of *Liza* which otherwise seem more highly evolved than other mugiline genera which have only small patches of teeth or none. In Mugilinae the lingual teeth occur in patches which share a common basal plate separated by soft tissue from other similar patches on basal plates. A common pattern is a series of small tooth-bearing plates along the antero–lateral edges of the tongue and a median central patch well back, opposite the teeth on the pteryoid.

Otoliths

Despite evidence from Morović (1953) and Erman (1960) that the Mediterranean species can be distinguished on the form of their otoliths, little use has been made of this character elsewhere. The otoliths of mullet are rather convoluted and require some degree of preparation for use in age studies. However, their form does seem to exhibit constant differences between species, though this has yet to be demonstrated for any great number of species.

The adipose eyelid

At one time the presence or absence of the adipose eyelid was regarded as a good generic distinction in mullets, particularly to distinguish *Mugil* from *Liza*. But there are a number of species which otherwise display characteristics of *Liza* which have a well-developed adipose eyelid. The occurrence of adipose eyelids of varying width within different genera suggests that the genera diverged after the development of the adipose tissue in the mugiline subfamily with subsequent independent trends to obsolescence of this characteristic. Although called an adipose eyelid, the tissue does not intrude on the eye in many species, particularly in many *Liza* spp. for example, but lies on the face in two triangular patches, one in front of the eye intruding to greater or less degree over the preorbital and the other lying behind the eye, but never reaching far back on the head, although its complete absence serves to distinguish some genera from others. The degree of intrusion over the eye has little generic significance, though it can be useful as an interspecific criterion, provided the fact that the coverage increases with age is recognized.

The pyloric caeca

The primitive number of pyloric caeca in mullet is two. This number is typical of the Agonostominae and the more primitive species of Mugilinae, including those of the genus *Mugil*. Other mugiline genera have varying numbers of caecae. The range within a species is limited, often being invariably the same number. In my own counts invariability has been the rule, but Daget & Iltes (1965) have claimed a considerable range for *Liza falcipinnis* and *L. grandiquamis* from West Africa.

Günther (1861) listed all the observations on pyloric caeca known to him, but obviously many species had been described without any observation on the caeca, and indeed many species were only known from skins. Generally pyloric caeca are simple unbranching sacs, but Luther (1975) has recorded that *Valamugil buchanani* has pyloric caeca which branch distally. The fact that different counts of pyloric caeca have been given for the same species, probably results from misidentification rather than reflecting a real variation within a species. For example Lagler, Bardach & Miller (1962) figure a stomach, alleged to be that of *Mugil cephalus*, featuring eight pyloric caeca. For the same species Rahimulla (1945) gave a count of five. I have examined undoubted *Mugil cephalus* from all parts of the world and the invariable count of pyloric caeca has been two.

Hotta and Tung (1966) in Japan and Luther (1975) in India have used the pyloric caeca as major factors in the identification of their local mullet species.

The vertebral column

The paraphysial stay is a bony bridge across the parapophyses of the trunk vertebrae. Luther (1975) studied the vertebrae of eight species of Indian mullets and found the three species of *Liza* lacked paraphysial stays. A fourth species often allocated to the same genus *Ellochelon vaigiensis* was found to possess a stay on the last trunk vertebra, a feature shared with the species of *Mugil* and *Valamugil* studies by Luther. Some individual intraspecific variation was noted by Luther who found that most of the species of *Mugil* and of *Valamugil* also had stays on the tenth vertebra. One specimen (out of 200) of *M. cephalus* also had a stay on the ninth vertebra and another had a stay only on the eleventh. Also one specimen (out of thirty-three) of *E. vaigiensis* had a stay on the tenth vertebra.

Luther (1975) also found that all the species of *Liza* examined by him had paired vertebral processes which projected postero-dorsally from the second vertebrae, but these were absent from *Mugil* spp., *Valamugil* spp. and *Ellochelon vaigiensis*. Paired short posterior zygophysial hooks occur on the second and third vertebrae of the Indian species of *Mugil* but on the second vertebrae only of *Valamugil* and *Ellochelon*. The complete absence of these hooks from the vertebrae of the species known in India as *Mugil cunnesius* has prompted Luther (1975) to place the species in a new genus *Osteomugil*.

Basioccipital processes are absent in *Mugil*, *Ellochelon* and *Liza*, but are present in *Valamugil* and *Osteomugil* (Luther, 1975).

The place of insertion of the first interneural spine (usually somewhere between the sixth and ninth vertebrae, combined with the number of pyloric caeca were used by Hotta & Tung (1966) to provide a key for identification of Japanese mugilids.

Biochemical factors

The contribution of biochemistry to taxonomy is a relatively new development. Its application is in its infancy and has probably been used more in plant taxonomy than in animal taxonomy (Alston & Turner, 1963; Swan, 1963). The majority of published studies tend to deal with the relationships of supra-specific groups. The use of biochemical data for interspecific discrimination is comparatively new.

Serological studies

Serum proteins have been studied in a number of groups including tuna (Sprague, 1967; Fujino & Kang, 1968*a*, *b*), sardines (Sprague & Vrooman, 1962), other Clupeoids (Sinderman & Honey, 1963) and eels (Sick, Westergaard & Frydenberg, 1962).

Sulya, Box & Gunter (1961) included both *Mugil cephalus* and *M. curema* in their study of the plasma proteins of fishes from the Gulf of Mexico. The electrophoretic patterns of the two species were very different. In the Black Sea Senkevich & Kulikova (1970) also found differences in the serum proteins from *M. cephalus. Liza saliens* and *L. aurata.* These authors also found some degree of change with age or maturity as well as individual differences in composition. Polymorphism within a population of *M. cephalus* was found by Hongskul (1968). He detected ten protein components by electrophoretic analysis, but only one band, identified by him as component III displayed genetic variability. An analysis by the Hardy-Weinberg technique indicated that the three phenotypes derived from a system of two alelles. The polymorphism was detected only in the population from the eastern shore of Australia. It was not discerned in the admittedly small sample from Western Australia which is believed to be reproductively isolated from the eastern stocks (Thomson, 1955).

Muscle protein

In studies on the electrophoresis of water soluble muscle proteins (myogen) Herzberg & Pasteur (1975) were able to distinguish five species of mugilid from the eastern Mediterranean. These studies also indicated a change of pattern with age. As juveniles all the species had a particular fast moving band in the pattern which decreased in intensity with growth and disappeared entirely in adults of *Mugil cephalus* and *Liza aurata.* Herzberg & Pasteur give a key to the identification of the mullet of the eastern Mediterranean based on the nature of their electrophoretic patterns.

Eye-lens protein

The electrophoretic patterns of eye-lens proteins have been used by Peterson & Shehadeh (1971) to distinguish genetically isolated mullet populations in the waters of the Hawaiian Islands. The fact that differences in proteins have been found intraspecifically as well as interspecifically, and also at higher levels indicates that the whole problem of biochemical differences is complex.

Physiological phenomena

Use of physiological phenomena is rarely attempted in taxonomic work, yet physiological performance is genetically controlled. The precision of use of physiological criteria is diminished by the phenomenon of acclimation.

Salinity tolerance

The variation in tolerance of low salinity between species of mullet was noted in the farm ponds of the Po River delta many years ago (DeToni, Bullo & Paoletti, 1892). Their results as quoted by Brunelli (1916) indicated that *M. cephalus* and *L. ramada* are the most tolerant to freshwater which agrees with the more recent observations of Bograd (1961). Brunelli (1916) does not indicate the size of the mullet whose tolerance limits are shown in his table which indicates that *Liza saliens* is more tolerant of lowered salinity than *L. aurata* which does not accord with the indications from distribution in coastal streams obtained by Bograd (1961).

The ability to acclimate to lowered salinities is not necessarily correlated with an ability to tolerate abrupt changes in salinity. Ganapati & Alikunhi (1952) found great individual variability in ability to adapt rapidly to large changes in salinity. Devanesan & Chacko (1943) were able to increase the survival rates of fry considerably by reducing the salinity gradually over three to five days.

The ability to withstand high salinities also varies between species. Simmons (1957) found live *Mugil cephalus* from waters up to 75‰ in the Laguna Madre of Texas, but *M. curema* did not occur in water more saline than 45‰.

Identification of young stages

There exist relatively few studies of the eggs, larvae and postlarvae of mullet, and several of the papers deal with only a single species. One of the first comparative studies was that of Brunelli (1916) who figured the postlarval stages of five species of mullet from the Mediterranean. The ova and larvae of *Mugil cephalus* and of *Chelon labrosus* have been figured and described by Sanzo (1936). These figures and descriptions were extended to include *Liza aurata, L. ramada, L. saliens* and *Oedalechilus labeo* by Vialli (1937) who reproduced Sanzo's figures of *M. cephalus* but added a new series to illustrate the post-metamorphosed young.

Morović (1957) also described and figured some stages of development of *M. cephalus* and *C. labrosus*. The most recent and perhaps the most accurate descriptions of the young stages of Mediterranean mullets have been produced by Yashouv & Berner–Samsonov (1970).

Besides these few comparative studies there are a number of investigations on the young stages of individual species of mullet. They suffer from a lack of identical observations, but are helpful in initiating the study of the young forms – e.g., Vodyanitsky (1936) for *L. aurata*; Pakrasi & Alikunhi (1952) for *Rhinomugil corsula*; Nair (1957) and Anderson (1958) for *M. cephalus*; Anderson (1957*a*) for *M. curema* and Anderson (1957*b*) for *Agonstomus monticola*).

Except for the diameter of the ova and the diameter of their contained oil globules, those ova which have been described are very similar to one another, being small and pelagic and none so far recorded having a diameter greater than 1.32 mm. There are differences in the reported size of the ovum of *Mugil cephalus* (Thomson, 1966) which may reflect a real variation by region or possibly may be an artificact produced by different observers or may even be due to misidentification. Usually mullet ova possess a single oil globule, though Chacko (1950) states that a globule is lacking in the ovum of *Valamugil seheli*.

The pigment pattern is an important aid in the identification of larval and post-larval mugilids (Yashouv & Berner-Samsonov 1970). Comparison of the published figures of the young stages of *M. cephalus*, however, raise doubts on whether the descriptions are sufficiently reliable or whether the patterns may change so rapidly as to be useless in practice. In the later larval stages the fin rays become apparent and may be used as an ancillary aid to identification.

Many of the facial characteristics which were described earlier are not useful in the newly hatched young. All newly-hatched mugilids have a straight edged and steeply oblique jawline, without the curves and visible ending to the maxilla which becomes apparent in the later stages of some species. The preorbital is also straight-edged without notches or grooves. The nostrils are close together high up on the face, the posterior nostril being over the anterior half of the eye.

Aquacultural use of taxonomy

Although there are a number of nomenclatural problems to be worked out, the identification of juvenile and adult mullet is readily achieved for practical purposes. But there is little guidance for either scientist or aquaculturist in attempts to identify larval and postlarval mullet. Once life histories are known in any region the time of arrival of fry in the estuaries may well be the best practical criterion of identification (Bograd, 1961).

Descriptions of the great majority of species at the larval and postlarval stages are lacking. Apart from the Mediterranean region few data are available. Even here the application of knowledge is tedious. The pigment patterns can only be discerned by microscopic examination. Electrophoretic analysis of the proteins in serum, plasma or muscle might enable samples from a homogeneous collection to be identified, but such analyses take time and require sophisticated equipment and a trained technician.

A practical sorting procedure for the aquaculturist might lie in the development of methods such as that available in India for the sorting of species of carp fry in mixed catches which makes use of the differential behaviour of the species as the oxygen content of the water declines (Alikunhi, Chaudhuri & Ramachandran, 1951).

There remains a range of other biochemical and physiological criteria which might provide a useful method of distinguishing species, but which like the protein methods mostly require time and more or less complex equipment for their measurement.

References

Alikunhi, K. H., Chaudhuri, H. & Ramachandran, V. (1951). A possible method for segregation of species from a mixed collection of carp fry. *Sci. Cult.* **17**, 220–2.

Alston, R. E. & Turner, B. L. (1963). *Biochem. Systematics.* Prentice-Hall, Englewood Cliffs, N.J.

Anderson, W. W. (1957*a*). Early development, spawning, growth, and occurrence of the silver mullet (*Mugil curema*) along the South Atlantic coast of the United States. *Fish. Bull. US Fish Wildl. Serv.* **57**(119), 397–414.

Anderson, W. W. (1957*b*). Larval forms of the freshwater mullet (*Agonostomus monticola*) from the open ocean off the Bahamas and South Atlantic coast of the United States. *Fish. Bull. US Fish Wildl. Serv.* **57**(120), 415–25.

Anderson, W. W. (1958). Larval development, growth and spawning of striped mullet (*Mugil cephalus*) along the South Atlantic coast of the United States. *Fish. Bull. US Fish Wild. Serv.* **58**(144), 501–9.

Bellini, G. (1916). Ricerche sul novellame dei Muggini con osservasioni e considerazioni sulla mugginicoltura. *R. Com. Talassogr. Ital. Mem.* **44**, 1–44.

Berg, L. S. (1947). *Classification of Fishes both Recent and Fossil.* J. W. Edwards, Ann Arbor, Michigan.

Bograd, L. (1961). Occurrence of *Mugil* in the rivers of Israel. *Bull. Res. Coun. Isr.* (*B*), **9**(4), 169–90.

Brunelli, G. (1916). Ricerche sul Novellame dei Muggini con osservazioni e considerazioni sulla mugginicoltura. *Mem.* R. Comitato. Talass. Italiano, **44**, 1–45.

Chacko, P. I. (1950). Marine plankton from waters around Krusadai Island. *Proc. Indian Acad. Sci.* (*B*), **31**(3), 162–74.

Daget, J. & Iltes, A. (1965). Poissons de côte d'Ivoire. *Mem. Inst. Fr. Afr. Noire*, **74**, 1–385.

DeToni, E., Bullo, G. S. & Paoletti, U. (1892). *Alcune notizie sul lago di Arqua Petrarca.* Antonelli.

Devansesan, D. W. & Chacko, P. I. (1943). On the possibility of culture of certain marine mullets in freshwater tanks. *Proc. Natl. Inst. Sci. India*, **9**, 249–50.
Ebeling, A. W. (1957). The dentition of eastern Pacific mullets with special reference to adaptation and taxonomy. *Copeia*, (3), 173–85.
Ebeling, A. W. (1961). *Mugil galapagensis*, a new mullet from the Galapagos Islands. *Copeia*, (3), 295–304.
Erman, F. (1960). A comparative morphological study on some Mugils' otoliths or sagitta. *Hidrobiol. Istanbul*, (*B*), **5**(1–2), 25–9.
Fujino, K. & Kang, T. (1968*a*). Transferrin groups of tunas. *Genetics*, **59**, 79–91.
Fujino, K. & Kang, T. (1968*b*). Serum esterase groups of Pacific and Atlantic tunas. *Copeia* (1), 56–63.
Ganapati, S. V. & Alikunhi, K. H. (1952). Experiments on the acclimatization of salt water fish to freshwater. *Proc. Indian Acad. Sci.* (*B*), **35**(2), 93–109.
Greenwood, P. H., Rosen, D. E., Weitzman, S. H. & Myers, G. S. (1966). Phyletic studies of teleostean fishes with a provisional classification of living forms. *Bull. Am. Mus. Nat. Hist.* **131**(4), 339–456.
Günther, A. (1861). *Catalogue of the Acanthopterygian fishes in the Collection of the British Museum*, vol. 3, pp. 1–586.
Gunter, G., Sulya, L. L. & Box, B. E. (1961). Some evolutionary patterns in fishes' blood. *Biol. Bull.* **121**(2), 302–6.
Herzberg, A. & Pasteur, R. (1975). The identification of grey mullet species by disc electrophoresis. *Aquaculture*, **5**, 99–106.
Hongskul, V. (1968). Serum polymorphism in grey mullet (*Mugil cephalus* L.). *Proc. Indo-Pac. Fish. Coun. Techn. Pap.* **18**, 1–9.
Hotta, H. & Tung, I. S. (1966). Identification of fishes of the family Mugilidae based on the pyloric caeca and the position of insertion of the first interneural spine. *Jap. J. Ichthyol.* **14**(1–3), 62–6 (in Japanese, English summary).
Jordan, D. S. (1907). *Fishes.* Holt & Co., New York.
Jordan, D. S. & Evermann, B. W. (1896). A descriptive catalogue of the fishes of North and Central America, part 1. *Bull. US Natl. Mus.* **47**, 1–240.
Lagler, K. F., Bardach, J. E. & Miller, R. R. (1962). *Ichthyol.* John Wiley & Sons, New York.
Luther, G. (1975). New characteristics for consideration in the taxonomic appraisal of grey mullets. *Aquaculture*, **5**, 107.
Morović, D. (1953). Sur la détermination des muges Adriatiques d'après la forme de l'otolithe sagitta. *Notes Inst. Oceanogr. Split*, **9**, 1–7.
Morović, D. (1957). Les Muges de l'Adriatique avec la bibliographie des muges. *Rib. Jugosl. Zagreb*, 1–26.
Nair, G. S. (1957). Notes on the early development of *Mugil cephalus* L. *Bull. Cent. Res. Inst. Univ. Kerala* (*C*), **5**(1), 77–84.
Pakrasi, B. B. & Alikunhi, K. H. (1952). On the development of the grey mullet *Mugil corsula* Hamilton. *J. Zool. Soc. India*, **4**(2), 123–40.
Peterson, G. L. & Shehadeh, Z. H. (1971). Subpopulations of the Hawaiian striped mullet *Mugil cephalus*: analysis of variations of nuclear eye-lens protein electropherograms and nuclear eye-lens weights. *Mar. Biol.* **11**(1), 52–60.
Rahimulla, M. (1945). A comparative study of the morphology, histology and probable functions of the pyloric caeca in Indian fishes, together with a discussion on their homology. *Proc. Indian Acad. Sci.* (*B*), **21**(1), 1–37.
Rosen, D. E. (1964). The relationships and taxonomic position of the half-beaks, killifishes, silversides and their relatives. *Bull. Am. Mus. Nat. Hist.* **127**(5), 219–67.

Sanzo, L. (1936). Contributi alla conoscenza dello sviluppo embrionale e post-embrionale nei Mugilidi. *Mem. R. Com. Talassogr. Ital.* **179**, 1–11.

Schultz, L. P. (1946). A revision of the genera of mullets, fishes of the family Mugilidae with descriptions of three new genera. *Proc. U.S. Natl Mus.* **96**, 377–95.

Senkivich, N. K. & Kulikova, N. I. (1970). Intraspecific differentiation of serum proteins in the blood of Black Sea mullet. *J. Ichthyol.* **10**, 538–44.

Sick, K., Westergaard, N. & Frydenberg, O. (1962). Haemoglobin pattern and chromosome number of American, European and Japanese eels (*Anguilla*). *Nature, Lond.* **193**, 1001–2.

Simmons, E. G. (1957). Ecological survey of the upper Laguna Madre of Texas. *Publ. Inst. Mar. Sci. Univ. Texas* **4**(2), 156–200.

Sinderman, C. J. & Honey, K. A. (1963). Electrophoretic analysis of the haemoglobins of Atlantic clupeoid fishes. *Copeia*, 534–7.

Sprague, L. M. (1967). Multiple molecular forms of serum esterase in three tuna species from the Pacific Ocean. *Hereditas*, **57**(11), 198–204.

Sprague, L. M. & Vrooman, A. M. (1962). A racial analysis of the Pacific sardine (*Sardinops caerulea*) based on studies of erythrocyte antigens. *Ann. N.Y. Acad. Sci.* **97**, 131–8.

Sulya, L. L., Box, B. E. & Gunter, G. (1960). Distribution of some blood constituents in fishes from the Gulf of Mexico. *Am. J. Physiol.* **199**(6), 1177–80.

Sulya, L. L., Box, B. E. & Gunter, G. (1961). Plasma proteins in the blood of fishes from the Gulf of Mexico. *Am. J. Physiol.* **200**(1), 152–4.

Swan, T. (1963). *Chemical Plant Taxonomy*. Academic Press, New York.

Thomson, J. M. (1954). The Mugilidae of Australia and adjacent seas. *Aust. J. Mar. Freshwater Res.* **5**(1), 70–131.

Thomson, J. M. (1955). The movements and migrations of mullet (*Mugil cephalus* L.). *Aust. J. Mar. Freshwater Res.* **6**(3), 328–47.

Thomson, J. M. (1966). The grey mullets. *Oceanogr. Mar. Biol. Ann. Rev.* **4**, 301–35.

Thomson, J. M. (1974). Features of systematic importance in the Mugilidae. In *Proceedings of the IBP/PM International Symposium on the Grey Mullets and their Culture, Haifa*, 2–8 June 1974.

Trewavas, E. & Ingham, S. E. (1972). A key to the species of Mugilidae (Pisces) in the northeastern Atlantic and Mediterranean, with explanatory notes. *J. Zool. London*, **167**(1), 15–29.

Troschel, F. A. (1865). Ueber das Gebiss der Fischgattung: *Mugil. Sitzungsber. Versuchsanst. Preuss. Rheinlandisches Westphalia*, **22**, 130.

Vialli, M. (1937). Uova larvae e stadi giovanili di Teleostei, Mugilidae. *Publ. Sta. Zool. Napoli.* (38), 433–56.

Vodyanitsky, V. A. (1936). Observations on pelagic eggs of fishes of the Black Sea. *Tr. Sevastop. Biol. Stn.* (5), 3–43.

Yashouv, A. & Berner-Samsonov, E. (1970). Contribution to the knowledge of eggs and early larval stages of mullets (Mugilidae) along the Israel coast. *Bamidgeh*, **22**(3), 72–89.

2. Means of identification of grey mullet fry for culture

LYKA ZISMANN

Introduction

To increase the efficiency in the culture of grey mullets in ponds, culturists prefer to stock the species which have been shown to have high growth and survival rates (Ben-Yami & Grofit, see Chap. 9). Once it has been determined which of the local species is best for pond culture, the problem which remains is how to recognize and select their fry from the others present. The work of building a key to identify the fry in the laboratory, and eventually in the field, is greatly simplified if the commercial-sized mullets have already been identified.

This chapter is intended, as a guide and aid to those interested, to build a regional key for the identification of young grey mullets, 20–100 mm in total length. Larger specimens can usually be identified by using keys for commercial-sized fish (Bograd, 1955; Ben-Tuvia, 1975; Ebeling, 1957; Trewavas & Ingham, 1972; Thomson, 1954).

Morphometric characters may or may not be obvious. They may change with size (Demir, 1971; Ebeling, 1957; Thomson, 1954) or remain constant. It is, of course, desirable that a key be based on morphological characters which can be seen and recognized easily and remain constant at all sizes. Such characters do not always occur in nature, therefore other, perhaps unconventional methods, should be considered.

The inherent differences between one organism and another begin inside the cells, and affect the highly complex processes of growth, development, behaviour and other traits of the particular species, as well as the morphology. The eggs of two species of grey mullet have been found to differ biochemically (Herzberg, 1975). Electrophoresis of the water-soluble muscle proteins has been used to build a tentative key for the six common grey mullets of the Mediterranean coast of Israel (Herzberg & Pasteur, 1975). The fry of various species of grey mullets have been found to appear at different seasons of the year (Yashouv & Berner-Samsonov, 1970), indicating different spawning seasons. Such information may aid in reducing the number of species of fry to be identified during a given period of the year.

Various aspects and problems of the systematics of grey mullets have been discussed by Schultz (1946), Smith (1935), Thomson (1954; also see Chap. 1) and by Trewavas & Ingham (1972). In this chapter, two groups of characters for the identification of fry are discussed together: characters used in routine work by the author over a period of many years, and characters used by other investigators working with grey mullet fry in various parts of the world. The former were selected by the author for use in field work or in a laboratory with unsophisticated equipment: a good dissecting binocular microscope, electricity, required glassware and simple dissecting equipment. The characters used by other investigators were evaluated by examination of the grey mullet species of the Mediterranean coast of Israel.

As an exercise in the application of the routine examination of the various characters, one sample of grey mullet (53.4–108.6 mm total length) from Tor, Gulf of Suez, was examined. A study of the fry and juveniles of this area has not yet been made (Ben-Tuvia, 1975). The key devised by Ben-Tuvia (1975) for commercial-sized grey mullets was used to identify several of the species of this sample.

The grey mullets of the Mediterranean shore of Israel

The investigations on the grey mullets of Israel progressed along three lines simultaneously: (1) the literature on the area was reviewed; (2) commercial-sized specimens taken by the fishermen were examined, specimens collected, preserved and identified; and (3) several series of the young stages were collected and studied. When a number of fry and juveniles could be recognized as belonging to various species, limited studies on stocking these species haphazardly were carried out by a number of kibbutzim.

By 1953, five of the common grey mullets of commercial size could be identified (Ben-Tuvia, 1953). By 1955, six of the commercial-sized mullets could be identified using a field key (Bograd, 1955), the fry and juveniles identified by a key (Perlmutter, Bograd & Pruginin, 1957) and the most suitable species for culture selected.

Although six different species of grey mullets could be recognized in the early stages of the investigations, only one or two could be named. The scientific names of all the species were found and confirmed as the studies progressed.

Understandably, the most common commercial-sized species were recognized and identified first. *Mugil cephalus* (L.) was the most easily identified, as it was the only common species with an adipose eyelid. *Oedalechilus labeo* (Cuvier) and *Chelon labrosus* (Risso, 1826) were easily identified by the high upper lip with the ornamentation characteristic of each species.

The descriptions of the pyloric caeca of the Mediterranean grey mullets by Heldt (1948) and the peculiar predorsal scales of *Liza saliens* (Risso) described by Popov (1930) solved the problem of separating *Liza saliens*, *Liza aurata* (Risso) and *Liza ramada* (Risso).

Ben-Tuvia (1966) recorded *Liza carinata* for the first time from the Israel Mediterranean coast; it had been reported by Wimpenny (1932) from the Egyptian Mediterranean coast as *Liza seheli* (Trewavas & Ingham, 1972).

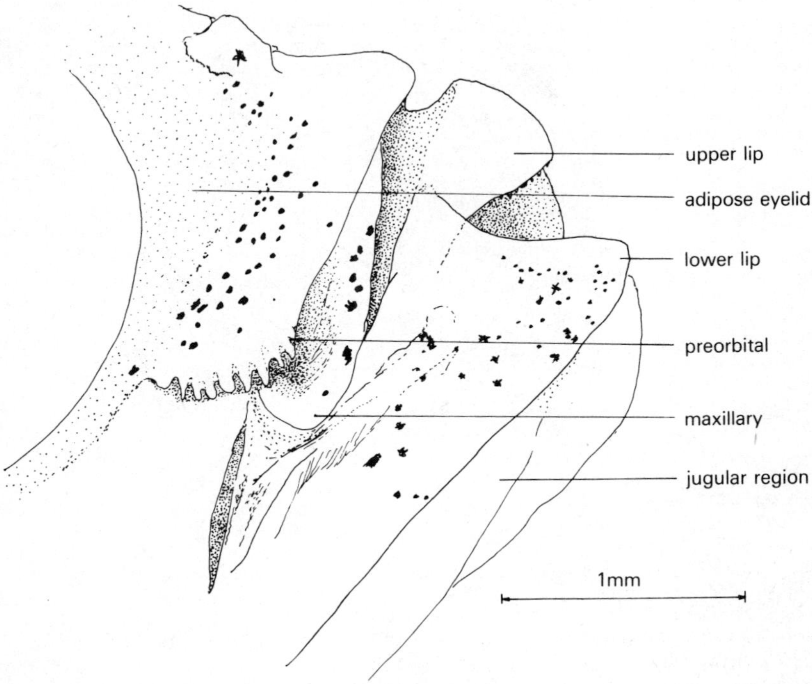

Fig. 2.1. Snout of *Chelon labrosus* (37.2 mm total length).

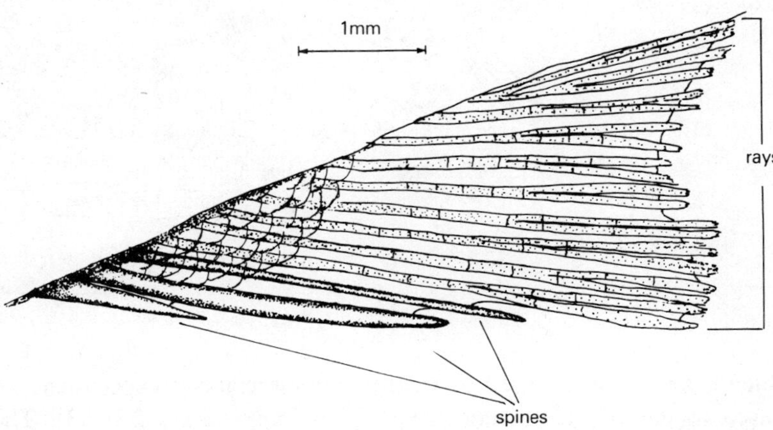

Fig. 2.2. Anal fin, drawn from a specimen of *Chelon labrosus* (34.0 mm total length).

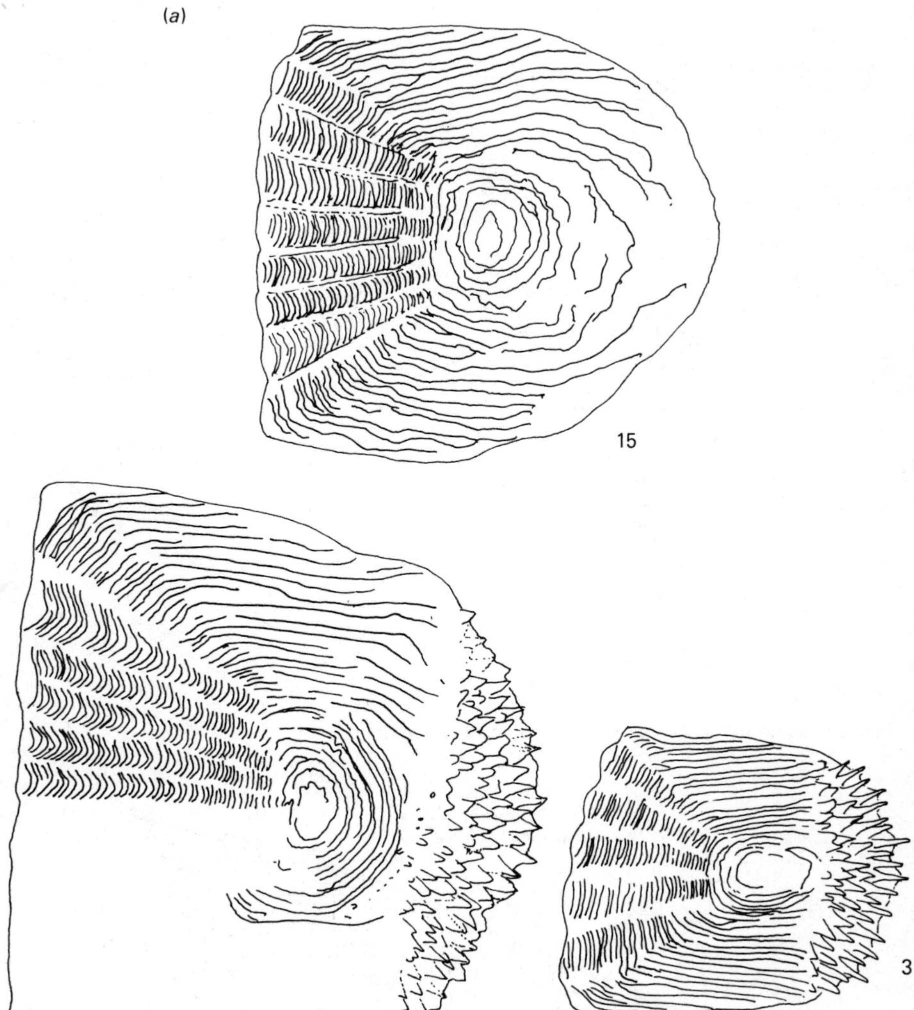

Some of the characters used to identify commercial-sized specimens were also used to identify the young (see keys in Appendixes 2.1–2.3). These characters included the number and arrangement of the pyloric caeca and the number of elements in the anal fin. However, these characters were insufficient and others were sought. Several series of sizes of the various species were collected and preserved in order to verify the identification of the fry and juveniles.

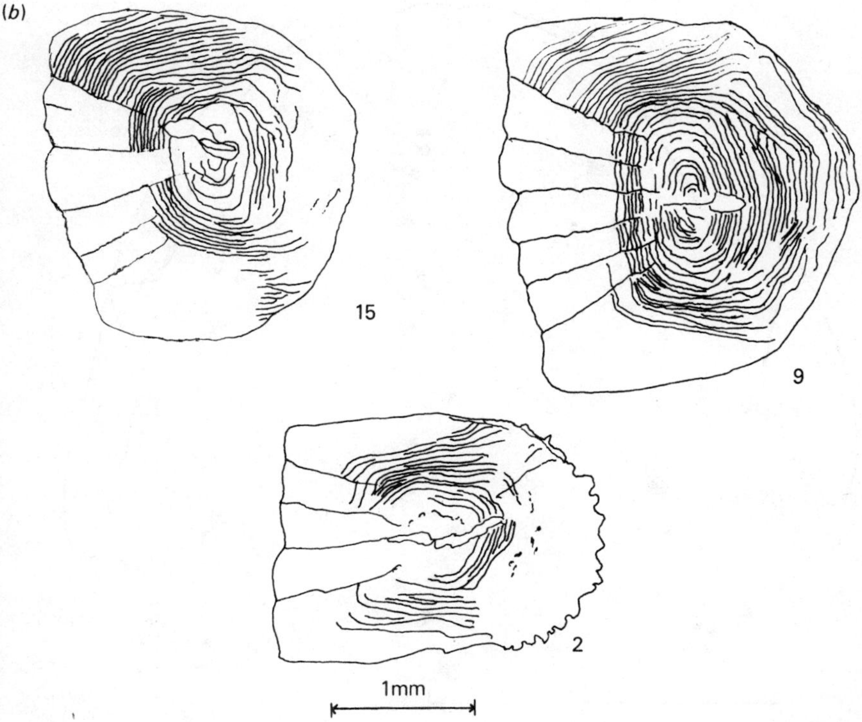

Fig. 2.3(*a*). Scales of *Chelon labrosus* (65.7 mm total length), nos. 3, 10 and 15, counting from the mid-ventral line upwards, along the transverse line from the anal opening towards the first dorsal. Scales no. 3 and 10 are ctenoid, no. 15, cycloid. (*b*). Scales from the transverse row of an unidentified mugilid species from the Gulf of Suez (77.6 mm total length). Scale no. 2 is digitated, the others are cycloid.

Monthly collections were made with a beach seine at various habitats (mouth of freshwater outlets to the sea, open beaches, protected beaches and a number of stations upstream from the sea) to find the variations in distribution seasonally and by habitat (Perlmutter *et al.*, 1957; Zismann, in preparation).

Terminology

Adults

Adipose eyelid. As Thomson (1954) points out, this is not a third eyelid but a fatty deposit around the eye, which at times may cover all or part of the iris. The adipose eyelid is transparent in live or very fresh specimens, but becomes dark and almost opaque in preserved specimens (Fig. 2.1). It becomes apparent in fry of 40–50 mm total length or over.

(a)

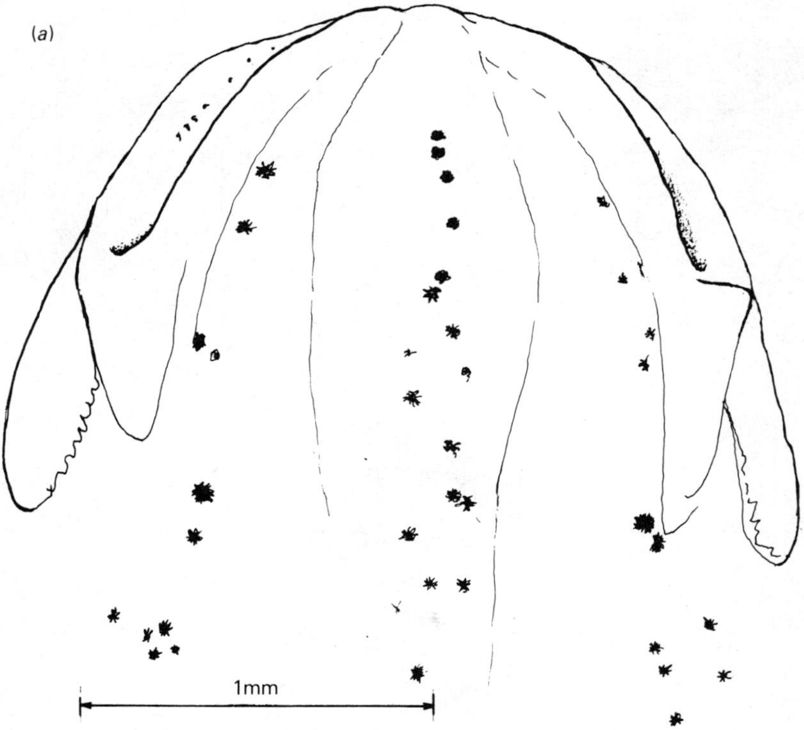

Fig. 2.4. Anterior jugular region of grey mullets to show variations in the shape of the lower jaw and pigment patterns: (*a*) *Mugil cephalus* (27 mm total length), no teeth visible on lower lip, (*b*) *Mugil cephalus* (42.5 mm total length), teeth visible on lower lip, (*c*) *Liza aurata* (40.8 mm total length), (*d*) *Liza saliens* (32.9 mm total length), (*e*) *Chelon labrosus* (42 mm total length), (*f*) *Oedalechilus labeo* (67.2 mm total length).

Anal fin elements. This refers to the total number of spines and rays in the anal fin. In young specimens, there are two spines and from nine to eleven rays. With growth (see below), the first ray in the Mugilinae changes to a third spine (Fig. 2.2).

Ctenoid scale. Scale with spines on the exposed portion (Fig. 2.3*a*).

Cycloid scale. Scale with the exposed edge smooth (Fig. 2.3*a*–*b*).

Digitated scale. Scale with the exposed edge thin, with scallop-like or finger-like extensions (Fig. 2.3*b*).

Jugular region. From the anterior interopercular area to the lower lip (Fig. 2.4).

Predorsal scales. Scales on the dorsal surface of the body between the first dorsal fin and the snout.

Preorbital. Small bone in front of the eye, usually denticulate. Its anterior edge

(*b*)

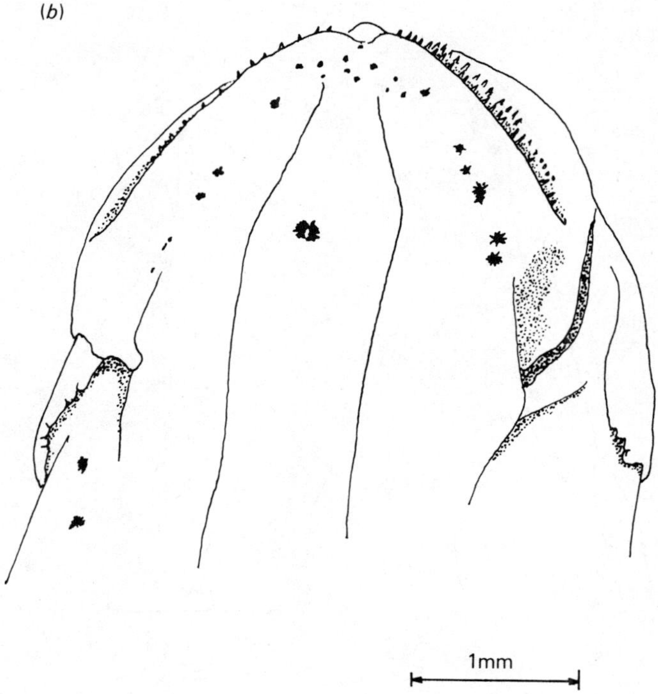

(*c*)

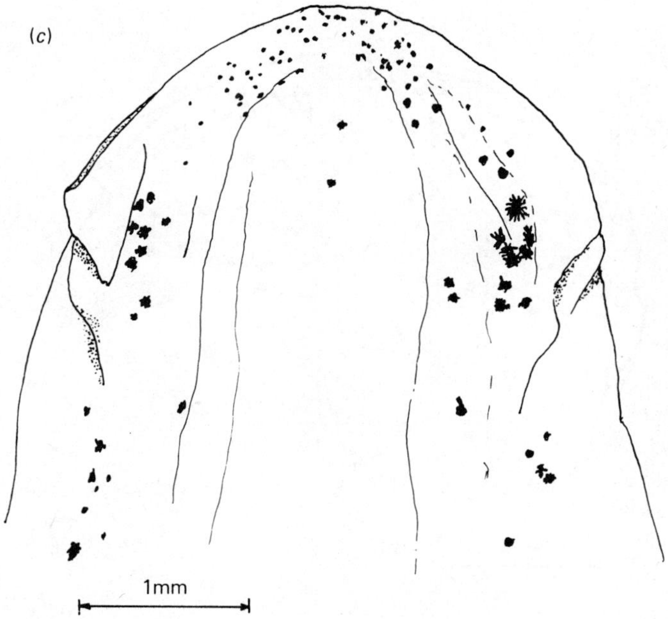

(*d*)

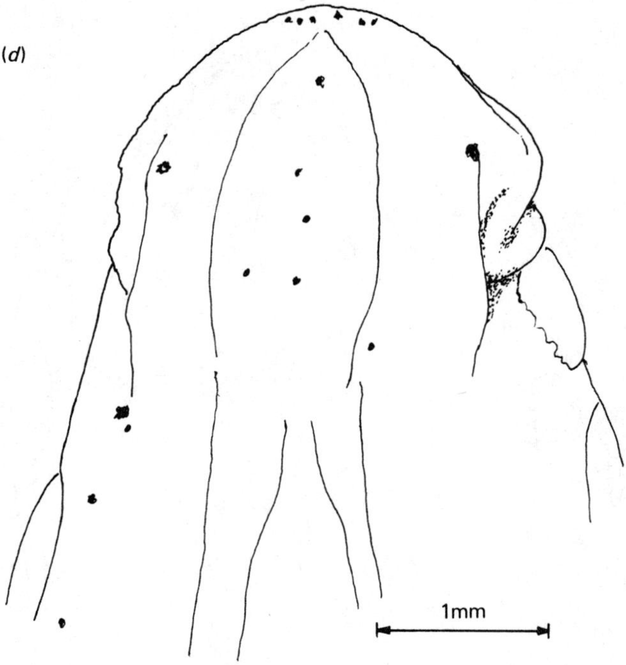

(*e*)

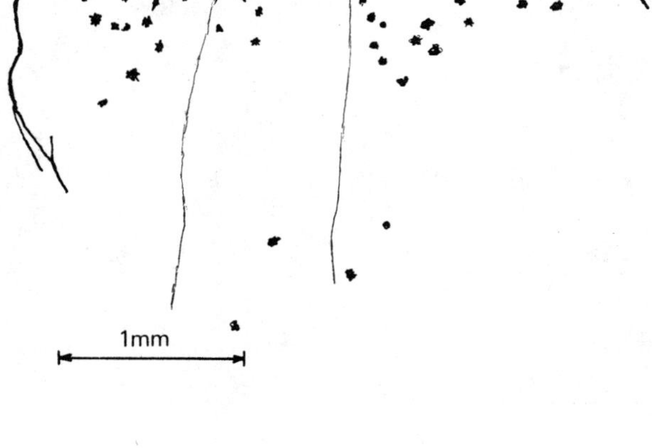

(f)

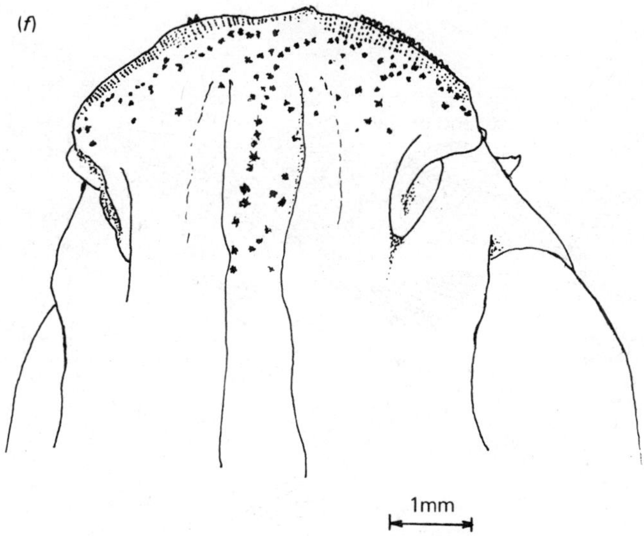

may be notched, curved or straight. The shape of the preorbital may change with growth (Thomson; Chap. 1 and Fig. 2.1).

Standard length (SL). Length of fish from tip of snout to end of vertebral column. On fish less than 30 mm in total length, the standard length was measured on the binocular platform, by means of vernier calipers. By using transmitted light, the junction between the end of the vertebral column and the caudal can be seen as a narrow translucent band. The standard length was then taken as the distance from the snout to the inner side of the band (Fig. 2.5).

Total length (TL). All lengths in this paper, unless stated otherwise, refer to total length. This is taken as the distance from the tip of the snout to the end of the caudal fin (Fig. 2.5).

Transverse row of scales. Taken as the row of scales beginning just in front of the anal opening and continuing towards the first dorsal fin.

Young stages

The terminology for describing the young stages of grey mullet varies in the literature, as pointed out by Demir (1971). Demir (1971) and Zismann, Berdugo & Kimor (1975) follow Hubbs' (1943) definition of 'postlarva' for the stage following the absorption of the yolk. At the beginning of the postlarval stage, the grey mullet are found off shore in the sea, and are usually captured by plankton or neuston nets. The specimens taken off shore in the Haifa Bay region are quite different in general colour and pigment pattern from the specimens of the same species and size-range caught close to the shore

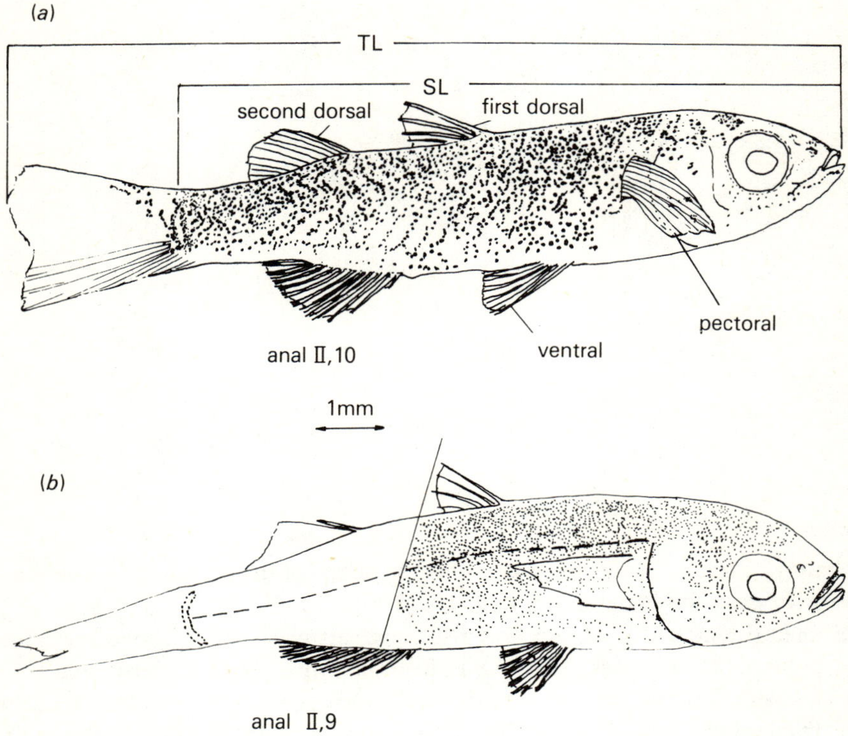

Fig. 2.5. Outline drawings of (*a*) *Liza aurata* and (*b*) *Mugil cephalus* (each 26.4 mm total length) to show differences in general outline of body and pigmentation. The pigment spots (drawn using a camera lucida) on *Mugil cephalus* are finer and of different distribution from those on *Liza aurata*. In *Mugil cephalus*, only part of the pigmentation is reproduced.

or in the mouths of freshwater outlets (Zismann, unpublished data). Demir (1971) discussed the problem of determining the end of the postlarval stage, and used the formation of the third anal spine to separate the postlarval stage from the juvenile. From the data I have collected, the range of sizes at which there are two spines in the anal fin (Table 2.1) may be as much as 30 mm. This parameter seems inadequate for use as a dividing line between two stages in the development (see section on anal fin) of a fish which changes rapidly in form and distribution by 100 mm total length.

In this paper, all young mullets, 20–60 mm in total length, are considered fry. As most of the characters used to identify the adult grey mullet can be distinguished in fish of 100 mm in total length or more, the size-range 60–100 mm is considered as the juvenile stage. All specimens and characters of grey mullets are based on material collected from the Mediterranean coast of Israel, unless stated otherwise.

Table 2.1. *Anal fin elements in fry and juveniles (20–79.9 mm total length) of seven grey mullets from the Mediterranean shore of Israel (number of fish per size-group per species)*

Size group (mm)	Anal fin elements		Species of grey mullet						
	Spines	Rays	*Mugil cephaus*	*Liza ramada*	*Liza aurata*	*Liza saliens*	*Liza carinata*	*Chelon labrosus*	*Oedalechilus labeo*
20.0–39.9	II	8	1	—	—	—	—	—	—
	II	9	80	1	1	2	—	1	—
	II	10	1	16	55	44	—	19	—
	II	11	—	1	2	—	—	—	—
	III	8	—	—	—	—	—	—	—
	III	9	—	—	2	—	—	4	—
	III	10	—	—	—	—	—	—	—
	III	11	—	—	—	—	—	—	1
Total			82	18	60	46	0	24	1
40.0–59.9	II	8	—	—	—	—	—	—	—
	II	9	11	—	1	5	—	1	—
	II	10	—	12	68	51	—	5	—
	II	11	—	—	—	—	—	—	—
	III	8	8	—	—	1	—	1	—
	III	9	—	11	52	22	—	33	—
	III	10	—	—	—	—	—	—	—
	III	11	—	—	—	—	—	—	4
Total			19	23	121	79	0	40	4
60.0–69.9	II	8	—	—	—	—	—	—	—
	II	9	2	—	—	—	—	—	—
	II	10	—	—	9	3	—	—	—
	II	11	—	—	—	—	—	—	—
	III	8	6	—	—	3	—	—	—
	III	9	—	8	35	14	2	14	—
	III	10	—	—	—	—	—	—	1
	III	11	—	—	—	—	—	1	4
Total			8	8	44	20	2	15	5
70.0–70.9	II	8	—	—	—	—	—	—	—
	II	9	1	—	—	—	—	—	—
	II	10	—	—	—	—	—	—	—
	II	11	—	—	—	—	—	—	—
	III	8	5	—	—	—	—	1	—
	III	9	—	6	33	16	1	5	—
	III	10	—	—	—	—	—	—	1
	III	11	—	—	—	—	—	—	6
Total			6	6	33	16	1	6	7
Total per species			115	55	258	161	3	85	17

Background information required

In Israel, there are seven species of grey mullets found along the Mediterranean shore (Ben-Tuvia, 1975), of which only two, *Mugil cephalus* and *Liza ramada*, are suitable for pond culture (Pruginin & Kitai, 1957). The veteran fish culturists identify these species quickly and quite accurately by means of a key based on data collected by biologists in Israel over a period of years and techniques obtained through practical experience in the field.

The first aspects investigated (once the fry could be recognized) were:

(1) Which are the common species of grey mullet fry available?

(2) In what season do the fry under 25 mm in length occur?

(3) Are these fry found in particular areas or habitats, if so, where?

To obtain this information, monthly samples were collected from a number of permanent stations chosen in various habitats (Perlmutter *et al.*, 1957; Zismann, in preparation). These collections also provided specimens of graduated sizes for each species of grey mullet, as well as material for studying the characters used to differentiate and identify the species.

Along the northern shore of Israel, it was found that of the five common species (*Mugil cephalus*, *Liza ramada*, *Liza aurata*, *Liza saliens* and *Chelon labrosus*), fry of 25 mm or less in length (those recently arrived from the open sea) appear during definite months of the year. *Mugil cephalus* of this size-range may appear as early as June (Table 2.2), but rarely in schools. Their numbers increase slowly during July and August, become common during September, usually reaching a peak during November and December, and decrease sharply in January. *Liza ramada* and *Liza aurata* fry both appear for the first time about the middle or end of January, and reach a peak towards the end of February and March. *Chelon labrosus* of less than 25 mm begin to appear in March, reach a peak in April and May and decrease in numbers in June. *Liza saliens* first appear in June and reach a peak about the end of September or October. As *Mugil cephalus* has the fastest rate of growth in the ponds (Perlmutter *et al.*, 1957), it is this species which is sought by the fish culturists. Small groups of fishermen begin to scout along the shore in and about freshwater outlets to the sea as early as July, in the hope of obtaining sufficient stock for the ponds by January, when these fry become scarce again. From July until December, fry of 30 mm or less are either *Mugil cephalus* or *Liza saliens*. *Chelon labrosus* of this size are rare by July as most are over 35 mm.

Yashouv & Berner-Samsonov (1970) have summarized the spawning seasons for the five common Mediterranean species of grey mullet as reported in the literature.

Veteran fish culturists in Israel now know that in the peak season of *Mugil cephalus* fry (October–December), the only other species of mullet fry numerous in the vicinity of freshwater outlets and upstream is *Liza saliens*.

Table 2.2. *Seasonal distribution of fry (< 250 mm total length) of seven grey mullet species from the Mediterranean shore of Israel*

	Species of grey mullet						
	Mugil cephalus	*Liza ramada*	*Liza aurata*	*Liza saliens*	*Liza carinata*	*Chelon labrosus*	*Oedalechilus labeo*
Jan.	II	II	III	I	0	—	0
Feb.	—	III	III	—	—	—	—
Mar.	—	III	II	—	—	I	—
Apr.	—	II	—	—	—	III	—
May	—	II	—	—	—	III	—
June	I	—	—	I	—	II	—
July	I	—	—	II	—	—	—
Aug.	II	—	—	II	—	—	—
Sep.	II	—	—	II	—	—	—
Oct.	III	—	—	III	—	—	—
Nov.	III	—	—	II	—	—	—
Dec.	III	—	—	II	—	—	—

I, rare; II, small numbers (< 20 per haul beach seine); III, large numbers (> 50 per haul beach seine); 0, no specimens.

In India, the young of *Mugil corsula* appear from March to September (Pakrasi & Alikunhi, 1952), while the fry of *Mugil tade* and *Mugil parsia* are collected for fish culture from November to February (Sarojini, 1951). Moore (1974) reported that along the south Texas coast, based on observations of ovary development and appearance of the fry, the spawning of *Mugil cephalus* and *Mugil curema* occurs at different seasons. Thomson (1966) summarized the information on the spawning seasons and time of occurrence of fry from various areas. The information available indicates that the grey mullets spawn during definite seasons, and these generally vary with the species. Knowledge of the season of occurrence of the fry desired for stocking may aid in reducing the number of species of fry which have to be distinguished one from the other.

Studies made along the northern shore of Israel also show that the fry of the various species of grey mullet vary in concentrations in different habitats (Bograd, 1961; Zismann, in preparation). This too is very helpful for fish culturists in obtaining stock only of the desired species.

Methods and equipment

Instructions on the various techniques and procedures involved in the preservation, handling and identification of fishes have been described and summarized by Lowe-McConnell (1968). His paper, although intended for work with freshwater fishes, is also applicable to grey mullets and other estuarine and marine fishes. The gear for the capture of young grey mullets for stocking is discussed by Ben-Yami & Grofit (Chapter 9).

The laboratory procedure followed by the author in examining fry involves the following procedures. The total and standard lengths of all grey mullets under 80 mm are measured with vernier calipers. Grey mullets of about 20 mm or less are measured on the microscope platform at a magnification of × 6. Weights are determined as part of the general routine work on all the fish. Counts of scales, pyloric caeca, anal fin elements and gill rakers were made by means of a Wild M-5 binocular. Small, fine, sharply pointed scissors, a pair of jewellers forceps (further sharpened on a stone), and fine insect pins (size 000) inserted into needle holders were used to dissect the fry. A spear-headed probe with two cutting edges (also sharpened on a stone) was used for both dissection and counting scales on small fish. Drawings were made by means of a camera lucida.

Observations taken

(1) Note the general pigments on the body – is there a pattern?

(2) Observe the head. Note the shape and colour of the upper lip, and dentition. Presence or absence of teeth may sometimes be discerned only at × 12 or × 25 magnification.

(3) Observe the jugular region. Is there a recurrent pigment pattern recognizable on all specimens of the same size-group? Note angle formed by the lower jaws.

(4) Find how many elements are in the anal fin.

(5) Count the number and examine the type of scales along the transverse row of scales.

(6) Count the number of scales along the lateral line.

(7) Dissect the fish along the mid-ventral line, avoiding the pelvic girdle. Note number, grouping and form of the pyloric caeca.

(8) Remove first gill arch for raker count, mount in a drop of water and glycerine.

(9) Remove tongue, mount in a drop of water and glycerine.

Once a set of diagnostic characters have been found sufficiently reliable to identify a species, only the data for these are recorded.

Examination of morphological characters

Anal fin

In grey mullets under 60 mm, the anal fin was examined under water. This can be done by filling a petri dish with water, or more simply, by putting a drop of water on the anal fin. When the fin is examined under water with transmitted light at a magnification of × 12 or × 25, the joints in the rays are clearly visible, while the spines which are unsegmented appear opalescent (Hubbs, 1944).

Lateral line count

The lateral line was taken as the first continuous row of scales from the free edge of the operculum, over the base of the pectoral fin, to the end of the vertebral column (Smith, 1935). In grey mullets fry (20–60 mm), these counts were made under the microscope. Under transmitted light, the hypural region appears as a light band; the first scale on the anterior side of the band was taken as the last scale of the lateral line. The scales were counted under reflected light.

Counts along the transverse row of scales

The number of rows of ctenoid scales was counted along the transverse row. The total number of ctenoid (or digitated) scales and the total number of scales in the row (from the mid-ventral to mid-dorsal) were recorded.

Pyloric caeca

The pyloric caeca are arranged in a whorl around the junction between the stomach and the duodenum; most are on the ventral side (Fig. 2.6). To avoid injuring the caeca during dissection, cut the abdomen from the anal opening along a line curving around the pelvic fins on the right side of the fish, and then back to the midline just behind the opercula. This avoids the pelvic girdle altogether. To count the pyloric caeca, the liver has to be lifted or removed; often, there is fatty and connective tissue surrounding the pyloric caeca as well. This tissue can be removed with forceps. When large numbers of caeca are found in a species, e.g. *Liza aurata*, the last caecum (counting from the right side of the fish to the left) may be near the dorsal side of the stomach. The stomach has to be partly lifted out and rotated, so that all the caeca are exposed. The pyloric caeca of the Mediterranean species of grey mullet are all unbranched. Luther (1975) noted that in India, there are two species (*Valamugil buchanani* and *Valamugil vaigiensis*) which have branched pyloric caeca, and suggests that discrepancies in recorded number of caeca may occur if one observer counts the bases and another counts the ends of the branches. He recommends that only the bases be counted. Data on pyloric caeca should therefore include: number, arrangement (one or more groups) and form (branched or unbranched).

Gill rakers

For gill raker counts, the first gill arch from the left side of the head was removed with fine forceps. The arch from the right side was taken only if the first was damaged (it was found that there was little variation between the two arches of the same specimen). The number of gill rakers on the upper and lower limbs were recorded separately, but combined for comparing the number of gill rakers per size-group per species.

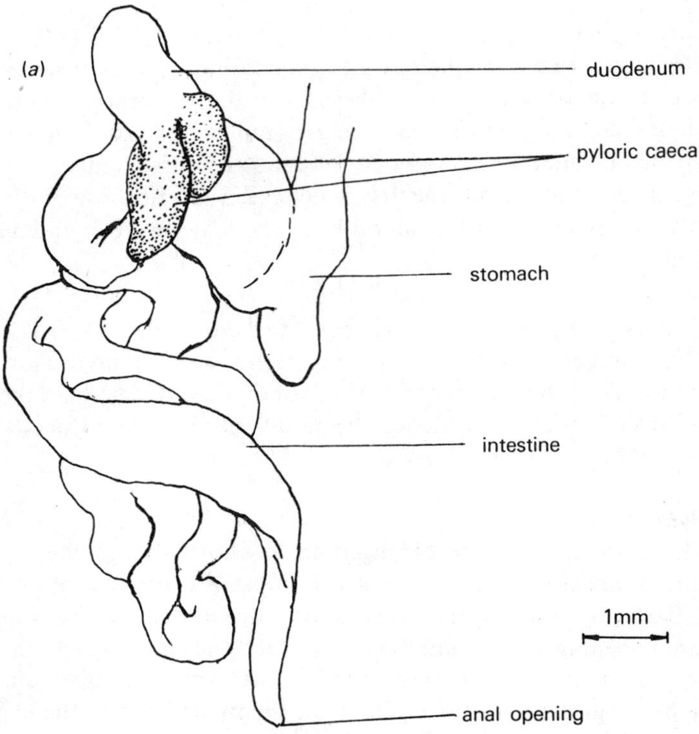

Fig. 2.6. Digestive tracts of three species of grey mullets drawn in outline, to show the variation in number and arrangement of pyloric caeca: (*a*) *Mugil cephalus* (27.2 mm total length), two caeca, (*b*) *Liza saliens* (32.9 mm total length), two groups of caeca, (*c*) *Oedalechilus labeo* (67.2 mm total length), one group of six caeca.

Teeth on tongue (*lingual teeth*)

If the teeth on the tongue are also examined, great care must be taken when removing the gill arch, as part of the lateral margin of the tongue may be torn off. The tongues of the fish were dissected and mounted only in water, with a drop of glycerin, to find whether or not, using very simple methods, differences among the species can be recognized. The tongues were examined under transmitted light, usually at the magnification of ×25 or ×50.

Characters used to identify grey mullet fry

As with most animals, the early stages of growth are accompanied by rapid changes in morphology, anatomy and relative proportions of the body, particularly in the head region (Thomson, 1954; also see Chap. 1). Of the various characters used to recognize species of grey mullet fry, some remain

(b)

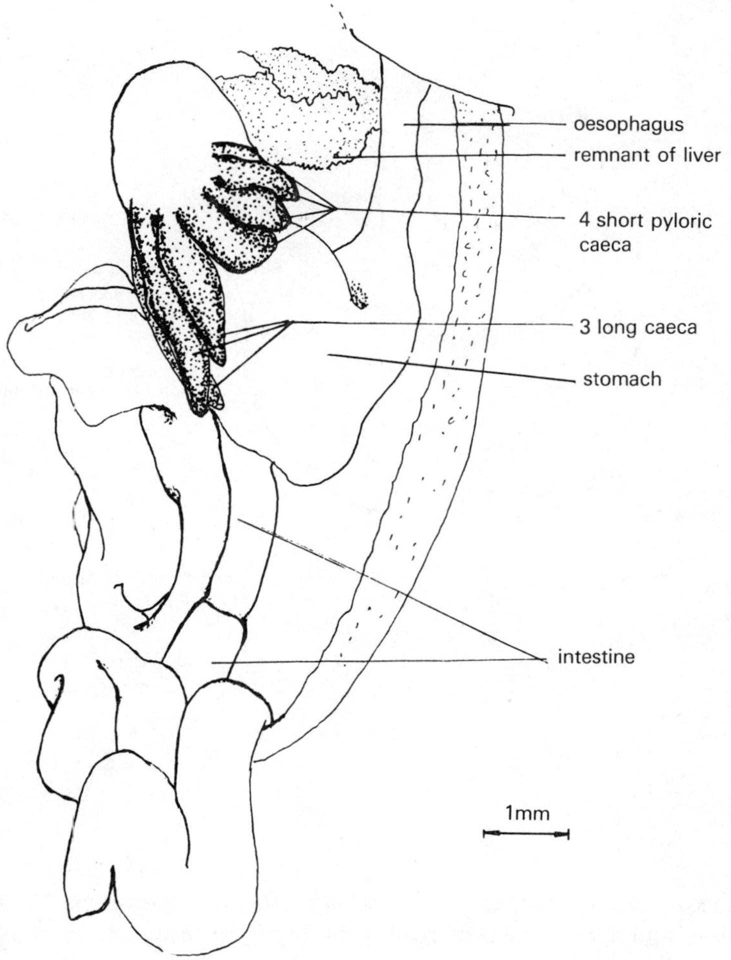

constant at all sizes, and some vary with the length of the fish. In Tables 2.1–2.7, the author has summarized data recorded during several studies on fry of grey mullet, indicating the extent of these variations at different sizes. The data on *Liza carinata* and *Crenimugil crenilabis* from the Red Sea were taken from one sample of fish for comparison only (see appendices).

Morphometric characters that remain constant

Pyloric caeca

The pyloric caeca can be counted in postlarvae of about 8.0 mm, but differences in size or grouping of large and small caeca can be noted in specimens of over 12 mm (Zismann, unpublished data). The range in numbers

(c)

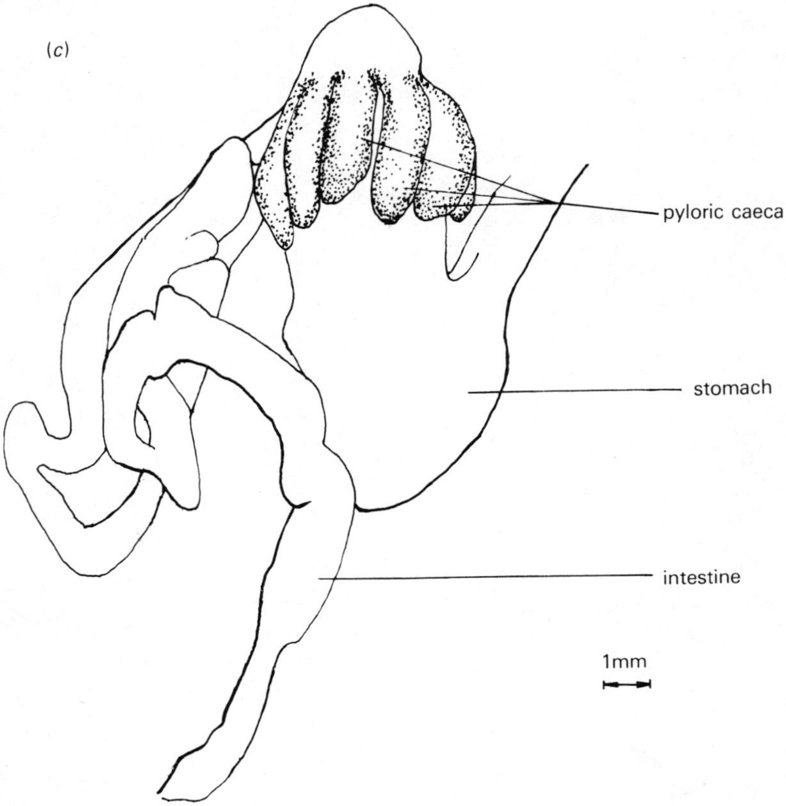

of caeca, however, appear to be the same for all size-groups. Data on the number and arrangement of pyloric caeca of the seven Mediterranean and two of the Red Sea species are given in Table 2.3. Data on the number of caeca found in various species of grey mullets observed by Ben-Tuvia (1975), Hotta & Tung (1966), Perlmutter *et al.* 1957), Trewavas & Ingham (1972) and Thomson (1954) are given in Table 2.4, where the species are arranged in order of the number of caeca.

The number of caeca, their arrangement and structure (branched or unbranched) can be observed in the field, with the aid of a pocket lens. The veteran fish culturists in Israel tear off the head of a specimen or two and squeeze the abdomen, so that the stomach and caeca are pressed forward and out through the opening left by the decapitation. With the naked eye, it can be seen whether the fish has two caeca, as in the sought-for *Mugil cephalus*, or two groups of caeca, as in *Liza saliens* which is thrown back. This character is useful in the field.

Table 2.3. *Number and arrangement of pyloric caeca for eight species of Israel grey mullets in number of fish per category*

Species	No. of groups		Number of pyloric caeca										Total no. fish
	1	2	2	3	4	5	6	7	8	9	10	11	
Mugil cephalus	×	—	141	—	—	—	—	—	—	—	—	—	141
Liza ramada	×	—	—	—	—	—	15	135	148	14	—	—	312
Liza aurata	×	—	—	—	—	—	1	21	244	227	15	1	509
Liza saliens	—	×	—	—	—	—	3	131	156	6	—	—	296
Liza carinata	×	—	—	—	—	5	—	—	—	—	—	—	5
Liza carinata[a]	×	—	—	—	—	13	4	—	—	—	—	—	17
Chelon labrosus	×	—	—	—	—	5	176	17	—	—	—	—	196
Oedalechilus labeo	×	—	—	—	—	11	126	23	2	—	—	—	162
Crenimugil crenilabis[a]	×	—	—	—	—	—	—	—	3	4	4	—	11

[a] Red Sea species.

Table 2. 4. *Pyloric caeca of grey mullet*

Species	No. of caeca	Author
Mugil cephalus	2	Hotta & Tung, 1966
		Perlmutter *et al.*, 1957
		Thomson, 1954
Liza argentea	2	Thomson, 1954
Aldrichetta forsteri	2	Thomson, 1954
Myxus elongatus	2	Thomson, 1954
Valamugil seheli	2	Thomson, 1954
Plicomugil labiosus	2–3	Ben-Tuvia, 1975
		Hotta & Tung, 1966
Liza macrolepis	5	Hotta & Tung, 1966
Liza subviridis	5	Ben-Tuvia, 1975
Liza dussumieri	5	Hotta & Tung, 1966
		Thomson, 1954
Liza carinata	5	Ben-Tuvia, 1975
		Hotta & Tung, 1966
Liza strongylocephalus	5–6	Hotta & Tung, 1966
Chelon labrosus	5–7	Perlmutter *et al.*, 1957
Liza haematocheila	5–7	Hotta & Tung, 1966
Liza ampinensis	5	Hotta & Tung, 1966
Oedalechilus labeo	5–7	Perlmutter *et al.*, 1957
Liza ramada	6–9	Perlmutter *et al.*, 1957
Liza formosae	8	Hotta & Tung, 1966
Liza saliens	3–4+4–5[a]	Perlmutter *et al.*, 1957
		Trewavas & Ingham, 1972
Liza aurata	8–9	Perlmutter *et al.*, 1957
Liza vaigiensis	10	Hotta & Tung, 1966
		Thomson, 1954
Valamugil seheli	8–9	Ben-Tuvia, 1975
Crenimugil crenilabis	8–10	Ben-Tuvia, 1975
		Hotta & Tung, 1966
Mugil georgii	22	Thomson, 1954

[a] Two size-groups of caeca.

Number of scales along the lateral line

The number of scales along the lateral line is constant within a given range for each of the species (Table 2.5). *Liza saliens*, with 42–49 scales, can be separated from *Liza carinata*, with 37–39 scales in the lateral line. In many cases, there is too much overlap for this character alone to be used. This is not a suitable character for field work, but if sufficient data are accumulated, and are used together with other characters, the number of scales along the lateral line may be useful in the field laboratory.

Number of elements in the anal fin

The Mugilidae are divided into two subfamilies: the Mugilinae and the Agonostominae (Thomson; also see Chap. 1). In the Mugilinae, the young

Table 2.5. *Number of scales in lateral line for eight species of Israel grey mullets (20–80 mm) in number of fish per category*

	Number of scales in lateral line																Total
Species	35	36	37	38	39	40	41	42	43	44	45	46	47	48	49	50	no. fish
Mugil cephalus	—	—	—	—	3	13	13	14	11	1	—	—	—	—	—	—	55
Liza ramada	—	—	—	—	—	2	4	9	6	2	3	—	—	—	—	—	26
Liza aurata	—	—	—	—	—	2	8	8	11	10	4	5	—	—	—	—	48
Liza saliens	—	—	—	—	—	—	—	8	8	20	17	10	1	1	1	—	66
Liza carinata	—	—	2	—	1	—	—	—	—	—	—	—	—	—	—	—	31
Liza carinata[a]	2	6	3	3	1	1	—	—	—	—	—	—	—	—	—	—	16
Chelon labrosus	—	—	—	1	5	5	7	4	6	3	—	—	—	—	—	—	31
Oedalechilus labeo	—	—	—	—	—	—	—	1	1	—	4	4	2	3	—	—	15
Crenimugil crenilabis[a]	1	—	3	2	3	1	—	—	—	—	—	—	—	—	—	—	7

[a] Red Sea species.

stages have only two spines in the anal fin, the first ray becoming a third spine during growth. In the Agonostominae, the adults remain with two spines in the anal fin (Hubbs, 1944). Rarely, in adults and fry, there is one ray less (or more) than usual in the full complement.

In fry, juveniles and adult grey mullets, the number of elements in the anal fin is useful for separating some species. *Mugil cephalus* has three spines and eight rays (or two spines and nine rays in the fry), while *Oedalechilus labeo* has three spines and eleven rays. *Liza ramada*, *Liza aurata*, *Liza saliens*, *Liza carinata* and *Chelon labrosus* have three spines and nine rays.

In specimens of about 30 mm (Anderson, 1957) *Agonostomus monticola* and *Mugil curema*, taken off the South Atlantic coast of the United States, each have the anal fin formula II, 10, (two spines, ten rays), while *Mugil cephalus* from the same location has II, 8. The first two species can be separated from each other by the tenth ray (in the fry). In *Agonostomus monticola*, the posterior branch of the last ray is not forked, while in *Mugil curema* and *Mugil cephalus*, both are forked.

In the size-range 20–39.9 mm of *Mugil cephalus* (Table 2.1), all the fish (82) have only two spines; at 40.0–59.9 mm, eight specimens of nineteen have three spines, the remainder having, two; at 60.0–69.9 mm, six out of eight specimens have three spines. Rarely, two spines may occur in specimens over 70 mm. All the specimens examined of *Liza ramada* and *Liza saliens* of 20.0–39.9 mm have only two spines, but in the 60.0–69.9 mm size-group, all the *Liza ramada* examined had three spines and seventeen of the twenty specimens of *Liza saliens* had three spines. Two specimens of *Liza aurata* (of 60), and the single specimen of *Oedalechilus labeo*, had three spines in the 20.0–39.9 mm size-range. All thirty-three specimens in the range 70.0–79.9 mm of *Liza aurata* and all specimens of *Oedalechilus labeo* from 20.0–79.0 mm had three spines.

In general, the Israeli specimens of grey mullets may have two or three spines from about 40 to 70 mm in total length.

As specimens with two and three spines are found in the same size-range and habitat at the same time, and both the two spine and three spine stages appear very similar to each other, the author rejects this character as a good dividing line between fry and juveniles. At the same time, no other clear-cut characteristic can be offered to separate these two stages; tentatively therefore, the total length of 60 mm is taken as the end of the fry stage and beginning of the juveniles.

Morphometric characters that change with size

The two morphometric characters examined which change with size are the number of gill rakers on the first gill arch (Table 2.6) and the number of rows of ctenoid scales along the transverse row (Table 2.7).

Gill rakers

For the 20.0–29.9 mm size group, there were no data available on *Liza carinata* or *Oedalechilus labeo*. For the other species, the ranges of numbers of gill rakers overlap to such an extent that this parameter is useless. At sizes 30.0–39.9 mm, the numbers of gill rakers still overlap for the six species. At the 40.0–49.9 mm size-range, it can be seen that the number of gill rakers for *Liza saliens* and *Oedalechilus labeo* is less than that for *Liza ramada* or for *Chelon labrosus*. In this size-range, there is still overlapping of the range of numbers of gill rakers with *Mugil cephalus* and *Liza aurata*. For the 60.0–69.9 mm size fish, *Oedalechilus labeo* has the smallest number of gill rakers, slightly overlapping the range found for *Liza saliens*, which has fewer gill rakers than *Mugil cephalus*, which in turn has fewer gill rakers than *Chelon labrosus*. The range in numbers of gill rakers overlaps for the remaining three species: *Liza ramada*, *Liza aurata* and *Liza carinata*. The numbers of gill rakers appear to be more useful for identification of juveniles and adults than for fry. This character is not suitable for field work, but may be useful in the field laboratory, especially for specimens over 50 mm.

Number of ctenoid scales along the transverse row

Observations made by the author on five grey mullets: *Mugil cephalus*, *Liza ramada*, *Liza aurata*, *Liza saliens* and *Chelon labrosus* (Table 2.7), show that the scales of specimens under 25 mm in length were cycloid; ctenoid scales found on larger fish are observed first along the belly. The number of longitudinal rows of ctenoid scales along the belly increases with the size of the fish, from the mid-ventral line upwards. A thorough study of the changing patterns over the whole body in the various species was not made. Counts of rows of ctenoid scales were made along the transverse row of scales. The total number of scales (from mid-ventral to mid-dorsal) was noted. For all species, the total number of scales along this transverse row varied from fourteen to sixteen, rarely, seventeen. In Table 2.7, only the number of rows of ctenoid scales is given.

In the size-range 40.0–59.9 mm, most of *Liza ramada* fry are almost completely covered by ctenoid scales, while *Mugil cephalus* have only a few rows along the belly. The other species have ctenoid scales along the belly and sides, while the upper part of the sides and back have cycloid scales. The high number of rows of ctenoid scales in many *Liza ramada* of less than 40 mm is worthwhile noting. The wide range in the number of rows of ctenoid scales does not permit the separation of these five species by this character alone. However, the presence of ctenoid scales distinguishes the postlarva of *Agonostomus monticola* from those of *Mugil cephalus* and *Mugil curema* (Anderson, 1957).

Table 2.6. *Range and average of total number of gill rakers for eight species of Israel grey mullets by size groups*

Size group (mm)		Species of grey mullet								
		Mugil cephalus	*Liza ramada*	*Liza aurata*	*Liza saliens*	*Liza carinata*	*Liza carinata*[a]	*Chelon labrosus*	*Oedalechilus labeo*	*Crenimugil labiosus*[a]
20.0–29.9	Min.	23	21	22	20	—	—	23	—	—
	Max.	36	39	36	31	—	—	39	—	—
	Av.	28.2	30.4	29.0	25.6	—	—	31.0	—	—
	SD	2.15	5.90	3.86	3.18	—	—	2.93	—	—
	No. of specimens	103	95	38	65	0	0	56	0	0
30.0–39.9	Min.	29	35	30	27	—	—	34	—	—
	Max.	41	51	46	37	—	—	49	—	—
	Av.	34.1	43.1	37.0	32.3	—	—	41.7	31.0	—
	SD	3.48	4.21	3.37	2.60	—	—	4.27	—	—
	No. of specimens	28	102	94	98	0	0	27	1	0
40.0–49.9	Min.	37	46	38	31	—	—	45	33	—
	Max.	49	66	60	45	—	—	59	35	—
	Av.	42.6	54.2	49.2	38.2	—	—	51.6	33.4	—
	SD	3.99	5.22	6.04	2.49	—	—	3.55	0.89	—

	Av.	49.8	66.2	63.2	43.3	—	69.7	59.3	37.3	66.6
	SD	2.08	3.54	4.01	2.52	—	1.53	3.62	2.37	3.5
	No. of specimens	5	13	24	31	0	3	12	73	5
60.0–69.9	Min.	53	68	61	43	65	71	58	32	68
	Av.	55.0	73.1	69.9	47.6	65.5	72.7	66.1	39.7	69.0
	SD	1.41	3.54	4.27	1.98	0.71	2.08	3.31	2.34	1.41
	No. of specimens	6	14	14	26	2	3	20	101	2
70.0–79.9	Min.	56	72	62	45	—	77	60	41	75
	Max.	66	83	75	56	—	78	78	47	80
	Av.	58.7	76.0	71.8	50.8	—	77.5	69.9	43.4	77.7
	SD	3.78	3.22	3.82	2.74	—	0.01	6.19	1.78	2.52
	No. of specimens	6	12	10	26	0	2	11	12	3
Total per species		162	261	247	336	2	8	158	192	10

[a] Red Sea species.

Table 2.7. *Number of rows of ctenoid scales from mid-ventral line of Israel grey mullet by size-group and species (in number of fish)*

Size group (mm)	No. of rows with ctenoid scales	Species of grey mullet						
		Mugil cephalus	*Liza ramada*	*Liza aurata*	*Liza saliens*	*Liza carinata*	*Chelon labrosus*	*Oedalechilus labeo*
20.0–39.9	0	44	1	—	2	—	1	—
	1–2	—	2	—	—	—	—	—
	3–4	—	6	3	2	—	2	—
	5–6	1	3	11	9	—	11	—
	7–8	—	4	8	4	—	10	1
	9–10	—	4	3	2	—	7	—
	11–12	—	12	—	—	—	—	—
	13–14	—	4	—	—	—	—	—
	15–16	—	1	—	—	—	—	—
	17–18	—	—	—	—	—	—	—
Total		45	37	25	19	0	31	1
40.0–59.9	0	4	—	—	—	—	—	—
	1–2	—	—	—	—	—	—	—
	3–4	5	—	—	—	—	1	—
	5–6	5	—	19	6	—	9	1
	7–8	3	—	12	41	—	11	2
	9–10	—	1	1	10	—	14	1
	11–12	—	—	9	—	—	2	1
	13–14	—	4	1	—	—	2	—
	15–16	—	4	—	—	—	—	—
	17–18	—	2	—	—	—	—	—
Total		17	11	42	57	0	39	5
60.0–79.9	0	2	—	—	—	—	—	—
	1–2	1	—	—	—	—	—	—
	3–4	—	—	1	—	—	—	—
	5–6	2	—	7	2	—	—	1
	7–8	4	—	10	16	—	1	4
	9–10	1	—	7	13	—	7	5
	11–12	—	1	6	1	—	4	1
	13–14	—	3	5	—	2	15	1
	15–16	—	4	—	—	—	1	—
	17–18	—	—	—	—	—	1	1
Total		10	8	36	32	2	29	13
Total per species		72	56	103	108	2	99	19

Other morphological characters

There are characters which cannot be counted or measured, and often are difficult to describe, even with the aid of illustrations.

General appearance and colour

Many of the grey mullet species appear very similar, superficially, to workers unacquainted with them. The mullets are all dark grey, or greyish brown or green along the back, the hues becoming lighter along the sides, and usually with a silver white belly. Sometimes, there are gold-coloured spots or patches on the head or body, varying in intensity with the species or condition of the specimen. They are all streamlined in outline with two dorsal fins, the first with the characteristic arrangement of four spines (Fig. 2.5).

The morphology of live or freshly-caught specimens is similar to that in properly preserved specimens.

In body outline, *Mugil cephalus* fry have a more blunt and rounded head and a deeper body than the fry of the other five common grey mullets found in and around freshwater outlets. Fry (20–60 mm) of *Liza ramada, Liza aurata* and *Liza saliens* are alike in body outline and colour, and are slimmer in outline than *Mugil cephalus* (Fig. 2.5). The fry of both *Chelon labrosus* and *Oedalechilus labeo* of over 40 mm are slender in form, but appear to have truncated snouts because of their high upper lips.

It is possible that *Liza carinata*, which has a deep body at about 60 mm in length, may also be deep-bodied as fry, but no specimens under 60 mm were available from the Mediterranean.

Colour in live or freshly preserved fry

The colours and pigment patterns in live fish may change under stress and with the type of habitat. These colours further change on preservation; the chromatophores become more obvious, the gold or blue fading.

There are also slight differences in colour patterns from the young stages to the commercial sizes. The gold patch, often quite obvious on the operculum of adult *Liza aurata*, is not found on fry, although a diffused gold colour may sometimes be seen on the whole operculum and on the sides of the body; sometimes this is lacking altogether. There are also differences in the pigmentation between fry taken in the mouth of an outlet or estuary and the same species and size-group taken upstream.

Live specimens of *Mugil cephalus* and *Liza ramada*, newly-arrived from the sea, are metallic silver in colour. These are usually found at the openings of freshwater outlets to the sea. Several hundred meters upstream, the fry of these two species are usually greyish or brownish along the back, light coloured on the sides and white below. The fry of the other three common species (*Liza aurata, Liza saliens* and *Chelon labrosus*), when they first appear along the shore or in the mouths of outlets, are silvery grey in colour, and often metallic blue on the back.

Fry kept in overcrowded tanks, or in water lacking oxygen, often have black and white patches over the body, and before death may appear very dark.

In the fry examined, the author has not seen the blue spot in the axillary area of the pectoral fin, often found in fresh specimens of *Mugil cephalus* over 150 mm.

Identification of fry

Colour of the fry, as a distinguishing character, has to be treated warily, and never as a sole identifying mark.

Colour patterns in preserved material

Mugil cephalus fry of about 30 mm or more, taken upstream, are brownish or bluish grey dorsally, fading to white along the belly without a marked pattern; the chromatophores are fine. In *Liza ramada*, the chromatophores are coarser (larger and smaller intermingled), and the pigment spots along the sides form a vague herring-bone pattern which is particularly dark along the mid-lateral line. The belly is white.

In *Liza aurata*, the herring-bone pattern in the 20–40 mm size group is more obvious than in the fry of *Liza ramada* or *Liza saliens* of this size, and the mid-lateral line is less obvious. The colour pattern of *Liza saliens* is very similar to that of *Liza ramada*. The fry of both *Chelon labrosus* and *Oedalechilus labeo* are, in general, darker in colour than the other four species; parallel stripes along the length of the body can be distinguished in *Chelon labrosus* at about 35 mm length and over. *Chelon labrosus* smaller than 35 mm have vague stripes, but the body is generally darker in colour than that of *Mugil cephalus*, *Liza ramada*, *Liza saliens* or *Liza aurata*.

The smallest *Oedalechilus labeo* obtained was 37.4 mm; all other specimens were about 46 mm or over. These all had dark blackish-grey stripes along the sides and a comparatively dark head. It is of interest that only two specimens were found in Bardawil Lagoon (Zismann & Ben-Tuvia, 1975). They were so light in colour that they were recognizable only by the upper lip and anal fin.

The smallest *Liza carinata* observed from the Mediterranean shore were 67.8 and 69.0 mm, so the colour pattern of fry is not known for this region. The juveniles are dark grey on the back; this colour fades gradually and smoothly to a silver white on the belly, with almost no definite markings along the sides.

The head

It has been pointed out by Schultz (1946), Thomson (1966; also see Chap. 1) and Ebeling (1957), that the features of the head are usually the most important characters for identifying the grey mullet species. This is also true for the fry and juveniles, although many of the features in the adults that are

Fig. 2.7. Outline drawings of the snouts of six species of Mediterranean grey mullets: (*a*) *Mugil cephalus* (27.2 mm total length), adipose eyelid barely covers edge of iris, (*b*) *Mugil cephalus* (42.5 mm total length), adipose eyelid covers about 1/3 iris, (*c*) *Liza aurata* (40.8 mm total length), note teeth in upper lip, (*d*) *Liza saliens* (32.9 mm total length), note teeth in upper lip, (*e*) *Liza carinata* (69.5 mm total length), teeth in upper lip not visible, (*f*) *Chelon labrosus* (37.2 mm total length), (*g*) *Oedalechilus labeo* (67.2 mm total length), horny scalloped ridges missing from both lips.

(a)

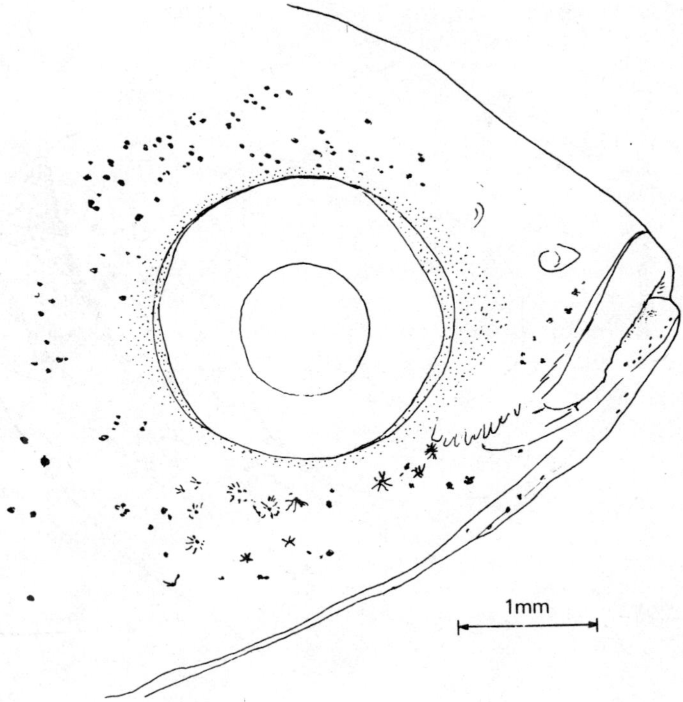

(b)

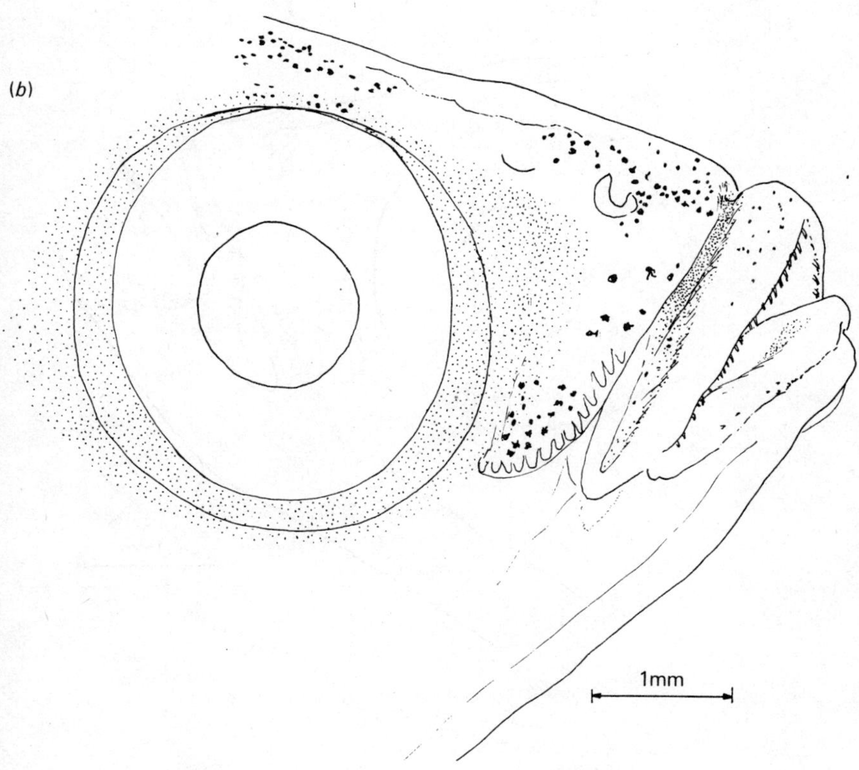

(c)

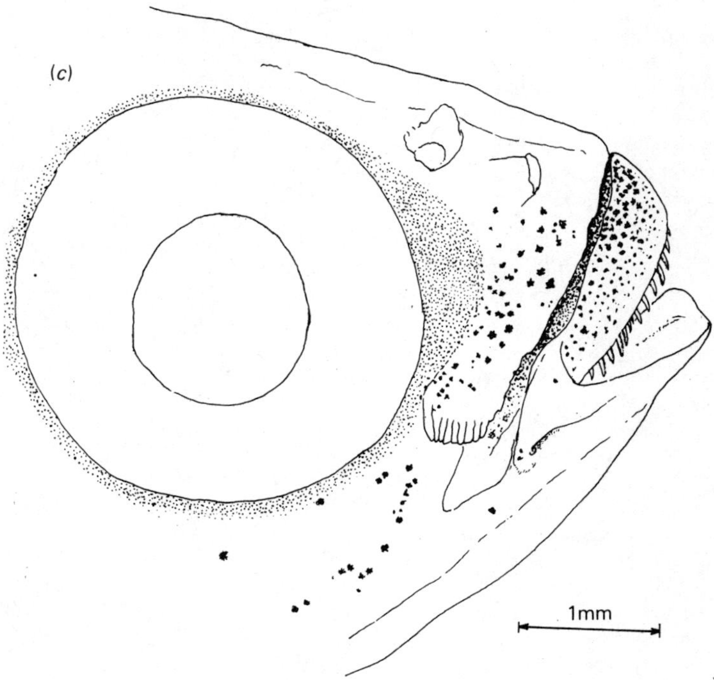

(d)

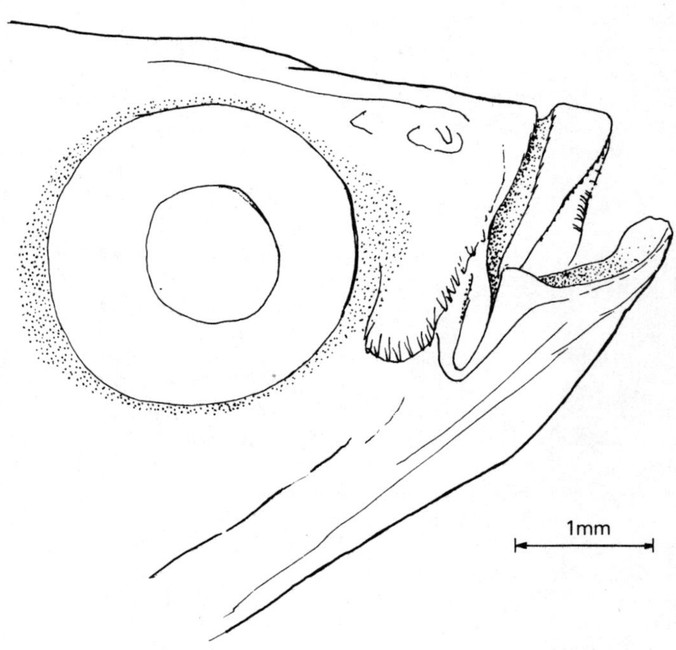

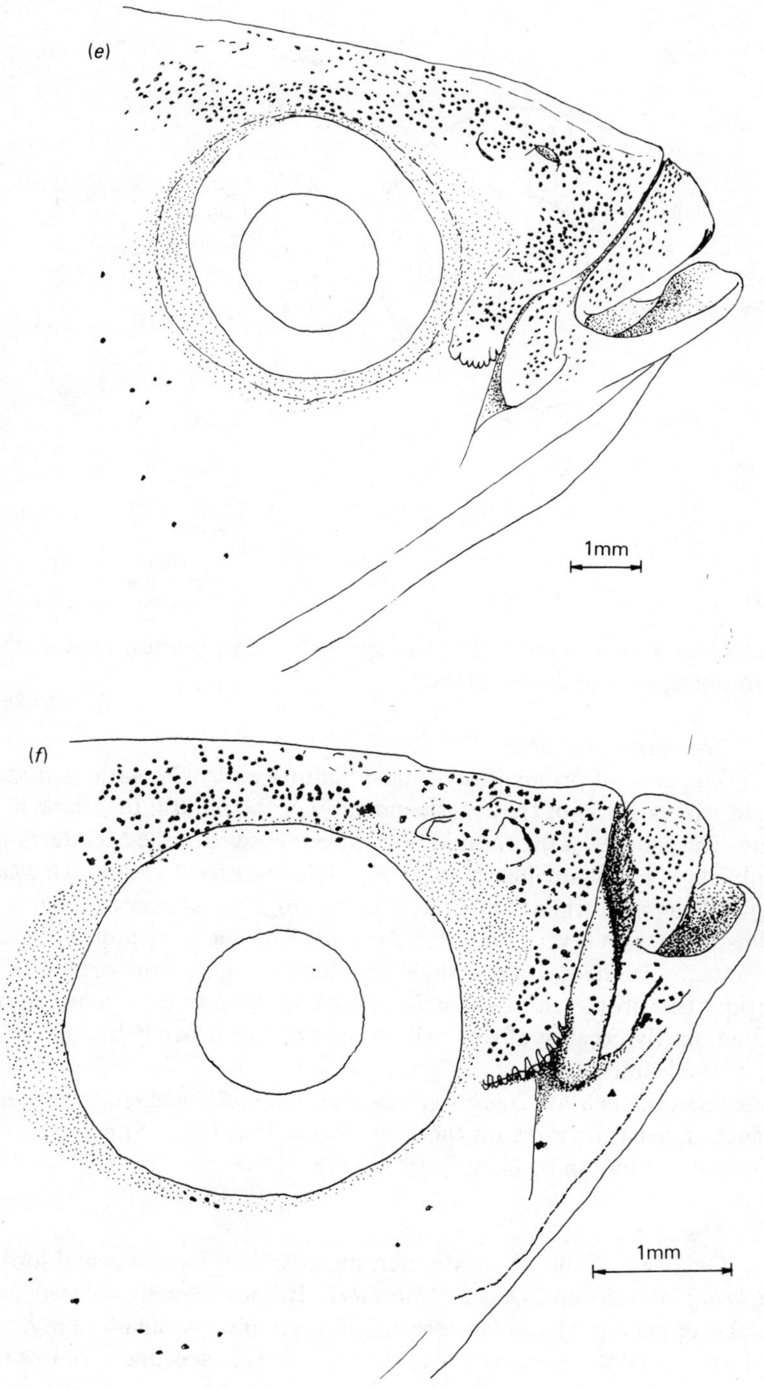
(e)
1mm
(f)
1mm

(g)

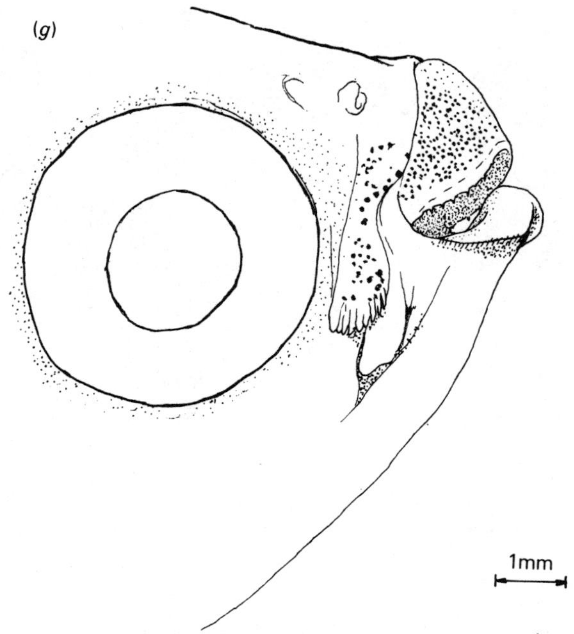

used as identification marks (adipose eyelids, ornamentation on the upper lips) do not appear at these stages.

The general profile

The general profile is, of course, influenced by the various parts of the head: the position of the eye, the position of the mouth, the shape of the lips, etc. The profiles of the snout of the various species of the Mediterranean grey mullets of Israel are illustrated in Fig. 2.7. The profile of *Mugil cephalus* is deep and rounded when compared with that of *Liza saliens* or *Liza ramada* (not illustrated, as it is very similar to *Liza saliens*), which are more elongated and pointed. *Liza aurata* has a slightly different angle in the profile of the upper lip; the profiles of *Chelon labrosus* and *Oedalechilus labeo* appear truncated, partly because of the higher upper lips, and partly because of the shape of the head.

Some species, such as *Crenimugil crenilabis* and *Rhinomugil squamipinis*, have such unusual features on the head region that they require very little additional information to identify them (Fig. 2.8).

The lips

Only two of the seven Mediterranean species have unusual-looking lips: *Chelon labrosus* and *Oedalechilus labeo*. In the commercial-sized specimens, the upper lip of *Chelon labrosus* is high and swollen-looking, with several rows of protuberances above the free edge. In specimens of less than

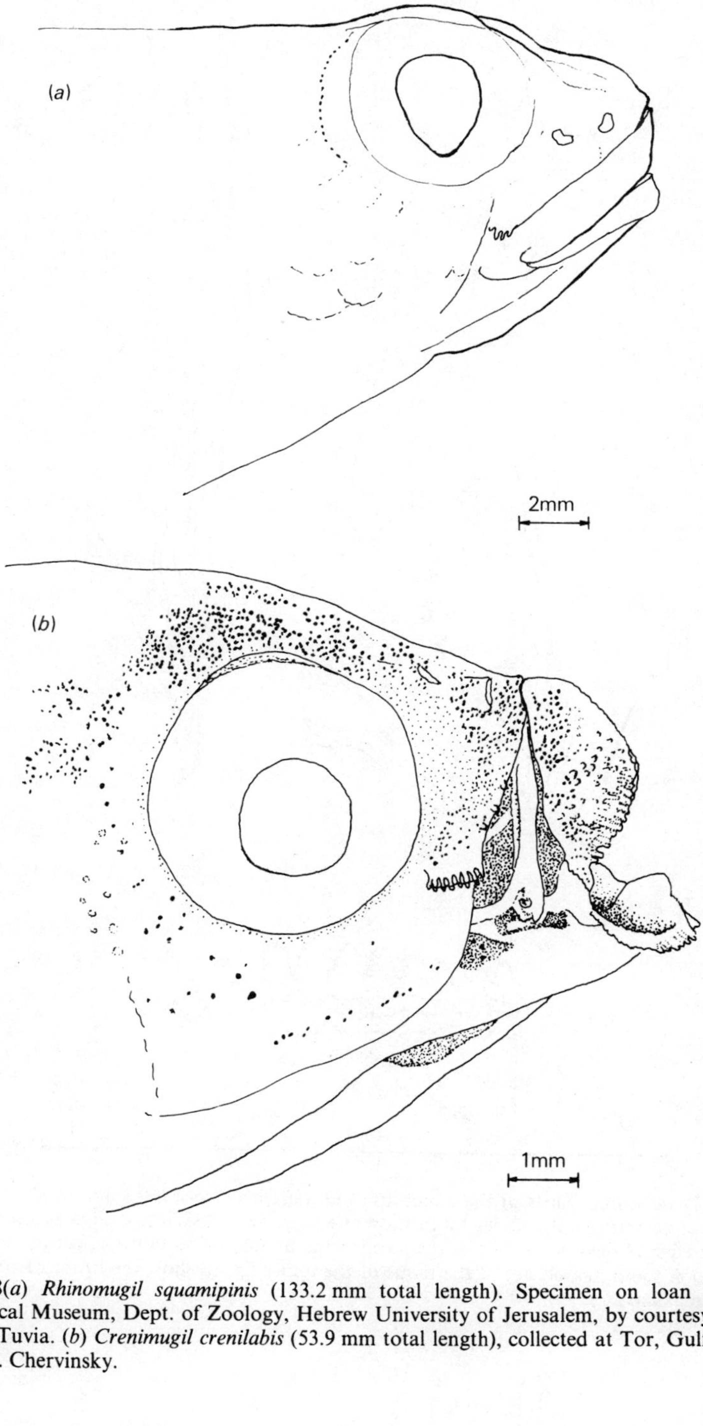

Fig. 2.8(*a*) ***Rhinomugil squamipinis*** (133.2 mm total length). Specimen on loan from the Zoological Museum, Dept. of Zoology, Hebrew University of Jerusalem, by courtesy of Prof. A. Ben-Tuvia. (*b*) *Crenimugil crenilabis* (53.9 mm total length), collected at Tor, Gulf of Suez, by Dr J. Chervinsky.

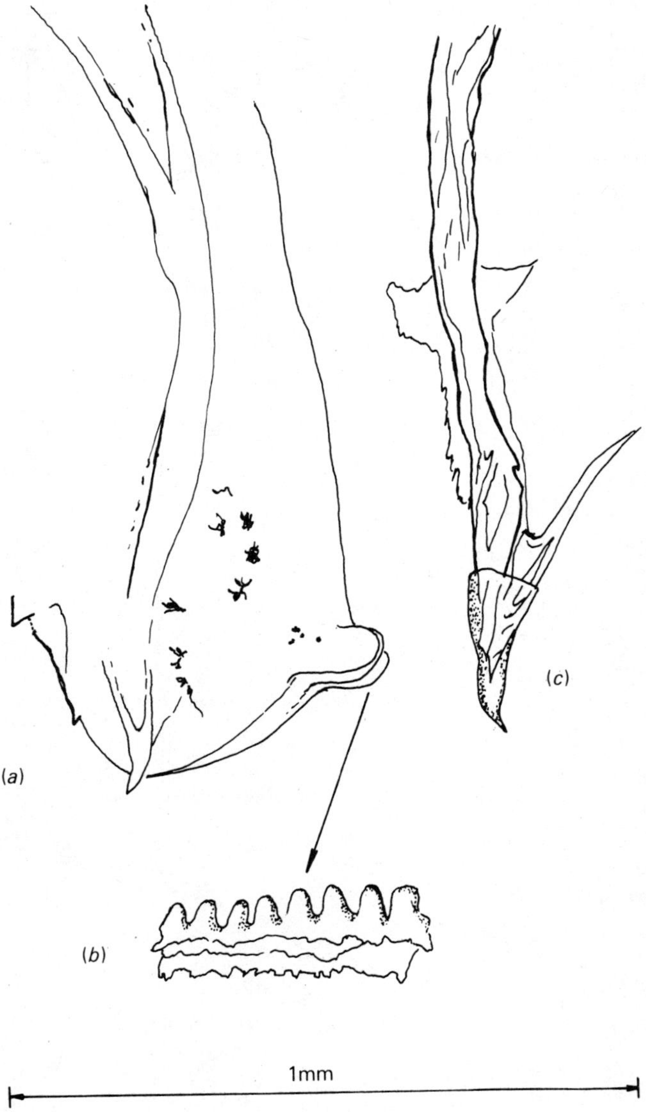

Fig. 2.9. Dissection of parts of the upper lip of *Oedalechilus labeo* (69.8 mm total length): (*a*) Sagital section through the upper lip to show the position of the horny ridge and teeth along the lower edge of the lip, (*b*) Part of the horny ridge broken off from the upper lip, view from above, (*c*) A tooth, teased out of the tissue of the upper lip, to show the ligament connecting it to the maxillary.

100 mm, the protuberances are not yet developed, but the upper lip is swollen in appearance. In *Oedalechilus labeo*, the high upper lip has a fringe of small projections along the free edge in commercial-sized specimens. In specimens below 100 mm, this fringe is not visible, but the peculiar, horny scallops along the edges of both the upper and lower lips (Fig. 2.9) are visible under low magnification. In the other five species, the upper lips are more-or-less narrow. In the Red Sea *Crenimugil crenilabis* (Fig. 2.8*b*), the peculiar folds of the upper and lower lips are obvious even in specimens of 53 mm.

The labial teeth

The teeth and their use in identification of genera and species have been described and discussed by Schultz (1946), Thomson (1954; also see Chap. 1) and Ebeling (1957). In the Agonostominae, the sessile teeth are attached directly to the jaws. The teeth of the Mugilinae do not conform to the general idea of 'teeth' in the sense that they are very small, bristle-like, and are not attached directly to the maxillary, but emerge along the edge of the lips and are attached to the maxillary indirectly by long ligaments (Fig. 2.9*c*).

Ebeling (1957) investigated the morphology and histology of mullet teeth from species of the eastern Pacific, and proposed a key based on these findings. He noted that the premaxillary (upper lip) teeth appeared before those in the lower lip, and that the number of teeth increased with the growth of the fish. The rows of replacement teeth are external to the functional teeth, and smaller teeth give way to larger teeth as the fish matures. The teeth are attached to the jaw by long fibres; there is a suture between the tip and the base, and the tips are easily broken. Wallace & van der Elst (1973) have used scanning electromicrographs to examine the dentition of 40 mm length fry.

In some species, such as *Crenimugil crenilabis*, there are no teeth. In others, such as *Liza carinata*, *Liza ramada* and *Liza saliens*, the teeth are so small that they are difficult to see even under magnification. The teeth of the upper lip in *Mugil cephalus* of 20–40 mm in length are visible in most specimens at a magnification of ×12. The teeth in the lower lip are visible in specimens over 30 mm in length (Fig. 2.4*a–b*). In the fry of *Liza ramada* and *Liza saliens*, the teeth are very small and are visible only under the magnification of ×6 or ×12.

In *Liza aurata* of 30 mm or less in length, the teeth in the upper lip are small, even under ×12 magnification. At sizes of over 40 mm, the teeth of *Liza aurata* are clearly seen at ×12 or even ×6 magnification. At sizes of about 50 mm or more, the teeth are visible, as a scintillating fringe, to the unaided eye, when examined in good light. Under the magnification of ×6, these teeth appear as fine, claw-like structures along the edge of the upper lip. In *Liza carinata* of 60 mm or more in length, the small teeth are visible only under magnification. In *Chelon labrosus* of 20–30 mm, the teeth are visible at a magnification of ×12. In specimens of over 50 mm, the teeth are visible at ×6.

The teeth in the lower lip do not protrude beyond the skin in *Liza ramada*, *Liza aurata*, *Liza saliens* and *Chelon labrosus*. In *Liza carinata*, no teeth were seen in the lower lip, even under the magnification of ×25.

When the upper and lower lips of *Oedalechilus labeo* are viewed from the side under a dissecting microscope, and the fish moved slightly from side to side, several rows of teeth can be seen in specimens of over 50 mm in length. In addition to the teeth, there is a row of evenly-spaced, horny projections near the free edge of the upper lips and along the edge of the lower lips (Fig. 2.9*a–b*). These projections were found on most of the specimens examined, but as they are very easily broken off, they are partly or entirely missing from the rest of the specimens. These projections are yellowish brown and opalescent when viewed under transmitted light.

A small section across the height of the upper lip was cut out with fine scissors; this section was further sliced with the spear-headed probe so that a slice of a little less than 1 mm was obtained. The position of the horny ridge in relation to the edge of the lip and the teeth could then be seen (Fig. 2.9*a–c*).

In the field, the presence or absence of teeth projecting beyond the skin of the lip can be determined with the aid of a good hand lens, in good light. In the field laboratory, the shape and arrangement of teeth can be examined by dissecting the upper lip, and mounting it in water on a slide, preferably with a cover slip, and viewing it by transmitted light.

One serious drawback in using the teeth as an identification character is that the teeth are easily broken. Replacement teeth grow in, but normally toothed species may have the teeth missing from a large portion of the lips, and so may appear to be without teeth or with very small teeth. On the other hand, a specimen in which the tissues have shrunk as a result of poor preservation, or in which the lip has been torn, may have the teeth exposed, so that it appears to have much longer teeth than normal.

The lingual teeth

As Thomson (Chap. 1) has pointed out, the lingual teeth of the Agonostominae are canine-like and scattered all over the tongue. In Mugilinae, the teeth occur in patches, which often share a common basal plate. Anderson (1958) examined the tongues of fry of grey mullet, and showed that the pattern of teeth on the tongue of *Mugil cephalus* changes with growth; he found no teeth on the tongue of *Mugil curema*. Demir (1971) found differences in size, number and pattern of the teeth in the two species, *Crenimugil labrosus* (*Chelon labrosus*) and *Liza auratus* (*Liza aurata*). The tongues of *Mugil cephalus*, *Liza ramada*, *Liza aurata*, *Liza saliens* and *Chelon labrosus*, of the size range of 20.0–30.7 mm, were dissected, mounted in water and glycerine, and the teeth patterns drawn (Fig. 2.10). On examination of tongues of

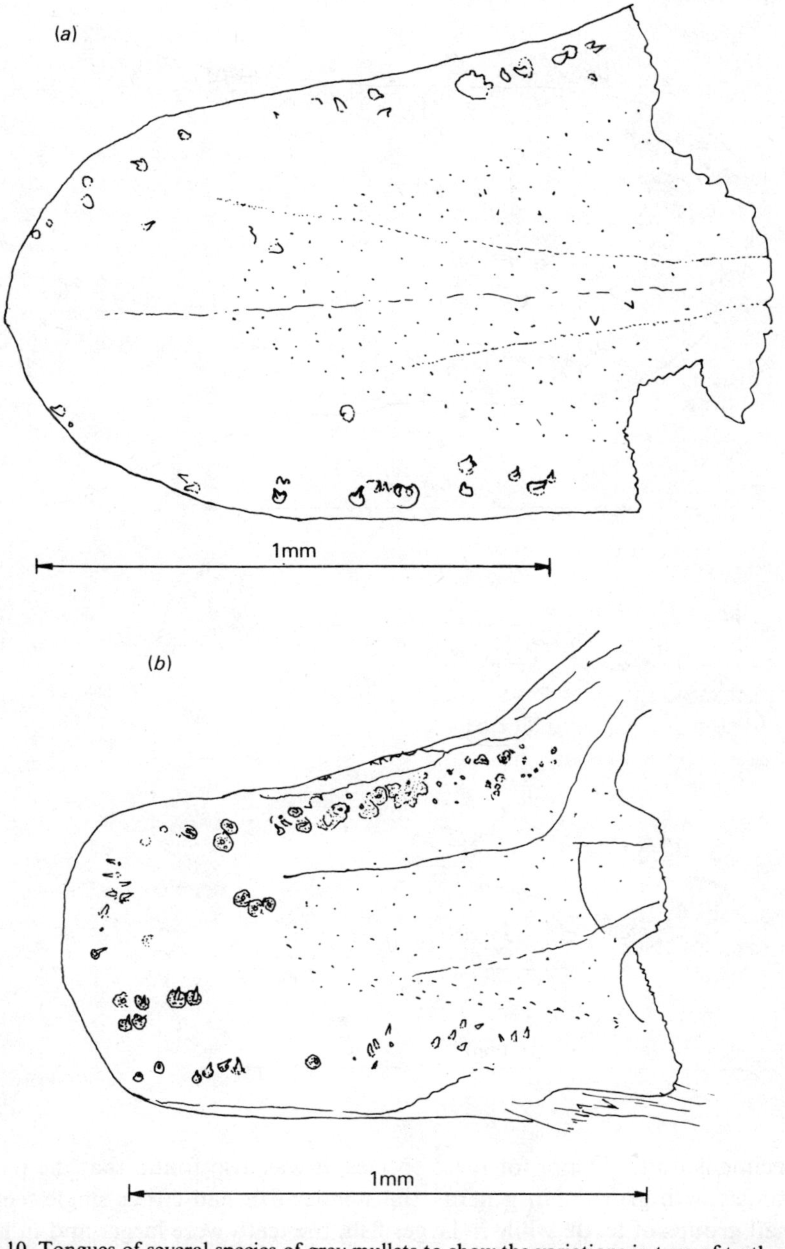

Fig. 2.10. Tongues of several species of grey mullets to show the variations in type of teeth and their distribution on the tongue: (*a*) *Mugil cephalus* (30.7 mm total length), (*b*) *Liza ramada* (23.4 mm total length), (*c*) *Liza aurata* (27.9 mm total length), (*d*) *Liza saliens* (20 mm total length), (*e*) *Chelon labrosus* (27.2 mm total length).

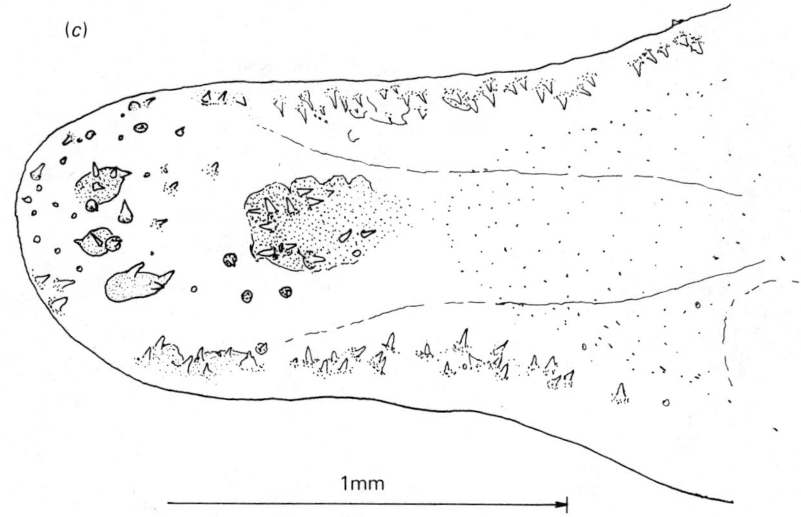

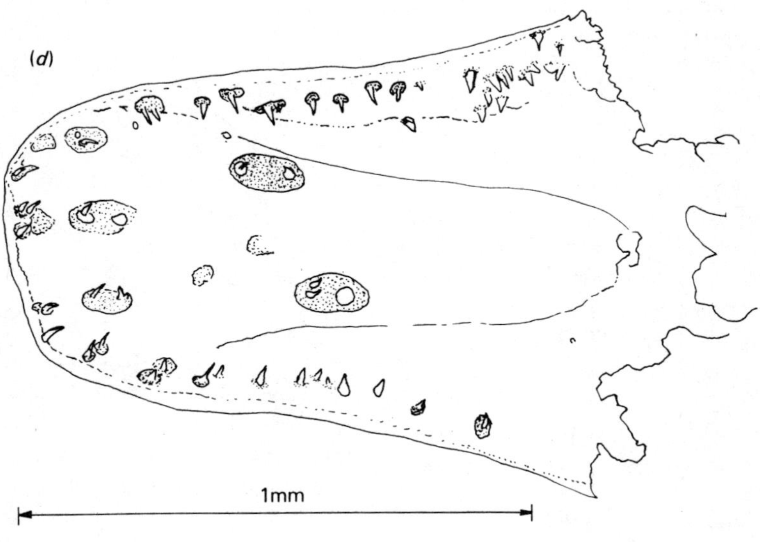

specimens up to 70 mm for these species, it was also found that the pattern changes with growth. In general, the smaller fish had either single teeth or small groups of teeth, while in larger fish, the teeth were larger and in larger groups on common bases. However, when several species of the same size-range are compared, differences in the pattern and type of teeth can be seen. This means of identification may be useful occasionally; for example, in Israel, it is sometimes very difficult to identify positively a specimen under

(e)

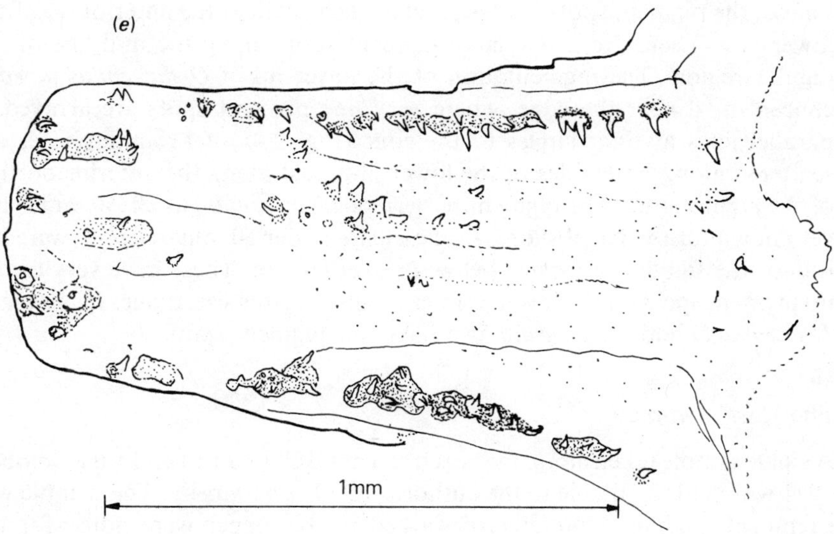

30 mm as being either *Liza aurata* or *Liza ramada*, especially if the upper lip is injured. The pattern of the lingual teeth for these two species differ quite markedly.

The jugular region

The outline of the lower jaw, and the pigment pattern along its edge and along the ventral, anterior part of the head, is one of the main set of external characters used to identify the seven species of fry and juveniles in Israel. It was found to be characteristic for each of the species.

The angle formed by the lower jaw (as well as the pigment pattern) is used to separate the fry of *Mugil cephalus* from those of the other species (Fig. 2.4). This angle was used also by Trewavas & Ingham (1972), who compared the ratio of the width to the length of the lower jaw, in order to separate the young of *Mugil capurii* from *Mugil cephalus* in the Atlantic.

The fine pigment spots on the lower jaw of *Mugil cephalus* fry outline the jaws, emphasizing the acute angle, and form a line along the centre of the jugular region. In *Liza aurata* (30–40 mm in length), the pigment spots form two slightly irregular rows, outlining the jaws in an arch. There may be a few scattered spots in the jugular area. Pigment spots are also concentrated at the corners of the mouth. In *Liza ramada* of about the same size, the pigment spots are concentrated near the anterior tip of the jaw, not forming a clear pattern, and at the corners of the mouth; the jugular region has very few spots. The fry of *Chelon labrosus* have numerous small, evenly scattered pigment spots across the lower lip and the anterior part of the jugular region. In *Liza saliens*, the pigment spots are few, and concentrated at the anterior tip of the lower jaw. There are a few large pigment spots along the mid-line of the jugular region. The pigmentation of the lower lip of *Oedalechilus labeo* is concentrated near the edge, where very fine pigment spots are arrayed in parallel lines at right angles to the edge (Fig. 2.4*f*). Chromatophores are scattered along both sides of the lower jaws and along the anterior portion of the jugular area. The pigment pattern of *Liza carinata* at 20–50 mm is not yet known for the Israel shore. *Liza carinata* under 50 mm in length were not found; the smallest were just below or over 60 mm. These have very few or no pigment spots on the lower jaws or jugular; at this size-range, *Liza saliens*, *Liza ramada* and *Liza aurata* also have few pigment spots.

The 'test' sample

A single sample taken along the beach at Tor (Gulf of Suez) on 18–19 October, 1971 was made available to the author by Dr J. Chervinsky. This sample was examined, and the characters described in this paper were noted for the specimens. The data are given in Appendixes 2.1–2.3.

First the sample was separated roughly into two groups: those with an adipose eyelid which covers part or all of the iris, and those with no adipose eyelid protruding over the iris. This could be seen with the unaided eye. Those with no adipose eyelid over the iris, or appearing to have none, were examined more closely; on the larger specimens, about 80 mm in length, the upper lip could be seen as rather high, with a peculiar edge. Under × 6 magnification, it was seen that both lips had many folds along the free edges. A few fish were different from the above-mentioned groups. In the subsample in which the adipose eyelid was seen, one specimen appeared different from the others, but very similar to *Mugil cephalus* of the Mediterranean. Tentatively, the subsamples were labelled:

species A – those with adipose eyelid, anal III, 9

species B – those with no adipose eyelid, but with crenulated lips, anal III, 9

species C – those with no adipose eyelid, no crenulated lips, anal III, 9

species D – one specimen only but external appearance similar to *Mugil cephalus*, anal III, 8

The fish were then measured, weighed and morphometric data recorded. Two specimens, each different from the other and from the above-listed groups, were set aside for the time being. Using Ben-Tuvia's (1975) key, species A was identified as *Liza carinata* and species B as *Crenimugil crenilabis* (this was confirmed by Prof. A. Ben-Tuvia). The specimen in species D appears to be *Mugil cephalus*, as it has two pyloric caeca, III, 8 in the anal fin, forty-one scales along the lateral line (which agrees with the range found in the Mediterranean, see Table 2.5) and had both cycloid and ctenoid scales and teeth in both upper and lower jaws. The acute angle of the lower jaws also agrees with this identification. Of the two remaining specimens, one is similar to *Liza aurata*, but the teeth appear slightly different, and there are seven pyloric caeca of almost equal size, divided into two groups. The second unknown specimen has a small adipose eyelid, with teeth along the edges of both lips. As these two were only single specimens, they were set aside for further study, when additional specimens will be available.

Discussion and conclusions

The study of morphometic and other characters in grey mullet fry (20–59 mm in total length) and juveniles (60–100 mm in total length) has shown that the keys devised for the identification of grey mullets of about 200 mm or more are not applicable for fry.

The rapid changes in the morphology, and the changes in the proportion of one part of the body to another, occurring during the early stages, make it necessary for the fry of the various species to be compared one with another only within narrow size-ranges. The smaller the fish, the more difficult it is to separate the species.

It is suggested that the length intervals taken for comparing fry of 20–60 mm be limited to 20 mm, that is, that the first size-group range from 20.0–39.9 mm and the second from 40.0–59.9 mm. For certain characters, such as the pigmentation on the jugular region, or visibility of the teeth in the upper lip, it may be necessary to subdivide the first group further, using 10 mm intervals.

Characters which remain constant at all sizes:

(1) Number and arrangement of pyloric caeca – can be used in field work.

(2) Number of scales along the lateral line – useful only in the laboratory.

(3) Total number of elements in the anal fin – useful in the laboratory.

Characters which change with size:

(1) Number of gill rakers on the first gill arch – not useful for fish below 50 mm total length; only for laboratory work.

(2) Number of ctenoid (or digitated) scales along the transverse row – only for laboratory work.

Of the other characters used for identification of fry, the general appearance and colour of the fry can be used for identification only after verification that they are characteristic under all conditions. The upper and lower lips, their shape, colour and dentition may be useful even in the field, if a good hand lens is used. Sophisticated means of differentiation of species by means of skeletal parts (Hotta & Tung, 1966; Luther, 1975), or the study of various parts by a scanning electron microscope (Wallace & van der Elst, 1973), are suitable for large institutes which have the complex equipment, trained personnel and the finance.

Upon the completion of studies, now in progress, on the immunological differentiation of mugilid species, a field test could be developed based on the results obtained (Herzberg, 1975, and personal communication).

Local efforts based on a world-wide standardized pattern of observations and recording of data may eventually lead to the production of keys for the identification of grey mullet fry in each oceanic area.

Appendix 2.1

Key for identification of Mediterranean grey mullets of Israel
150 mm total length or longer

1*a* Adipose eyelid covering all or part of iris 2
1*b* Adipose barely covering edge of iris or not at all 3
2*a* Anal fin III, 8; pyloric caeca 2 *Mugil cephalus*
2*b* Anal fin III, 9; pyloric caeca 5–6 *Liza carinata*
3*a* Upper lip high with ornamentation (papillae or projections) 4
3*b* Upper lip narrow, with no ornamentation, teeth may or may not be visible 5
4*a* Anal fin III, 11; pyloric caeca usually 6; upper lip high with projections along free edge *Oedalechilus labeo*
4*b* Anal fin III, 9; pyloric caeca usually 6; upper lip high with 2–3 rows of papillae along lower half of lip *Chelon labrosus*
5*a* Predorsal scales with several grooves in each; pyloric caeca in 2 groups: long and short *Liza saliens*
5*b* Predorsal scales with single groove 6
6*a* Posterior margin of naked patch at tip of snout reaches at least to anterior nares; large scales border patch; tip of pectoral fin almost reaches pupil of eye when fin is folded forwards; teeth in upper lip visible to eye; pyloric caeca 6–11, usually 8 or 9 *Liza auratus*
6*b* Posterior margin of naked patch at tip of snout does not reach anterior nares; several rows of small scales border patch; tip of pectoral fin does not reach, or barely reaches posterior rim of eye when fin is folded forwards; teeth in upper lip barely or not visible to naked eye; pyloric caeca 6–9, usually 7 or 8 *Liza ramada*

Appendix 2.2

Key for identification of Mediterranean grey mullet fry of Israel
20–60 mm total length

1*a* Pyloric caeca 2; anal fin elements 11; angle of mandibles acute; in specimens over 30 mm, teeth in both jaws *Mugil cephalus*
1*b* Pyloric caeca more than 2; angle of mandibles obtuse 2
2*a* Anal fin elements 14; pyloric caeca usually 6; heavy pigmentation on both lips and anterior jugular area *Oedalechilus labeo*
2*b* Anal fin elements 12 3
3*a* Pyloric caeca in two size groups (long plus short) *Liza saliens*
3*b* Pyloric caeca in one group almost equal in size 4
4*a* Numerous fine, black pigment spots across anterior end of jugular and lower lip; upper lip appears thick, with small teeth visible at magnification of $\times 12$ or $\times 6$; lateral line scales 38–44; gill rakers 53–65 at 50.0–59.9 mm length *Chelon labrosus*

4b Few black pigment spots around anterior part of jugular and lower lip; upper lip narrow — 5

5a Upper lip with teeth, visible to eye in specimens over 35 mm total length; pyloric caeca 6–11, usually 8 or 9; body with herring-bone pattern or stripes; lateral line 40–46 — *Liza aurata*

5b Upper lip with no teeth visible to eye, may be visible at magnification of ×6 or greater — 6

6a Lateral line count 35–40; pyloric caeca 5–6; no teeth visible in upper lip — *Liza carinata*

6b Lateral line count 40–45; pyloric caeca 6–9, mostly 7 or 8; teeth in upper lip may be visible at magnification of ×6 or ×12 — *Liza ramada*

Appendix 2.3

Data on specimens of grey mullets from test sample taken at Tor, Red Sea, 18–19 October 1971

Specimen no.	TL (mm)	SL (mm)	Wt. (g)	Anal fin	Lat. line	Ctenoid or digitated scales in transverse row/total	Magnification at which teeth visible	No. of pyloric caeca	No. of gill rakers	Portion of iris covered by adipose eyelid
Species A. Adipose eyelid over part of iris, lips without ornamentation, ctenoid and cycloid scales[a]										
1	58.4	46.1	2.19	III, 9	36	9/14	—	5	46, 1, 24	$\frac{1}{4}$
2	69.4	53.4	2.63	III, 8	38	11/13	—	6	45, 1, 26	$\frac{1}{3}$
3	72.5	57.9	4.10	III, 9	37	11/14	12×	5	50, 1, 27	$\frac{1}{2}$
4	73.6	58.0	3.81	III, 9	39	15/15	12×	5	48, 1, 28	$\frac{1}{2}$
5	108.6	85.8	14.91	III, 9	37	14/14	12×	5	55, 1, 32	All
6	65.3	50.9	2.92	III, 9	37	9/14	12×	5	47, 1, 27	$\frac{1}{3}$
7	66.5	53.2	3.00	III, 9	36	11/14	12×	6	45, 1, 25	$\frac{1}{3}$
8	96.3	78.5	10.86	III, 9	36	14/14	6×	6	53, 1, 33	$\frac{1}{2}$–$\frac{3}{4}$
9	59.2	47.2	2.20	III, 9	37	5/14	12×	5	43, 1, 26	$\frac{1}{3}$
10	56.8	44.6	1.64	III, 9	36	11/14	12×	5	43, 1, 24	$\frac{1}{4}$

Species B. Adipose eyelid does not cover iris, both lips with crenulations, digitated and cycloid scales, small axillary scale in fish over 54 mm.[b]										
1	79.9	63.7	6.24	III, 9	40	4/15	No teeth	10	44, 1, 35	—
2	77.0	61.2	4.66	[c]	37	3/14	No teeth	9	45, 1, 32	—
3	71.2	54.7	4.12	III, 9	41	4/14	No teeth	10	42, 1, 32	—
4	63.2	51.1	3.31	III, 9	37	3/15	No teeth	9	40, 1, 27	—
5	61.7	49.0	2.86	III, 9	39	3/15	No teeth	8	42, 1, 27	—
6	58.9	50.2	2.48	III, 9	39	3/15	No teeth	8	42, 1, 28	—
7	59.0	47.2	2.54	III, 9	38	2/16	No teeth	10	41, 1, 27	—
8	58.8	47.1	2.48	III, 9	38	3/15	No teeth	9	40, 1, 26	—
9	54.9	43.4	1.91	III, 9	37	3/15	No teeth	10	37, 1, 25	—
10	53.4	42.6	1.56	III, 9	39	2/15	No teeth	8	37, 1, 25	—
Species C. snout much shorter than in species A or B, adipose eyelid over anterior part of iris only, scales cycloid or with digitated posterior edge, no teeth visible in upper lip										
1	91.7	72.0	7.75	III, 9	39	16/16	0 at 25 ×	7	47, 1, 33	$\frac{1}{4}$
2	84.8	66.2	5.73	III, 9	37	15/15	0 at 25 ×	6	46, 1, 31	$\frac{1}{4}$
3	76.8	60.3	5.50	III, 9	37	4/15	0 at 25 ×	8	41, 1, 29	$\frac{1}{3}$
4	75.5	60.9	4.67	III, 9	39	6/15	0 at 25 ×	7	42, 1, 28	$\frac{1}{3}$
5	64.3	51.2	2.85	III, 9	37	6/14	0 at 25 ×	7	—	$\frac{1}{8}$
Species D. Adipose eyelid covers iris, acute angle formed by lower jaws, ctenoid and cycloid scales										
1	84.9	69.4	8.54	III, 8	41	6/15	6 ×	2	—	All

[a] 18 specimens (56.0 – 108.6 mm TL.) not tabulated.
[b] 3 specimens (51.0 – 53.6 mm TL) not tabulated.
[c] Injured.

References

Anderson, W. W. (1957). Larval forms of the fresh-water mullet (*Agonostomus monticola*) from the open ocean off the Bahamas and South Atlantic coast of the United States. *Fish. Bull. US Fish Wildl. Serv.* **57**(120), 415–25.

Anderson, W. W. (1958). Larval development, growth and spawning of striped mullet (*Mugil cephalus*) along the South Atlantic coast of the United States. *Fish. Bull. US Fish Wildl. Serv.* **58**(144), 501–19.

Ben-Tuvia, A. (1953). Mediterranean fishes of Israel. *Bull. Sea Fish. Res. Stn. Isrl.* **8**, 1–40.

Ben-Tuvia, A. (1966). Red Sea fishes recently found in the Mediterranean. *Copeia*, (2), 254–75.

Ben-Tuvia, A. (1975). Mugilid fishes of the Red Sea with a key to the Mediterranean and Red Sea species. *Bamidgeh* **27**(1), 14–20.

Bograd, L. (1955). Identification of the grey mullets of the Israel Mediterranean coast. *Commer. Fish. Bull. Isr.*, (5), 22–4. (in Hebrew.)

Bograd, L. (1961). Occurrence of *Mugil* in the rivers of Israel. *Bull. Res. Coun. Isr. (B)*, **9**(4), 169–90.

Demir, N. (1971). On the occurrence of grey mullet postlarvae off Plymouth. *J. Mar. Biol. Assoc. UK*, **51**(2), 235–46.

Ebeling, A. W. (1957). The dentition of Eastern Pacific mullets with special reference to adaptation and taxonomy. *Copeia*, (3), 173–84.

Heldt, H. (1948). Contribution à l'étude de la biologie des Muges des lacs Tunisiens. *Bull. Stn. Océanogr. Salammbô*, (41), 1–35.

Herzberg, A. (1975). The use of carp (*Cyprinus carpio*, 1758) instead of rabbit as an antibody-donor in immunological research on fishes. *Anim. Blood Grps. Biochem. Genet.* **6**, 215–20.

Herzberg, A. & Pasteur, R. (1975). The identification of grey mullet species by disc electrophoresis. *Aquaculture*, (5), 99–106.

Hotta, H. & Tung, I. S. (1966). Identification of the fishes of the family Mugilidae based on the pyloric caeca and the position of insertion of the first interneural spine. *Jap. J. Ichthyol.* **14**(1–3), 62–6 (in Japanese, English summary).

Hubbs, C. L. (1943). Terminology of early stages of fishes. *Copeia*, (4), 260.

Hubbs, C. L. (1944). Fin structure and relationships of the phallostethid fishes. *Copeia*, (2), 69–79.

Lowe-McConnell, R. H. (1968). Identification of freshwater fishes. In *Methods of Assessment of Fish Production in Fresh Waters*, ed. W. E. Ricker, IBP Handbook no. 3, pp. 46–77. Blackwell Scientific Publications, Oxford & Edinburgh.

Luther, G. (1975). New characters for consideration in the taxonomic appraisal of grey mullets. *Aquaculture*, **5**, 107 (abstract).

Moore, R. (1974). General ecology distribution and relative abundance of *Mugil cephalus* and *Mugil curema* on the South Texas coast. *Contrib. Mar. Sci.* **18**, 241–55.

Pakrasi, B. & Alikunhi, K. M. (1952). On the development of the grey mullet *Mugil corsula* Hamilton. *J. Zool. Soc. India*, **4**(2), 123–40.

Perlmutter, A., Bograd, L. & Pruginin, J. (1957). Use of the estuarine and sea fish of the Family Mugilidae for pond culture in Israel. *Proc. Tech. Pap. Gen. Fish. Coun. Mediterr.* (4), 289–304.

Popov, A. M. (1930). Mullets of Europe (Mugilidae) with description of a new species from the Pacific waters of the USSR. *Tr. Sevastop. Biol. Stn.* (2), 47–125.

Pruginin, J. & Kitai, H. (1957). Cultivating mullets as a secondary fish in the carp ponds. *Bamidgeh*, **9**(4), 70–5.

Sarojini, K. K. (1951). The fishery and biology of the Indian grey mullets – a review. *J. Zool. Soc. India*, **3**(1), 159–79.

Schultz, L. P. (1946). A revision of the genera of mullets, fishes of the Family Mugilidae with descriptions of three new genera. *Proc. US Natl. Mus.* **96**(3204), 377–95.

Smith, J. L. B. (1935). The fishes of the Family Mugilidae. *Ann. S. Afr. Mus.* **30**(19), 587–644.

Thomson, J. M. (1954). The Mugilidae of Australia and adjacent seas. *Aust. J. Mar. Freshwater Res.* **5**(3), 70–131.

Thomson, J. M. (1966). The grey mullets. In *Oceanography and Marine Biology*, ed. H. Barnes, vol. 4, pp. 301–35. George Allen & Unwin, London.

Trewavas, E. & Ingham, S. E. (1972). A key to the species of Mugilidae (Pisces) in the northeastern Atlantic and Mediterranean, with explanatory notes. *J. Zool. Lond.* **167**(1), 15–29.

Wallace, J. & van der Elst, R. (1973). Identification of juvenile mullet. *Bull. S. Afr. Assoc. Mar. Biol. Res.* (10), 26–7.

Wimpenny, R. S. (1932). Faune et Flore de la Mediterranée. Liste des fiches composant les cahiers I a XVI classes par ordre zoologique avec le nom de l'auteur de chacune d'elles. *Rap. P-V. Reun. Comm. Int. Explor. Sci. Mer Mediterr. Paris*, (*NS*) **7**, 361.

Yashouv, A. & Berner-Samsonov, E. (1970). Contribution to the knowledge of eggs and early larval stages of mullets (Mugilidae) along the Israel coast. *Bamidgeh* **22**(3), 72–89.

Zismann, L. & Ben-Tuvia, A. (1975). Distribution of juvenile mugilids in the hypersaline Bardawil Lagoon, January 1973 – January 1974. *Aquaculture*, **6**(2), 143–61.

Zismann, L., Berdugo, V. & Kimor, B. (1975). The food and feeding habits of early stages of grey mullets in the Haifa Bay region. *Aquaculture*, **6**, 59–75.

3. FAO identification sheets

FAO

MUGIL

1974

FAO SPECIES IDENTIFICATION SHEETS

FAMILY: MUGILIDAE

FISHING AREAS 57,71
(E Ind. Ocean)
(W Cent. Pacific)

MUGILIDAE

Grey mullets

Elongate fishes, usually with *broad, flattened head,* blunt snout and cylindrical or little compressed body. *Mouth rather small, terminal or inferior;* premaxillae protractile; teeth small, feeble, hidden or absent. Eyes often partly covered by fatty tissue (adipose eyelid). *No lateral line. 2 short dorsal fins, the first with 4 slender spines; pectoral fins set rather high on body;* pelvic fin base about equidistant between pectoral fin base and origin of first dorsal fin; 3 spines in anal fin; caudal fin moderately forked, emarginate or truncate. Scales large or moderate, often with one or more striae to give appearance of longitudinal streaks down sides; modified scales may be present below first dorsal fin and above pectoral and pelvic fins (axillary scales).

Colour: in life, blue/green, green or olive on back, silvery on sides and belly, often with 3 to 9 longitudinal streaks on back, sides and belly; fins hyaline or dusky.

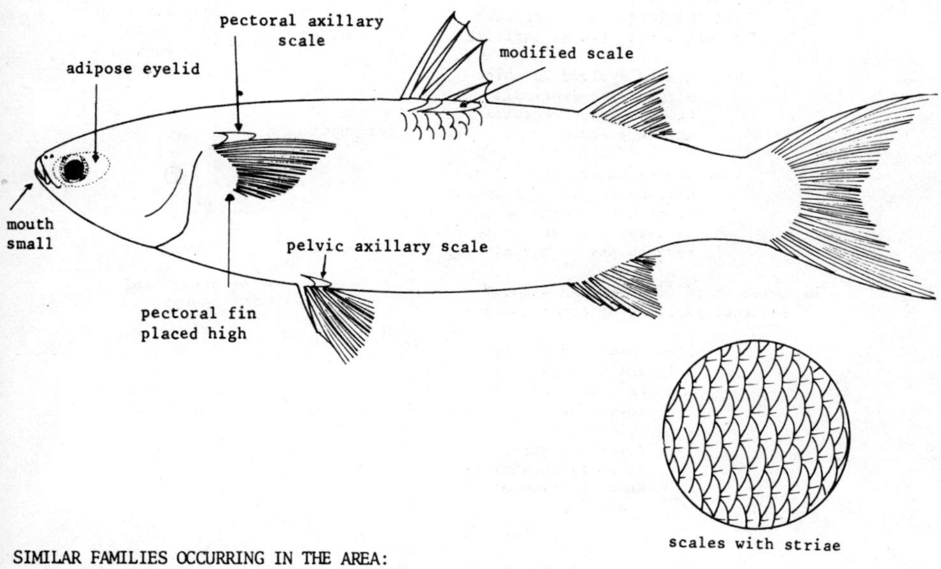

SIMILAR FAMILIES OCCURRING IN THE AREA:

Atherinidae: have silvery stripe along sides, larger eyes and soft anal fin rays usually more than 10 (usually less than 10 in Mugilidae).

FAO Sheets | MUGILIDAE | Fishing Areas 57,71

Key to Genera

1 a. Fleshy lobes between arms of lower jaw (Fig. 1) ..... *Cestraeus*

1 b. No such fleshy lobes (Fig. 2)

2 a. Spine on gill cover above pectoral fin base .... *Sicamugil*

2 b. No spine on gill cover

3 a. Head concave between eyes, the latter projecting above this level; anterior nostril at level of eye centre or lower .... *Rhinomugil*

3 b. Head concave between eyes, the latter not projecting above this level; anterior nostril above level of eye centre

4 a. Lower third of upper lip bearing enlarged papillae or crenellations

5 a. Preorbital deeply notched (Fig. 3); lips with a single row of horny projections .... *Oedalechilus*

5 b. Preorbital not or but little notched (Fig. 4); lips with a multiserial row of papillae

6 a. Scales cycloid but hind margin with denticulations; large pectoral axillary scale .... *Crenimugil*

6 b. Scales either ctenoid or cycloid but without denticulations on hind margin; pectoral axillary scale rudimentary or absent .... *Chelon*

4 b. Lower third of upper lip without enlarged papillae or crenellations

7 a. Lower lip thick; small symphysial knob at front of lower jaw .... *Aldrichetta*

7 b. Lower lip thin; large symphysial knob at front of lower jaw

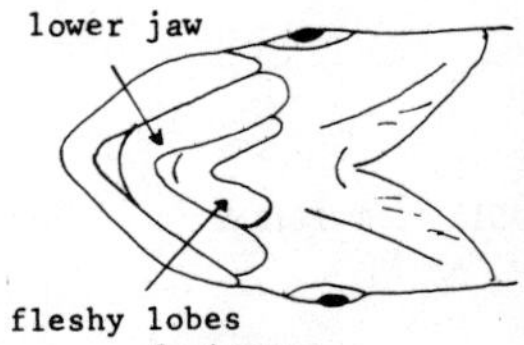

Cestraeus
underside of head

Fig. 1

Liza
underside of head

Fig. 2

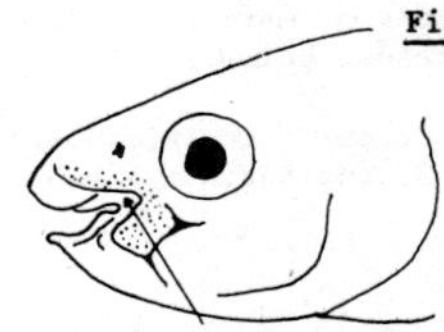

Oedalechilus

Fig. 3

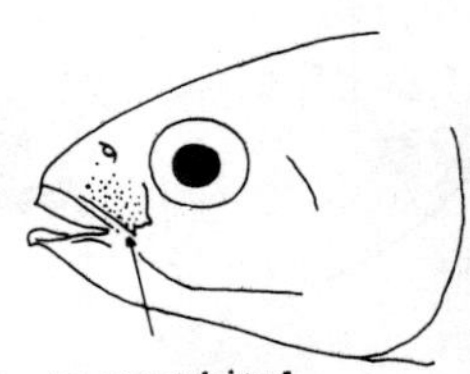

Fig. 4

8 a. Hind tip of maxilla not curved down below tip of premaxilla (Fig. 5); adipose eyelid to pupil in adults .... *Mugil*

8 b. Hind tip of maxilla curved down below tip of premaxilla (Fig. 6); adipose eyelid absent or to iris only

9 a. Teeth present on vomer and palatines (roof of mouth); maxilla tip not greatly curved downward .... *Myxus*

9 b. Teeth absent on vomer and palatines (roof of mouth); maxilla tip strongly curved down at corner of mouth (Fig. 6)

10 a. Scales ctenoid or cycloid but with no digitations on hind margin; tip of maxilla apparent when mouth closed ...... *Liza*

10 b. Scales cycloid, hind margin with digitations; maxilla tip hidden beneath tendon .. *Valamugil*

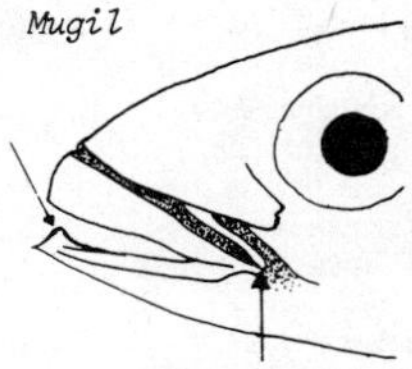

Fig. 5

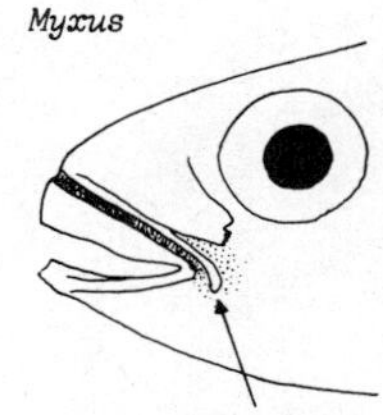

Fig. 6

FAO Sheets MUGILIDAE Fishing Areas 57,71

List of Species occurring in the Area
(Code numbers are given for those species for which Identification Sheets are included)

Aldrichetta forsteri	MUGIL Aldr 1
Cestraeus oxyrhynchus	
Cestraeus plicatilis	
Crenimugil crenilabis	
Crenimugil heterocheilos	
Liza argentea	MUGIL Liza 1
Liza carinata	
Liza macrolepis	
Liza melinoptera	
Liza parmata	
Liza parsia	
Liza subviridis	MUGIL Liza 2
Liza tade	MUGIL Liza 3
Liza vaigiensis	MUGIL Liza 4
Mugil cephalus	MUGIL Mugil 1
Myxus elongatus	MUGIL Myx 1
Myxus petardi	
Oedalechilus labiosus	
Rhinomugil nasutus	
Rhinomugil squamipinnis	
Sicamugil cascasia	
Sicamugil hamiltoni	
Valamugil buchanani	
Valamugil cunnesius	MUGIL Vala 1
Valamugil engeli	
Valamugil georgii	
Valamugil seheli	MUGIL Vala 2
Valamugil speigleri	MUGIL Vala 3

MUGIL Liza 1

1974

FAO SPECIES IDENTIFICATION SHEETS

MILY: MUGILIDAE

FISHING AREAS 57,71
(E Ind. Ocean)
(W Cent. Pacific)

Liza argentea (Quoy & Gaimard, 1824)

'NONYMS STILL IN USE: *Gracilimugil ramsayi* Whitley, 1941

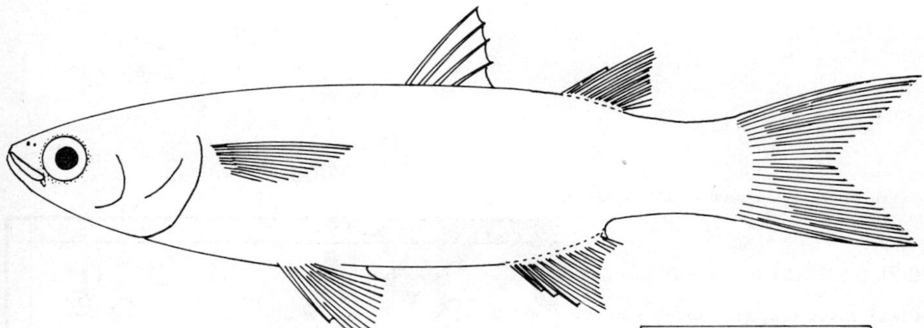

RNACULAR NAMES:

FAO: En - Ramsay's grey mullet
Fr -
Sp -

NATIONAL:

STINCTIVE CHARACTERS:

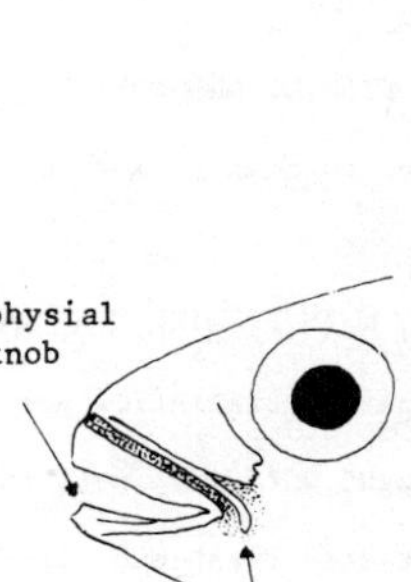

Liza

Body slender, head flattened on top but relatively narrow, its ength 26 to 28% of standard length; *fatty (adipose) tissue only a m around eye;* lips thin; a large symphysial knob at front of wer jaw; *posterior tip of upper jaw strongly curved down and ill visible when mouth closed;* a single row of fine teeth in th jaws. Origin of first dorsal fin nearer to caudal fin base an to snout tip; origin of second dorsal at vertical from midy along anal fin base; *pectoral fins short, about 3/4 of head ngth, with no axillary scale;* anal fin with 10 soft rays; udal fin deeply forked; scales in lateral series 35 to 38.

Colour: light brown on back, silvery on sides and belly; is purple with gold flecks; a patch of bright gold at upper sterior corner of operculum.

DISTINGUISHING CHARACTERS OF SIMILAR SPECIES OCCURRING IN THE AREA:

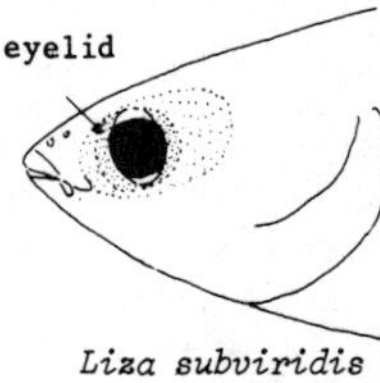

Liza vagiensis: caudal fin truncate, not forked.

Liza subviridis, L. tade: adipose tissue covering eye except for pupil (only a rim in *L. argentea*); also, usually 9 soft anal fin rays.

Mugil species: adipose tissue covering eye except for pupil; also, maxilla not curving down behind mouth corner.

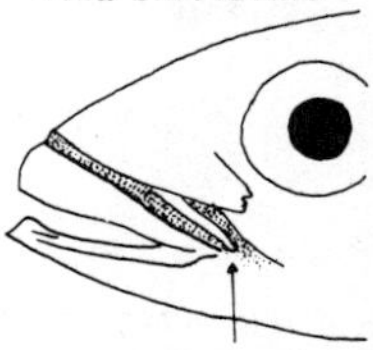

Valamugil species: longer pectoral fins (reaching to level of first dorsal fin).

Aldrichetta species: lower lip thick, and only a small symphysial knob at front of lower jaw.

Myxus species: no adipose tissue around eye.

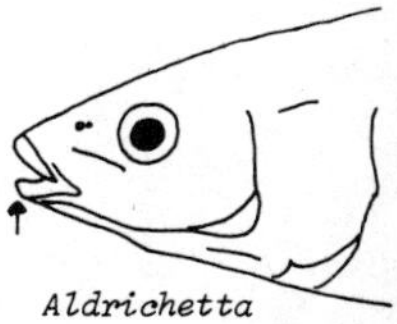

SIZE:

Maximum: 30 cm; common: 15 to 22 cm.

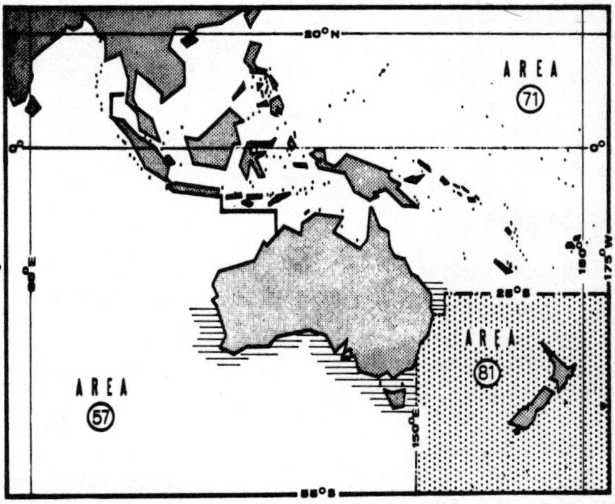

GEOGRAPHICAL DISTRIBUTION AND BEHAVIOUR:

Southern Australia, from mid-Queensland to Geraldton on west coast.

Schools occur in shallows, bogs and lower estuaries, and more saline lagoons. Spawning takes place in the sea.

Feeds on minute bottom-living organisms, including small crustaceans and filamentous algae.

PRESENT FISHING GROUNDS:

Shallow coastal waters, estuaries.

CATCHES, MAIN FISHING GEAR AND PRINCIPAL FORMS OF UTILIZATION:

Separate statistics are not reported for this species.

Caught with gill nets and beach seines.

Marketed fresh and filleted.

FAO SPECIES IDENTIFICATION SHEETS

FAMILY: MUGILIDAE

FISHING AREAS 57,71
(E Ind. Ocean)
(W Cent. Pacific)

Liza subviridis (Valenciennes, 1836)

SYNONYMS STILL IN USE: *Mugil dussumieri* Valenciennes, 1836
Mugil javanicus Bleeker, 1852
Mugil sundanensis Bleeker, 1853

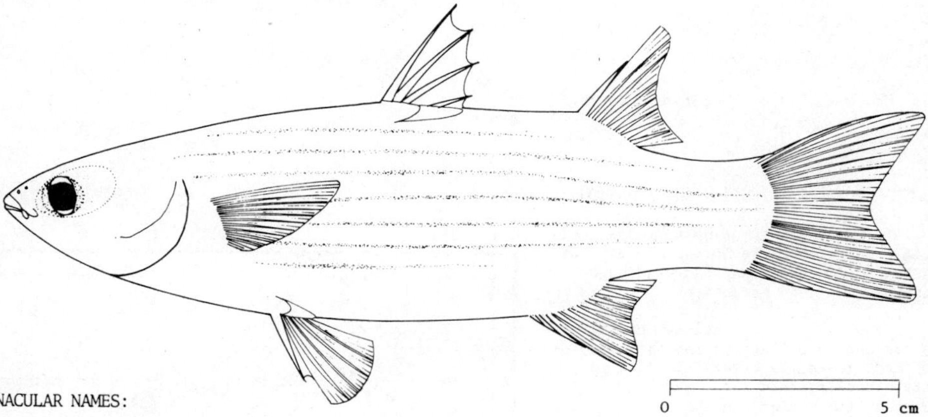

VERNACULAR NAMES:

FAO: En - Greenback grey mullet
Fr -
Sp -

NATIONAL:

DISTINCTIVE CHARACTERS:

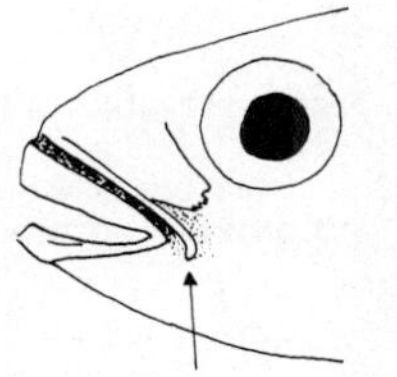

maxilla tip
Liza

Body rather stout, head broad and flattened on top, its length 23 to 25% of standard length; *fatty (adipose) tissue covering eye except for pupil.* Upper and lower lips thin; a large symphysial knob at front of lower jaw; *posterior tip of upper jaw strongly curved down and still visible when mouth closed;* several rows of teeth in upper lip, a single row in lower lip, the latter absent in adults. Origin of first dorsal fin nearer to snout tip than to caudal fin base; origin of second dorsal fin behind vertical from origin of anal fin; *pectoral fins very short, about 3/4 of head length, with no axillary scale;* caudal fin slightly forked. Scales in lateral series 30 to 32.

Colour: back green/grey, flanks and belly silvery; often 3 to 7 blackish longitudinal stripes along flanks; pectoral fins not blackish.

DISTINGUISHING CHARACTERS OF SIMILAR SPECIES OCCURRING IN THE AREA:

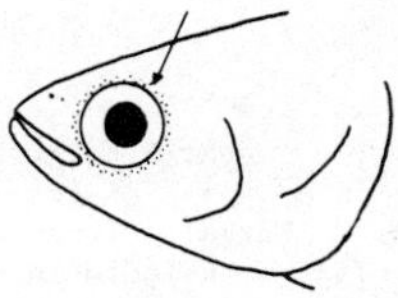

Liza argentea

Liza vaigiensis: pectoral fins as long as head and blackish; scales in lateral series 24 to 27 (30 to 32 in *L. subviridis*); also, caudal fin almost truncate (slightly forked in *L. subviridis*).

Liza tade: head very short (19 to 23% of standard length; 23 to 25% in *L. subviridis*) and with distinctive bulge at sides; also, caudal fin forked.

Liza argentea, L. vaigiensis and *Valamugil seheli:* adipose tissue only a rim around eye.

Liza parsia: corner of mouth reaches to vertical from posterior nostril (to vertical from anterior nostril in *L. subviridis*).

Mugil species: maxilla not curved down and hidden when mouth closed.

Aldrichetta species: lower lip thick, and only a small symphysial knob at front of lower jaw.

Myxus species: no adipose tissue around eye.

Other mugilid species: large axillary scale usually present above pectoral fin.

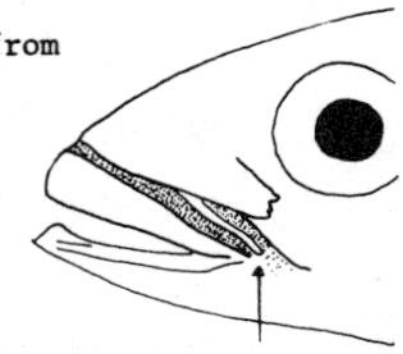

maxilla tip
Mugil

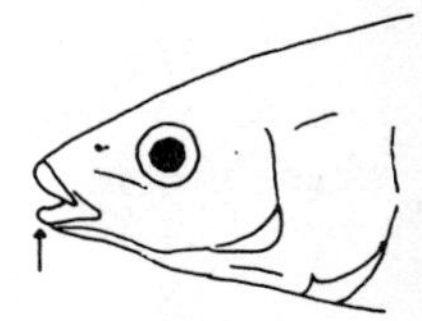

Aldrichetta sp.

pectoral
axillary scale

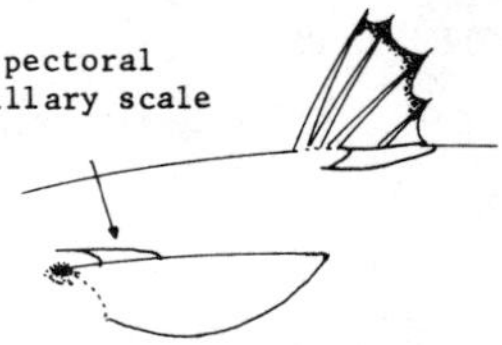

Valamugil speigleri

SIZE:

Maximum: 40 cm; common: 25 cm.

GEOGRAPHICAL DISTRIBUTION AND BEHAVIOUR:

Throughout most of northern part of area and southward to northern coasts of Australia (recorded as *Liza planiceps*); also, westward to Karachi.

Schools occur in shallow coastal waters and enter lagoons and estuaries to feed, juveniles often occurring in rice fields and mangrove swamps. Spawning takes place in the sea.

Feeds on minute bottom-living organisms and on organic matter contained in mud and sand; perhaps also on floating algae.

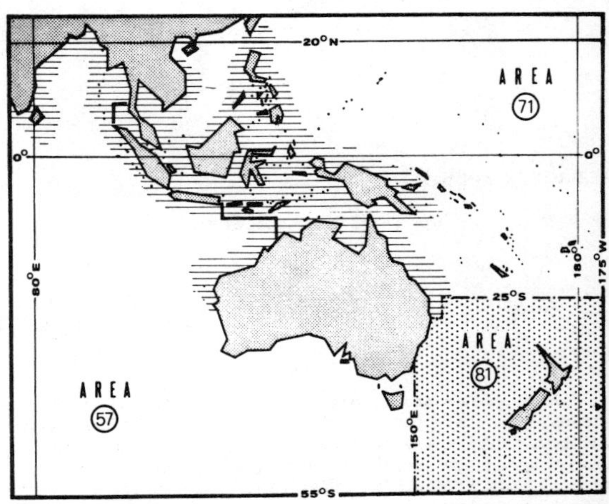

PRESENT FISHING GROUNDS:

Shallow coastal waters, estuaries, and mangrove swamps.

CATCHES, MAIN FISHING GEAR AND PRINCIPAL FORMS OF UTILIZATION:

Separate statistics are not reported for this species. The total reported catch of unclassified grey mullets in 1972 was:

area 57 (Eastern Indian Ocean): 1 700 tons (Australia: 1 200 tons)
area 71 (Western Central Pacific): 2 300 tons (Australia: 1 500 tons)

Caught with gill nets, beach seines and cast nets.

Marketed fresh and salted; also boiled (Thailand) and canned or frozen (Australia). The roe is often marketed as a salted product.

FAO SPECIES IDENTIFICATION SHEETS

FAMILY: MUGILIDAE

FISHING AREAS 57,71
(E Ind. Ocean)
(W Cent. Pacific)

Liza tade (Forsskål, 1775)

SYNONYMS STILL IN USE: *Mugil planiceps* Valenciennes, 1836

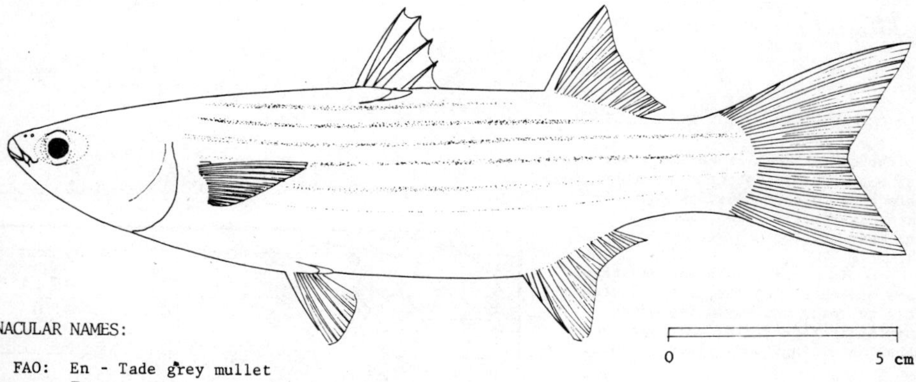

VERNACULAR NAMES:

FAO: En - Tade grey mullet
Fr -
Sp -

NATIONAL:

DISTINCTIVE CHARACTERS:

Body rather stout, head short, broad and flattened on top, its length 19 to 23% of standard length, *with distinctive bulge at sides; fatty (adipose) tissue covering eye except for pupil.* Upper and lower lips thin; a large symphysial knob at front of lower jaw; *posterior tip of upper jaw strongly curved down and still visible when mouth closed;* a row of fine straight teeth (and 4 to 8 rows of smaller teeth) on upper lip, a row of sparse cilia on lower lip. Origin of first dorsal fin nearer to tip of snout than to base of caudal fin. Origin of second dorsal fin behind vertical from origin of anal fin; *pectoral fins very short, 18 to 20% of standard length, about 3/4 of head length, with axillary scale very small or absent;* caudal fin forked. Scales in lateral series 31 to 33.

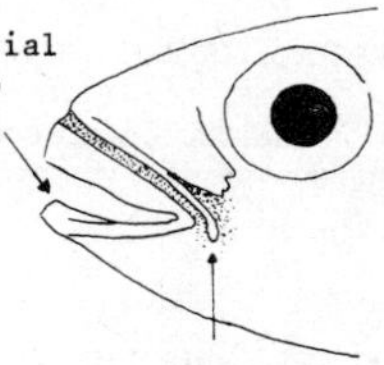

Liza

Colour: back olive, flanks and belly silvery; often 5 to 9 dark longitudinal stripes along flanks.

DISTINGUISHING CHARACTERS OF SIMILAR SPECIES OCCURRING IN THE AREA:

Liza subviridis: head longer (23 to 25% of standard length; 19 to 23% in *L. tade*) and without distinctive bulge at sides; also, caudal fin almost truncate.

Liza argentea, L. vaigiensis and *Valamugil seheli:* adipose tissue only a rim around eye; also, pectoral fins about equal to head length and blackish in *L. vaigiensis* (3/4 of head length in *L. tade*).

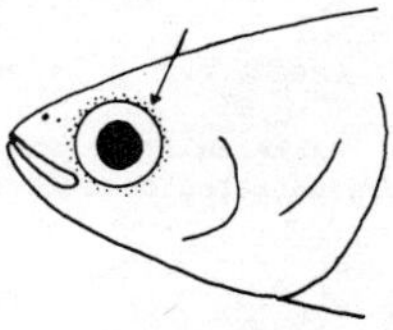
Liza argentea

Mugil species: maxilla not curved down and hidden when mouth closed.

Aldrichetta species: lower lip thick, and only a small symphysial knob at front of lower jaw.

Myxus species: no adipose tissue around eye.

Other mugilid species: large axillary scale usually present above pectoral fin.

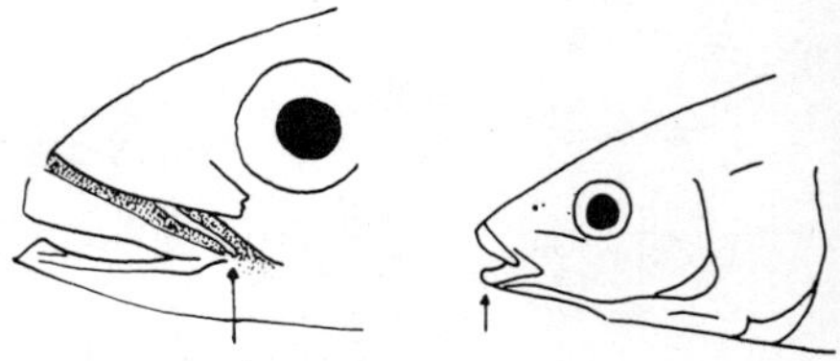

maxilla tip *Mugil*   *Aldrichetta*

large pectoral axillary scale

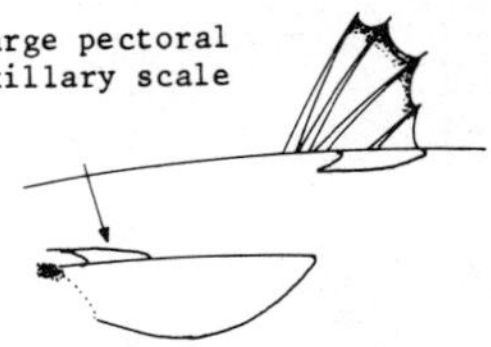

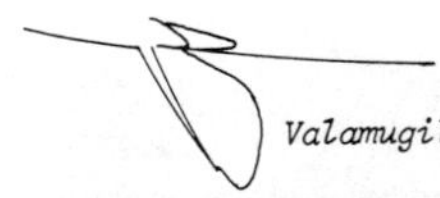

Valamugil speigleri

SIZE:

Maximum: 47 cm; common: 30 cm.

GEOGRAPHICAL DISTRIBUTION AND BEHAVIOUR:

Throughout most of northern part of area, but probably not from Australian coasts (previous records, as *Mugil planiceps*, refer to *Liza subviridis*); also, westward to Bombay.

Schools occur in shallow coastal waters and enter lagoons, estuaries and rivers to feed, the juveniles often occurring in rice fields and mangrove swamps. Spawning takes place in the sea.

Feeds on minute bottom-living organisms and on organic matter contained in mud and sand; perhaps also on floating algae.

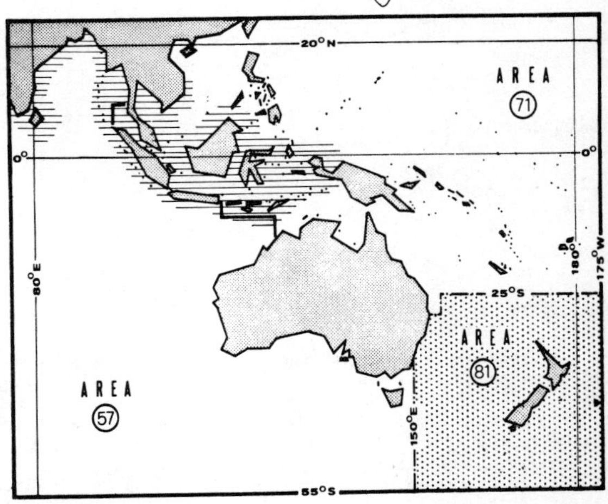

PRESENT FISHING GROUNDS:

Shallow coastal waters, estuaries, and mouths of rivers.

CATCHES, MAIN FISHING GEAR AND PRINCIPAL FORMS OF UTILIZATION:

Separate statistics are not reported for this species. The total reported catch of unclassified grey mullets in 1972 was:

area 57 (Eastern Indian Ocean): 1 700 tons (Australia: 1 200 tons)
area 71 (Western Central Pacific): 2 300 tons (Australia: 1 500 tons)

Caught with gill nets, beach seines and cast nets.

Marketed fresh and salted; also boiled (Thailand) and canned or frozen (Australia). The roe is often marketed as a salted product.

FAO SPECIES IDENTIFICATION SHEETS

FAMILY: MUGILIDAE

FISHING AREAS 57,71
(E Ind. Ocean)
(W Cent. Pacific)

Liza vaigiensis (Quoy & Gaimard, 1824)

SYNONYMS STILL IN USE: *Mugil vaigiensis* Quoy & Gaimard, 1824

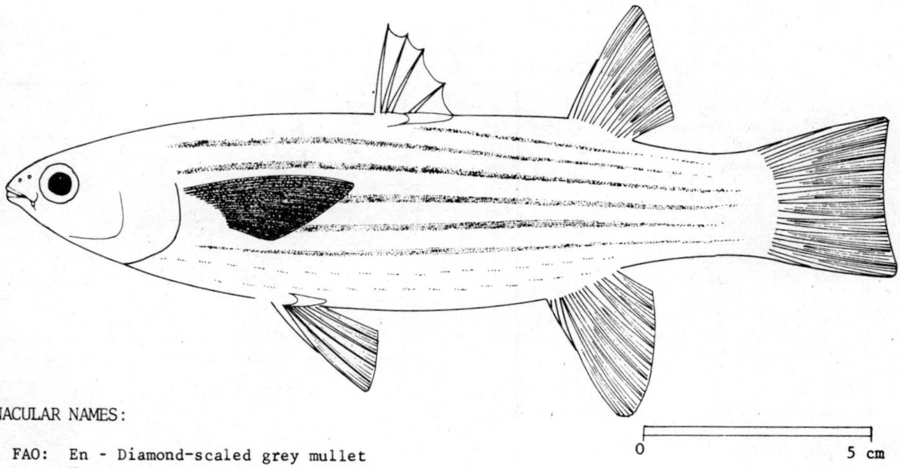

VERNACULAR NAMES:

FAO: En - Diamond-scaled grey mullet
Fr -
Sp -

NATIONAL:

DISTINCTIVE CHARACTERS:

Body rather stout, head broad and flattened on top, its length 24 to 27% of standard length; *fatty (adipose) tissue only around rim of eye*. Upper and lower lips thin; a large symphysial knob at front of lower jaw; *posterior tip of upper jaw strongly curved down and still visible when mouth closed;* adults without teeth, juveniles with sparse row of teeth in each lip. Origin of first dorsal fin nearer to base of caudal fin than to snout tip; origin of second dorsal fin behind vertical from origin of anal fin; *pectoral fins about equal to head length, with no axillary scale; caudal fin almost truncate*. Scales in lateral series 24 to 27.

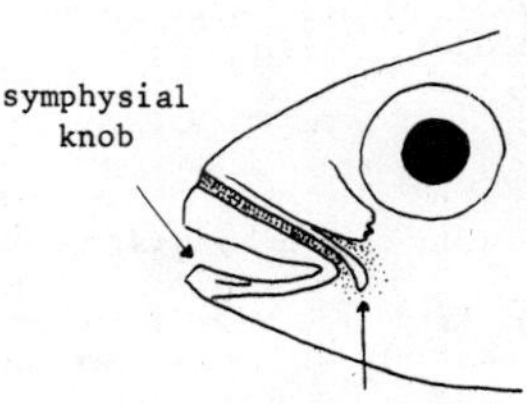

Liza

Colour: back dark green, flanks lighter, belly silvery; usually 6 longitudinal spotted bands along flanks, the second and fifth the most conspicuous. *Pectoral fins blackish*, other fins with dusky margins.

DISTINGUISHING CHARACTERS OF SIMILAR SPECIES OCCURRING IN THE AREA:

Liza subviridis, L. tade: adipose tissue covering eye except for pupil; pectoral fin ¾ of head length and not blackish; scales in lateral series 30 to 33 (24 to 27 in *L. vaigiensis*).

Other mugilid species: caudal fin more or less forked and axillary scale usually present above pectoral fin.

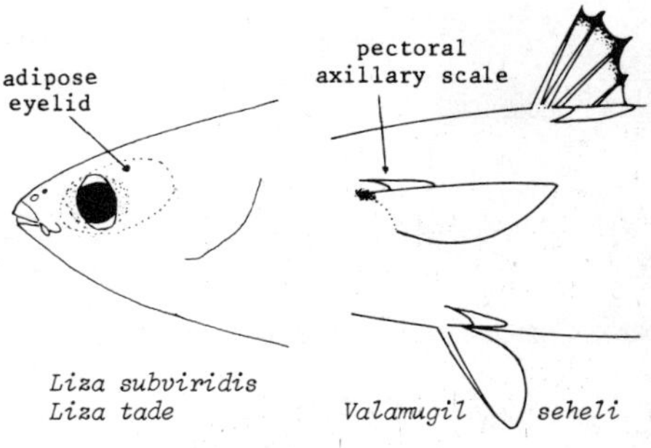

SIZE:

Maximum: 45 cm; common: 30 cm.

GEOGRAPHICAL DISTRIBUTION AND BEHAVIOUR:

Throughout most of northern part of area and southward to northern coasts of Australia; also, westward to South Africa and northward to Japan.

Schools occur in shallow coastal waters and enter lagoons, estuaries and rivers, the juveniles often occurring in rice fields and mangrove swamps. Spawning takes place in the sea.

Feeds on minute bottom-living organisms and on organic matter in mud and sand; perhaps also on floating algae.

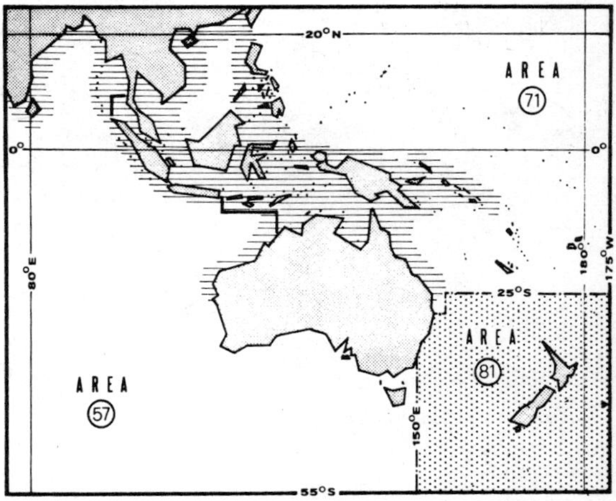

PRESENT FISHING GROUNDS:

Shallow coastal waters, estuaries, and mouths of rivers.

CATCHES, MAIN FISHING GEAR AND PRINCIPAL FORMS OF UTILIZATION:

Separate statistics are not reported for this species. The total reported catch of unclassified grey mullets in 1972 was:

area 57 (Eastern Indian Ocean): 1 700 tons (Australia: 1 200 tons)
area 71 (Western Central Pacific): 2 300 tons (Australia: 1 500 tons)

Caught with gill nets, beach seines and cast nets.

Marketed fresh and salted; also boiled (Thailand) and canned or frozen (Australia). The roe is often marketed as a salted product.

FAO SPECIES IDENTIFICATION SHEETS

FAMILY: MUGILIDAE

FISHING AREAS 57,71
(E Ind. Ocean)
(W Cent. Pacific)

Mugil cephalus Linnaeus, 1758

SYNONYMS STILL IN USE: *Mugil japonicus* Schlegel, 1845
Mugil galapagensis Ebeling, 1961

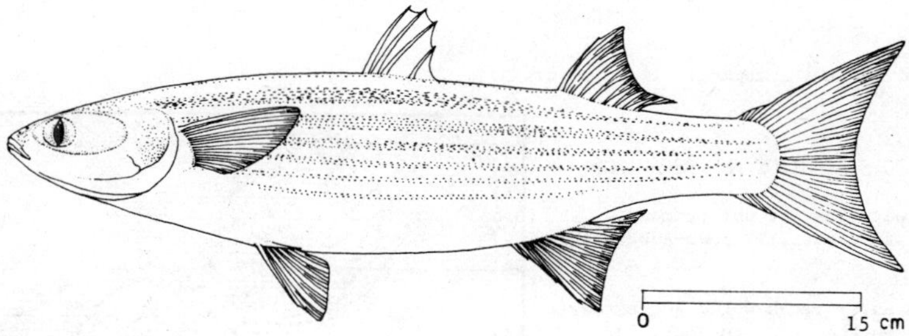

VERNACULAR NAMES:

FAO: En - Flathead grey mullet
Fr -
Sp -

NATIONAL:

DISTINCTIVE CHARACTERS:

Body rather stout, head broad and flattened on top, its length 27 to 29% of standard length; *fatty (adipose) tissue covering most of eye;* lips thin; a large symphysial knob at front of lower jaw; posterior tip of upper jaw not curved down and hidden when mouth closed; several rows of teeth in upper lip; one row or sometimes more in lower lip. Origin of first dorsal fin nearer to snout tip than to caudal fin base; origin of second dorsal fin behind vertical from origin of anal fin; *pectoral fins short, not reaching first dorsal fin origin, with an axillary scale; anal fin with 8 soft rays;* caudal fin forked. Scales in lateral series 38 to 42.

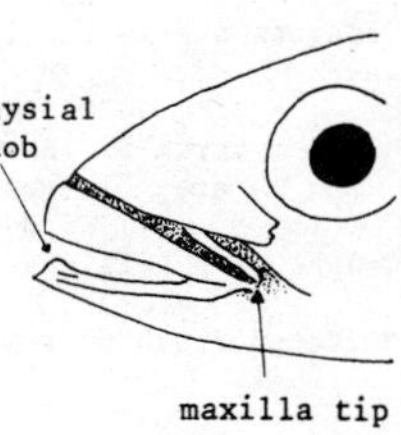

Mugil

Colour: olive green on back, silvery on sides shading into white below; 6 to 7 indistinct longitudinal brown bars on flanks.

DISTINGUISHING CHARACTERS OF SIMILAR SPECIES OCCURRING IN THE AREA:

Liza and *Myxus* species: adipose tissue only covering rim of eye or at most to iris border, no axillary pectoral scale, and maxilla curving down behind mouth corner; also, usually 9 to 11 soft anal rays (8 in *Mugil cephalus*, but 8 in *L. vaigiensis* as also in *Aldrichetta*).

Valamugil species: longer pectoral fins (reaching to level of first dorsal fin).

Aldrichetta species: lower lip thick, and only a small symphysial knob at front of lower jaw.

Myxus species: no adipose tissue around eye.

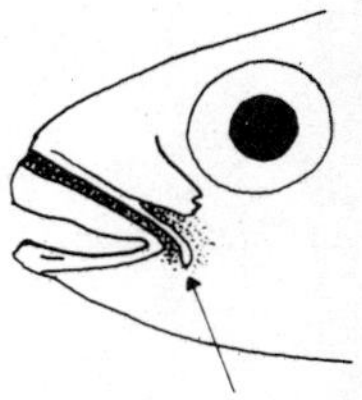

maxilla tip
Liza

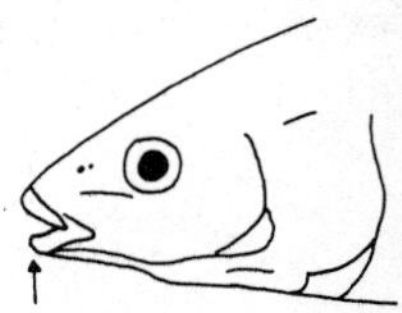

Aldrichetta

SIZE:

Maximum: 60 cm; common: 14 to 32 cm.

GEOGRAPHICAL DISTRIBUTION AND BEHAVIOUR:

Throughout area but not common in tropical areas (especially Indo-Australian archipelago).

Schools enter fresh water and estuaries but spawning in the sea; commonly leaps from the water.

Feeds on microscopic organisms and organic detritus in bottom muds.

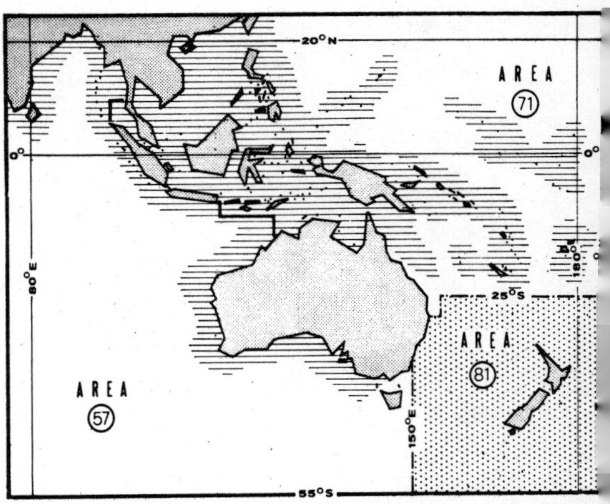

PRESENT FISHING GROUNDS:

Shallow estuaries and coastal lakes, as well as sea beaches on the migration to the spawning grounds.

CATCHES, MAIN FISHING GEAR AND PRINCIPAL FORMS OF UTILIZATION:

Separate statistics for this species are reported only by Australia. The Australian catch in 1972 was:

area 57 (Eastern Indian Ocean): 452 tons
area 71 (Western Central Pacific): 1 300 tons

Caught with gill nets, beach seines and shore-line traps.

Marketed fresh or sometimes frozen. Roe also sold fresh or smoked.

FAO SPECIES IDENTIFICATION SHEETS

FAMILY: MUGILIDAE

FISHING AREAS 57,71
(E Ind. Ocean)
(W Cent. Pacific)

Myxus elongatus (Günther, 1861)

SYNONYMS STILL IN USE: None

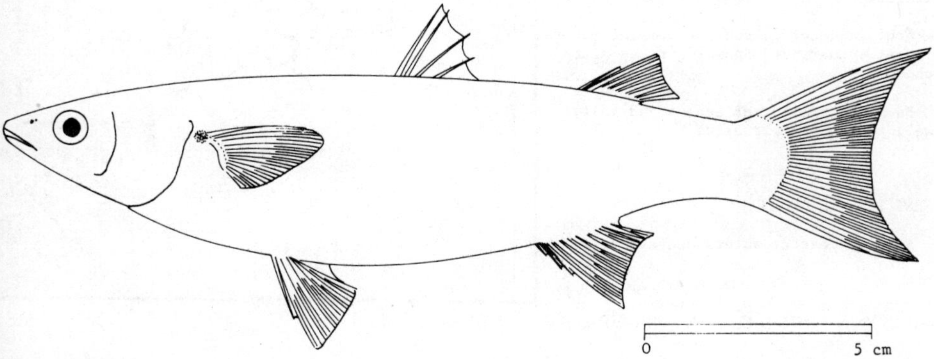

VERNACULAR NAMES:

FAO: En - Sand grey mullet
Fr -
Sp -

NATIONAL:

DISTINCTIVE CHARACTERS:

Body slender, head slightly convex and rather broad, its length 25 to 26% of standard length; *no fatty (adipose) tissue around eye;* lips thin; a large symphysial knob at front of lower jaw; hind tip of upper jaw hidden when mouth closed; *a single row of spatulate teeth in upper lip,* a marginal row in lower lip and also lateral patches near mouth corners. Origin of first dorsal fin nearer to caudal fin base than to snout tip; origin of second dorsal fin behind vertical from origin of anal fin; *pectoral fins very short,* less than ¾ of head length, *with no axillary scale;* caudal fin deeply forked. Scales in lateral series 43 to 46.

Colour: olive green on back, silver on sides, white below; fins with dusky edges except anal fin which has white edge; iris yellow; a dark axillary spot at base of pectoral fins.

DISTINGUISHING CHARACTERS OF SIMILAR SPECIES OCCURRING IN THE AREA:

Myxus petardi: ciliate teeth in upper lip (spatulate in *M. elongatus*) and more scales in lateral series (47 to 50; 43 to 46 in *M. elongatus*).

Liza and *Mugil* species: adipose tissue around at least rim of eye.

Valamugil species: longer pectoral fins (reaching to level of first dorsal fin).

Aldrichetta species: lower lip thick, and only a small symphysial knob at front of lower jaw; also, 58 to 64 scales.

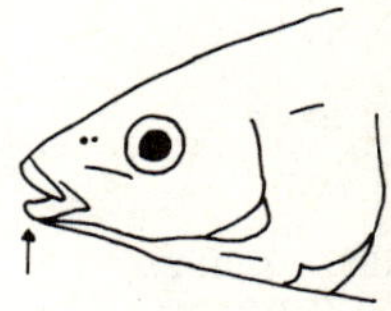

Aldrichetta

SIZE:

Maximum: 27 cm; common: 15 to 20 cm.

GEOGRAPHICAL DISTRIBUTION AND BEHAVIOUR:

Southern coasts of Australia, from southern Queensland to south western Australia.

Schools occur in shallow coastal waters and lower estuaries. Spawning takes place in the sea.

Feeds on small crustaceans and molluscs as well as microscopic algae.

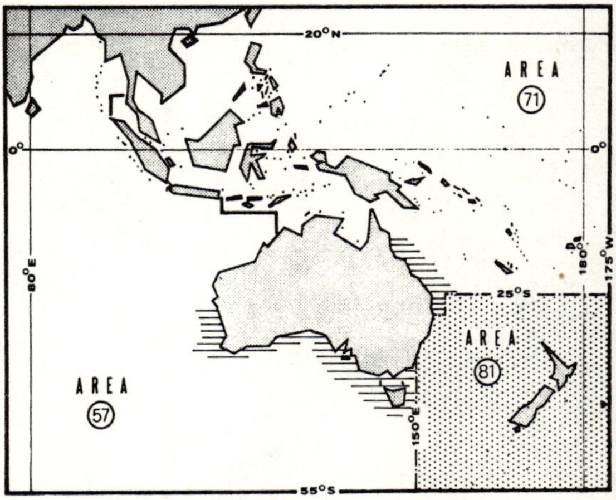

PRESENT FISHING GROUNDS:

Shallow coastal waters and estuaries.

CATCHES, MAIN FISHING GEAR AND PRINCIPAL FORMS OF UTILIZATION:

Separate statistics are not reported for this species.

Caught with gill nets and beach seines.

Marketed fresh or filleted.

FAO SPECIES IDENTIFICATION SHEETS

FAMILY: MUGILIDAE

FISHING AREAS 57,71
(E Ind. Ocean)
(W Cent. Pacific)

Valamugil cunnesius (Valenciennes, 1836)

SYNONYMS STILL IN USE: *Mugil strongylocephalus* Richardson, 1846
Mugil longimanus Günther, 1861

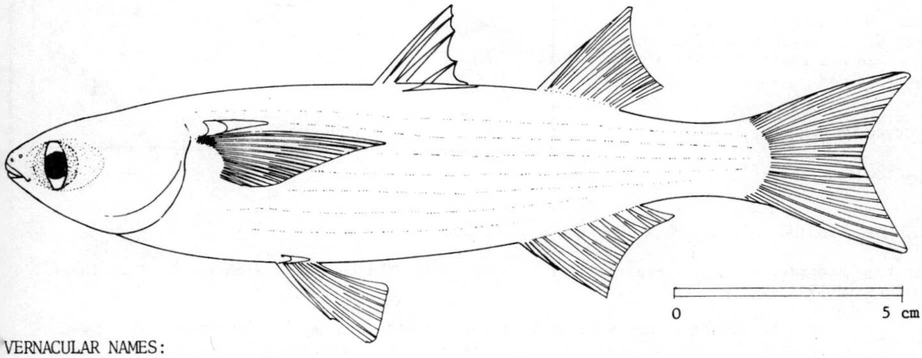

VERNACULAR NAMES:

FAO: En - Longfin grey mullet
Fr -
Sp -

NATIONAL:

DISTINCTIVE CHARACTERS:

Body elongate, rather slender, head moderate; *fatty (adipose) tissue covering eye except for pupil.* Upper lip fleshy, lower lip thin; *posterior tip of upper jaw hidden when mouth closed;* fine teeth on both lips, but more sparse on upper. Origin of first dorsal fin a little nearer to tip of snout than to base of caudal fin; origin of second dorsal fin a little behind vertical from origin of anal fin; *pectoral fins equal to or longer than head, reaching to below 3rd or 4th rays of first dorsal fin, with axillary scale nearly half the length of fin;* caudal fin forked. Scales in lateral series 30 to 35.

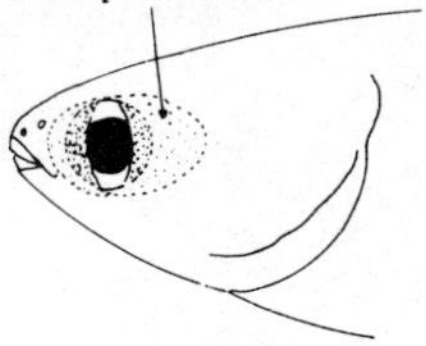

V. cunnesius

Colour: back blue, flanks and belly silvery; pectoral fins with dark mark at axil.

DISTINGUISHING CHARACTERS OF SIMILAR SPECIES OCCURRING IN THE AREA:

Valamugil speigleri: outer margin of first dorsal fin black; also, scales in lateral series 37 to 40 (30 to 35 in *V. cunnesius*).

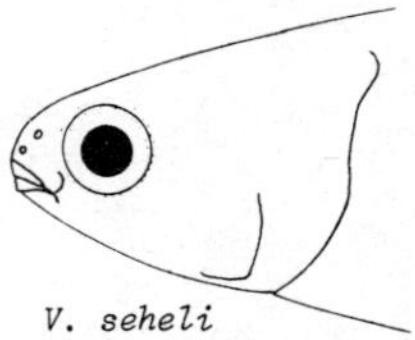

V. seheli

Valamugil seheli, V. buchanani: fatty (adipose) tissue only around rim of eye.

Other mugilid species: shorter pectoral fins, not reaching beyond first dorsal fin origin.

SIZE:

Maximum: 35 cm; common: 15 to 20 cm.

GEOGRAPHICAL DISTRIBUTION AND BEHAVIOUR:

Throughout most of northern part of area and southward to northern Australia; also, westward to Bombay.

Schools occur in shallow coastal waters and enter lagoons, estuaries and rivers to feed, the juveniles often occurring in rice fields and mangrove swamps.

Spawning takes place in the sea.

Feeds on minute bottom-living organisms and on organic material contained in mud and sand; perhaps also on floating algae.

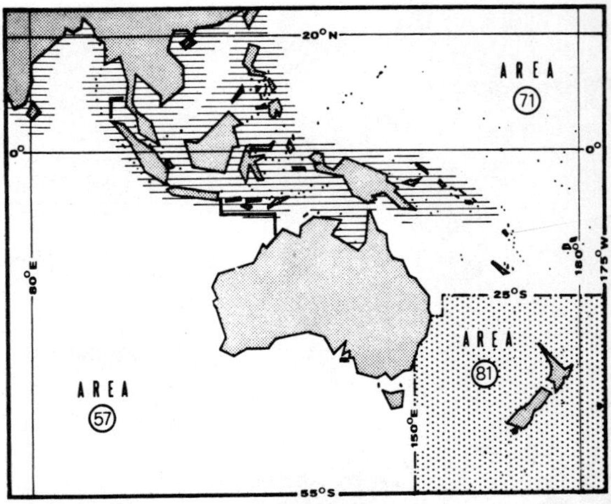

PRESENT FISHING GROUNDS:

Shallow coastal waters, estuaries, and mouths of rivers.

CATCHES, MAIN FISHING GEAR AND PRINCIPAL FORMS OF UTILIZATION:

Separate statistics are not reported for this species. The total reported catch of unclassified grey mullets in 1972 was:

area 57 (Eastern Indian Ocean): 1 700 tons (Australia: 1 200 tons)
area 71 (Western Central Pacific): 2 300 tons (Australia: 1 500 tons)

Caught with gill nets, beach seines and cast nets.

Marketed fresh and salted; also boiled (Thailand) and canned or frozen (Australia). The roe is often marketed as a salted product.

FAO SPECIES IDENTIFICATION SHEETS

FAMILY: MUGILIDAE

FISHING AREAS 57,71
(E Ind. Ocean)
(W Cent. Pacific)

Valamugil seheli (Forsskål, 1775)

SYNONYMS STILL IN USE: *Mugil axillaris* Valenciennes, 1836
Mugil caeruleomaculatus Lacepede, 1803

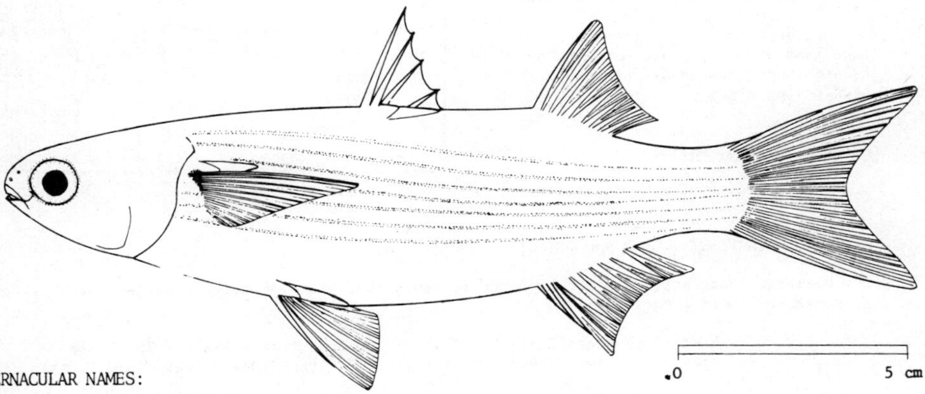

VERNACULAR NAMES:

FAO: En - Bluespot grey mullet
Fr -
Sp -

NATIONAL:

DISTINCTIVE CHARACTERS:

Body elongate, rather slender, head moderate; *fatty (adipose) tissue only around rim of eye.* Upper lip fleshy, lower lip thin; *posterior tip of upper jaw hidden when mouth closed;* fine teeth on both lips, but shorter and more sparse in upper. Origin of first dorsal fin about equidistant between tip of snout and base of caudal fin; origin of second dorsal fin on vertical from anal fin origin; *pectoral fins equal to head length or a little shorter, just reaching to vertical from first dorsal fin origin, with axillary scale 1/3 the length of fin;* caudal fin forked. Scales in lateral series 38 to 42.

Colour: back blue, flanks and belly silvery; pectoral fins yellow, with dark blue spot at axil.

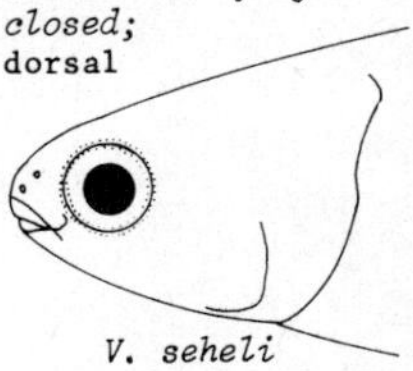

V. seheli

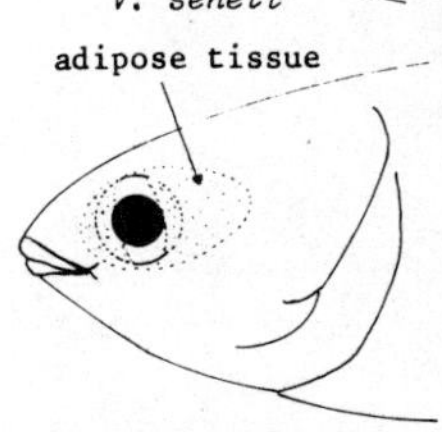

V. speigleri

DISTINGUISHING CHARACTERS OF SIMILAR SPECIES OCCURRING IN THE AREA:

Valamugil speigleri, V. cunnesius, V. engeli: fatty (adipose) tissue covering most of eye.

Valamugil buchanani: fewer scales in lateral series (32 to 35; 38 to 42 in *V. seheli*).

Other mugilid species: shorter pectoral fins not reaching to vertical from first dorsal fin origin.

SIZE:

Maximum: 45 cm; common: 20 to 30 cm.

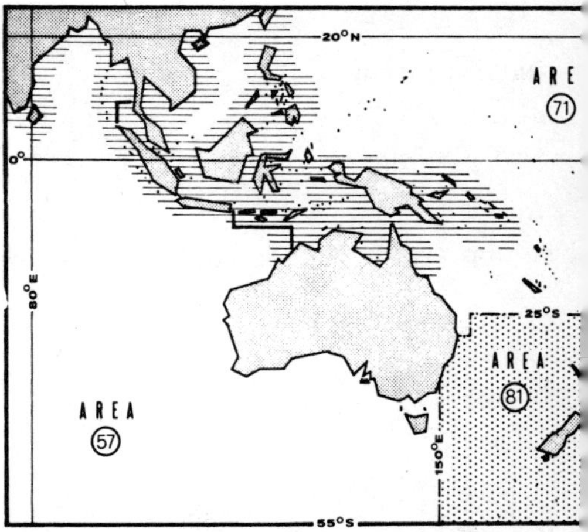

GEOGRAPHICAL DISTRIBUTION AND BEHAVIOUR:

Throughout most of northern part of area and probably southward to tropical waters of Australia; also westward to South Africa.

Schools occur in shallow coastal waters and enter lagoons, estuaries and rivers to feed, juveniles often occurring in rice fields and mangrove swamps.

Spawning takes place in the sea.

Feeds on minute bottom-living organisms and on organic matter contained in mud and sand; perhaps also on floating algae.

PRESENT FISHING GROUNDS:

Shallow coastal waters, estuaries, and mouths of rivers.

CATCHES, MAIN FISHING GEAR AND PRINCIPAL FORMS OF UTILIZATION:

Separate statistics are not reported for this species. The total reported catch of unclass grey mullets in 1972 was:

area 57 (Eastern Indian Ocean): 1 700 tons (Australia: 1 200 tons)
area 71 (Western Central Pacific): 2 300 tons (Australia: 1 500 tons)

Caught with gill nets, beach seines and cast nets

Marketed fresh and salted; also boiled (Thailand) and canned or frozen (Australia). The r often marketed as a salted product.

FAO SPECIES IDENTIFICATION SHEETS

FAMILY: MUGILIDAE

FISHING AREAS 57,71
(E Ind. Ocean)
(W Cent. Pacific)

Valamugil speigleri (Bleeker, 1858)

SYNONYMS STILL IN USE: None

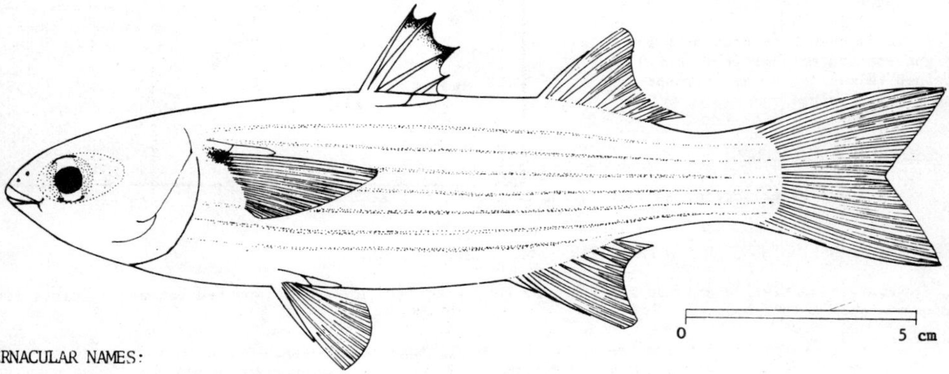

VERNACULAR NAMES:

FAO: En - Speigler's grey mullet
Fr -
Sp -

NATIONAL:

DISTINCTIVE CHARACTERS:

Body elongate, rather slender, head moderate; *fatty (adipose) tissue covering eye except for pupil.* Upper lip fleshy, lower lip thin; *posterior tip of upper jaw hidden when mouth closed;* fine teeth on both lips, but more sparse on upper. Origin of first dorsal fin nearer to tip of snout than to base of caudal fin; origin of second dorsal fin behind origin of anal fin; *pectoral fin a little shorter than head length, reaching somewhat beyond vertical from first dorsal fin origin, with axillary scale almost half the length of fin;* caudal fin forked. Scales in lateral series 37 to 40.

Colour: back green, flanks and belly silvery; pectoral fin with black spot at axil; *margin of first dorsal fin black;* other fins dusky.

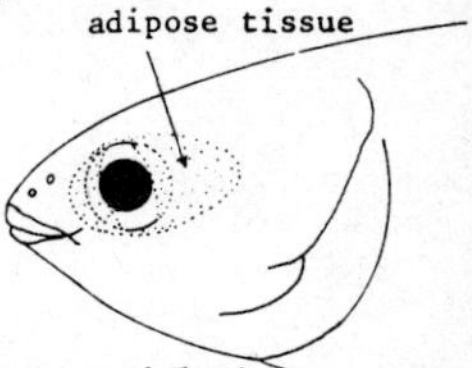

V. speigleri

DISTINGUISHING CHARACTERS OF SIMILAR SPECIES OCCURRING IN THE AREA:

Valamugil seheli, V. buchanani: fatty (adipose) tissue only around rim of eye; also, margin of first dorsal fin not black.

Valamugil cunnesius: outer margin of first dorsal fin not black; also, fewer scales in lateral series (30 to 35; 37 to 40 in *V. speigleri*).

Other mugilid species: shorter pectoral fins, not reaching beyond first dorsal fin origin.

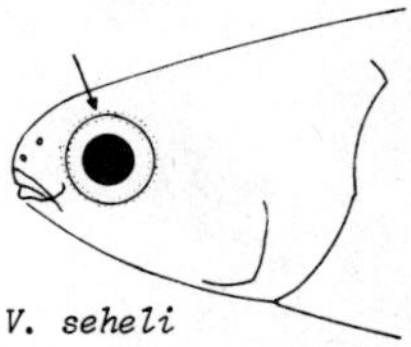

V. seheli

SIZE:

Maximum: 35 cm; common: 15 to 20 cm.

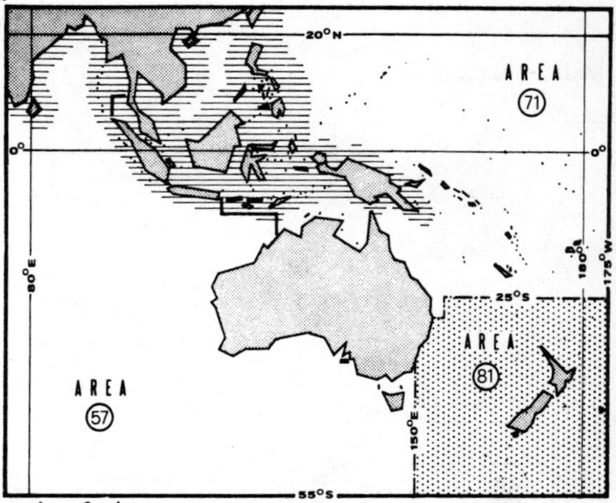

GEOGRAPHICAL DISTRIBUTION AND BEHAVIOUR:

Throughout most of northern part of area, but perhaps not southward to Australian coasts.

Schools occur in shallow coastal waters and enter lagoons, estuaries and rivers to feed, juveniles often occurring in rice fields and mangrove swamps.

Spawning takes place in the sea.

Feeds on minute bottom-living organisms and on organic material contained in mud and sand; perhaps also on floating algae.

PRESENT FISHING GROUNDS:

Shallow coastal waters, estuaries, and mouths of rivers.

CATCHES, MAIN FISHING GEAR AND PRINCIPAL FORMS OF UTILIZATION:

Separate statistics are not reported for this species. The total reported catch of unclassified grey mullets in 1972 was:

area 57 (Eastern Indian Ocean): 1 700 tons (Australia: 1 200 tons)
area 71 (Western Central Pacific): 2 300 tons (Australia: 1 500 tons)

Caught with gill nets, beach seines and cast nets

Marketed fresh and salted; also boiled (Thailand) and canned or frozen (Australia). The roe is often marketed as a salted product.

FAO SPECIES IDENTIFICATION SHEETS

FAMILY: MUGILIDAE

FISHING AREAS 57,71
(E Ind. Ocean)
(W Cent. Pacific)

Aldrichetta forsteri (Valenciennes, 1836)

SYNONYMS STILL IN USE: *Agonostomus forsteri:* Waite, 1921

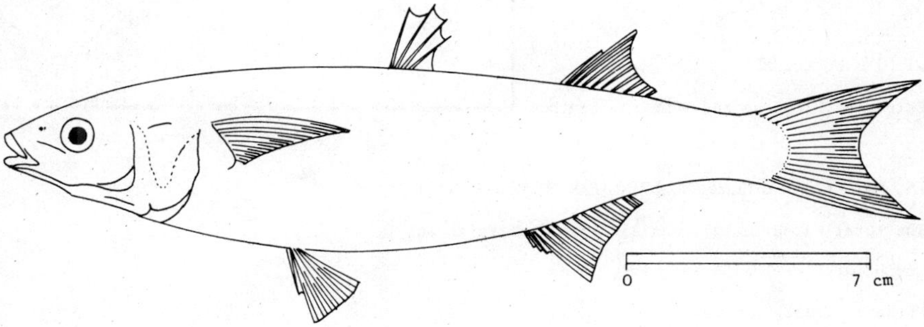

VERNACULAR NAMES:

FAO: En - Yellow eye grey mullet
Fr -
Sp -

NATIONAL:

DISTINCTIVE CHARACTERS:

Body slender, elongate, head somewhat convex on top, its length 23 to 26% of standard length; *no fatty (adipose) tissue on head; lips thick,* posterior tip of upper jaw not visible when mouth closed; several rows of teeth in both lips, sessile in jaws. Origin of first dorsal fin nearer to caudal fin base than to snout tip. Second dorsal fin origin behind vertical from origin of anal fin; pectoral fins of moderate length, 78 to 80% of head length, with no axillary scale; caudal fin forked. *Scales in lateral series 58 to 64.*

Colour: olive/brown on back, silvery to yellowish white on sides and belly; fins with brown margins; iris yellow.

DISTINGUISHING CHARACTERS OF SIMILAR SPECIES OCCURRING IN THE AREA:

All other mugilid species in area: lower lip thin, with large symphysial knob at front of lower jaw; also, fewer scales (less than 58 in lateral series).

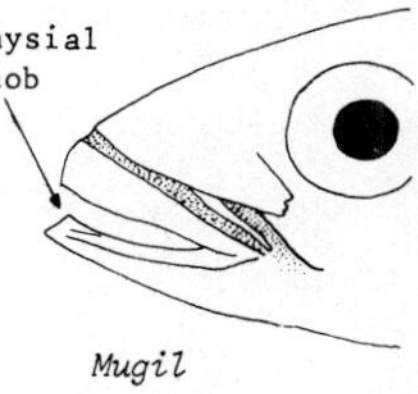

Mugil

SIZE:

Maximum: 40 cm; common: 15 to 20 cm.

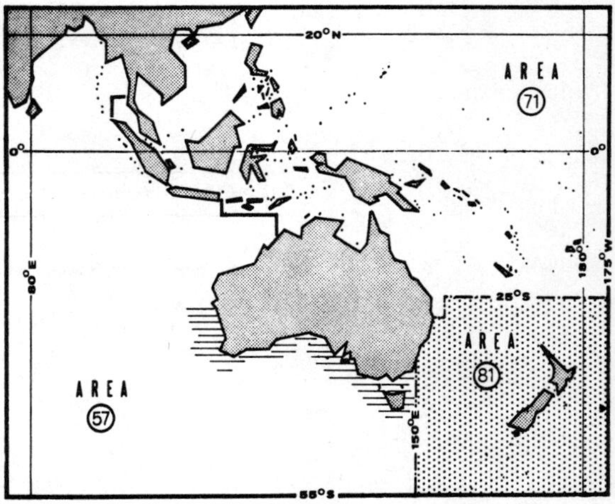

GEOGRAPHICAL DISTRIBUTION AND BEHAVIOUR:

Southern coasts of Australia, from Shark Bay on the west coast to southern New South Wales on the east; also, extending southward to New Zealand.

Shoals occur in shallow estuaries and close inshore, spawning takes place in the sea.

Feeds on small crustaceans and molluscs, as well as filamentous algae and diatoms.

PRESENT FISHING GROUNDS:

Estuaries, lagoons and some sea beaches.

CATCHES, MAIN FISHING GEAR AND PRINCIPAL FORMS OF UTILIZATION:

The total catch in 1972 was 250 tons (Australia only).

Caught with gill nets and beach lines.

Marketed fresh.

FAO SPECIES IDENTIFICATION SHEETS

FAMILY: MUGILIDAE

Fishing Area 37
(Medit. and Black Sea)

Mugil cephalus Linnaeus, 1758

SYNONYMS STILL IN USE: None

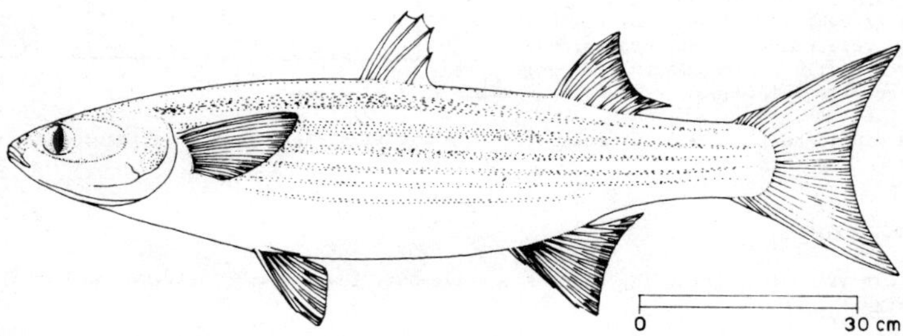

VERNACULAR NAMES:

FAO - En : Flathead grey mullet
Fr : Muge cabot
Sp : Pardete

NATIONAL -

ALBN:		ISRL:	Kifon gedol harosh	ROMN:	Laban
ALGR:		ITAL:	Cefalo	SPAN:	Pardete
BULG:	Kefal	LEBN:	Bouri ram	SYRI:	
CYPR:	Kephalos	LIBY:	Bouri	TUNS:	Bouri
EGYP:	Bouri	MALT:	Mulett tal-iswed	TURK:	Haskefal
FRAN:	Muge cabot	MONC:	Müsaru	USSR:	Loban
GREC:	Képhalos	MORC:	Bouchakfa	YUGO:	Cipal batas

DISTINCTIVE CHARACTERS AND DIAGNOSIS:

Body elongate, slightly compressed from side to side; head short and flattened with a broad terminal mouth; teeth very small, hardly visible; *upper lip thin (its greatest depth less than half the eye diameter) and smooth (without tubercules); a thick adipose lid covers most of the eye; two dorsal fins, the first short, with 4 slender spines; anal fin usually with 8 soft rays;* back bluish-grey; belly silvery, often with grey lengthwise stripes.

Other field characters: scales large and adherent; no external lateral line; a gizzard-like stomach with thick walls.

DISTINCTION FROM MOST SIMILAR SPECIES OCCURRING IN THE AREA:

Other Mediterranean and Black Sea mugilids differ from *M. cephalus* by the absence of a thick adipose lid over part of the eye and by a higher number of anal rays (9 to 11).

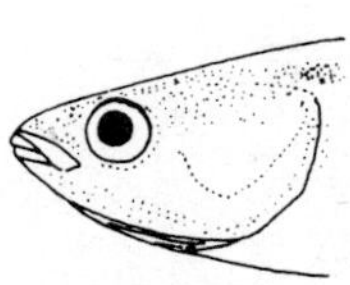

Mugil sp.

SIZE:

Maximum: 120 cm; common: 30 to 50 cm.

GEOGRAPHICAL DISTRIBUTION AND BEHAVIOUR:

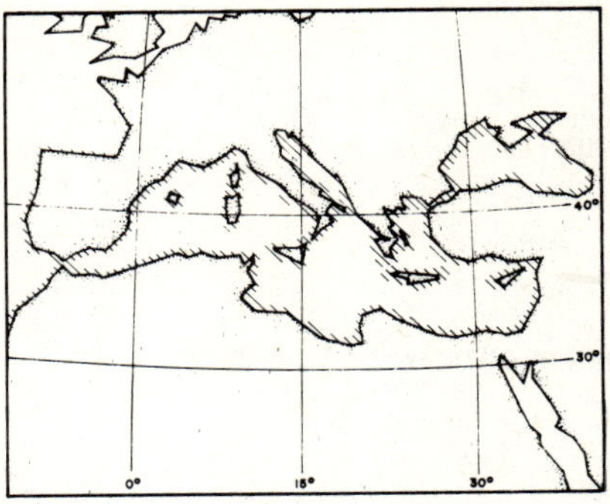

Cosmopolitan species inhabiting the warm belt of the Atlantic, Indian and Pacific Oceans and adjacent seas; very common in coastal waters of the Mediterranean and the Black Sea, including the Azov Sea; in the eastern Atlantic it occurs as far north as the Bay of Biscay.

Schools of *M. cephalus* occur mostly in shallow waters especially in coastal lagoons of varying salinity; a fast swimmer, leaping out of the water when disturbed; enters estuaries and rivers for feeding but spawning takes place in the sea only; juveniles often concentrate near the outflow of freshwater streams. In some countries fry of *M. cephalus* is brought into fresh- or brackish-water ponds and reservoirs for culture purposes, generally together with other fish. Promising results were recently obtained with induced spawning in confinement.

Feeds on minute bottom-living organisms or on algae floating near the surface; also on organic matter contained in mud and sand.

PRESENT FISHING GROUNDS:

Shallow coastal waters including hypersaline lagoons, lower arms of rivers and brackish water lakes and reservoirs.

CATCHES, MAIN FISHING GEAR AND PRINCIPAL FORMS OF UTILIZATION:

All mugilid species are included in a single statistical category. Countries reporting on this category are: Algeria, Egypt, France, Greece, Israel, Italy (1970: 6 900 tons), Libya (1969: 2 500 tons), Malta, Romania, Spain, Turkey (1969: 3 700 tons), USSR and Yugoslavia, the catches in the area reported for 1970 totalling 17 000 tons.

Caught with gill nets, trammel nets, beach seines, cast nets, handlines and occasionally with purse seines.

Marketed fresh, frozen or salted; the roe is often marketed as a salted caviar-like product.

FAO SPECIES IDENTIFICATION SHEETS

FAMILY: MUGILIDAE

Fishing Area 37
(Medit. and Black Sea)

Mugil chelo Cuvier, 1829

SYNONYMS STILL IN USE: *Crenimugil labrosus* (Risso, 1826)
Liza provensalis (Risso, 1826)

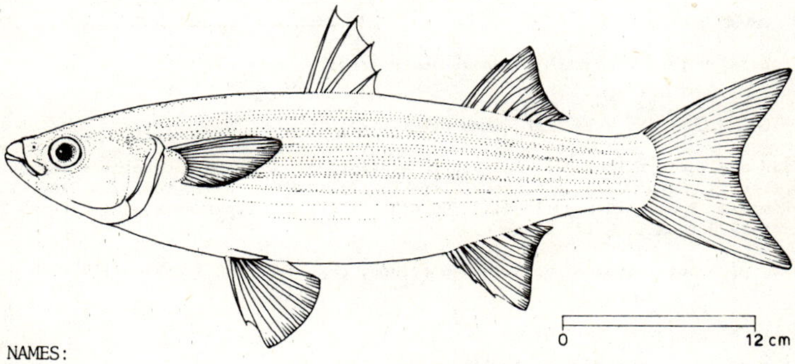

VERNACULAR NAMES:

FAO - En : Thicklip grey mullet
Fr : Muge à grosses lèvres
Sp : Lisa

NATIONAL -

ALBN:		ISRL:	Kifon belut hassafa	ROMN:	
ALGR:		ITAL:	Cefalo bosega	SPAN:	Lisa
BULG:		LEBN:	Bouri sailoun	SYRI:	
CYPR:	Kephalos	LIBY:	Bouri	TUNS:	Kahlayoun
EGYP:	Gabayesh	MALT:	Kaplat	TURK:	Kefal
FRAN:	Mulet à grosses lèvres	MONC:	Müsaru	USSR:	
GREC:	Velanitsa	MORC:	Mulet	YUGO:	Cipal putnik

DISTINCTIVE CHARACTERS AND DIAGNOSIS:

Body elongate, slightly compressed from side to side, head short and flattened with a broad terminal mouth; teeth very small, hardly visible; *upper lip thick* (its greatest depth more than half the eye diameter); *on the lower edge of the lip, a series (2 to 5 rows) of small tubercules;* anal fin usually with 9 soft rays; *the eye is not covered by an adipose lid;* colour of the back bluish-grey; belly silvery, often with grey lengthwise stripes.

Other field characters: two dorsal fins, the first short with 4 slender spines; scales large and adherent; no external lateral line; a gizzard-like stomach with thick walls.

DISTINCTION FROM MOST SIMILAR SPECIES OCCURRING IN THE AREA:

Other Mediterranean mugilids differ from *M. chelo* by the absence of tubercules on the upper lip.

SIZE:

Maximum: 60 cm; common: 20 to 40 cm.

GEOGRAPHICAL DISTRIBUTION AND BEHAVIOUR:

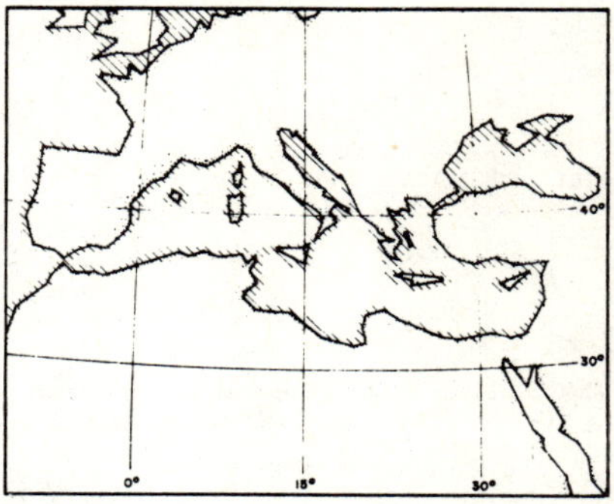

Common throughout the Mediterranean and the Black Sea; also occurs in the eastern Atlantic from Scotland and Norway to Morocco.

Schools of *M. chelo* occur mostly in shallow waters, often in brackish and coastal lagoons of varying salinity; a fast swimmer, leaping out of the water when disturbed; enters rivers and estuaries for feeding but spawning takes place in the sea.

Feeds on minute bottom-living and planktonic organisms.

PRESENT FISHING GROUNDS:

Shallow coastal waters, brackish lagoons and reservoirs.

CATCHES, MAIN FIGHING GEAR AND PRINCIPAL FORMS OF UTILIZATION:

All mugilid species are included in a single statistical category. Countries reporting on this category are: Algeria, Egypt, France, Greece, Israel, Italy (1970: 6 900 tons), Libya (1969: 2 500 tons), Malta, Romania, Spain, Turkey (1969: 3 700 tons), USSR and Yugoslavia, the catches in the area reported for 1970 totalling 17 000 tons.

Caught with gill nets, trammel nets, beach seines, cast nets and occasionally with purse seines and handlines.

Marketed fresh, frozen or salted.

FAO SPECIES IDENTIFICATION SHEETS

FAMILY: MUGILIDAE

Fishing Area 37
(Medit. and Black Sea)

Mugil capito Cuvier, 1829

SYNONYMS STILL IN USE: *Liza ramada* (Risso, 1826)

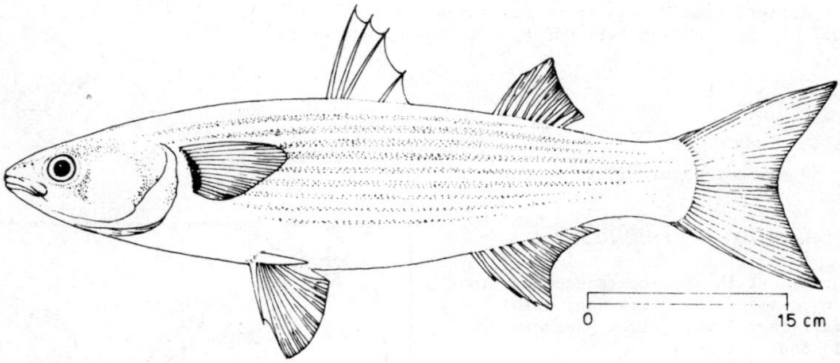

VERNACULAR NAMES:

FAO - En : Thinlip grey mullet
Frr : Muge porc
Sp : Morragute

NATIONAL - ALBN:		ISRL:	Kifon matzui	ROMN:	
ALGR:		ITAL:	Cefalo calamita	SPAN:	Morragute
BULG:		LEBN:	Bouri dahban	SYRI:	
CYPR:	Kephalos	LIBY:		TUNS:	Bitoum
EGYP:	Tobar	MALT:	Mulett tal-incarrat	TURK:	Pulatarina
FRAN:	Mulet porc	MONC:	Müsaru	USSR:	
GREC:	Mavraki	MORC:	Mulet	YUGO:	Cipal balavac

DISTINCTIVE CHARACTERS AND DIAGNOSIS:

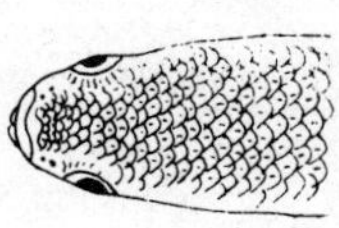

Body elongate, slightly compressed from side to side; head short and flattened with a broad terminal mouth; teeth, very small, hardly visible; *upper lip thin (its greatest depth less than half the eye diameter) and smooth (without tubercles);* anal fin usually with 9 soft rays; *scales on the top of the head extending forward to the anterior nostrils, (almost to the upper lip); eye is not covered by a thick adipose lid;* colour of the back bluish-grey, belly silvery, often with grey lengthwise stripes.

Other field characters: two dorsal fins, the first short with 4 slender spines; scales large and adherent; no external lateral line; pectoral fin short; when folded forward its extreme tip at most reaches the rear edge of the orbit; a gizzard-like stomach with thick walls.

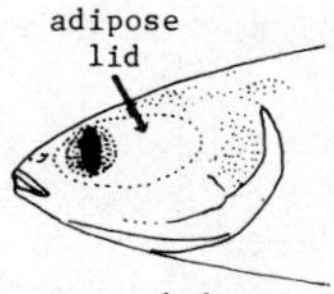

M. cephalus

DISTINCTION FROM MOST SIMILAR SPECIES OCCURRING IN THE AREA:

Mugil cephalus differs from *M. capito* by the presence of a distinct adipose lid around the eye, and of usually 8 rays in the anal fin.

M. saliens: differs from *M. capito* by its more slender body, by having 2 to 3 grooves (instead of 1) on the scales of the top of the head and the back (in small fish this is visible only under a magnifying glass); also by a comparatively long pectoral fin (when folded forward its extreme tip reaches well past the rear edge of the orbit).

scales (schematic)

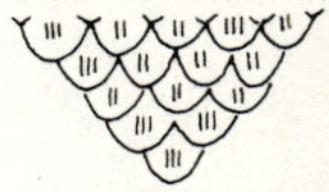

M. saliens

M. auratus differs from *M. capito* by the scales on the top of the head not extending forward beyond the level of the posterior nostrils; also by a comparatively long pectoral fin; (when folded forward, its extreme tip reaches well beyond the rear edge of the orbit).

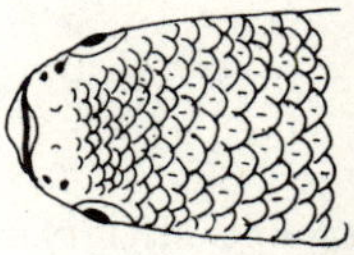

M. auratus

M. chelo: differs from *M. capito* by the presence of small tubercules on the upper lip.

M. labeo: differs from *M. capito* by its thick upper lip (its depth greater than half of the eye diameter) and by the high number of rays (11) in the anal fin.

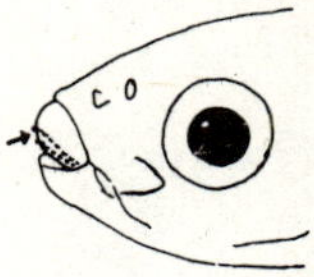

M. chelo

SIZE:

Maximum: 60 cm; common: 20 to 40 cm.

GEOGRAPHICAL DISTRIBUTION AND BEHAVIOUR:

Common throughout the Mediterranean and along the northwestern coast of the Black Sea; also occurs in the eastern Atlantic, from southern Norway to Natal (South Africa).

Schools of *M. capito* occur mostly in shallow water, especially in brackish and coastal lagoons of varying salinity; a fast swimmer, leaping out of the water when disturbed; enters estuaries and rivers for feeding but spawns in the sea; juveniles often concentrate in the vicinity of freshwater outflows.

Feeds on minute bottom-living or planktonic organisms; also on suspended organic matter.

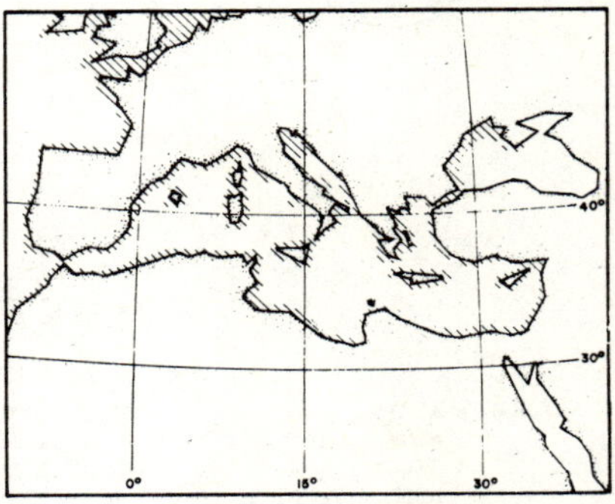

PRESENT FISHING GROUNDS:

Shallow coastal waters, including lagoons, lower arms of rivers, brackish water lakes and reservoirs.

CATCHES, MAIN FISHING GEAR AND PRINCIPAL FORMS OF UTILIZATION:

All mugilid species are included in a single statistical category. Countries reporting on this category are: Algeria, Egypt, France, Greece, Israel, Italy (1970: 6 900 tons), Libya (1969: 2 500 tons), Malta, Romania, Spain, Turkey (1969: 3 700 tons), USSR and Yugoslavia, the catches in the area reported for 1970 totalling 17 000 tons.

Caught mainly with gill nets, trammel nets, beach seines, cast nets and occasionally with purse seines.

Marketed fresh, frozen and salted.

FAO SPECIES IDENTIFICATION SHEETS

FAMILY: MUGILIDAE

Fishing Area 37
(Medit. and Black Sea)

Mugil saliens Risso, 1810

SYNONYMS STILL IN USE: *Liza saliens* (Risso, 1810)

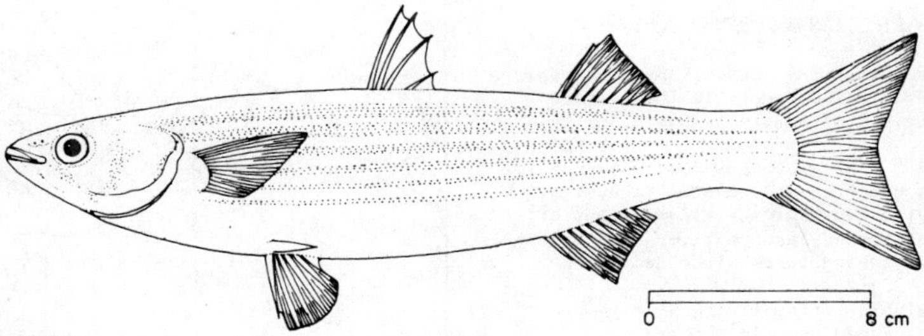

VERNACULAR NAMES:

FAO - En : Leaping grey mullet
Fr : Muge sauteur
Sp : Galúa

NATIONAL - ALBN:		ISRL:	Kifon haruz	ROMN:	Ostreinos
ALGR:		ITAL:	Cefalo verzelata	SPAN:	Galúa
BULG:	Ilarija	LEBN:	Bouri toubaraii	SYRI:	
CYPR:	Kephalos	LIBY:		TUNS:	Karshou
EGYP:	Garan	MALT:	Mulett buri	TURK:	Kefal
FRAN:	Mulet sauteur	MONC:	Müsaru	USSR:	Ostronos
GREC:	Găstros	MORC:	Mulet	YUGO:	Cipal skocac

DISTINCTIVE CHARACTERS AND DIAGNOSIS:

Body slender and elongate, slightly compressed from side to side; head short and flattened, with a broad terminal mouth; *eye not covered by an adipose lid;* teeth very small, hardly visible; *scales on the top of head and back with 2 or 3 grooves* (in small fish these are visible only under a magnifying glass); *upper lip thin* (its greatest depth less than half the eye diameter) *and smooth* (without tubercles); anal fin usually with 9 soft rays; colour of the back bluish-grey, belly silvery, often with grey lengthwise stripes.

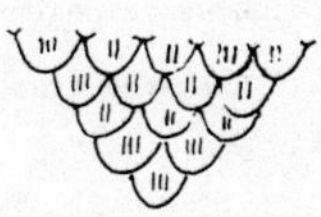
scales
(schematic)

Other field characters: 2 dorsal fins, the first short, with 4 slender spines; scales large and adherent; no external lateral line; pectoral fin comparatively long (when folded forward, its extreme tip reaches well beyond the rear edge of the orbit); a gizzard-like stomach with thick walls.

DISTINCTION FROM MOST SIMILAR SPECIES OCCURRING IN THE AREA:

Mugil cephalus differs from *M. saliens* by the presence of a distinct adipose lid around the eye and of usually 8 rays in the anal fin.

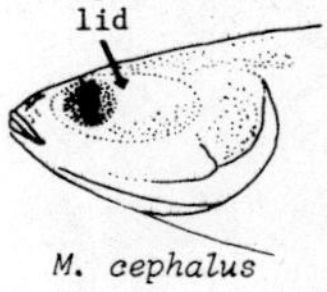

M. cephalus

M. capito and *M. auratus* differ from *M. saliens* by having 1 groove only (instead of 2 to 3) on the scales of the top of the head and the back (in small fishes this is visible only under a magnifying glass); *M. capito* also differs by a shorter pectoral fin (when folded forward, its extreme tip does not reach the rear edge of the orbit).

scales (schematic)

M. capito
M. auratus

M. chelo differs from *M. saliens* by the presence of small tubercles on the upper lip.

M. labeo differs from *M. saliens* by a thick upper lip (its depth greater than half the eye diameter) and by the high number (11) of rays in the anal fin.

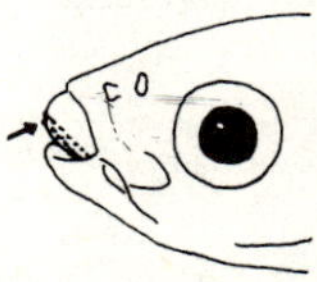

M. chelo

SIZE:

Maximum: 40 cm; common: 15 to 30 cm.

GEOGRAPHICAL DISTRIBUTION AND BEHAVIOUR:

Common throughout the Mediterranean and the Black and Azov Seas; also occurs in the eastern Atlantic from the Bay of Biscay to South Africa.

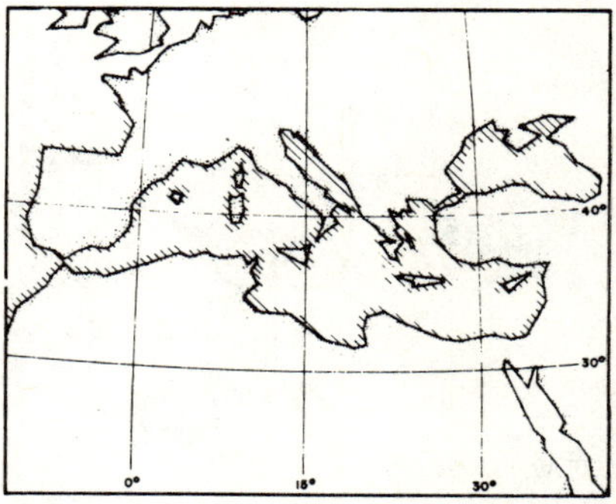

Schools of *M. saliens* occur mostly in shallow waters, often in brackish and coastal lagoons of varying salinity; a fast swimmer, leaping out of the water when disturbed; enters rivers and estuaries for feeding but spawning takes place in the sea.

Feeds on minute bottom-living and plankton organisms and suspended organic matter.

PRESENT FISHING GROUNDS:

Shallow coastal waters including lagoons, lower arms of rivers, brackish water lakes and reservoirs.

CATCHES, MAIN FISHING GEAR AND PRINCIPAL FORMS OF UTILIZATION:

All mugilid species are included in a single statistical category. Countries reporting on this category are: Algeria, Egypt, France, Greece, Israel, Italy (1970: 6 900 tons), Libya (1969: 2 500 tons), Malta, Romania, Spain, Turkey (1969: 3 700 tons), USSR and Yugoslavia, the catches in the area reported for 1970 totalling 17 000 tons.

Caught with gill nets, trammel nets, beach seines, cast nets and hand lines; occasionally also with purse seines.

Marketed fresh, frozen and salted.

FAO SPECIES IDENTIFICATION SHEETS

FAMILY: MUGILIDAE

Fishing Area 37
(Medit. and Black Sea)

Mugil auratus Risso, 1810

SYNONYMS STILL IN USE: *Liza aurata* (Risso, 1810)

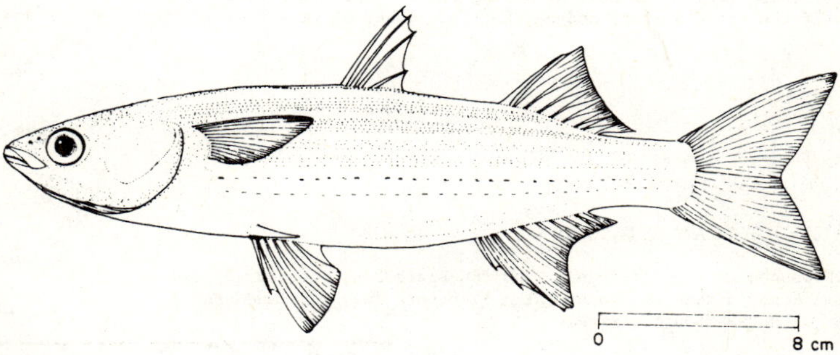

VERNACULAR NAMES:

FAO - En : Golden grey mullet
Fr : Muge doré
Sp : Galupe

NATIONAL -					
ALBN:		ISRL:	Kifon happaz	ROMN:	Singhil
ALGR:		ITAL:	Cefalo dorato	SPAN:	Galupe
BULG:	Platerina	LEBN:	Bouri chilau	SYRI:	
CYPR:	Kephalos	LIBY:	Mulett tal-misluta	TUNS:	Saffraya
EGYP:	Halili	MALT:		TURK:	Altinbas kefal
FRAN:	Mulet doré	MONC:		USSR:	Singil
GREC:	Mixinári	MORC:	Mulet	YUGO:	Cipal zlatac

DISTINCTIVE CHARACTERS AND DIAGNOSIS:

Body slender and elongate, slightly compressed from side to side; head short and flattened with a broad, terminal mouth; *teeth small but clearly visible ; upper lip thin, (its greatest depth less than half the eye diameter) and smooth (without tubercles); anal fins usually with 9 soft rays; scales on the top of the head not extending beyond the level of the posterior nostrils; each scale is provided with one groove; eye not covered by a thick adipose lid*; colour of the back bluish grey, belly silvery, often with grey lengthwise stripes; *a golden blotch on the gill cover.*

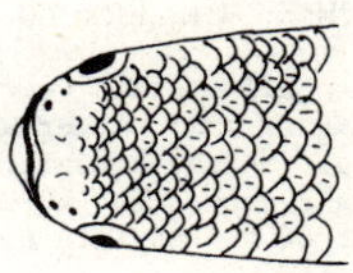

Other field characters: two dorsal fins, the first short with 4 slender spines; scales large and adherent; no external lateral line; pectoral fin comparatively long (when folded forward its extreme tip reaches well beyond the rear edge of the orbit); a gizzard-like stomach with thick walls.

DISTINCTION FROM MOST SIMILAR SPECIES OCCURRING IN THE AREA:

Mugil cephalus differs from *M. auratus* by the presence of a distinct adipose lid around the eye and usually 8 rays in the anal fin.

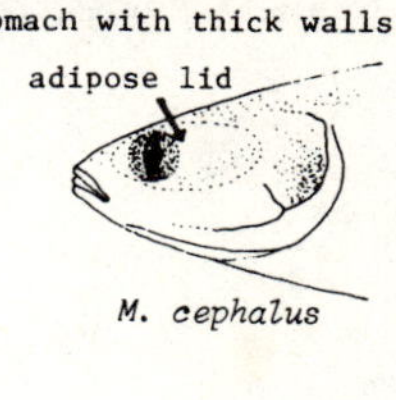

M. cephalus

scales (schematic)

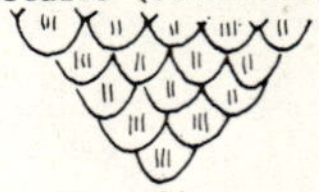

M. saliens

M. saliens differs from *M. auratus* by having 2 to 3 grooves (instead of 1) on the scales of the top of the head and the back (in small fishes this is visible only under a magnifying glass).

M. capito differs from *M. auratus* by the scales on the top of the head extending to the level of the anterior nostrils (almost to the upper lip), by the comparatively short pectoral fin (when folded forward, its extreme tip at most reaches the rear edge of the orbit) and by the absence of a golden blotch on the gill cover.

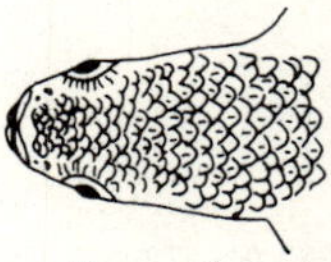

M. capito

M. chelo differs from *M. auratus* by the presence of small tubercles on the upper lip.

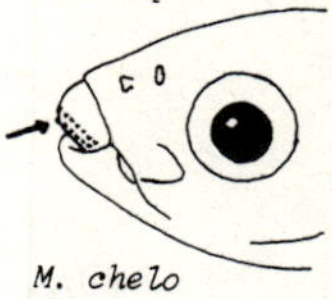

M. chelo

M. labeo differs from *M. auratus* by having a thick upper lip (its depth greater than half the eye diameter) and by the high number of rays (11) in the anal fin.

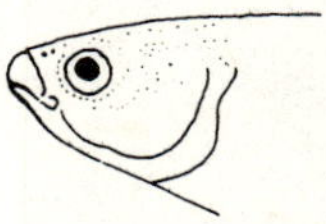

M. labeo

SIZE:

Maximum: 50 cm; common: 15 to 40 cm.

GEOGRAPHICAL DISTRIBUTION AND BEHAVIOUR:

Common throughout the Mediterranean and the Black Sea; introduced also into the Caspian Sea; in the eastern Atlantic it occurs from Scotland and southern Norway to Natal (South Africa).

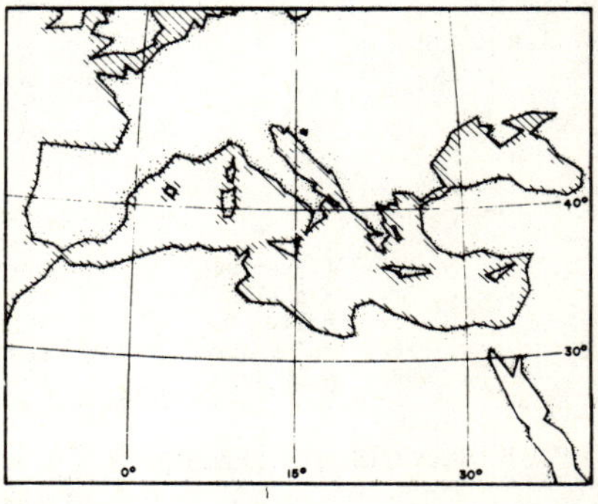

Schools of *M. auratus* occur mostly in shallow water, especially in coastal lagoons of varying salinity; a fast swimmer, leaping out of the water when disturbed; enters rivers and estuaries for feeding, but spawns in the sea; juveniles often concentrate in the vicinity of freshwater outflows.

Feeds on minute bottom-living and planktonic organisms.

PRESENT FISHING GROUNDS:

Shallow coastal waters, including lagoons, lower arms of rivers and brackish water lakes and reservoirs.

CATCHES, MAIN FISHING GEAR AND PRINCIPAL FORMS OF UTILIZATION:

All mugilid species are included in a single statistical category. Countries reporting on this category are: Algeria, Egypt, France, Greece, Israel, Italy (1970: 6 900 tons), Libya (1969: 2 500 tons), Malta, Romania, Spain, Turkey (1969: 3 700 tons), USSR and Yugoslavia, the catches in the area reported for 1970 totalling 17 000 tons. It is estimated that *M. auratus* makes up about 80 per cent of the mugilid catches in the Black and Azov Seas.

Caught with gill nets, trammel nets, beach seines, cast nets and occasionally with purse seines and handlines.

Marketed fresh, frozen and salted.

4. Sexuality and biology of reproduction in grey mullets

J. BRUSLE

Sexuality

Mullets are *heterosexual* fishes. No externally observable characters are known to distinguish males from females, but in schools during the spawning period the males are generally smaller and more slender. Evidence of differentiation and maturation of gonads has been shown for different mugilid species in many ways:

Macroscopic observation

In the majority of fishes, it has been found (Le Dantec, 1955; El Maghraby, Hashem & El Sedfy, 1974*b*) that there was a constant difference between the right and the left lobes of the gonads; the left gonad is smaller than the right one and the difference in size varies between 12.8% and 15% (El Maghraby, *et al.*, 1974*b*).

Juvenile

Brulhet (1975) working on *Mugil cephalus ashenteensis* reports that it is impossible, at the microscopic level, to distinguish between male and female gonads until stage II is reached, and Abraham (1963) states that the sexes of *M. cephalus* can be distinguished only when fishes are about 28 cm long, at 6–7 months of age, in Israel, the weight of the gonads at this time being about 0.05 g. Discrimination of sex in juvenile gonads is still not easy in *M. chelo* and *M. auratus* of less than 18 cm FL, and in *M. capito* under 19 cm FL (Thong, 1969*a*), sexual differentiation not being evident before age class II+ in *M. chelo* and *M. capito* (Cassifour, 1975).

Female

Several ovarian stages have been described, in *M. dobula* (= *cephalus*) by Kesteven (1942) stages I to VII; in *M. chelo* by Erman (1961), 0 to V, in

* The fish have been named as in 'Fiches FAO d'identification des espèces Mediterranée et Mer Noire', FAO, Roma, 1973. *Mugil auratus* = *Liza aurata* (Golden grey mullet) – FAO no. 31 – GPCM no. 31; *Mugil capito* = *Liza ramada* (Thin-lipped grey mullet) – FAO no. 29 – GPCM no. 29; *Mugil cephalus* (Flat-head grey mullet, striped mullet) – FAO no. 27 – GPCM no. 27; *Mugil chelo* = *Crenimugil labrosus* (Thick-lipped grey mullet) – FAO no. 28 – GPCM no. 28; *Mugil saliens* = *Liza saliens* (Leaping grey mullet) – FAO no. 30 – GPCM no. 30.

M. cephalus by Ezzat (1965), I to VII; in *M. cephalus*, *M. capito* and *M. saliens* by Rafail (1968): I, immature; II, developing; III, ripe and IV, spent stage; in *Aldrichetta forsteri* by Webb (1973), I to V; in *M. cephalus ashenteensis* by Brulhet (1975); I to VI; in *M. curema* and *M. cephalus* by Moore (1974), 0 to V and in *M. capito* by El Maghraby *et al.* (1974*b*).

Male

After rough examination, seven (I to VII) testicular stages have been recorded in *M. dobula* by Kesteven (1942) and five (I to V) in *A. forsteri* by Webb (1973). The stages of gonad maturity in *M. capito* were standardized according to the shape of testis and the condition of milt inside: I, immature; II, mature; III, ripening; IV, nearly ripe; V, ripe; and VI, spent, by El Maghraby *et al.* (1974*b*).

Histological examination

Many detailed descriptions of male and female gametogenesis have been recorded.

Undifferentiated gonad

Cytological evidence of sexual differentiation begins in *M. cephalus* at lengths over 15 cm standard length (SL) and Stenger (1959) gives an account of 'germ cells' present in 'protogonial masses' of 20 mm fry. Primary germ-cells have no single morphological characteristic making them unique and identifiable. They are usually referred to as large cells, sometimes as the largest cells of the embryo, measuring 9 to 12 μm in diameter, with an ovoid shape and relatively clear aspect (Stenger, 1959). Thong (1969*a*) set distinction in *M. auratus*, *M. chelo* and *M. capito* between germ-cells, 3.5 μm in diameter (max. 5 μm) and 'ovogonies' and 'spermatogonies', polygonal cells 8–10 μm with the same morphological appearance. A similar account is given for *M. capito* and *M. chelo* by Cassifour (1975). Cytological information on primordial germ-cells, ovogonies and spermatogonies are lacking; the problem of ultrastructural analysis of precocious germinal differentiation is at present being investigated (Brusle, in preparation).

Ovarian structure

Oogenesis and vitellogenesis stages have been described in different mugilid species, essentially *M. cephalus* (Erman, 1959; Stenger, 1959; Leray & Stahl, 1961; Abraham, 1963; Abraham, Blanc & Yashouv, 1966, 1968; Ochiai & Umeda, 1969; Shehadeh & Ellis, 1970; Livni, 1971; Kuo, Nash & Shehadeh, 1974*b*; Kuo & Nash, 1975; Timoshek & Shilenkova, 1974). Other descriptions of microscopical examinations of ovarian lamellae have been made by Thong (1969*a*) in *M. auratus*, *M. chelo* and *M. capito*, and by Donato & Contini (1974*a*, *b*) in *M. chelo*.

Oogenetic stages were derived from average oocyte diameter in *M. cephalus* by Abraham (1963): A, up to 30 μm (= 'early perinucleolus stage); B, up to 60 μm (= 'late perinucleolus stage'); C, up to 75 μm (= 'yolk vesicle stage'); D, up to 130 μm (= 'primary yolk stage'); E, up to 375 μm and over, up to 600 μm (= 'secondary' and 'tertiary yolk stages'). Stage B represents the state of the ovary during the latent period and there is rapid progress from stage B to E in the course of 1–1½ months. A, B and C are considered (Abraham *et al.*, 1966) as 'previtellogenetic stages' and D to E as the 'vitellogenetic period'. Stage B (previtellogenetic oocyte) is characterized with a 'vitelline body' (Stenger, 1959) or persistent 'yolk nucleus' (Abraham *et al.*, 1966; Livni, 1971), which is described as having (*a*) intranuclear, (*b*) crescent layer and (*c*) intracytoplasmic phases and disintegrates in stage C (Abraham *et al.*, 1966, 1968).

Frequency distribution histograms for oocyte diameter have been studied by Abraham (1963), Ezzat (1965), Abraham *et al.* (1966) and Kuo & Nash (1975: p. 21, Fig. 1). There was one size group of ova seen in each ovary of *M. dobula* by Kesteven (1942). From analysis of the frequency distribution of oocyte diameter there was evidence that oocytes of *M. cephalus* appear to develop in synchrony (Shehadeh, Kuo & Milisen, 1973*b*). So, the validity of using extruded intraovarian oocytes for *in vivo* assessment of ovarian maturity was established by Shehadeh and his co-workers. Accounting for variations in oocyte diameters among samples removed experimentally with a cannula from different ovarian locations indicates that mullet oocytes develop in synchrony and that *in vivo* samples taken from any area in the ovary would be representative of the entire ovary. From the general homogeneity of frequency distribution of oocyte diameter or unimodal type (Grant & Spain, 1975) and from the simultaneity in the evolution of oocyte age classes, Timoshek & Shilenkova (1974) and Cassifour (1975) conclude a 'non-intermittent spawning'. Oogenesis of *M. cephalus* in Israel is completed within one month and ripening of oocytes is simultaneous indicating a single spawning per season (Abraham, 1963).

Cytochemical examination of previtellogenetic and vitellogenetic oocytes of *M. chelo* have shown that lipid yolk co-exists with protein yolk (Donato & Contini, 1974*a* and *b*). Reaching a critical oocyte diameter (690 or 730 μm) has been observed to precede effective spawning (Shehadeh, Kuo & Milisen, 1973*a*) or to follow gonadotrophin treatment (Shehadeh, Madden & Dohl, 1973); beyond this size atresia occurred and atrophy took place in *M. cephalus* females. 'Atretic' follicles within the ovary were studied by Stenger (1959), atretic oocytes being characterized by destruction of the *zona radiata* and phagocytosis of the fatty and protein yolk elements by follicular cells. Degenerative stages (*a* to *e*) have been recognized by Abraham *et al.* (1966) and oocytes atresia and resorption described by Timoshek & Shilenkova (1974), Zhitenev, Kalinin & Abayev (1974) and Cassifour (1975).

Testicular constitution

Seminiferous tubules with sperm cysts have been described in *M. cephalus* by Stenger (1959) and Thierberger-Abraham (1967), in *M. cephalus*, *M. capito* and *M. auratus* by Leray (1968), Thong (1969*a*), and in *M. chelo* by Cassifour (1975). Primordial germ-cells and spermatogonies are recognized in juvenile and latent testes (Leray, 1968) as well as in spent gonads, in the wall of the cyst epithelium (Stenger, 1959). Stage I spermatocytes (15 μm in diameter), stage II spermatocytes (5 μm) and spermatids (2.4 μm) characterize maturing testes (Thong, 1969*a*; Cassifour, 1975).

Gonad weight analysis

In *M. dobula* gonad index varies from 1 to 7 for fish with undeveloped or developing gonads and from 11 to 12 for fish with ripe gonads (Kesteven, 1942). The gonosomatic index (GSI) values are calculated as the gonad weight/body weight × 100, being greater than 20% in female specimens at the peak of the spawning season. Since the gonad of a gravid female may weigh a quarter of the total weight of the fish, there is a length-weight ratio difference between nearly ripe and recently spent females (Kennedy & Fitzmaurice, 1969, in *M. chelo*). The male gonads of this thick-lipped mullet do not exceed about $4\frac{1}{2}$% of the total weight and so, the males do not vary so much in weight at the spawning season. Non-spawning females do not undergo any weight loss at this time. In *M. cephalus* from Japan (Tamura, 1974) ovarian weight is 20% of the body weight and testicular weight is only 11%. The heaviest gonad weight recorded by Kesteven (1942) was 230 g but, in *M. cephalus asthenteensis* Brulhet (1974) found a maximum weight for the ovary of 566 g and only 34.3 g for maximum testicular weight. GSI values in *M. cephalus* females are given as 7.66–16.53% by Shehadeh, Kuo & Nash (1973), 19 and 20% by Yashouv (1969) and over 20% by Grant & Spain (1975). Other GSI values are quoted by Kuo & Nash (1975) for *M. cephalus* (Fig. 4.1), by Gandolfi & Orsini (1969) in *M. saliens*, by Edimar (1973) in *M. curema*, by Abraham *et al.* (1966) and Abraham, Yashouv & Blanc (1967) in *M. capito*, in which the index during the spawning season is much higher in fish brought from freshwater ponds (= 15.3) and from Lake Tiberias (= 19.3) than those caught in their natural habitat (= 12.5), see Abraham *et al.* (1966: p. 163, Fig. 1), but the higher GSI indicates a process of oocyte atresia at a very high rate in fish unable to ovulate because of inability to make the seaward migration and the result of which is an accumulation of the products of atresia in the ovaries (Abraham *et al.*, 1966). Many authors give similar accounts of RGS ('rapport gonosomatique') in different mugilid species (Morovic, 1963; Ezzat, 1965; Brulhet, 1974; Farrugio, 1975; Landret, 1975; Cassifour, 1975) with a fixed mathematical relationship between gonad weight and total body weight or standard length

(Farrugio, 1975: p. 70, Fig. 11). Gonadosomatic indices are useful to measure experimental maturation and artifically induced spawning after pituitary treatment (Shehadeh & Ellis, 1970).

Ripe roes of grey mullets (*M. cephalus*) are of great value in markets where, dried and salted eggs like 'caviar' (with claimed aphrodisiac virtues) are sold as 'boutargue' or 'poutargue', in countries such as Corsica (Belloc, 1938; Sardinia (Paolucci, 1917), Algeria (Gruvel, 1926), Tunisia (Heldt, 1948; Davidson, 1963), the Adriatic region (Morovic, 1963), the eastern Mediterranean Sea (Erman, 1957; Acara & Gozenalp, 1959), the south of the USSR (Berg, 1965) and western Africa (Brulhet, 1974, 1975; Landret, 1975).

Conditions for sexual maturation

The gametogenetic process may be dependent on environmental regulators or ecological factors at certain stages of maturity. Interaction of environmental regulators and the endocrine system indicate that influences on the reproductive cycle are reflected more readily by changes in the ovaries rather than in the testes. The reproductive cycle is under the dual control of an internal physiological rhythm and of an external seasonal rhythm which includes the factors of temperature, salinity and photoperiod.

Abraham (1963) noted that, in Israel, oogenesis of *M. cephalus* begins when water temperatures are highest, i.e., 27 °C off the Mediterranean coast and 30 °C in Lake Tiberias, and that oocytes ripen as temperatures begin to decrease: they are 'fall spawners'; but earlier ripening in Lake Tiberias suggests that in warmer waters *M. cephalus* may spawn twice a year. In the same fashion, in *M. capito*, the timing of oogenesis coincides with a peak of water temperature (Abraham, 1963). Evidence was given by Kuo *et al.* (1974*b*) that temperature regulates vitellogenesis towards functional maturity; at 17 °C, oocyte maturation was incomplete and only limited yolk deposition occurred in stage III oocytes. The combination of a constant temperature of 21 °C and retarded photoperiod is the most effective means of inducing completion of vitellogenesis of oocytes to functional maturity (Kuo *et al.*, 1974*b*).

Specimens of *M. cephalus* confined to freshwater do not ovulate and throughout the year the GSI is about 1. In females prevented from their seaward migration, the normal course of development of the gonads is arrested (no oocyte beyond stage B, 100 μm diameter) and the fish do not reproduce (Abraham *et al.*, 1966). According to Abraham *et al.* (1966, 1967) *M. cephalus* and *M. capito* cannot reproduce in fresh water and oocytes remain in the previtellogenetic stage. The gonadotropin contents of the pituitary of this species held in fresh water was considerably lower than that of pituitary from seawater fish (Blanc-Livni & Abraham, 1969). Abraham *et al.* (1966) assumed that the arrest of oocyte development in *M. cephalus* from fish-ponds was the result of the low osmotic values of the water to which the fish were confined, but the fact that *M. capito* in fishponds and both species under the same salinity conditions prevailing in Lake Tiberias had oocytes which reached stage E, led them to eliminate the hypotonicity factor as an explanation of this phenomenon. Other factors, like crowded conditions prevailing in the fish-ponds, have been considered. Although captive males of *M. cephalus* complete spermiogenesis, the vitellogenesis of ovarian oocytes in captive females maintained in running sea water 32 parts per thousand in Hawaii has been observed only up to the tertiary yolk globule stage (Kuo, Shehadeh & Nash, 1973). These studies suggested that salinity may influence gametogenesis but probably does not function as an environmental synchronizer in the timing of reproduction.

Many experiments, initiated early in the natural refractory period or primary oocyte stage, have indicated that photoperiod and temperature are the most significant regulators of the reproductive cycle. A retarded photoperiod plays a dominant role in stimulating oocyte growth. So, in nature in Hawaii (Kuo *et al.*, 1974*b*; Kuo & Nash, 1975), vitellogenesis in the ovaries of *M. cephalus* begins shortly before day-length reaches its annual minimum (10 h 50 min.) In experimental conditions, the onset of vitellogenesis is determined directly by the effective photoperiod and vitellogenesis is stimulated within 49–62 days, by a short and constant retarded photoperiod of six hours of light and eighteen of darkness.

It may be concluded that the controlling mechanism of vitellogenesis in grey mullets is the marked endogenous activity of the pituitary gland which itself is initiated by the external environmental factor of daylight. Many aspects of endocrinology of vitellogenesis, oocyte maturation and spawning (pituitary and gonadotropin inductions) have been detailed for mullets by Tang (1964), Abraham *et al.* (1967), Gandolfi & Orsini (1968, 1969), Blanc & Abraham (1968) Yashouv (1969), Shehadeh & Ellis (1970), Lumare & Villani (1972), Shehadeh *et al.* (1973*a*), Shehadeh, Madden & Dahl (1973*d*), Kuo *et al.* (1973*b*, 1974*a*, *b*), Zhitenev *et al.* (1974), Cassifour & Chambolle (1975), Kuo & Nash (1975) and Liao (1975). The importance of exogenous factors in controlling annual reproduction cycles has been demonstrated by De Vlaming (1972). There are numerous instances in which disturbance to the spawning migration and sharp temperature fluctuations have caused egg resorption (Zhitenev *et al.*, 1974). Although photoperiod has been credited as the dominant environmental regulatory factor in many previous investigations, photoperiodism seems to be temperature sensitive in the majority of Teleost species and in particular fashion in the family Mugilidae. So, in experimental conditions the spawning season of *M. cephalus* could be prolonged throughout the year and primary objectives for aquaculture could be achieved (Kuo *et al.*, 1974*b*).

Table 4.1. *Hermaphroditism or intersexuality in grey mullets*

Authors	Locality	Species	Number	Evidence
Orlandi (1902)	Ligurian Sea	*M. chelo*	1 (unknown length)	Ovotestis
Kesteven (1942)	Australia	*M. dobula* (= *cephalus*)	1 or 2 each season	—
Johnson (1954) (see Stenger, 1959)	Florida	*M. cephalus*	Not known	Oocyte-like cells
Stenger (1959)	Florida	*M. cephalus*	1 (SL = 250 mm)	Ovotestis
			Several	Oocyte-like cells
Moe (1966)	Florida	*M. cephalus*	2	Ovotestis
Gandolfi, Caronna & Orsini (1969)	Venice	*M. saliens*	4 (of 342 specimens)	Ovotestis
Gandolfi & Orsini (1969)	Venice	*M. saliens*	2 (of 170 specimens)	Ovotestis
Thong (1969*a*)	Brittany	*M. chelo*	1 (CFL = 276 mm) of 492 specimens	Ovotestis
		M. capito	1 (CFL = 301 mm) of 183 specimens	Ovotestis
Brusle & Brusle (1974)	Tunisia	*M. cephalus*	1 (TL = 375 mm) of 114 specimens	Testicular intersexuality
		M. capito	2 (TL = 255 and 280 mm) of 200 specimens	

Hermaphroditism or intersexuality

Mugilids are heterosexual fishes and gonochorism is the general rule, but occasional individuals may be hermaphrodites. Some cases of abnormal hermaphrodites or intersexual mullets have been noted in different species and localities (Table 4.1).

The first description of an *ovotestis* with ovarian lobes and testicular areas, in a specimen of *M. chelo* was found by Orlandi (1902). Then, Kesteven (1942, p. 61) wrote of the Australian mullet (*M. dobula*) 'hermaphroditic roes in otherwise apparently normal fish are not very rare, one or two being found in each season'. Johnson (1954, quoted by Stenger, 1959), from the presence of oocyte-like cells in young fish, suggested that Florida grey mullets (*M. cephalus*) might be hermaphroditic, and Stenger (1959) found evidence of oocyte-like cells in immature mullet and in spent testes, and of a testis-ovary; but the presence of oocyte-like cells in immature and ripe male mullet suggests intersexuality rather than protogyny. Moe (1966), studying the same species, is interested in the juxtaposition of male and female areas in two ovotestes which are considered like ovaries with secondary testicular development (and possible self-fertilization?). Histological examination of four hermaphroditic gonads in a group of 342 specimens of *M. saliens* (Gandolfi & Orsini, 1969) show some different testicular and ovarian lobes, some development of testicular tissue in a lateral appendix of the ovarian gonad, and, finally, a few oocytes scattered in the testicular area. One hermaphroditic sample of *M. chelo* with a proximal female region and a distal male area, like progressive masculinization from anterior to posterior, and one ovotestis in *M. capito*, have been reported in Brittany by Thong (1969*a*). Evidence of a few oocytes (in the previtellogenetic stage) inside spermatogenetic cysts of unripe, mature and spent males of *M. cephalus* and *M. capito* from Tunisian lakes is interpreted (Brusle & Brusle, 1974) as testicular intersexuality. Are permanent primordial germ cells or ovogonies present in male gonads of grey mullets? A hypothesis for such a latent germinal bipotentiality is being investigated currently by the author.

Sex-ratio

Information on the sex-ratio in mugilid species is very inconsistant. This variability means that sex separation in different regions and in various populations of grey mullets is not significant (Table 4.2).

Dominance of females was observed in most species: *Aldrichetta forsteri* (Thomson, 1957*a* – 1.2 female for each male; Webb, 1973: p. 55, Table 13), *M. cephalus* (Erman, 1959; Ezzat, 1965; Brusle, unpublished), *M. cephalus ashenteensis* (Brulhet, 1974, 1975; Landret, 1975), *M. curema* (Mefford, 1955 quoted by Thong, 1969*a*; Edimar, 1973), *M. chelo* (Brusle unpublished), *M. capito* (Ezzat, 1965; Brusle unpublished), *M. saliens* (Gandolfi *et al.*, 1969; Gandolfi & Orsini, 1970; El Zarka & El Sedfy, 1970 – 1. 45 female for each male) and *M. tade* (Pillay, 1954 quoted by Thong, 1969*a*).

Males are considered as the dominant sex in other mugilid populations: *M. auratus* (Ezzat, 1965; Thong, 1971), *M. cephalus* (Kesteven, 1942 – 2 males for each female; Broadhead, 1953; Rakaj, 1960; Jhingran, 1958, all quoted by Thong, 1969*a*; Hendricks, 1961; Nikolskii, 1961; Tung, 1968; Peterson & Shehadeh, 1971; Brulhet, 1974, 1975; Tamura, 1974), *M. chelo* (Erman, 1961) and *M. saliens* (Ezzat, 1965).

It was suggested by El Zarka & El Sedfy (1970) that the following factors might be responsible for sex composition: (*a*) segregation of the sexes through various periods of the year including segregation resulting from sex differences in age and size maturity, (*b*) gear selectivity in relation to sex differences in morphology and in physiological activity and (*c*) differences in natural and fishing mortality between the sexes. So, they suggest that in Lake Qarun, (El Zarka & El Sedfy, 1970: p. 24, Table 7) the higher percentage of females (*M. saliens*) throughout the year and especially during the breeding season (sex-ratio is 4 in August for an annual mean of 1. 45) may be due to their greater activity which in turn increases their chances of being caught.

Having analysed the conditions affecting sex composition, that are quoted in the scientific literature, the different factors which can affect results may be considered as follows.

Table 4. 2. *Sex-ratio in grey mullets*

Authors	Locality	% males	% females	No. in sample and conditions
		Aldrichetta forsteri		
Thomson (1957*a*)	Australia	43.6	56.3	1353
Webb (1973)	New Zealand	21	78.9	1564
		(ratio differs with the month, pl. 4: 1 male/8.5 females in III, I, male/7.6 females in VI)		
		Mugil auratus		
Ezzat (1965)	Mediterranean: Berre	61.5	38.8	Age group III
		59.5	40.5	Age group IV
		51.8	48.2	Age group V
		54.5	54.4	Age group VI
Thong (1971)	Brittany: Gulf of Morbihan	60	40	29 (CFL > 180 mm)
		Mugil capito		
Ezzat (1965)	Mediterranean: Berre	50	34	TL = 250–300 mm
		45	55	TL = 300–50 mm
		32	68	TL = 350–400 mm
		59	40.5	TL = 400–50 mm
El Maghraby, Hashem & El Sedfy (1973)	UAR: Lake Borullus	38	62	1439
El Maghraby *et al.* (1974*a*, *b*)	Lake Borullus	65	35	During the day, specimens of small size, 170–200 mm
		30	70	Sunset to sunrise, specimens of greater, size, 200–340 mm
Brusle (unpublished)	Tunisian lakes	16	84	200
		(% males increases during spawning migration)		
		Mugil cephalus		
Kesteven (1942)	Eastern Australia (*M. dobula*)	63.5	36.5	3107
Erman (1959)	Bosphorus and Sea of Marmara	43	57	758
Hendricks (1961)	Salton Sea	75	—	In a sample of very old mullets (13–16 yr.)
Nikolskii (1961)	Black Sea	Males more abundant than females in spawning schools		
Ezzat (1965)	Mediterranean: Berre	—	Females > males (% females decreases after spawning)	—
Tung (1968)	Taiwan	—	31–35	—
Peterson & Shehadeh (1971)	Hawaii	53	46	113

Table 4. 2. (*cont.*)

Authors	Locality	% males	% females	No. in sample and conditions
Brulhet (1974, 1975)	Mauritania (*M. cephalus ashenteensis*)	0 to 29	100 to 71	Large fishes in schools with sex separation
		76.2	23.8	One sampling 1974 in Nouadhibou
Tamura (1974)	Japan			
	Isla Kaba	83	16.6	6887
	Tomie	57.1	42.8	—
Landret (1975)	Western Africa (*M. cephalus ashenteensis*)	40	60	—
Brusle (unpublished	Tunisian lakes	7.9 (% males increases during spawning migration)	92.1	114
		Mugil chelo		
Erman (1961)	Dardanelles	54.5	45.5	310
Karvounaris (1963)	Italy: Paola	42	40	318 (18% undetermined)
Brusle (unpublished)	Tunisian lakes	16.6	83.3	66
		Mugil curema		
Edimar (1973)	Venezuela: Laguna La Restinga	31.5 (% varies according to the month)	68.4	1074
		Mugil saliens		
Ezzat (1965)	Mediterranean: Berre	males > females (% males decreases with cold winter as mortality increases)	—	—
Gandolfi & Orsini (1970)	Venetian lagoon	60.6	39.4	SL = 130–140 mm (826)
		45.7	54.3	SL = 170–180 mm (826)
		0	100	SL = 250–269 mm (826)
Gandolfi *et al.* (1969)	Venetian lagoon	19.2	80.7	338
El Zarka & El Sedfy (1970)	UAR: Lake Qarun	40.8	59.1	628
			43(Mar.)	Variability with month
			52(June)	
			66(July)	
			80(Aug.)	
			27(Sep.)	
			71(Nov.)	

Ecological conditions

The cold winter, in Berre, in 1963, was responsible for higher mortality rate for males of *M. auratus* and *M. saliens* (Ezzat, 1965). Captures of *M. capito*, in Lake Borullus (El Maghraby *et al.*, 1974*b*) differ diurnally between 65% of males to 70% females, from sunset to sunrise; extreme darkness is particularly favourable for the swimming activity and migrations of females.

Fishing activities

Kesteven (1942) in *M. dobula* describes a sex-ratio which occurs due to differential fishing, in Queensland, the males are often rejected at the net and only females taken to market.

Sex differences in size and age

Ezzat (1965) observed that, in Berre, *M. capito* females are dominant, but the number of males increases with greater size; this is different from findings with *M. auratus* in which males are less frequent among older specimens. In the Venetian lagoon (Gandolfi & Orsini, 1970: p. 86: Table 1), the life span of females seems to be longer than that of the males and this displaces the sex-ratio in the former's favour in the larger size classes (with 100% females in specimens of standard length > 250 mm, as in the protandric condition).

Spawning migrations

Nikolskii (1961) stated that males are more abundant than females in *M. cephalus* spawning schools in the Black sea. This differs from other species in which the female frequency increases during the breeding period (*M. saliens:* 80% in August, 43% in March, from El Zarka & El Sedfy, 1970; *M. curema*: 80–90% in June–July, 50% in December, from Edimar, 1973). In *Aldrichetta forsteri* from New Zealand (Webb, 1973, pl. 4), significant predominance of females which migrate in schools, are related to two reproductive periods (1 male/8.5 females in March, 1 male/7.6 females in June). Sex separation in migrating schools was observed off the west African coast by Brulhet (1974, 1975). Females are absent at the beginning of the migration, dominant (from 71 to 100% in the population) throughout the year and less frequent in some particular samples (23.8% in Nouadhibou in 1974). After spawning, *M. cephalus* females are fewer in Berre (Ezzat, 1965). In the 'bordigues' of Tunisian lakes (Tunis, Ischkeul), the number of males (*M. cephalus, M. capito*) is higher during the seaward migration while females are dominant throughout the year (Brusle, unpublished). In the same country, ripe females of *M. chelo* are unknown in the lakes and supposed to live indefinitely in the sea once sexually mature (Farrugio, 1975).

Knowledge about sex-composition in juvenile populations of grey mullets, which affects their collection, is lacking. This problem is to be investigated in spite of difficulties such as absence of sexual discrimination until 180 mm (Thong, 1969*a*) or 200–250 mm (Ezzat, 1965).

First sexual maturity

The age at which mullets reach sexual maturity is highly variable and is related to the rate of growth which may vary greatly depending on the ecological factors and the food supply. Some published estimates of size and age at first maturity vary (Table 4.3), being due to differences in interpretation of age-recording structures. Very different estimates of length for age arise from examination of otoliths or of scales. Variability is also found in relation to local geographic conditions. Thong (1969*a*) has suggested that a condition factor, $K = W/L^3 \times 100$, may represent sexual maturity, because weight increases in relation to the development of gonads. So, in *M. cephalus ashenteensis*, Landret (1975) gives an account of variation in *K* (from mean value 1.29 to 1.32–1.42) when fish attain 310 mm in length, and concludes that this is evidence of first sexual maturity in the beginning of the fourth year. However, Farrugio (1975) demonstrates that maximal value of *K* is not only a function of gonadal development but also related to increase in body weight due to seasonal food supply and he suggests that *K* is not a good criterion for determining sexual maturity.

There is a rough correspondence between reported size at maturity and the local mean sea temperature, such that the youngest ages and smallest minimum size at maturity have been recorded from the warmer waters and the greatest size and age at first maturity from the coldest waters. So, in Israel (Abraham, 1963), sexual maturity in *M. cephalus* is reached off the Mediterranean coast in fish three years old (TL = 270 mm), and sooner (before the end of the second year, TL = 570 mm) in the warmer waters of Lake Tiberias where growth is faster. The situation is the same for *M. capito* from Lake Tiberias with sexual activity being found at the age of less than one year (Abraham, 1963), or, from UAR, in fish one year old (Faouzi, 1938; El Maghraby *et al.*, 1974*b*). Maturity of *M. capito* is less precocious in the northern Mediterranean sea (at three years in Berre, Ezzat, 1965) and in Brittany (Thong, 1969*a*, *b*) or in the Bay of Biscay (Cassifour, 1975). Evidence of temperature influence is supplied by Linder *et al.* (1975) in Galveston Bay (Texas). Pond cultivated striped mullet (*M. cephalus*), receiving heated effluents from a power plant, had fully developed gonads at the age of one year (SL = 160 mm).

It has been observed generally that sexual maturity occurs at a greater size and age in higher than lower latitudes. So, in Irish waters, (Kennedy & Fitzmaurice, 1969) and in southern England (Hickling, 1970), *M. chelo* seems to be a long living, slow growing and also a late maturing fish (males at

Table 4.3. *First sexual maturity in grey mullets*

Authors	Locality	Length	Age (years)
		Aldrichetta forsteri	
Vebb (1973)	New Zealand	220–240 mm (TL)	—
		Mugil auratus	
'omazo (1940, quoted by Berg (1965)	Black Sea	—	3–4 (males), 5–6 (females)
Jikolskii (1961)	Black Sea	—	3–4 (males), 5–6 (females)
Borysenko (quoted by Berg, 1965)	Black Sea	280–370 mm (TL) (males), 350–430 mm (TL) females)	
:zzat (1965)	Mediterranean: Berre	270 mm (TL) (males), 325 mm (TL) (females)	End 3rd year (males) End 4th year (females)
'hong (1969*a*, *b*; 1971)	Brittany	210–220 mm (FL)	3
		Mugil capito	
Vimpenny (1934)	UAR	> 210 mm	2 (cl. 1+) (some samples: 0+ = 1 year)
aouzi (1938)	UAR	130 mm (males), 190 mm (females)	1
l Maghraby *et al.* (1974*b*)	UAR: Lake Borullus	130 mm (males) 170 mm (females)	After 1st winter
braham (1963)	Israel		
	Mediterranean coast	275–310 mm	—
	Lake Tiberias	440–550 mm	Less than 1 year
zzat (1965)	Mediterranean: Berre	330 mm (TL) (males), 370 (TL) (females)	5 (males), 6 (females)
arrugio & Quignard (1973); Farrugio (1975)	Tunisian lakes	First males at 210 mm (SL) all at 230 mm (SL) (males), 255 mm (SL) (females)	Beginning 3rd yr.
'hong (1969*a*, *b*)	Brittany	250–60 mm (FL)	3 (age class III+)
assifour (1975)	Bay of Biscay: Arcachon	—	Age class III+
		Mugil cephalus	
acot (1920, quoted by Linder *et al.* 1975)	Florida (E)	235 mm (SL)	2
ubbs (1921, quoted by Ezzat, 1965)	Florida	330 mm (males), 350 mm (females)	2
roadhead (1953, quoted by Thomson, 1963)	Florida (W)	230–90 mm (males), 240–310 mm (females)	3
unter (1945)	Texas	200–355 mm (males), 250–350 mm (females)	—
ill (1944, quoted by Stenger, 1959)	Salton Sea	412 mm (males), 562 mm (females)	—
enger (1959, quoted by Thomson, 1963)	Florida (E)	236 mm (males), 250 mm (females)	2 (males), 3 (females)
ireman (1965, quoted by Moore, 1974)	Gulf of Mexico	—	3

Table 4.3. (*cont.*)

Authors	Locality	Length	Age (years)
Cech & Wohlschlag (1973)	Gulf of Mexico	—	1–2
Arnold & Thompson (1958)	Gulf of Mexico	240 mm (males), 258 mm (females)	—
Marquez (1974)	Mexico (Tamiahua Lagoon)	330 mm (SL)	3
Linder *et al.* (1975)	Texas Bay (Galveston 'thermal effluents')	160 mm (SL)	1
Wimpenny (1934)	UAR	> 300 mm (1 male specimen 130 mm)	> 15 months
Faouzi (1938)	UAR	310 mm (males), 330 mm (females)	2
Morovic (1954, 1957)	Adriatic:	310 mm (TL)	3rd year during 8th mo
	Venice	> 330 mm (TL)	—
Denizci (1958)	Istanbul	470 mm (TL) (1 female specimen)	5
Abraham (1963)	Israel		
	Mediterranean coast	470 mm (TL)	3
	Lake Tiberias	570 mm (TL) (females)	Before the end of 2nd y
Morovic (1963)	Pantan Lagoon	330–50 mm (TL) (males), 360–80 mm (TL) (females)	4
Erman (1959)	Sea of Marmara Bosphorus	400 mm (TL) (males), 415 mm (TL) (females)	5
Farrugio (1975)	Tunisian lakes	270 mm (SL) (males), first females at 280 mm (SL), all by 355 mm (SL)	End 2nd year
Slatenenko (1956, quoted by Erman, 1959)	Black Sea	312 mm (males), 404 mm (females)	6–7 (males), 7–8 (femal
Nikolskii (1961) } Berg (1965)	Black Sea	—	6–7 (males), 7–8 (femal
Kesteven (1942)	Eastern Australia (*M. dobula*)	310 mm (males), 330 mm (females)	End 3rd year
Thomson (1951)	Western Australia	310–50 mm	3
Thomson (1957, quoted by Farrugio, 1975)	Western Australia	310 mm (males), 340 mm (females)	4
Ezzat (1965)	Mediterranean: Berre	> 330 mm (TL) (males), > 360 mm (TL) (females)	End 3rd year
Brulhet (1974, 1975)	Mauritania (*M. ceph. ashenteensis*)	390 mm (SL) (mean, 430 mm) (males), 460 mm (SL) (mean, 500 mm) (females)	3
Landret (1975)	Mauritania	282 mm (FL) or 345 mm (FL)	End 3rd or 4th year
Hotta (1955, quoted by Thomson, 1963)	Japan (Nagasaki Pref.)	< 400 mm (males), > 450 mm (females)	—
Tamura (1974)	Japan (Nagasaki Pref.)	310–410 mm (males), 430–550 mm (females)	—
Ochiai & Umeda	Japan	270–420 mm (SL) (males),	—

Table 4.3. (*cont.*)

Authors	Locality	Length	Age (years)
(1969)	(Kochi Pref.)	320–560 mm (SL) (females)	
		Mugil chelo	
Le Dantec (1955)	Bay of Biscay: Certes	380 mm (SL)	3rd winter
Kennedy & Fitzmaurice (1969)	Irish waters	> 385 mm (FL) (1 specimen 355 mm FL) (males), > 410 mm (FL) (2 specimens 409–416 mm FL) (females)	9–10 (males), 11–12 (females)
Thong (1969*a*, *b*)	Brittany	220–30 mm (FL)	3
Hickling (1970)	Southern England	> 350 mm (males), > 380 mm (females)	9 (males), 11 (females)
Cassifour (1975)	Bay of Biscay: Arcachon	—	6 (males), 7 (females)
Rossignol (1951)	Morocco	—	3rd winter
Karvounaris (1963)	Italy: Paola	> 300 mm	2–3
Erman (1961)	Dardanelles	340 mm (males), 360 mm (females)	4
Morovic (1963)	Adriatic: Pantan	340 mm (males), 350 mm (females)	4
Farrugio & Quignard (1973) & Farrugio (1975)	Tunisian lakes	First males at 220 mm (SL), and all at 260 mm (SL), first females at 290 mm (SL), and all at 310 mm (SL)	—
		Mugil cunnesius	
Sarojini (1958, quoted by Gopalakrishnan, 1974)	India	131 mm (SL)	—
		Mugil curema	
Anderson (1957)	South Florida	189 mm (males), 209 mm (females)	1
Edimar (1973)	Venezuela	285 mm (males), 303 mm (females)	1
		Mugil macrolepis	
Jhingran & Natarajan (1969, quoted by Gopalakrishnan, 1974)	India	212 mm (males), 337 mm (females)	2 (males), 3 (females)
		Mugil saliens	
Ezzat (1965)	Mediterranean: Berre	First males at 210 mm (TL), and all at 250 mm(TL), first females at 220 mm (TL), and all at 280 mm (TL)	—
Gandolfi & Orsini (1970)	Venetian lagoon	150 mm (SL) (males), first females at 210 mm (SL), all at 250 mm (SL)	2 (males), 3 (females)
El Zarka & El Sedfy (1970)	UAR: Lake Qarun	> 130 mm (TL) (at > 150 mm males spent)	2

SL, standard length; FL, fork length; TL, total length.

380 mm FL, age nine years and females at 450 mm FL, age 11–12 years, in the former station and 350–80 mm and 9–11 years old in the latter area). Maturity is known to be more precocious in Tunisia (males mature at 220–60 mm SL and females at 290–310 mm SL, Farrugio & Quignard, 1973; Farrugio, 1975), in Morocco (during the third winter, Rossignol, 1951) and in the Bay of Biscay (Le Dantec, 1955).

There are some sex-differences and the males are generally more precocious (one year) in maturity than the females. So, in *M. saliens* from the Venetian lagoon, sexual maturity is first reached by most males in the second year while the females ripen in the third year (Gandolfi & Orsini, 1970). In relation to this fact, the life span of the females seems to be longer than that of the males and this displaces the sex-ratio in the favour of the females in the larger size classes.

Biology of reproduction

Although the Mugilidae are of circumglobal distribution in the temperate and tropical zones, there is little information available on the actual area and time of spawning of many of the numerous species, because fully ripe females are generally lacking.

So, Breder (1940: *M. cephalus* in Florida), Dekhnik (1953, quoted by Libosvarsky & Darrag, 1975*a*: *M. auratus* in the Black sea), Erman (1959: *M. cephalus* in the Sea of Marmara), Bograd (1961: different species in Israel), Perlmutter, Bograd & Pruginin (1957: in Israel), Abraham *et al.* (1966: in Israel), Morovic (1957: *M. cephalus* in the Adriatic), Farrugio & Quignard (1973: *M. chelo* in Tunisia), Moore (1974: *M. curema* in Texas) were not able to find any running ripe females during the spawning period. The paucity of ripe spawners *M. cephalus* was also noted in the estuaries, bays and coastal waters of Hawaii in which spawning schools are yet to be located (Shehadeh & Ellis, 1970).

In the absence of knowledge of the spawning grounds of mullets in different regions, many methods are used to determine the reproductive season.

Histological examination of the gonads and histograms have been used by Abraham *et al.* (1966), with gonadal-stages recognized (Moore, 1974).

GSI values have also been used (Morovic, 1963; Abraham *et al.*, 1966; Thong, 1969*a*, *b*; Farrugio & Quignard, 1973; Cassifour, 1975). So, the loss in weight between ripe and spent gonads in *M. capito* from Lake Borullus was 6–12.2% in males and 11.1–20.5% in females (El Maghraby *et al.*, 1974*b*).

Noting when spent females are captured also provides useful information. Postspawning fishes have been recorded, shortly after the spawning period, in bays and coastal waters: *M. cephalus* in Japan (Ochiai & Umeda, 1969), in the Black Sea (Borcea, 1934), in south Texas (Moore, 1974), *M. chelo* in the Dardanelles (Erman, 1961), in Irish areas (Kennedy & Fitzmaurice, 1969), near the Scilly Isles (Hickling, 1970), *M. capito* in the UAR (Rafail, 1968), *M. saliens* in Lake Qarun (El Zarka & El Sedfy, 1970). They vary from 82.2% of females in June, at the peak of the spawning season, to 27% in July and 16% in September). The question of the spent fish swimming back into lakes and estuaries is not quite clear. If, after spawning, some fishes return to the rivers they left (Thomson, 1966), do all the total spent spawners return into freshwater habitats, a question posed by Libosvarsky & Darrag (1975*a*) or do they spend their whole lives in sea water, as suggested by Thomson (1950, 1953, 1955) for *M. dobula* in Australia and Farrugio & Quignard (1973) for *M. chelo* in Tunisia?

The reproductive season may also be deduced from collections of planktonic eggs and larvae (Gourret, 1890; Sanzo, 1936; Dekhnik, 1953, quoted by Libosvarsky & Darrag, 1975*a*; Caldwell & Anderson, 1959) and lastly, from fry appearance and of size distribution of fry (Gunter, 1945; Le Dantec, 1955; Bograd, 1961; Demir, 1971).

However, Arnold & Thompson (1958) reported on *M. cephalus* collected in the wild, from which eggs could be extruded with the slightest abdominal pressure and milt ran freely from the males. Submature specimens and nearly ripe females have been caught in some cases (Heldt, 1948*a*, in Tunisia; Anderson, 1957, in Florida; Yang & Kim, 1961, in Korea; Perlmutter *et al.*, 1957, in Israel; Ezzat, 1965, in Berre; Tung, 1968, in Taiwan; El Zarka, 1970, in UAR; Fitch, 1972, in California and Franks *et al.*, 1972, in the Gulf of Mississippi).

Sometimes spawning behaviour (mating and fertilization) have been observed (Breder, 1940; Nikolskii, 1961; Arnold & Thompson, 1958).

Conflicting reports on the reproductive history of Mugilidae are common (Table 4.4). Indeed, as shown by Avanesov (1972), when one compares the main conclusions of investigations on the reproduction of mullet, in a given region, one may readily note certain differences in determination of the time of the beginning of spawning and the location of the main spawning grounds relative to the shores and depth (Table 4.4). The effect of temperature on the intensity of spawning is also evaluated in differing ways.

Spawning grounds

Though mullets grow to a large size and sometimes become submature in lakes, coastal lagoons and ponds, spawning has generally not been observed in enclosed waters, like 'limans' of the Black Sea (Berg, 1965), the Tunisian lakes (Heldt, 1948*a*), the Salton Sea (Hendricks, 1961), coastal lagoons near Odessa (Zaytsev, 1960, quoted by Timoshek & Shilenkova, 1974), the Lake of Tiberias and ponds in Israel (Abraham *et al.*, 1966).

Gonads do not ripen fully unless the fish get to the open sea and final stages of maturation must occur after the fish have left the inland freshwaters. So,

Table 4.4. *Period of spawning of grey mullets*

Locality	Authors	Period	Evidence
	Aldrichetta forsteri		
Australia	Thomson (1957*a*, *b*,)		—
Victoria, Tasmania, Western Australia		XII → II (summer) (winter) } two races	
New Zealand	Webb (1973)	{VI–VII XI → II (sexual maturity)	Spent females: III–IV
	Mugil auratus		
Eastern Atlantic			
Brittany: Bay of St Michel	Thong (1967)	end V → ?	—
Brittany	Thong (1969*a*)	late VI → early IX	GSI
Brittany: Gulf of Morbihan	Thong (1971)	VI → early IX	GSI
Plymouth	Demir (1971)	late VIII–IX	Postlarvae (between 6 mm and 21 to 27 mm): mid X–mid XI
Northern Mediterranean			
Gulf of Marseille	Gourret (1890)	IV–V	Fry (4.5–6 mm): VI
Gulf of Marseille	Marion (1890*a*, *b*)	III	Eggs and fry (17–20 mm): late V–early VI
Gulf of Marseille	Holt (1899)	VII	1 postlarva (14 mm): on 10/VIII
Gulf of Marseille	Leray (1968)	IX → XI	Gonadal stages
Berre Lagoon	Ezzat (1965)	VIII	Fry: IX–X–XI
Corsica: Biguglia	Belloc (1938)	V–VI–VII	—
Italy	Brunelli (1916, quoted by Serbetis, 1939)	II–III	Fry
Naples, Fusaro	Mazzarelli (1922, quoted by de Angelis, 1967)	late XI–late X	—
Messina	Sanzo (1931, quoted by Demir, 1971)	XII–I–II	Eggs and larvae
Naples	Roule (1917, quoted by Demir, 1971)	XI–XII	Postlarvae (7 mm) and juveniles (60 mm)
Spain: Castellon	Gavard (quoted by Belloc, 1938)	V–VII	—
Mar Menor	Lozano-Rey (1947)	late IX–early XI	—
Eastern Adriatic	Morovic (1957)	II	Eggs and fry
	Morovic (1963 quoted by Thong, 1971)	V–VII	GSI
Southern Mediterranean			
Tunisia	Heldt (1948*a*)	IX–X	Ripe specimens
Algiers Bay	Marinaro (1971)	X	Eggs and larvae
Eastern Mediterranean			
Israel: Haifa	Perlmutter *et al.* (1957)	I–II	Fry (15–35 mm)

Table 4.4. (*cont.*)

Locality	Authors	Period	Evidence
Israel	Bograd (1961)	XII–I–II	Fry (19 mm): I
Israel	Abraham *et al.* (1966)	mid XI–mid XII	GSI
Israel	Yashouv & Berner-Samsonov (1970)	X	Planktonic eggs and early larval stages: X
Black Sea and Caspian Sea			
Black Sea: Romanian shores	Borcea (1934)	Late summer (IX)	1 ripe female and 1 ripe male (IX/1933)
Black Sea	Banarescu (1964)	IX–X	—
	Tomazo (1940, quoted by Yashouv & Berner-Samsonov, 1970)	Early VIII–early X	—
Black Sea	Ilin (1949, quoted by Yashouv & Berner-Samsonov, 1970)	VII → X	—
Black Sea	Tanarenko (1950, quoted by Yashouv & Berner-Samsonov, 1970)	early VI → early XI	—
Black Sea	Tanarenko (1951, quoted by Yashouv & Berner-Samsonov, 1970)	late VII → mid IX	—
Black Sea	Nikolskii (1961)	IX–X	—
Black Sea	Vodyanitsky & Kazanova (1954, quoted by Demir, 1971)	mid VI → X	—
Black Sea (NW)	Zambriborsch (1962, quoted by Thong, 1971)	VI → IX	—
Black Sea (NW) (20 km from Odessa)	Malyatskii & Borisenko (quoted by Berg, 1965)	VII–IX	Fry: VIII/1938 and IX/1939
Black Sea (NW)	Savchuk (1967)	VII → IX	Fry: X–XI
Middle and South Caspian	Avanesov (1972)	Middle: mid VII–early VIII (peak: late VIII–early IX) South: ? → X	Eggs and larvae
	Mugil capito		
Eastern Atlantic			
Bay of Biscay: Certes	Le Dantec (1955)	XI–XII	Fry: III
Bay of Biscay: Arcachon	Cassifour (1975)	VIII → XI	GSI
Brittany	Thong (1969*a*, *b*)	late VII → early XI	GSI
South Devon	Hickling (1970)	VIII–IX	2 ripe females: VIII–IX 1 spent female: X
Northern Mediterranean			
Corsica: Biguglia	Belloc (1938)	X	—
Gulf of Marseille	Leray (1968)	XI–XII	Gonadal stages
Berre Lagoon	Ezzat (1965)	IX–X–XI (peak: IX, females until II)	Fry: X–XI
Marseille	Albertini-Berhaut (1975)	III–IV	Fry (15 mm) IV/1971 and V/1972

Table 4.4. (*cont.*)

Locality	Authors	Period	Evidence
Spain	Lozano-Rey (1947)	X–XII	—
Italy: Gulf of Naples	Lo Bianco (1888, 1909)	X	Pelagic eggs: X, and fry (20 mm): XI → IV
Italy	Brunelli (1916, quoted by Serbetis, 1939)	II–III	Fry
Italy: Fusaro, Naples	Mazzarelli (1922, quoted by De Angelis, 19767)	mid VIII–late IX	—
Adriatic	Morovic (1963, quoted by Albertini-Berhaut, 1975)	II–III	Fry: II–IV
Eastern Adriatic	Morovic (1957)	XII	Eggs and fry
Eastern Mediterranean			
Israel	Perlmutter *et al.* (1957)	XII	3 ripe specimens, 1 spent female: XII/1955 Fry (15–20 mm): I
Israel	Bograd (1961)	Early winter (XII)	Fry (10.4 mm): I
Israel	Abraham (1963)	XII	GSI, gonadal stages
Israel	Abraham *et al.* (1966)	XII–I	GSI
Israel	Thierberger-Abraham (1967)	XII	Spermatogenetic activity
Israel	Yashouv (1969)	Mid XI → early I	Fry
Israel	Yashouv & Berner-Samsonov (1970)	IX	Planktonic eggs and larval stages: IX
UAR	Paget (1922, quoted by Rafail, 1968)	XI	—
UAR	Wimpenny (1934)	X (females)–XI–XII	Ripe specimens
UAR: Lake Qarun	Wimpenny (1936)	X	Eggs X/1934 (*Mugil saliens*?)
UAR	Faouzi (1938)	X → I	Fry (10 mm): XI
Mex canal	El Zarka & Kamel (1967)	X → XII or I	Fry: XII → VI (peak: II)
UAR: Mediterranean coast	Rafail (1968)	X → I (peak: XI–XII)	{ 79 ripe specimens 2 spent females: XII
UAR: Lake Edku	El Zarka (1970)	XII → V	Fry
UAR: Lake Borullus	El Maghraby *et al.* (1973)	X → XII	Ripe specimens
UAR: Lake Borullus	El Maghraby *et al.* (1974*a*)	X → I	Spent specimens: II
UAR: Lake Borullus	El Maghraby *et al.* (1974*a*, *b*)	{ (*a*) XI–XII (*b*) X → I	Fry Gonadal stages
UAR: Lake Borullus	Hashem *et al.* (1973)	XI–XII	Ripe specimen
UAR: Lake Borullus	Libosvarsky & Darrag (1975*a*)	X → I	Running milt males
Southern Mediterranean			
Tunisia	Heldt (1948*a*)	X–XI	Ripe specimens
Tunisia	Farrugio & Quignard (1973) Farrugio (1975)	(X) XI–XII (I)	GSI
Tunisia	Brusle (unpublished)	X–XI (XII)	Gonadal stages
	Mugil cephalus		
Eastern Atlantic			
Mauritania (*M. cephalus ashenteensis*)	Brulhet (1974, 1975)	XI → II	GSI Gonadal stages

Table 4.4. (*cont.*)

Locality	Authors	Period	Evidence
Mauritania	Landret (1975)		
Senegal			
North Cap Vert		XI–XII (decreasing)	GSI
Senegal mouth		V–VI	
South Cap Vert		VI → IX	
Western Atlantic			
North Carolina	Jacot (1920, quoted by Anderson, 1958)	IX → XII (peak: X–XI)	—
North Carolina	Higgins (1928, quoted by Anderson, 1958)	XI–XII	Ripe specimens and fry: spring
North Carolina	Taylor *et al.* (1951, quoted by Anderson, 1958)	XI–XII	—
North Carolina and Florida	Anderson (1958)	Late XI → II (peak: XII)	Postlarvae (20 mm = 3–4 week old): XI
Georgia	Anderson (1958)	X (peak: XII)	—
Western Florida	Kilby (1955, quoted by Anderson, 1958)	Cedar Bay: X–V Bayport: XII–VII	—
Western Florida	Breder (1940)	XI → I–II	Gonadal stages
North Florida	Broadhead (1953, quoted by Anderson, 1958)	Late autumn–early XII	—
Florida	Broadhead & Mefford (1954, quoted by Thomson, 1963)	X → II (peak: XI–I)	—
Maryland	Stenger (1959)	X → I (peak: end XI–XII)	Gonadal stages
Texas	Gunter (1945)	Late X → early I (peak: XI–XII)	—
Texas	Hoese (1965, quoted by Moore, 1974)	Autumn–winter	—
Texas	Moore (1974)	Autumn–winter	Gonadal stages
Gulf of Mexico	Arnold & Thompson (1958)	XII	Ripe specimens, 4 males, 3 females: XII/1956
Gulf of Mississippi	Franks *et al.* (1972)	XII	1 ripe female: XI/1968
Eastern Pacific			
California	Fitch (1972)	Late XI	1 ripe female (late XI/1971)
Colorado River	Johnson & McClendon (1970)	XII–I	31 postlarvae (120 miles up-stream): III/1966
Salton Sea	Hendricks (1961, quoted by Fitch, 1972)	XII–I	Gonadal condition
Western Mexico	Campa de Guzman (1973)	IV?	2 postlarvae: on 30/V
Hawaii	Peterson & Shehadeh (1971)	Late XII–late I	—
Hawaii	Shehadeh (1973)	XII → II	Artificial spawning
Hawaii	Kuo *et al.* (1973*b*)	XII → II	Artificial spawning
Hawaii	Nash (quoted by Anon., 1973)	II → IV	—
Hawaii	Kuo *et al.* (1974*b*)	XII → III	Gonadal stages
Hawaii	Nash *et al.* 1974)	XII → III	Induced spawning

Table 4.4. (*cont.*)

Locality	Authors	Period	Evidence
Hawaii	Kuo & Nash (1975)	I–II	Induced spawning +GSI
Hawaii	Sylvester & Nash (1975)	XII → IV	Eggs and larvae
Western Pacific			
Korea (SW)	Yang & Kim (1962)	IV → VI	Ripe specimens
Hong-Kong	Bromhall (1954)	XI–XII–I	Week 3–4 XI One month later (full moon) 1 spent female: XII/1953 Fry: late I
Taiwan	Tung (1968)	XII–I	Ripe specimens
Taiwan	Liao *et al.* (1974, see Shehadeh *et al.* 1973*b*)	XII–I	Induced spawning
Japan: Kochi Pref.	Ochiai & Umeda (1969)	Late X–early XI	Ripe females: XI Spent females: XI
Japan: Mie Pref.	Tamura (1974)	X → I	Eggs and larvae
Australia			
Eastern Australia (SW) (*Mugil dobula*)	Kesteven (1942)	III → VIII (autumn and early winter)	Gonadal stages
Western Australia (*Mugil dobula*)	Thomson (1951)	III → IX	—
Western Australia	Thomson (1963)	V → VIII (winter)	GSI
Western Australia	Grant & Spain (1975)	V → VII (winter)	GSI
India	Jacob & Krishnamurthi (1948), quoted by Bromhall, 1954)	X → V	—
India	Nair (1957)	I → V	Fry (10–12 mm): II → VI
India: Lake Chilka	Kowtal (1967)	X → I	—
India: Lake Chilka	Gopalakrishnan (1974)	X → V	—
Northern Mediterranean			
Corsica: Biguglia	Belloc (1938)	VIII–IX (X)	—
Berre Lagoon	Ezzat (1965)	Early VII → late IX (peak: VIII, some specimens late X)	Ripe specimens GSI Fry: IX–XI
Marseille	Leray (1968)	VII → IX	Gonadal stages
Spain: Mar Menor	Navarro (1927)	VI–VII	1 ripe female: VI/1926
Spain: Mar Menor	Lozano-Rey (1947)	VIII–IX	—
Italy	Brunelli (1916, quoted by Serbetis, 1939)	IX → XII	Fry
Sardegna	Paolucci (1917)	VIII–IX	—
Messina	Sanzo (1936)	IX–X	Artificial breeding: IX/1932

Table 4.4. (*cont.*)

Locality	Authors	Period	Evidence
Adriatic			
Eastern	Morovic (1957)	VII–IX	Eggs and fry
Pantan Lagoon	Morovic (1963)	VII–IX	GSI
Eastern Mediterranean			
Istanbul	Denizci (1958)	VI	Ripe specimens
Bosphorus	Erman (1959)	V → X	Spent females: X/1957
Israel	Bograd (1961)	VI → X	Fry (15 mm): late VI → XII
Israel	Abraham (1963)	(IX)X–XI–XII (and twice IV and XI in Lake of Tiberias)	GSI Gonadal stages
Israel	Abraham *et al.* (1966)	X—XI (XII)	GSI
Israel	Thierberger-Abraham (1967)	IX → XI	Spermatogenetic stages
Israel	Yashouv (1969)	X	GSI Induced spawning
Israel	Yashouv & Berner-Samsonov (1970)	IX	Planktonic eggs Early larval stages
Israel	Blanc & Abraham (1968)	X–XI	—
UAR	Wimpenny (1934)	VIII → XI	Ripe specimens
UAR	Faouzi (1938)	VI → X	Fry (10 mm): VII
UAR: Lake Maryut	Paget (1929, quoted by Navarro, 1927)	VII → X	Fry: mid VIII → late XII
UAR: Mediterranean coast)	Rafail (1968)	IV → XI (peak: VIII–X)	Ripe specimens
UAR: Mex canal	El Zarka & Kamel (1967)	IX–X	Fry: VIII → I
UAR: Lake Edku	El Zarka (1970)	VI → XI	Fry and juveniles
UAR: Lake Borullus	Hashem *et al.* (1973)	VI → IX	Ripe specimens
Southern Mediterranean			
Tunisia	Heldt (1948*a*)	Mid VIII–late IX (VII in Lake Ischkeul)	Ripe specimens
Tunisia	Farrugio (1975)	(VIII) IX–X	GSI
Tunisia	Brusle (unpublished)	(VII: males) VII–IX	Gonadal stages
Algeria	Dieuzeide, Novella & Roland (1955)	IX	—
Algeria: Algiers Bay	Marinaro (1971)	IX–X	Planktonic eggs
Black Sea	Arnold (1896, quoted by Yashouv & Berner-Samsonov, 1970)	VI–VII–VIII	—
Black Sea	Popov (1930, quoted by Morovic, 1963)	early IV	—
Black Sea	Tomazo (1940, quoted by Berg, 1965)	Late VI → early IX	Fry (15–20 mm: mid VII
Black Sea	Ilin (1949, quoted by Yashouv & Berner-Samsonov, 1970)	Late V → mid X	—
Black Sea	Taranenko (1950, quoted by Yashouv & Berner-Samsonov, 1970)	Late V → early IX	—

Table 4.4. (*cont.*)

Locality	Authors	Period	Evidence
Black Sea	Taranenko (1951, quoted by Yashouv & Berner-Samsonov, 1970)	VII–VII	—
Black Sea	Dekhnik (1953, quoted by Morovic, 1963)	V?	—
Black Sea: Romanian shores	Borcea (1934)	VIII	1 spent female: VIII/1933
Black Sea: Romanian shores	Banarescu (1964)	IX	—
Black Sea: Romanian shores	Berg (1965)	VI → IX	Spent females: X
Black Sea: Romanian shores	Vodyanitski & Kazanova (1954, quoted by Thomson, 1966)	V → VIII	—
Black Sea: Romanian shores	Malyatski (quoted by Berg, 1965)	VIII	Planktonic eggs: VIII/1938
	Mugil chelo		
Eastern Atlantic			
Nordmark	Duncker (1960)	Early summer	Planktonic eggs
Irish waters	Kennedy & Fitzmaurice (1969)	V	Spent females: VI
Channel coast & Southern Ireland	Wheeler (1969)	IV	—
Plymouth	Demir (1971)	Late V–early VI → late VII	Postlarvae (14.8–21.4 mm): X–XI
Southern England: Scilly Islands	Hickling (1970)	I → IV	Spent females: II Eggs: IV
English coast	Cunningham (1891)	IV	Larvae (10.5 mm): V
Brittany: Roscoff	Le Gall & Cantacuzene (1955)	Summer	Larvae (32 mm): VIII (Batz Island)
Brittany: Bay of St Michel	Thong (1967)	XII → II	—
Brittany: La Rance, Bay of Morlaix, Gulf of Morbihan)	Thong (1969*a*, *b*)	Mid II → mid IV	GSI
Bay of Biscay: Certes	Le Dantec (1955)	I → IV	Gonadal stages Appearance of fry: autumn
Bay of Biscay: Certes	Cassifour (1975)	II → V Gonadal stages	GSI
Northern Mediterranean			
Spain	Lozano-Rey (1947)	II–III–IV	—
Gulf of Marseille	Gourret (1890)	VI	Eggs and fry
Corsica: Biguglia	Belloc (1938)	I (II–III)	—
Berre Lagoon	Ezzat (1965)	V–VI	Mature specimens: VI Fry: V–late IX
Italy	Brunelli (1916, quoted by Serbetis, 1939)	V	Fry
Italy: Messina	Sanzo (1936)	mid II–III–mid IV	Planktonic eggs: II

Table 4.4. (*cont.*)

Locality	Authors	Period	Evidence
Italy: Paola	Karvounaris (1963)	XII–I–II	—
Adriatic: Trieste	Graeffe (1888, quoted by Morovic, 1963)	XI–XII	—
Eastern Adriatic	Morovic (1957, 1963)	II–III	GSI
Eastern Mediterranean			
Dardanelles	Erman (1961)	Early XII → mid III	GSI Gonadal stages
Israel	Bograd (1961)	II–III–early IV	Fry (17–19 mm): III–IV
Israel	Abraham *et al.* (1966)	Late XII–I–II	GSI
Israel	Yashouv & Berner-Samsonov (1970)	X	Planktonic eggs & Early larval stages: II
Southern Mediterranean			
Tunisia	Heldt (1948*a*)	XII–I (early III)	Mature specimens
Tunisia	Farrugio & Quignard (1973) Farrugio (1975)	II–III–IV (males)	GSI (no mature females)
Tunisia	Brusle (unpublished)	I–II	Submature males: II Submature females: XII–I
Algeria	Dieuzeide *et al.* (1955)	XII → III	—
	Mugil cunnesius		
India	Gopalakrishnan (1974)	V → VIII	—
	Mugil curema		
Western Atlantic			
North Carolina	Jacot (1920, quoted by Anderson, 1957)	Mid IV → mid VIII (peak: V)	Young specimens (20 mm)
South Florida	Anderson (1957)	Late III or early IV IX (peak: IV–V–VI)	6 ripe females: IV/54 Fry: III–IV
Gulf of Mexico	Caldwell & Anderson (1959)	IV–V	Larval stages (7–9 mm): V/1954
Texas	Hoese (1965, quoted by Moore, 1974)	Spring–early summer	—
Texas	Moore (1974)	Spring–early summer	Gonadal stages
Venezuela: Campoma Lagoon	Rojas (1972)	VII	1 ripe female
Venezuela: Campoma Lagoon	Edimar (1973)	VII–VIII	GSI Gonadal stages
	Mugil labeo		
Italy: Messina	Sanzo (1938)	VII–VII–early IX	Ripe specimens Eggs and larvae
	Mugil macrolepis		
India: Mandapam coast	Luther (1963, quoted by Jhingran & Natarajan, 1974)	VII–VIII	—
	Mugil saliens		
Northern Mediterranean			
Languedoc: Thau Lagoon	Gourret (1897)	X	Seaward migration: VIII

Table 4.4. (*cont.*)

Locality	Authors	Period	Evidence
Berre Lagoon	Ezzat (1965)	Late VI–early VII (peak: VIII)	Fry: VIII
Spain	Lozano-Rey (1947)	VI–VII	—
Italy	Brunelli (1916, quoted by Serbetis, 1939)	III → VII	Fry
Italy	Bullo (1891, quoted by Gandolfi & Orsini, 1970)	VII–VIII	—
Western Adriatic			
Gulf of Trieste	Graeffe (1888, quoted by Gandolfi & Orsini, 1969)	X–XI	—
Chioggia	Varagnolo (1964*a*, *b*, *c*)	Late V → early VIII	Pelagic eggs: V → IX
Venetian lagoon	Gandolfi & Orsini (1970)	Late V → early VIII	GSI
Eastern Adriatic	Morovic (1957)	VI–VII	Eggs and fry
Eastern Mediterranean			
Israel	Bograd (1961)	V–VI–VII	Fry (13–15 mm): VI → X
Israel	Abraham *et al.* (1966)	VIII–mid IX	GSI
Israel	Yashouv & Berner-Samsonov (1970)	V → VIII	Planktonic eggs: V → VIII
UAR: Lake Menzaleh	Wimpenny (1934)	(1) V (males dominant) (2) VII → XII (females)	Ripe specimens
UAR: Lake Qarun	El Zarka (1963*a*)	IV → VII	Ripe specimens (males ripe since I) spent females: IV Fry: mid IX
UAR: Mediterranean coast	Rafail (1968)	IV → X (peak: VIII)	768 ripe specimens Spent specimens: XI
UAR: Lake Qarun	Aleem (1969)	IV → VII	—
UAR: Lake Edku	El Zarka (1970)	VI → XI	Fry and juveniles
UAR: Lake Qarun	El Zarka & El Sedfy (1970)	(1) III → VI (peak: VI) (2) VII → XII (peak: XI)	Spent females: V → IX (peak: VI)
UAR: Lake Borullus	El Maghraby *et al.* (1973)	(1) spring (2) autumn	Ripe specimens
Southern Mediterranean			
Tunisia	Heldt (1948*a*)	mid VI–mid VIII	Ripe specimens
Tunisia	Brusle (unpublished)	VII	Gonadal stages
Black Sea and Caspian Sea			
Romanian shores	Borcea (1934)	VIII	Ripe female: VII/1933
Black Sea	Tomazo (1940, quoted by Yashouv & Berner-Samsonov, 1970)	VIII–IX	—
Black Sea	Ilin (1940, quoted by Yashouv & Berner-Samsonov, 1970)	VIII–IX	—

Table 4.4. (*cont.*)

Locality	Authors	Period	Evidence
Black Sea	Taranenko (1951, quoted by Yashouv & Berner-Samsonov, 1970)	Late VI → VIII	—
Black Sea	Nikolskii (1961)	VI → VIII	—
Romanian shores	Banarescu (1964)	IX	—
Danubian Delta	Brezeanu & Zinevici (1971)	VII–VIII	Larval stages (17–18 mm)
Southern Caspian			
Turkmenian waters	Avanesov (1972)	Late V–early VI (peak: VIII) Late VII–early VII (northwards)	Eggs and larvae
	Mugil tade		
India	Gopalakrishnan (1974)	V–VI IX	—

different species of mullet migrate to the sea for breeding and generally spawn in salt water; indeed, the eggs are presumed to need waters of high salinity for fertilization and this is believed to be an adaptation for ensuring the most favourable conditions for the development of the eggs and larvae (Nikolskii, 1963, quoted by El Maghraby *et al.*, 1974*b*).

Spawning migrations

This means the movement (named '*appel du large*' by Roule, quoted by Heldt, 1931) of the ripe fish from the feeding grounds in lakes and estuaries to the spawning grounds in the sea. The fish form schools and migrate to the sea; these schools may consist of immature fish but generally consist of mature, nearly ripe, specimens. In Australian waters (Kesteven, 1942) *M. dobula* males outnumber females and are often one year younger than these females with which they travel; so, a considerable number of fish proceeds northwards to warmer waters in the course of the spawning migration (Kesteven, 1942; Sidhu, Vale, Shipton & Murray, 1972).

Various types of movement have been observed, such as from lakes to the sea for *M. capito* in Lake Borullus (Libosvarsky & Darrag, 1975*a*) in which captures increase greatly during the spawning migration (Hashem, El Maghraby & El Sedfy, 1973), *M. cephalus*, *M. tade* and *M. macrolepis* in India (Kurian, 1975), *M. cephalus* and *M. capito* in the Lakes of Tunis and Ischkeul, with a greater number of males during this period (Brusle, unpublished). Movement may take place downstream, from rivers or estuaries to the sea as in *M. dobula* in Australia (Thomson, 1955) and *M. cephalus* in southern mainland China (Tseng, 1975), or from bays and coastal areas to the open sea as in *M. cephalus* and *M. curema* in south Texas (Moore, 1974). Migration may also take place in the sea, e.g., *M. cephalus* moving northwards in the Adriatic (Morovic, 1963), southwards in Taiwan (Tung, 1968; Tseng, 1975) and in Mauritania (Brulhet, 1974, 1975) parallel to the shore where *M. cephalus ashenteensis* moves, over some ten metres, searching for the brackish waters of the Senegal mouth.

Spawning migration seems to be possible at more than one time in a year: Wallace (1975) gave details of two annual seaward spawning migrations for *M. cephalus* on the eastern coast of South Africa; so it seems also to be affected and stimulated by environmental factors.

An optimal temperature is supposed to induce spawning movements which for (*M. cephalus*) takes place at 17.1–20.6 °C in Japan (Tamura, 1974) and at 21–23 °C in Taiwan (Tung, 1968). It is known that mobility of fish increases with the water temperature; the temperature of Lake Borullus contributes to the increased activity of *M. capito* observed by Libosvarsky & Darrag (1975*b*). These authors also have shown that *M. capito* captures in the beach area of Lake Borullus vary with weather conditions and certain atmospheric phenomena (wind velocity, barometric pressure and air temperature); and they noted that ordinary barometric changes appear to be a sufficient stimulus for extensive activity of fishes. The mass movement of fish from the estuaries seems frequently to be triggered by strong offshore wind (Anon., 1951; Taylor, 1951, both quoted by Thomson, 1966). In Australia, it appears that the intensity of the wind is the most influential factor, as fish (*M. dobula*) frequently move out on the first day of a strong blow and more schools move to the sea when westerly winds are blowing than at other times (Thomson, 1955); Kesteven (1942) has explained the hypothesis that mullet in Australia swim against the ocean current (offshore wind) during the spawning migration so that the eggs and fry drift back to repopulate the rivers from which the migrating schools have come. Evidence of the dominant influence of the wind has been also demonstrated in Berre for several species (Ezzat, 1965) and in Lake Borullus (El Maghraby *et al.*, 1974*b*; Libosvarsky & Darrag, 1975*b*) in which the schools of *M. capito* began to move towards the sea only when a northwest wind blew, driving a strong water current from the sea to the lake; so, an increased catch of fish is frequently allied with such an atmospheric disturbance.

The effect of photoperiod rhythm was observed by Libosvarsky & Darrag (1975*b*) while working in Lake Borullus. They give an account of night captures being superior to those during the day. It was also observed (El Maghraby *et al.*, 1974*b*) that extreme darkness is particularly favourable for the migration of female *M. capito* and the catch collected from the set nets in Lake Borullus increased on a moonless windy night. Schools moving to the sea for spawning differed in their sex composition and sizes from day to night, while the migration of males took place during the day.

It is suggested (Wallace, 1975) that organic cues derived from seawater inflow and from fresh-water drainage might help to orientate the fish seawards.

Other factors have also been suggested as having an effect on migration. Roule (1915, quoted by Thomson, 1955) believed that the determining factor is the relative level of dissolved oxygen in waters, but experimental evidence is lacking.

Spawning migration is dependent on endocrinological mechanisms and seems to be regulated by the activity of the thyroid and pituitary glands (Stahl, Seite & Leray, 1960; Leray & Carlon, 1963; Olivereau, 1968; Leray, 1971, Blanc-Livni & Abraham, 1970; Cassifour, 1975), and in relation to osmo-regulation. So, submature specimens (*M. cephalus* and *M. capito*) are in a condition of hypothyroid activity when they migrate from freshwaters or brackish waters to the sea, like a case of antagonism of the thyroid gland and the gonads (Leray, 1971). Mesoadenohypophysis cell activity was compared (Blanc & Abraham, 1968) in *M. cephalus* from freshwater (gonadotropin activity very weak = 7%, without FSH and LH production), from Kinnereth Lake (secretory production = 25% in FSH but LH is lacking) and from seawater (gonadotropin production = 40% , FSH and LH). The spawning season is characterized, as shown by Abraham (1974), by intense stimulation of both synthesizing and releasing neurosecretory influence on the gonadotropin-secreting cells. During this period the gonadotropin cells of seawater specimens were characterized by a large number of A and B type axons of the proximal *pars distalis* in the pituitary of *M. cephalus*, pointing to the possibility that neurosecretory cells may be stimulatory.

When migration to the sea is not possible or is prevented, females show arrest of normal gonadal development and their gonadosomatic index is low (Abraham *et al.*, 1966, 1967). Stages of atretic oocytes were found in fish which did not ovulate because they were confined to freshwater. Some cases of sickness and mortality among females have been observed (*M. cephalus* and *M. capito*) in Tunisian Lakes (Heldt, 1931 and 1948*b*). So, there is evidence that mullet are, as asserted by Thomson (1955), 'generatively marine fish' (and trophically freshwater or brackish fish). A permanent marine habitat has been adopted, it is suggested by Thomson (1955), in *M. dobula* from Australia, because freshwaters are very restricted in extent in certain areas. It was also suggested by Farrugio & Quignard (1973), who have never found ripe and spent females of *M. chelo* in the Lake of Tunis, that this species spend its whole life in marine waters after their first sexual maturity, as asserted previously by Thong (1969*a*) for *M. auratus*, *M. capito* and *M. chelo* along the shores of Brittany. Indeed, a sojourn in freshwater does not appear to be essential though many fish are found there (Thomson, 1955); for this author, believes that there is some evidence to support the hypothesis that the majority of mature fish do not migrate every spawning season, but at greater intervals.

However, it was shown that, in some cases, that spawning takes place in interior brackish waters, in lakes and lagoons. The only species that can successfully reproduce in Lake Qarun (UAR) is *M. saliens* (El Zarka, 1963*a*, *b*; El Zarka & Kamel, 1965; Aleem, 1969, 1972; El Zarka & El Sedfy, 1970; Alkholy & Abdelmalek, 1972) and not *M. capito* as asserted by Faouzi (1936), Wimpenny (1936) and Wimpenny & Faouzi (1935), because of misidentification suggested by El Zarka (1963*a*). Other Mugilidae seem able to spawn in similar conditions, deduced for *M. capito* in Berre (Ezzat, 1965, from larvae 7, 9 and 12 mm collected in the lagoon), *M. cephalus* in Biguglia Corsica), in waters with the same hydrological conditions as in the sea (Belloc, 1938), but there is no spawning of some species in hypersaline Salton Sea (Hendricks, 1961). Spawning in brackish waters of estuaries has been reported by Yang & Kim (1961) in the mouth of the Yungsan River (SW Korea), by Brulhet (1974, 1975) and Landret (1975) in the mouth of the Senegal river, by Webb (1973*a*) in the Avon-Heathcote estuary (New Zealand) and by Wallace (1975) in the mouth of Lake St Lucia (on the east coast of South Africa).

Spawning in fresh water seems not to be possible, except in the case of *Rhinomugil corsula* in India, which is reported to breed in rivers and enclosed waters (Kurian, 1975), and *M. cephalus ashenteensis* which is said to spawn in the Senegal River, 50 km upstream from St Louis (Landret, 1975). Lastly, Johnson & McClendon (1970) presented a fairly convincing set of circumstances for *M. cephalus* spawning in fresh water, i.e., they captured thirty one postlarval mullet 120 miles upstream from the mouth of the Colorado River and by using growth data of Anderson (1958) they demonstrated that fish could not have hatched in saline waters and then arrived in this location while only 28–40 mm length.

Location of spawning grounds

There is conflicting information on local spawning areas, some in close vicinity to the sea coast, others off-shore in the open sea. Locations of spawning grounds remain controversial.

There is evidence for spawning grounds being located in littoral zones. Breder (1940) attested that spawning of *M. cephalus* took place in the bays on sandy grounds; he reported observations on spawning behaviour in a small tidal creek on the west coast of Florida, but females with residual eggs were lacking and so there is doubt about this hypothesis. Yang & Kim (1961) have collected ripe running *M. cephalus* in coastal waters, in a fairly well protected area of SW Korea, 7.5–9 m deep in brackish waters and, at low tide, a salinity of 5.5–7.8‰. Berg (1965) states that Borisenko observed spawning of *M. auratus* near the shore, 20 km from Odessa. Spawning of *M. macrolepis* (Jhingran & Natarajan, 1966, 1974) and of *M. cephalus* (Kowtal, 1967) take place in the vicinity of the mouth of Lake Chilka, that of *M. chelo* near the Scilly Islands (Hickling, 1970) or on the shores of Brittany (Thong, 1969*a*), and that of *M. saliens* not far from the Venetian Lagoon (Gandolfi & Orsini, 1970) or limited to the neighbourhood of Lake Edku/sea connections; nearly full and mature males and females obviously do not need to go far away to the Mediterranean to spawn (El Zarka, 1970).

Gunter (1945) attested also that *M. cephalus* does not spawn in Chesapeake Bay (Texas) but in the vicinity because small fry are present as soon as spring arrives, and Dekhnik (1953, quoted by Libosvarsky & Darrag, 1975), from egg distribution, assumed that *M. auratus* spawned along the coast of the Black Sea, over depths up to some 50–60 m. Finally, Timoshek & Shilenkova (1974) give evidence of the main spawning grounds of Black Sea mullet as relatively close to the shore, because of the ecological characteristics of these fish and other authors who they quote had advanced weighty arguments in support of inshore spawning. Moreover, evidence that deep waters, in North Queensland are some distance offshore and that females in running condition were trapped at river mouths, favours the suggestion of sublittoral sheltered spawning grounds (Grant & Spain, 1975).

There is also evidence for off-shore spawning. From captures of fully matured or running ripe females, there was evidence of spawning grounds off the coast, in the open sea, near the edge of the continental shelf, at various depths: 20 fathoms for *M. curema* in southern Florida (Anderson, 1957), 50 fathoms for *M. auratus* in the Black Sea (Dekhnik, 1953, quoted by Libosvarsky & Darrag, 1975), 5–40 m for *M. saliens*, 5–7 miles off the shore of the Turkmenian sector of the Caspian Sea (Avenesov, 1972), 30 fathoms for *M. cephalus* in California (Franks *et al.*, 1972), 40 miles SW of California (Fitch, 1972), and lastly 575 fathoms for *M. curema*, 55 miles offshore in the western Gulf of Mexico (Caldwell & Anderson, 1959).

Indications of breeding in deeper seas, far off the coast, were presented by Gopalakrishnan (1974) for *M. tade* in India, Moore (1974) for *M. cephalus* and *M. curema*, in south Texas, because the final stages of maturation must occur after the fish have left the bays and fish are absent from coastal areas during spawning activity. Malyatskii (quoted by Berg, 1965), provides evidence for *M. cephalus*, from eggs collected in the pelagic part of the Black Sea, also Wallace (1973) for *M. cephalus*, in South Africa and Zaytsev (1960, 1963, 1964, quoted by Libosvarsky & Darrag, 1975), plus Babayan & Zaytsev (1964, quoted by the same authors) for Black Sea mullets. In the mid and south Caspian Sea, intensive spawning of *M. auratus* beginning in the middle and ending in the southern part, was noted at a considerable distance (50–60 miles) from the shore, at depths of 400–600 m (Avanesov, 1972). This fish does not spawn less than 30 miles from the coast, while in the eastern part of the south Caspian sea, spawning extends over a wide area approximately 40 miles from the coast. Additional evidence has been given by the presence of fertilized eggs and early larvae (*M. cephalus*) in the north central part of the Gulf of Mexico (Arnold & Thompson, 1958), over the continental slope, at 500–900 fathoms and 40–50 miles south-eastern of the Mississippi delta; ripe males and females in the process of spawning activity were collected in surface waters where the total depth was 750 fathoms.

Hypotheses concerning geographic subspecies and local races have been offered, from meristic considerations, by Cadenat (1954) in western Africa, by Thong (1969*a*) in Brittany and by Landret (1975) in Mauritania. Peterson & Shehadeh (1971) gave convincing evidence, from nuclear eye-lens protein electrophoregrams, for a genetic polymorphism of mullet in the Hawaiian Islands, and Senkevich & Kulikova (1970), from electrophoretic mobility of the serum proteins of Black Sea mullets, for intra-specific (or racial) differentiation useful for investigation of subpopulations. However, Kesteven (1942) assumed that *M. dobula* spawning occurs in Australia over a wide continuous area with mixed populations; and Thomson (1951) thought that there was no raciation in grey mullets of this region, but distinction between A (eastern Australia) and B (western Australia) populations was presented by Blackburn (1956). So, more precise knowledge from spawning areas would be very interesting for confirmation of hypothetical geographical subpopulations, isolated in very local spawning grounds frequented by geographically separate schools with limited genetic interchange.

Spawning period

Mullet spawn in different geographical areas at different times of year. Conflicting reports on the reproductive period of many species (Table 4.4), in each region, are due to several difficulties. It may be that ripe fish are scanty or that captures are lacking. So, in Israel (Bograd, 1961), when no ripe fish of any of the species were found in the investigations, estimated the spawning periods from the time of the first appearance of the fry; presumptive spawning periods were estimated as taking place a month or more before the entrance of the first of the smallest size fry.

The frequency of ovulations vary in individual fish and in the population as a whole. As shown in the section on sexuality, from analysis of distribution of oocyte diameter, there was evidence that ova lie within one size group only, with general homogeneity of frequency (of unimodal type), and develop in synchrony (Kesteven, 1942; Timoshek & Shilenkova, 1974; Shehadeh *et al.*, 1973*b*; Cassifour, 1975; Grant & Spain, 1975). These authors conclude a non-intermittent and single spawning per season because of the production of only one set of ova per year, but the occurrence of partly spent ovaries suggested that shedding of the ova may sometimes spread over some period of time (Thomson, 1957*a*, *b*). Moreover, there is some evidence that some females spawn only in alternate years (Thomson, 1955, Kennedy & Fitzmaurice, 1969, where it was apparent that only some of the large females of *M. chelo* spawned in any given year), or that others might reproduce more than once in a season, e.g., twice a year for *M. cephalus* from South Africa (Wallace, 1975) and in *Aldrichetta forsteri* from New Zealand (Webb, 1973). But are these the same fish producing more than one batch of eggs, or rather different fish inside the same population? From the evidence of partially spawned out females, Moore (1974) suggests that individual mullet (*M. cephalus* in Texas) may spawn more than once during their long spawning season and may return to the coast between successive spawnings, but a clear answer is difficult because experiments are generally lacking; mullet held in artificial conditions of photoperiod and temperature control were induced to spawn only seven months after their first spawning (Shehadeh, *et al.*, 1973*a*).

In the UAR, two spawning periods, in a year, one in the spring, the second one in autumn, have been recorded in *M. saliens* by Wimpenny (1934), El Zarka & El Sedfy (1970) and by El Maghraby *et al.* (1973); it does not mean necessarily that each fish has two periods of maturity but rather that spawning of the whole population spreads over a considerable duration, with two peaks, in June and November. Intermittent spawning was still suggested by Savchuk (1967) because variability in fry and fingerlings (*M. auratus*), and by many observers in Black Sea mullet (Tomazo, 1940, quoted by Berner-Samsonov, 1970; Ilin, 1949; Ilin and Taranenko, 1950; Smirnov, 1950; Tkacheva, 1952; Zaytsev, 1960), but little is known about the individual and collective behaviour of fish during this period.

Generally spawning is a phenomenon of long duration. So, the spawning period of *Aldrichetta forsteri* is prolonged over several months in Australia (Thomson, 1957*b*). A spawning spread over several months is also seen in *M. chelo* from southern England (Hickling, 1970), in *M. cephalus* from Berre (Ezzat, 1965), in *M. capito* from the UAR (Rafail, 1968), in *M. saliens* from the Venetian Lagoon (Gandolfi & Orsini, 1970) and from Lake Qarun (El Zarka, 1963*a*; El Zarka & El Sedfy, 1970), and lastly, in *M. cephalus* and *M. saliens* from the Caucasian coasts of Crimea in which average body size and weight of migratory fry form heterogenous populations (Savchuk, 1973). It was assumed that some spawning process are different, of limited duration (*M. cephalus* in North Carolina mentioned by Higgins, 1928), and relatively abrupt and concentrated in one month (*M. cephalus* in north-eastern Australia reported by Grant & Spain, 1975).

The duration of spawning may seem more or less prolonged according to the sex examined. Males generally mature sooner than females; so, in Lake Quarun, running ripe males (*M. saliens*) are caught in January and mature females only in April (El Zarka, 1963*a*). Such a male precocity has been shown in other species such as *M. chelo* (Erman, 1961; Hickling, 1970; Farrugio & Quignard, 1973). Moreover, spermatogenetic activity is longer than ovarian maturity; so, in Berre, ovarian ripeness (*M. capito*) decreases and stops in November but mature males are still observed in February and March (Ezzat,

1965). Thierberger-Abraham (1967) found that male *M. cephalus* have a longer spawning peak than females, so fertilization is assured; it is evident that, for propagation of the species, male and female reproductive activity must synchronize, but it is of great importance that male maturation was closely parallel to that of females during the time of spawning.

Some geographical races of mullet may have particular spawning periods. So, in yellow-eye mullet (*Aldrichetta forsteri*) of Australian waters, western fish are winter spawners and eastern fish, summer spawners (Thomson, 1957*a*, *b*).

Variations in precocity, duration and importance of spawning are observed according to climatic variations. Spawning is dependent on the conditions which vary from year to year and from region to region (Table 4.4). Temperature seems to be one preponderant factor in inducing spawning activity. So, spawning of *M. curema* begins with the rise of water temperature in early spring (20 °C in late March) in South Florida, and rising temperature stimulates spawning activity and appears to be of more importance than absolute temperature since spawning continues into summer when water temperatures are much warmer (Anderson, 1957), and this author indicates (1958) that *M. cephalus* in North Carolina spawns during autumn or winter when temperatures are falling.

Spawning of each species takes place according to an optimal temperature: 19.5–22 °C for *M. cephalus* in Korea (Yang & Kim, 1961), 25–29 °C for *M. saliens* of the Caspian Sea (Avanesov, 1972), at the moderate temperature of autumn for *M. capito* in Lake Borullus (El Maghraby *et al.*, 1974*b*). In laboratory conditions, Yashouv & Berner-Samsonov (1970) have shown that temperatures below 22 °C for *M. cephalus* and below 18 °C for *M. capito* arrest the development of eggs and causes them to perish. So, the sharp drop of water temperature along the Israel coast in the surface 30 m layer, in November, corresponds to the end of spawning season of *M. cephalus* and the further drop below 18 °C in January corresponds to the end of spawning of *M. capito*.

Such a temperature variation may be responsible for differences observed in the intensity of spawning. Thus, in the Caspian Sea, *M. saliens* spawned less intensively in 1956 when compared with 1940 (Probatov & Tereschenk, 1951, reported by Avanesov, 1972) and *M. auratus* less in 1969 than in 1968 (Avanesov, 1972). The period of spawning may still vary both in time and in space; so, in Australia, there is a progressive ripening of *M. dobula* northwards; the gradual decline in temperature in the summer-autumn period could initiate the gonad development and the change occurring with the onset of westerly winds could provide critical temperature initiating the spawning migration (Kesteven, 1942).

Photoperiod has yet to be considered. Observations by Hartley (1940), Dekhnik (1953, see Morovic, 1963), Arnold & Thompson (1958), Varagnolo

(1964*a*, *b*), suggest that spawning takes place at night, just before dawn. Only eggs (*M. auratus* in stage II–III of development have been collected during the day by Avanesov (1972).

A tidal or lunar influence was reported by Bromhall (1954) for *M. cephalus* in Hong-Kong, with two spawnings, one during the full moon period at the end of November and the other a month later, at the full moon of the end of December. Lastly, *M. cephalus* is supposed to breed in India during the post-monsoon period (Kowtal, 1967).

Since there are so many various and conflicting reports on the reproductive history of mullet species in each region, with more or less controlled conditions, the hope of Oren (1974) is to see a 'travelling team' able to follow the reproduction of the mullets from one geographical area to another and to repeat experiments on reproduction several times in a year and in different areas.

Fecundity

Estimations of the numbers of eggs produced by female Mugilidae vary greatly according to species, region, period and technical procedures used (Tables 4.5, 4.6 & 4.7); thus, a considerable variability has been shown in different populations of mullets.

Absolute fecundity (Table 4.5) in the number of ripening ova in a female before spawning and relative fecundity (Table 4.6) is the number of eggs per unit weight or length of fish. The detailed results may then be incorporated into the mathematical expression of fecundity (Table 4.7), by using the proportional weight or length estimation. Such results are presented by Ezzat (1965) for Berre, Edimar (1973) for Venezuela, Farrugio & Quignard (1973: p. 575, Table 4; p. 576, Fig. 6) in Tunis, El Maghraby *et al.* 1974*b*: p. 49, Fig. 3; p. 51, Fig. 4) for the UAR, and Grant & Spain (1975: p. 188, Fig. 9) for Queensland.

The fecundity of all mullets generally increases with the increase in size of the females; thus, fecundity and quality of the sexual products is seen to alter as the growth rate of the spawners changes in three species of Black Sea mullet (Vladimirov, 1973), and absolute fecundity increases at a rate of about three times the length of the fish, as in *M. capito* from Lake Borullus (El Maghraby *et al.*, 1974*b*). Increase in the growth rate of females and the accompanying increase in their fecundity are considered as an adaptative response to an improvement in the living conditions of populations. Some complexity of the phenomena is evident: for equality in length, the fecundity of *M. chelo* is less than that of *M. capito*, but the number of eggs per gram of body weight is greater in the latter species than in the former, as seen in the Lake of Tunis (Farrugio & Quignard, 1973, see Plate 7). It was found that the number of eggs of *M. capito* in Lake Borullus increases with the increase of age; results also show (El Maghraby *et al.*, 1974*b*) that the variation in fecundity between fishes

Table 4.5. *Absolute fecundity in grey mullets*

Locality	Authors	Absolute fecundity values	Evidence
		Aldrichetta forsteri	
Australia	Thomson (1957*a*)	125000	Specimens 24.5 cm
		630000	Specimens 39.1 cm
		Mugil auratus	
Black Sea	Tomazo (1940, quoted by Berg, 1965)	150000–927000	—
Black Sea	Nikolskii (1961)	157700–926000	—
Black Sea	Berg (1965)	1200000–210000	—
Berre Lagoon	Ezzat (1965)	324043	Specimens 400 mm (TL)
		325678	Specimens 370 mm (TL)
		Mugil capito	
			Specimens
Berre Lagoon	Ezzat (1965)	4502391	460 mm (TL)
		2500003	400 mm (TL)
		2600255	397 mm (TL)
		1091123	394 mm (TL)
Lake Borullus	El Maghraby (1974*b*)	45568	170 mm
		316828	340 mm
		102339	Age group I
		169415	Age group II
		280172	Age group III
Southern England	Hickling (1970)	Cf. table 12, p. 629; Table 4.6	
		Mugil cephalus	
Australia	Tosh (1903, quoted by Kesteven, 1942)	2000000–2500000	2 specimens
Australia (*M. dobula*)	Kesteven (1942)	1275000–2781000	—
Australia	Grant & Spain (1975)	1572000–2781000	—
India	Jacob & Krishnamurthi, 1948, quoted by Thomson, 1963)	1320000	Specimens 500 mm
Hawaii	Shehadeh *et al.* (1973*a*)	340000–795000	(Induced spawning)
Hawaii	Nash *et al.* (1974)	1000000 (effectively released)	(From relative fecundity evaluation)
Taiwan	Tung (1968)	700000–3000000	—
SW Korea	Yang & Kim (1962)	3790000	Specimen 787 mm (5 years old)
Mauritania (*M. ceph. ashenteensis*)	Brulhet (1974)	4000000–6000000	—
Mauritania and Senegal	Landret (1975)	2322400	Specimens 500 mm (FL)
Black Sea	Nikolskii (1961)	3089000–7206000	—
Black Sea: Danubian delta	Popescu (1961)	5065800–5085440 (mean prolificity from 5 specimens: 5072993)	—
Black Sea	Berg (1965)	5000000–7000000	Specimens 520 mm (13 years old)

Table 4.5. (*cont.*)

Locality	Authors	Absolute fecundity values	Evidence
		Mugil chelo	
			Specimens
Berre Lagoon	Ezzat (1965)	177591 175336	442 mm (TL) 440 mm (TL)
Brittany	Thong (1969*a*)	528320	230 mm (FL)
Southern England	Hickling (1970)	372000–387000 646000–685000 +table 12 p. 629; Table VI	470 mm (TL) 560 mm (TL)
		Mugil macrolepis	
India	Jhingran & Natarajan (1974)	152000–676200	—
		Mugil saliens	
Black Sea: Danubian delta	Popescu (1961)	1175000–2347000 (mean prolificity: 1 625 668)	11 specimens
Berre Lagoon	Ezzat (1965)	210560 180238	Specimens 315 mm (TL) Specimens 307 mm (TL)

Table 4.6. *Relative fecundity in grey mullets*

Locality	Authors	Relative fecundity
	Mugil capito	
Lake Borullus	El Maghraby *et al.* (1974*b*)	728 eggs/g body weight (age group I) 890 eggs/g body weight (age group II) 992 eggs/g body weight (age group III)
Southern England	Hickling (1970)	see Hickling (1970: Table 12, p. 629)
	Mugil cephalus	
Hawaii	Shehadeh *et al.* (1973*a*)	648 ± 62 eggs/g body weight
Hawaii	Nash *et al.* (1974)	849 ± 62 eggs/g body weight
	Mugil chelo	
Brittany	Thong (1969*a*)	395,600 eggs/kg body weight
Southern England	Hickling (1970)	see Hickling (1970: Table 12, p. 629)
	Mugil curema	
Venezuela: Laguna La Restinga	Edimar (1973)	1369 eggs/g body weight (mean prolificity, from statistical evaluation)

Table 4.7. *Mathematical expression of fecundity in grey mullets (using proportional weight or length estimation)*

Mugil capito

El Maghraby *et al.* (1974*b*, see Plates 8 & 9) – Lake Borullus (UAR)

$\log f$ (absolute) $= -1.9351 + 2.9414 \log L$ (TL)

$\log f$ (relative) $= 0.7863 + 1.8080 \log L$ (TL)

$\log f$ (absolute) $= 2.6462 + 1.1577 \log W$ (TW)

$\log f$ (relative) $= 2.3972 + 0.1803 \log W$ (TW)

Mugil cephalus

Grant & Spain (1975, see Plate 10) – Queensland

$f = 0.009\ L^{3.16}$

L, length in mm; standard error of the exponent = 0.29

Mugil curema

Edimar (1973) – Venezuela

$f = 0.0255\ L^{2.95}$

Mugil capito and Mugil chelo

Farrugio & Quignard (1973, see Plate 7) – Tunisia

f, fecundity; L, length (TL, total length); W, weight (TW, total weight)

of the same age is greater than the variation between fishes of the same weight or length. This is due, mostly, to the wide length and weight range of different age groups.

Greatest prolificity occurs in *M. cephalus*; such a characteristic seems very important for fish culture in every region (Israel, Korea, Hawaii...) in which grey mullets are cultivated.

Spawning behaviour, mating and fertilization

There still exists a considerable gap in our knowledge of spawning behaviour because this activity generally takes place at night (Hartley, 1940; Dekhnik, 1953; Perceva-Ostroumova, 1951, all quoted by Nikolskii, 1961; Arnold & Thompson, 1958; Varagnolo, 1964*a*, *b*; Avanesov, 1972); for example, eggs of *M. saliens* have been collected by Varagnolo (1964*b*), which had been deposited between 10.00 and 11.00 p.m. Spawning is pelagic: ova and sperm are indiscriminately ejected and fertilization takes place at the surface, over more or less deep waters (15 to 900 fathoms).

Breder (1940) stated that spawning of *M. cephalus* took place, in the Bay of Florida, on sandy grounds. Several males (usually 4–5) surround one female which is larger than the males. Arnold & Thompson (1958) working in the Gulf of Mexico, observed, before dawn, a compact school of approximately 200 striped mullet swimming within a few feet of the surface. Small groups of three to six fish swam close to the surface in an erratic manner, each female surrounded by several males, noticeably smaller and more slender, which maintained a position slightly behind the female. When spawning, the males stay at the side of the females, close to the tail and fertilize the eggs as soon as they are deposited. Such a spawning behaviour was observed during submarine diving by Farrugio (1975). In captivity, Yashouv (1969) observed the fish moving slowly above the bottom of the tanks, showing no signs of spawning when suddenly, after 5.0 p.m., the surface of the water was covered with eggs.

In laboratory conditions fertilization is known only after natural spawning (Sanzo, 1932, 1936, *M. cephalus*; Vodyanitski, 1934, quoted by Berg, 1965, *M. auratus*; Anderson, 1957, *M. curema*; Yang & Kim, 1962, *M. cephalus*). Artificially induced spawning and artificial fertilization have been studied for many years in Israel (Yashouv, 1969), in Italy (Lumare & Villani, 1972), in Taiwan (Tang, 1964; Liao *et al.*, 1971, quoted by Shehadeh *et al.*, 1973*b*; Liao, 1975) and in Hawaii (Shehadeh & Ellis, 1970; Shehadeh *et al.*, 1973*a*) to increase the available supply of fish.

Hines & Yashouv (1971) have shown that spermatozoa of *M. capito* are completely inactive in coelomic fluid and become active in less than eight seconds when diluted with sea water. Sperm activity is increased as the osmolar value approaches that of normal sea water. Thus, they become suitable for successful fertilization (at 18–23 °C). Fertilization of *M. cephalus* is usually successful to over 87% in Korea (Yang & Kim, 1962) and to over 90% in Hawaii (Nash *et al.*, 1974) but less successful in Taiwan – 32% – (Tang, 1964).

Eggs

Eggs, from ovarian extrusion or from planktonic collection, are characterized according to mean diameter and the number and size of oil droplets. Some variability in these characters (Table 4.8) is due to specific and geographical variations and also to natural or experimental conditions in the process of extrusion and cropping.

Mullet eggs are spherical and transparent, the surface of the egg's shell is smooth without any sculpturing, and one or several oil globules are present. The mean egg diameter of *M. cephalus* is about 930 μm and eggs are characterized by a single large oil droplet of uniform diameter 330 μm (Kuo *et al.*, 1973*a*; Nash *et al.*, 1974). The former authors inferred that oil droplets in the oocytes may be directly related to premature inducement by manual extrusion or to incomplete inducement following hormonal injection. Spontaneous release of eggs by the female produces eggs with a single oil globule, which are invariably fertilized and develop naturally, producing viable larvae (Nash & Kuo, 1975). Moreover, it is possible that the difference in the size of the eggs from a single female is of greater importance for the survival of larvae than the difference in the size of the eggs from different females (Vladimirov, 1973).

Pelagic eggs from many species have been collected in various areas:

Table 4.8. *Egg characteristics of grey mullets*

Authors	Locality	Ovum diameter (mm)	Oil globule diameter (mm)	Months observed
		Mugil auratus		
Sanzo (1935, quoted by Marinaro, 1971)	Messina	1.08–1.14	0.32–0.36	I–II
Krizanovsky & Potepyaev (1937, quoted by Thomson, 1966)	Black Sea	0.61–0.71	0.36	—
Vodianitsky & Kazanova (1954, quoted by Yashouv & Berner-Samsonov, 1970)	Black Sea	0.87–0.98	0.32–0.36	—
Yashouv & Berner-Samsonov (1970)	Israel	0.87–0.98	36.75% of egg diameter	X
Marinaro (1971)	Algiers	0.95–0.98	0.30–0.35	X
Avanesov (1972)	Caspian Sea	0.80–0.90 (average: 0.84)	—	—
		Mugil capito		
Wimpenny (1936)	Lake Qarun	0.66–0.80	0.20–0.30	IX
Yashouv (1969, quoted by Yashouv & Berner-Samsonov, 1970)	Israel: fish ponds	0.78–1.03	—	—
Yashouv & Berner-Samsonov 1970)	Israel	0.66–1.03	33.91% of egg diameter	IX
		Mugil cephalus		
Kawakami (1917, quoted by Thomson, 1966)	Japan	0.65–0.78	—	—
Nakano (1918, quoted by Thomson, 1966)	Japan	0.91–1.08	—	—
Sanzo (1936)	Messina	0.72	0.28	IX–X
Belloc (1938)	Corsica: Biguglia	0.72	—	—
Vodianitsky & Kazanova (1954, quoted by Marinaro, 1971)	Black Sea	0.60–0.72	0.28	VI–VIII
Hotta (1955, quoted by Thomson, 1963)	Japan	0.65–0.85	—	—
Yang & Kim (1962)	Korea	1.00	0.80	—
Tung (1968)	Taiwan	0.62–0.68	—	XII–I
Yashouv (1969, quoted by Yashouv & Berner-Samsonov, 1970)	Israel: fish ponds	0.66–0.72	—	—
Tang (1964)	Taiwan	0.93 ± 0.032	0.38 ± 0.020	—
Ling (1969, quoted by Yashouv & Berner-Samsonov, 1970)	Taiwan	0.93–0.95	0.38	—
Yashouv & Berner-Samsonov (1970)	Israel	0.66–1.08	36.1–40.8% of egg diameter	IX
Marinaro (1971)	Algier	0.72–0.78	0.26–0.31	IX–X
Avanesov (1972)	Caspian Sea	0.62–0.90 (average: 0.73)	—	—
Kuo *et al.* (1973*a*)	Hawaii	0.93	0.33	—
Sylvester, Nash & Emberson (1975)	Hawaii	0.921–0.931	—	—

Table 4.8. (*cont.*)

Authors	Locality	Ovum diameter (mm)	Oil globule diameter (mm)	Months observed
		Mugil chelo		
Sanzo (1936)	Messina	1.32	0.44	II–III
Belloc (1938)	Corsica: Biguglia	1.32	0.44	—
Erman (1961)	Dardanelles	0.90–1.02 (ovarian eggs)	—	XII–I
Hickling (1970)	Southern England	0.70–0.84 (ovarian eggs)	—	I–III
Yashouv & Berner-Samsonov (1970)	Israel	1.32	33.3% of egg diameter	X
Marinaro (1971)	Algier	0.98–1.20	0.35–0.45	XI
Cassifour & Chambolle (1975)	Bay of Biscay: Arcachon	1.25	—	—
		Mugil curema		
Anderson (1957)	South Florida	0.77–0.86 (mean: 0.82)	0.27–0.32 (mean: 0.30)	—
		Mugil labeo		
anzo (1938)	Messina	0.76	0.28–0.30	—
		Mugil saliens		
Perceva-Ostroumova (1951, quoted by Yashouv & Berner-Samsonov, 1970)	Caspian Sea	0.60–0.71	0.25–0.36	—
Dekhnik (1954, quoted by Yashouv & Berner-Samsonov, 1970)	Black Sea	0.77–0.88	0.28–0.37	—
Vodianitsky & Kazanova (1954, quoted by Marinaro, 1971)	Black Sea	0.76–0.86	0.31–0.40	VIII–IX
Varagnolo (1964*c*)	Venetian lagoon: Chiogga	0.72 (0.67, later)	0.27	VI–VII–VIII–IX
Yashouv & Berner-Samsonov (1970)	Israel	0.60–0.88	42.8% of egg diameter	V → VIII

M. capito in Naples (Lo Bianco, 1888), *M. auratus* in Marseille (Marion, 1890*a*; Gourret, 1891), *M. curema* in Florida (Anderson, 1957), *M. saliens* in the Adriatic Sea (Venetian Lagoon, Varagnolo, 1964*a*, *b*), *M. labeo* in the Straight of Messina (Sanzo, 1938), *M. cephalus* in the Mediterranean Sea (Sanzo, 1936; Lumare & Villani, 1972), in the Black Sea (Dekhnik, 1953, quoted by Stenger, 1959), in Taiwan (Tang, 1964). A comparative description of eggs of Mediterranean species was presented by Vialli (1937) and two keys, for eggs and larval stages from Israel and Algerian fishes were published by Yashouv & Berner-Samsonov (1970) and by Marinaro (1971). Further information is required. Nikolskii (1961) assumed that eggs of mullet which spawn in marine waters are pelagic but those of species which spawn in fresh water are demersal. Some sinking movements have been recorded under natural and experimental conditions, but little information is available.

Eggs and larvae of *M. cephalus* have been taken in plankton tows at a depth of 150 to 200 m (Hotta, 1955, quoted by Sylvester & Nash, 1975) and eggs of *M. saliens* were caught in the South Caspian by Avanesov (1972) at a depth of 600 m, when the temperature of the sea had reached 20 °C. Sanzo (1936) observed that all fertile eggs sank soon after fertilization, and Yashouv (1969) reported that eggs of *M. cephalus* sank, as their specific gravity being increased, towards the end of the incubation period (after twenty two hours). Lastly Kuo *et al.* (1973*a*) recorded that the majority of eggs which sank to the bottom of tanks were unfertilized or undeveloped. Fertilized eggs, maintained in suspension by strong aeration and circulation of the sea water do not sink and develop conveniently (Nash *et al.*, 1974); so that fertilization of more than 90% of the eggs takes place with hatching rates of 90–95% (Kuo *et al.*, 1973*a*).

Incubation

Kuo *et al.* (1973*a*) reported successful incubation of *M. cephalus* eggs between 22 and 24 °C, at 32% salinity in well aerated sea water, and Nash *et al.* (1974) showed that 25 °C and above was definitely inhibitory to successful incubation, mortality above this temperature being, usually, over 90% and often total. Highest survival is through incubation within the temperature range 21–24 °C with an optimum at 22 °C. Liao *et al.* (1972) used temperatures of 18–23 °C and Berner-Samsonov (1970) stated that temperatures below 22 °C arrested the development of the eggs and caused them to perish.

Relations between temperature, mortality of eggs and survival of larvae have been described for Hawaiian *M. cephalus* (Sylvester & Nash, 1975). Eggs were incubated experimentally between 1.0 and 40.4 °C and mortalities were lowest (4.2%) within the range 22.7–23.3 °C. The length of the incubation period was approximately 40–49 h within the optimum temperature range of 22.7–23.3 °C. Hatching time was inversely related to temperature and the equation $Y = 3.6674 - 1.524OX$ describes the relation between incubation time and temperature over the range of 11–30 °C. High mortality (92.1%) at the temperature at which the parent fish were spawned in captivity (24 °C) suggested that the temperature to which eggs were first exposed would be a major factor contributing to high mortality if prolonged. Eggs incubated in oxygen concentrations above 5.0 ppm had survival rates ranging from 81 to 98% and eggs held in oxygen concentrations below 4.5 ppm the rate ranged from 0 to 21%. The greatest increase in survival occurred between 4.5 and 5.0 ppm, and salinity for the incubation of eggs of Hawaiian striped mullet lies within the range of 30–32‰ (Sylvester *et al.*, 1975).

From these observations, authors concluded that mullet eggs are preadapted over a wide thermal range to accommodate temperature changes as they sink vertically through the water. Such a high tolerance to temperature changes during incubation and hatching has an obvious selective advantage for fish that spawn in areas subjected to wide temperature variations.

Hatching

Under laboratory conditions, the eggs of *M. cephalus* and *M. capito* hatched within 36–44 h in temperatures of 22–23 °C (Yashouv & Berner-Samsonov (1970); the first prelarvae appeared after 36–38 h but the peak of hatching took place after 40–44 h (Yashouv, 1969). Hatching occurs in *M. curema* about 40 h after fertilization (Anderson, 1957), and in *cephalus* 36 h following fertilization (Kuo *et al.*, 1973*a*) or 34–38 h at 23–24.5 °C in salinities of 30.1–33.8‰ (Liao *et al.*, 1970, reported by Kuo *et al.*, 1973*a*,) and in less than two days in Messina (Sanzo, 1936).

Hatching time of *M. cephalus* was prolonged further at lower temperatures: 49–54 h at 22.5 °C; 60–65 h at 21 °C (Liao, 1969 and Liao *et al.*, 1970, quoted by Kuo *et al.*, 1973*a*); 59–64 h at 20–24 °C and salinity 24.39–35.29‰ (Tang, 1964); and 60 h at 18.2–19.5 °C (Yang & Kim, 1962). Yashouv *et al.* (1969) recorded that eggs from hybrid female *M. cephalus* × male *M. capito* hatched after 58–60 h at 21–22 °C.

Hatching of *M. cephalus* was 87% successful in Korea (Yang & Kim, 1962) and higher rates (90–95%) were obtained when egg development occurred in water with strong aeration (Kuo *et al.*, 1973*a*) whereas the rate was less than 10% in Taiwan (Tang, 1964).

Embryonic stages

Embryonic development of mullet was studied by Sanzo (1936), Yashouv (1959), Yang & Kim (1962), Tang (1964), Kuo *et al.* 1973*a*), Nash *et al.* (1974) in *M. cephalus*, Anderson (1957) in *M. curema* and Sanzo (1938) in *M. labeo*.

A key to development was presented by Yashouv & Berner-Samsonov (1970), by Tang (1964) and by Kuo *et al.* 1973*a*: p. 461, Table 1). Development proceeds rapidly and gastrulation is reached in *M. cephalus* twelve hours after spawning (Nash *et al.*, 1974).

Larval stages (or juvenile phase)

Size of larvae upon hatching differs according to species and according to authors (Table 4.9). A key to larval stages in five species of mullet along the Israel coast was presented by Yashouv & Berner-Samsonov (1970).

Some larvae have been collected in natural plankton (Cunningham, 1891, also Cassifour & Chambolle, 1975, *M. chelo*; Anderson, 1958, *M. curema*;

Table 4.9. *Size of larvae upon hatching in grey mullets*

Authors	Locality	Size (mm)
	Mugil auratus	
Vodianitsky & Kazanova (1954, quoted by Yashouv & Berner-Samsonov, 1970)	Black Sea	2.35
	Mugil capito	
Yashouv (1969, quoted by Yashouv & Berner-Samsonov, 1970)	Israel	3.6–3.7
	Mugil cephalus	
Sanzo (1936)	Messina	2.4
Vodianitsky & Kazanova (1954, quoted by Yashouv & Berner-Samsonov, 1970)	Black Sea	2.0–2.05
Liao (1969, quoted by Yashouv & Berner-Samsonov)	Taiwan	3.01
Yashouv (1969, quoted by Yashouv & Berner-Samsonov)	Israel	3.4–3.6
Yang & Kim (1961)	Korea	2.7
Tang (1964)	Taiwan	2.2
Liao *et al.* (1971, quoted by Kuo *et al.* 1973*a*)	Taiwan	2.08–3.40
Lumare & Villani (1972)	Italy	2.2
Kuo *et al.* (1973*a*)	Hawaii	2.65 ± 0.23
Nash *et al.* (1974)	Hawaii	2.6
	Mugil chelo	
Sanzo (1936)	Messina	4.2
Cassifour & Chambolle (1975)	Bay of Biscay: Arcachon	3.8
	Mugil curema	
Anderson (1957)	South Florida	1.63–1.76 (average, 1.69)
	Mugil labeo	
Sanzo (1938)	Messina	2.20
	Mugil saliens	
Perceva-Ostroumova (1951, quoted by Yashouv & Berner-Samsonov, 1970)	Caspian Sea	1.7

Demir, 1971, *M. chelo* and *M. auratus*; Brezeanu & Zinevici, 1971, *M. saliens*; Johnson & McClendon, 1970, *M. cephalus*). Others have been produced after artificial spawning and fertilization (Sanzo, 1936, 1938; Yang & Kim, 1962; Tang, 1964; Yashouv, 1969; Lumare & Villani, 1972; Liao *et al.*, 1974).

Postlarvae may be regarded as 'juvenile' at about 50 mm total length. Perceva-Ostroumova (1951, quoted by Avanesov, 1972) distinguished the following stages: prolarvae, up to 4 mm long; larvae, 4.1–8 mm; fingerlings, 8.1–11 mm; underyearlings, 11 mm or longer.

The sequence of spawning mullet species is followed by a sequence of the approach of their fry to protected areas along the coast or in estuaries, as in

Israel (Yashouv & Berner-Samsonov, 1970). However, larvae have been studied mainly in hatcheries (Nash & Kuo, 1975) and problems of growth, nutrition (larval foods), salinity, temperature and bacterial contamination, have been discussed recently by Nash & Kuo (1975). Information on the controlled rearing of larvae, under laboratory conditions, is scant, despite the economic importance of the problem for projects in fish culture.

Growth

Larval growth values have been recorded in a few cases: 2.5 mm for *M. cephalus* at two days old in Italy (Lumare & Villani, 1972), 3.9, 4.0 and 4.1 mm for the same species at 1, 2 and 3 days old in Korea (Yang & Kim, 1962), 2.8 mm for *M. labeo* at five days old in Messina (Sanzo, 1938) and 3.1 mm for *M. cephalus* at eight days old in the same region (Sanzo, 1936). Anderson (1958) assumed a period of 3–4 weeks for striped mullet postlarvae to attain a size of about 20 mm (SL). The relationship between age and size is important because fry are generally used as indicators of the spawning period. Because the fish occur in pure monospecific schools as in Israel (Perlmutter *et al.*, 1957), in coastal shallow areas, 15–20 days after the start of spawning as in Berre (Ezzat, 1965), spawning periods and spawning grounds may be evaluated on the basis of the size distribution of these fry. Thus, the problem of larval growth is of great interest. Larval growth during the forty-two days larval period of *M. cephalus* was monitored in Hawaii at 22 °C and 32‰ salinity by Kuo *et al.* 1973*a*: p. 465, Fig. 3); larval length increased from 2.65 to 3.36 mm during the first day and no obvious growth increment was evident from the second day to the ninth day, this gradual increase became prominent on the ninth day and is coincident with the onset of larval feeding (8th day) and intensification of feeding (9–12th day). Mean length on the 42nd day was 17.69 mm (27.00 mm from Liao *et al.*, 1971, reported by Kuo *et al.*, 1973*a*).

Feeding

Larval feeding is the major problem in rearing mullet fry (Kuo *et al.*, 1973*a*; Nash & Kuo, 1975). No food was taken during the first few days, and food was offered on the third day according to a feeding schedule for larvae and postlarvae proposed by Nash *et al.* 1974: p. 22, Fig. 6). Food uptake during the first fifteen days of larval life was followed by Kuo *et al.* (1973*a*): initial feeding was carried out three days after hatching but the yolk sac was not totally absorbed till the fifth day and the larvae presumably survived on this endogenous food reserve during the first four days of the larval life. Liao *et al.* (1971) reported that three day old larvae fed when offered oysters veligers (< 70 μm) and Kuo *et al.* (1973*a*) assumed that ingestion of wild zooplankton, smaller than 150 μm in size, began on the fifth day; moderate uptake of *Artemia* nauplii took place at this time also and increased by the eleventh day, intensive feeding being apparent on the ninth day after hatching *Gymnodinium* was not observed in any of the larvae during the fifteen day check and the larvae began to feed on benthic diatoms on the forty-second day (Kuo *et al.* 1973*a*).

Vertical distribution and specific gravity

Observations on vertical distribution of larvae during the first ten days of larval life was carried out by Kuo *et al.* (1973*a*). Newly hatched larvae were inactive and not immediately free-swimming; they tended to remain suspended in the water column in an inclined position with the ventral side oriented towards the water surface (Kuo *et al.*, 1973*a*: p. 462, Fig. 1). Sustained larval activity tended to increase after the second day, then a change in larval specific gravity (from 1.005 to 1.047 g/cc) would have a significant effect on their vertical distribution. So, vertical migrations were not photoperiodic reactions, but apparently were controlled by an internal rhythm (Kuo *et al.*, 1973*a*; Nash *et al.*, 1974), and larvae responded to morphological or internal physiological changes. The first sinking (at two and a half days) is probably related to the rapid absorption of the yolk sac and the resulting change in specific gravity; the second descent at eight days is accounted for by a sudden increase in the specific gravity of the larvae on the seventh day.

Survival

Most larvae died a few days after hatching: there are reports of death at 2 days (*M. cephalus*, Yang & Kim, 1961; Tang, 1964), 36 h (*M. curema*, Anderson, 1957), 7–8 days (*M. cephalus*, Kuo *et al.*, 1973*a*). Survival is limited to 4 days for *M. cephalus* in Taiwan (Tang, 1964) and in Israel (Yashouv, 1969); one week for the hybrid *M. cephalus* × *M. capito* (Yashouv *et al.*, 1969), 22 days in Messina (Sanzo, 1936), 30 days in the laboratory at Dor (Yashouv, 1967). Generally this occurs because of a lack of an adequate food supply for the postlarvae. However, emergent larvae of grey mullet (*M. cephalus*) have been reared successfully through the 42 day larval period in Hawaii (Kuo *et al.*, 1973*a*; Nash *et al.*, 1974). Ninety per cent of larval mortality occurred during the initial 10 days of larval life (Kuo *et al.*, 1973*a*), the highest survival to 12 days (19.9%) occurred between 24.5 and 25.3 °C. (Sylvester & Nash, 1975).

The relationships between larval survival, temperature and initial feeding has been demonstrated by Kuo *et al.* (1973*b*) and Nash *et al.* (1974). There was evidence that, in experimental conditions, survival is dependent on tank size and larval density. So, in *M. cephalus*, mortalities which followed the second sinking were always greater than those associated with the first descent (Kuo *et al.*, 1973*a*); the reasons for these high mortalities are not apparent but could possibly be attributed to mechanical damage brought about by prolonged contact with a solid surface, i.e., the tank bottom. Moreover, exposure of larvae to direct sunlight during the first five days is lethal and larvae up to 4 weeks old are vulnerable to problems resulting from ingestion of the empty cysts or shells of *Artemia* (Nash *et al.*, 1974).

The 'critical' periods involving very high mortalities were clearly indicated by Kuo *et al.* (1973*a*). The first coincided with the opening of the mouth (2nd–3rd day) and the second preceded and overlapped the second growth phase (8th–12th day). These two critical periods accounted for 25% and 75% of total mortalities during the 42 day larval period and were always preceded by the larvae sinking to the bottom. Liao *et al.* 1971, quoted by Kuo *et al.*, 1973*a*) reported two similar critical periods (at the 3rd and the 11–13th days) with a total mortality of 98.5% and a final survival rate of 0.31%. Finally, Liao *et al.* (1972, quoted by Sylvester & Nash, 1975) described a good survival rate of mullet larvae at 23.0–24.5 °C, but little survival at 20 °C. This result suggests that optimum temperatures for egg development and hatching may occur at levels which are inefficient or inimical for larval growth and development. This problem is of the greatest importance for aquaculture. Minimal mortalities for larvae were shown to occur between 18·9 and 25.3 °C, although some larvae survived temperatures as low as 15.9 °C and as high as 29.1 °C (Nash *et al.*, 1974). Moreover, larvae exposed to salinities from 24–36‰ for periods up to 96 h had their greatest overall survival (97–100%) at 26‰ (Sylvester *et al.*, 1975). Significant increase in survival for larval rearing was above oxygen level of 5.4 ppm (at 32‰ salinity), and median survival occurred between 6·4–7·9 ppm for 96 h (Sylvester *et al.*, 1975). These authors have demonstrated in Hawaiian striped mullet (*M. cephalus*) that optimal larval survival was evident from 26–28‰ while median egg survival would occur between 28–30‰; so, they conclude that mullet eggs have much less tolerance to salinity variation than do mullet larvae which seem better adapted to brackish waters and estuarine existence.

The viability of the fish in the early stages of life and the abundance of progeny are considered by Vladimirov (1973) as a function of the growth rate of the spawner and as an indicator of their quality. So, the interrelationships between the growth rate of spawners, the age of spawners, and the survival and abundance of the progeny is an important constituent of the mechanisms of the natural self-regulation of abundance in fish populations. A density dependent reduction in survival may undoubtedly occur, owing to deterioration in environmental conditions in the early stages of life because of overpopulation of the spawning grounds or reduction of feeding when the number of larvae increases. Ichthyologists pay, at the present, great attention to the study of spawning conditions and egg incubation, feeding conditions of larvae and juveniles, predators, and the effect of the number of spawners and fishing, as important factors in population dynamics. Grey mullets constitute good experimental material for research of this type.

Migration of fry

Many papers deal with the periodic migration of fry of different species of mullet, from the sea to coastal areas, bays, gulfs, estuaries, lakes and lagoons, which function as 'nursery habitats' (Wallace, 1973) during the fattening period.

Fry (0 age group) of various lengths (10–40 mm, or more), about one month after hatching (Faouzi, 1938) or at 2–3 months old (Thomson, 1953) spend their juvenile lives within estuaries in brackish waters (Kesteven, 1942; Thomson, 1963; Wallace, 1973) and generally, with a seasonal periodicity, move into freshwater rivers, as in Israel (Perlmutter *et al.*, 1957), enter canals or lake-sea connections, as in the UAR (El Zarka & Kamel, 1965; El Zarka, 1970; El Maghraby *et al.*, 1974*a*; El Sedfy & Libosvarsky, 1974; Rafail & Hamid, 1974) and spread gradually over large areas of lakes to feed and grow. Heldt (1929) called this inward movement '*appel du lac*'. Penetration of fry occurs in hypersaline mangrove swamps in Puerto-Rico (*M. curema*, Austin, 1971, quoted by Moore, 1974).

Thus, important migrations of fry are observed periodically and controlled in the Venetian lagoons and in the River Po delta ('fishing *valli*' or '*valli di pesca*': De Angelis, 1960, 1967; Gandolfi, 1973), in the '*limans*' of the Black Sea (Antipa, 1931, 1933; D'Ancona, 1934) and in the northwestern part of the Black Sea in which young migrate into shallow waters to feed, in search of fattening areas (Savchuk, 1968).

Each species of fry varies from month to month and according to its own spawning period; the extended period of breeding of some species (*M. cephalus*, *M. saliens*) is responsible for the prolonged feeding migration of the young, as in the Black Sea (Savchuk, 1968).

The entrance of fry into inland waters and their abundance in lakes depends on the success of spawning and the chances of survival of different developmental stages. Environmental factors must be taken into account such as the effect of salinity decrease. The distribution of mullet fry is greatly affected by the degree of dilution and the tolerance of the species to, this as shown by El Zarka (1970) in Lake Edku. Adaptability of young golden-grey mullet was demonstrated by Chervinsky (1975). Though this movement (called 'osmoregulatory migration' by Myers, 1949, reported by Thomson, 1955) is evident, each year, for littoral populations, certainly it would not appear to be absolutely necessary for mullet to enter freshwaters, as shown by Thomson (1955).

The effect of temperature should also be taken into account. Fry seem to prefer shallow waters because they are always warmer than the sea and the appearance of fry is generally influenced by water temperature. Zambriborsch (1949, quoted by El Zarka, 1970) stated that the high water temperature of coastal lagoons of the Black Sea during summer, stimulated the development of diatoms, a basic food supply for the fry and young mullet.

The amount of food available should also be considered. As asserted by Thomson (1955), mullet are 'trophically brackish water fish' and their fry movements are 'trophic migrations'. The favoured diet of mullet is both detritus and living benthic diatoms which are very abundant in the areas inhabited by fry and young mullet.

Others environmental factors also can have an effect. De Angelis (1960) mentioned that dissolved oxygen, alkalinity and the presence of dissolved nitrates, were important factors. Kristensen (1971) supposed that an 'attractive component', present in brackish waters, may be responsible for inward migration and Creutzberg (1961) considered that the attractiveness of inland waters is due to an 'odorous substance' that can be removed by a charcoal filter, a specific odour to which fishes show an innate response, but a true demonstration of such an influence has never been shown using mullet fry. However, it seemed probable to Thomson (1955) that, if all the fishes which have participated in the spawning migration are not seen to return to their former habitats, at least some of their progeny may do so. Finally, Grant & Spain (1975) demonstrated the influence of wind in the coastal migration of fry in Queensland; the wind causes a local south-westerly current bringing inshore the larvae and juveniles which were spawned in offshore waters.

The maintenance of the lake populations of mullet and the stability of the fishery stock, depend on the recruitment fry, as demonstrated in Lake Borullus by Libosvarsky & Darrag (1975*a*, *b*). So, Wimpenny (1934) and El Zarka & Koura (1967) proposed prohibition of fishing in a belt around the sea-lake connection in the UAR, to protect the spawning and young mullet necessary to stock the lakes; such a protective measure to secure successful recruitment of fishing stock in this area also is asserted by El Zarka (1970).

Lastly, the development of natural fish culture ('valliculture' in Italy, De Angelis, 1960, 1967) and especially of mullet fish farming in ponds (as in Israel, Sarig, 1969; Chervinsky, 1975) depended greatly, until recently, on fry collection in natural habitats. The decrease of mullet production, in Israel in 1973 (to 57 tons, Sarig, 1974) is due to the continuing difficulties in catching fry along the Mediterranean; this fry decrease is because of occasional dry winters (Sarig, 1966) but mainly because of freshwater and marine pollution due to domestic and industrial wastes (Oren, 1971). So, availability of young mullet for fish farming will be due, in a few years, for the most part, to artificial reproduction and successful breeding and rearing of fry, as asserted by Oren (1975).

References

Less recent references are quoted from:

Pillay, S. R. (1972). A bibliography of the grey mullets, family Mugilidae *FAO Fisheries Technical paper no.* 109, *Rome.*

Alvarez-Lajonchere, L. A. (1974). Adiciones a la bibliografia de la familia Mugilidae (Teleostei) publicada por FAO en 1972. *Ciencias Inv. marinas, Univ. La Habana,* **8**(13), 1–116.

Abraham, B. (1963). See Lajonchere: 001.

Abraham, M. (1974). Ultrastructure of the cell types and of the neurosecretory innervation in the pituitary of *Mugil cephalus* (L.) from freshwater, the sea and a hypersaline lagoon. 2. The proximal pars distalis. *Gen. Comp. Endocrinol.* **24**, 121–32.

Abraham, M., Blanc, N. & Yashouv, A. (1966). See Pillay: 0003.

Abraham, M., Blanc, N. & Yashouv, A. (1968). See Pillay: 0004.

Abraham, M., Yashouv, A. & Blanc, N. (1967). See Pillay: 0005.

Acara, A. & Gozenalp, F. (1959). see Pillay: 0007.

Albertini-Berhaut, J. (1975). Biologie des stades juvéniles de Téléostéens Mugilidae *Mugil auratus* R. 1810, *M. capito* Cuv. 1828 et *M. saliens* R. 1810. *Aquaculture,* **5**, 179–97.

Aleem, A. A. (1969). See Lajonchere: 010.

Aleem, A. A. (1972). Effect of river outflow management on marine life. *Mar. Biol.* **15**, 200–8.

Alkholy, A. A. & Abdel-Malek, S. A. (1972). Food and feeding habits of some Egyptian fishes in Lake Quarun. 1. *Tilapia zillii* according to different localities. *Bull. Inst. oceanogr. Fish., UAR,* **2**, 187–201.

Anderson, W. W. (1957). See Pillay: 0028.

Anderson, W. W. (1958) See Pillay: 0030.

Anon. (1973). Nouvelles des Instituts de recherche. *Bull. Aquacult. FAO,* **5**, 3–12.

Antipa, G. (1931). Les bases biologiques de la production des pêcheries dans la région NW. de la Mer Noire. *P-V. Comm. Int. Explor. Sci. Mer. Mediterr.* **6**, 357–74.

Antipa, G. (1933). La vie dans la Mer Noire. *Ann. Inst. Oceanogr. Monaco,* **13**, 53–90.

Arnold, E. L. & Thompson, J. R. (1958). See Pillay: 0035.

Avanesov, E. M. (1972). Present spawning conditions of Mullets (genus *Mugil*) in the Caspian Sea. *J. Ichthyol.* **12**, 419–5.

Banarescu, P. (1964). See Lajonchere: 073.

Belloc, G. (1938). See Pillay: 0070.

Berg, L. S. (1965). Freshwater fishes of the USSR and adjacent countries. *Isr. Progr. Scient. Transl., Jerusalem.*

Blackburn, M. (1956). Real and apparent changes in size of marine animal populations during the Australian fisheries investigations. *J. Cons. Int. Expl. Mer,* **21**, 284–95.

Blanc, N. & Abraham, M. (1968). See Pillay: 0086.

Blanc-Livni, N. & Abraham, M. (1969). See Pillay: 0087.

Blanc-Livni, N. & Abraham, M. (1970). See Lajonchere: 097.

Bograd, L. (1961). See Pillay: 0169.

Borcea, I. (1934). See Pillay: 0175.

Breder, C. M. (1940). See Pillay: 0190.

Brezeanu, G. & Zinevici, V. (1971). See Lajonchere: 106.

Bromhall, J. D. (1954). See Pillay: 0201.

Brulhet, J. (1974). Situation et perspective des pêcheries du mulet jaune de Mauritanie. *Pêche Marit.*, **1159**, 702–6.
Brulhet, J. (1975). Observations on the biology of *Mugil cephalus ashenteensis* and the possibility of its aquaculture on the Mauritanian coast. *Aquaculture*, **5**, 271–81.
Brusle, S. & Brusle, J. (1974). Intersexualité testiculaire chez les muges méditerranéens *Mugil cephalus* et *Mugil ramada. Bull. Soc. Zool. Fr.* **100**, 249.
Brusle, S. & Brusle, J. (1977). Les Muges de Tunisie: pêche lagunaire et biologie de la reproduction de trois espèces (*Mugil capito, Mugil cephalus* et *Mugil chelo*) des lacs d'Ischkeul et de Tunis. *Rapp. P.-V. Comm. Int. Explor. Mer Méditerr.* **24**, 101–30.
Brusle, S. & Brusle, J. (1977). Infrastructure des cellules germinales précoces de *Mugil auratus*, Poisson Téléostéen marin. *Biol. Cell.* **30**, 4a.
Brusle, S. & Brusle, J. (1978). Early sex differentiation in *Mugil* (*Liza*) *auratus* Risso, 1810 (Teleostei, Mugilidae). An ultrastructural study. *Ann. Biol. anim. Bioch. Bioph.* **18**, 871–5.
Brusle, S. & Brusle, J. (1978). An ultrastructural study of early germ cells in *Mugil* (*Liza*) *auratus* R.1810 (Teleostei, Mugilidae). *Ann. Biol. anim. Bioch. Bioph.* **18**, 1141–53.
Brusle, S. (1979). Sex-ratio chez diverses espèces de Muges des milieux lagunaire et marin du littoral Languedoc Roussillon. *Rapp. P.-V. Comm. int. Explor. Mer Méditerr.* **25–26**, 67–9.
Caldwell, D. K. & Anderson, W. W. (1959). See Pillay: 0214.
Campa de Guzman, S. (1973). Larvas de peces colectadas en la costa suroeste de Baja California durante mayo-junio de 1973. *Inst. Nac. Pesca*, INP/SC: 4.
Cassifour, P. (1975). Contribution à l'étude de la biologie des Mugilidés des réservoirs à poissons du bassin d'Arcachon. Thesis no. 1208, Bordeaux.
Cassifour, P. & Chambolle, P. (1975). Induction de la ponte par injection de progesterone chez *Crenimugil labrosus* (R.), poisson téléostéen, en milieu saumâtre. *J. Physiol.* **70**, 565–70.
Cech, J. J. & Wohlschlag, D. E. (1973). Respiratory response of the striped mullet *Mugil cephalus* (L.) to hypoxic conditions. *J. Fish Biol.* **5**, 421–8.
Chervinski, J. (1975). Experimental raising of golden grey Mullet (*Liza aurata* R.) in saltwater ponds. *Aquaculture*, **5**, 91–8.
Creutzberg, F. (1961). On the orientation of migratory elvers (*Anguilla vulgaris*) in a tidal area. *Neth. J. Sea Res.* **1**, 257–338.
Cunningham, J. T. (1891). See Pillay: 0267.
D'Ancona, U. (1934). Faune et flore des eaux saumâtres. *Rap. P.-V. Comm. Int. Explor. Sci. Mer Mediterr.* **8**, 167–4.
Davidson, A. (1963). *Seafish of Tunisia and the Central Mediterranean.* Brit. Chamb. Comm. Tunis.
De Angelis, R. (1960). See Pillay: 0291.
De Angelis, R. (1967). See Pillay: 0292.
Demir, N. (1971). See Lajonchere: 160.
Denizci, R. (1958). See Pillay: 0302.
De Vlaming, V. L. (1972). Environmental control of Teleost reproductive cycles: a brief review. *J. Fish. Biol.* **4**, 131–40.
Dieuzeide, R., Novella, M. & Roland, J. (1955). See Pillay: 0318.
Donato, A. & Contini, A. (1974*a*). Vitellogenetic process and cytochemical aspects in the growing oocytes of the benthonic Teleost, *M. chelo. Act Histochem.* **49**, 132–41.
Donato, A. & Contini, A. (1974*b*). Further data on the yolk chemical nature of the Teleosts *M. chelo* and *Belone belone. Boll. Zool.* **41**, 261–8.

Duncker, G. (1960). *Die Fische der Nordmark*. De Gruyter and co., Hamburg.
Edimar, C. L. A. (1973). Algunos aspectos de la biologia de la Lisa *Mugil curema* en aguas hipersalinas de Nororiente de Venezuela. *Mem. Soc. scienc. nat. contrib.* **51**, 223–38.
El Maghraby, A. M., Hashem, M. T. & El Sedfy, H. M. (1973). Some biological characters of *Mugil capito* (Cuv.) in Lake Borullus. *Bull. Inst. Oceanogr. Fish.* **3**, 55–82.
El Maghraby, A. M., Hashem, M. T. & El Sedfy, H. M. (1974*a*). Species composition and distribution of the Mullet fry in Lake Borullus. *Bull. Inst. Oceanogr. Fish. ARE*, **4**, 3–31.
El Maghraby, A. M., Hashem, M. T. & El Sedfy, H. M. (1974*b*). Sexual maturity, spawning migration and fecundity of *M. capito* in Lake Borullus. *Bull. Inst. Oceanogr. Fish. ARE*, **4**, 35–56.
El Sedfy, H. M. & Libosvarsky, J. (1974). Some effects of Aswan High Dam on water and fishes of Lake Borullus, ARE. *Zool. Listy*, **23**, 61–70.
El Zarka, S. El-D. (1963*a*). See Pillay: 0340.
El Zarka, S. El-D. (1963*b*). Acclimatization of *Solea vulgaris* in Lake Quarun, Egypt. *J. Cons. Int. explor. méditerr.* **28**, 126–36.
El Zarka, S. El-D. (1970). Studies on the distribution, growth, and abundance of migrating fry and juveniles of Mullet in a brackish coastal lake (Edku) in the UAR. *Stud. Rev. Gen. Fish. counc. Mediterr.* **46**, 5–19.
El Zarka, S. El-D. & El Sedfy, H. M. (1970). See Pillay: 0345.
El Zarka, S. El-D. & Kamel, F. (1967). See Pillay: 0343.
El Zarka, S. El-D. & Koura, R. (1967). See Pillay: 0344.
Erman, F. (1957). See Pillay: 0349.
Erman, F. (1959). See Pillay: 0352.
Erman, F. (1961). See Pillay: 0354.
Ezzat, A. (1965). See Pillay: 0369.
Faouzi, H. (1936). See Pillay: 0374.
Faouzi, H. (1938). See Pillay: 0375.
Farrugio, H. (1975). *Les muges (Poissons, Téléostéens) de Tunisie. Répartition et pêche. Contribution à leur étude systématique et biologique.* Thesis, Univ. Montpellier.
Farrugio, H. & Quignard, J. P. (1973). Biologie de *Mugil* (*Liza*) *ramada* R. 1826 et de *Mugil* (*Chelon*) *labrosus* R. 1826 (Poissons, Téléostéens, Mugilidés) du lac de Tunis. *Bull. Inst. Océanogr. Pêche Salammbô* **2**, 565–78.
Fitch, J. E. (1972). See Lajonchere: 193.
Franks, J. S. *et al.* (1972). A study of nektonic and benthic fauna of the shallow Gulf of Mexico off the State of Mississippi. *Gulf Res. Rep.* **4**, 1–148.
Gandolfi, G. (1973). Primi dati sul popolamento ittico nelle acque interne del Delta padano. *Ateneo Parmense, Acta. Nat.* **9**, 409–17.
Gandolfi, G. , Caronna, E. & Orsini, P. (1969). See Lajonchere: 205.
Gandolfi, G. & Orsini, P. (1968). See Lajonchere: 202.
Gandolfi, G. & Orsini, P. (1969). See Lajonchere: 203.
Gandolfi, G. & Orsini, P. (1970). See Lajonchere: 204.
Gopalakrishnan, V. (1974). Taxonomy and biology of tropical finfish for coastal aquaculture in the Indo-Pacific region. In *Coastal Aquaculture in the Indo-Pacific Region* ed. T. V. R. Pillay, Fishing News (Books) Ltd., London, 120–49.
Gourret, P. (1890). See Lajonchere: 218.
Gourret, P. (1897). Les étangs saumâtres du Midi de la France et leurs pêcheries. *Ann. Mus. Hist. Nat. Marseille*, **5**, 1–386.
Grant, C. J. & Spain, A. V. (1975). Reproduction, growth and size allometry of *M. cephalus* L. (Pisces: Mugilidae) from North Queensland inshore waters. *Austr. J. Zool.* **23**, 181–201.

Gruvel, A. (1926). La pêche maritime en Algérie. *Bull. Stn. Aquaculture Pêche Castigl.* **2**, 151–7.
Gunter, G. (1945). See Pillay: 0504.
Hashem, M. T., El Maghraby, A. M. & El Sedfy, H. M. (1973). The grey mullet fishery of Lake Borullus. *Bull. Inst. Oceanogr. Fish.* **3**, 29–54.
Heldt, H. (1929). Le lac de Tunis (partie nord); résultats des pêches au filet fin. *Bull. Stn. Oceanogr. Salammbô*, (11), 5–74.
Heldt, H. (1931). See Pillay: 0517.
Heldt, H. (1948*a*). See Pillay: 0518.
Heldt, H. (1948*b*). See Pillay: 0519.
Heldt, H. & Heldt, H. (1943). Etudes sur le thon, la daurade et les muges. *Broch. Stn. Oceanogr. Salammbô*, (1), 3–48.
Hendricks, L. J. (1961). See Pillay: 0524.
Hickling, C. F. (1970). See Pillay: 0545.
Hines, R. & Yashouv, A. (1971). See Lajonchere: 250.
Holt, E. W. L. (1899). See Pillay: 0554.
Jhingran, V. G. & Natarajan, A. V. (1974). Role of Chilka Lake fisheries in the development of coastal aquaculture in Eastern India. In *Coastal Aquaculture in the Indo-Pacific. Region, ed. S. R. Pillay*, 109–54.
Johnson, D. W. & McClendon, E. L. (1970). See Pillay: 0613.
Karvounaris, D. (1963). See Pillay: 0679.
Kennedy, M. & Fitzmaurice, P. (1969). See Pillay: 0689.
Kesteven, G. L. (1942). See Pillay: 0691.
Kowtal, G. V. (1967). Occurrence and distribution of pelagic fish eggs and larvae in the Chilka Lake during the years 1964 and 1965. *Indian J. Fish.* **14**, 198–214.
Kristensen, I. (1971). Preference of euryhaline species for brackish and superhaline waters. *Vie et Milieu*, suppl. 22, 811–26.
Kuo, C. M. & Nash, C. E. (1975). Recent progress on the control of ovarian development and induced spawning of the grey mullet (*M. cephalus*). *Aquaculture*, **5**, 19–29.
Kuo, C. M., Nash, C. E. & Shehadeh, Z. H. (1974*a*). A procedural guide to induce spawning in grey mullet (*Mugil cephalus* L.). *Aquaculture*, **3**(1), 1–14.
Kuo, C. M., Nash, C. E. & Shehadeh, Z. H. (1974*b*). The effects of temperature and photoperiod on ovarian development in captive grey mullet (*Mugil cephalus* L.). *Aquaculture*, **3**(1), 25–43.
Kuo, C. M., Shehadeh, Z. H. & Milisen, K. K. (1973*a*). A preliminary report on the development, growth and survival of laboratory reared larvae of the grey mullet, *Mugil cephalus* L. *J. Fish Biol.* **5**, 459–70.
Kuo, C. M., Shehadeh, Z. H. & Nash, C. E. (1973*b*). Induced spawning of captive grey mullet (*Mugil cephalus* L.) females by injection of human chorionic gonadotropin (HCG). *Aquaculture*, **1**(4), 429–432.
Kurian, C. V. (1975). Mullets and mullet fisheries of India. *Aquaculture*, **5**, 114.
Landret, J. P. (1975). Contribution à l'étude de *Mugil cephalus* et comparaison avec d'autres espèces de Mugilidae. Thesis, Paris.
Le Dantec, J. (1955). See Pillay: 0728.
Le Gall, T. & Cantacuzene, A. (1955). Inventaire de la faune marine de Roscoff Poissons. *Trav. stat. biol. Roscoff, suppl. 8*, 1–67.
Leray, C. (1968). Etude de l'histophysiologie de l'hypophyse chez le muge et la carpe: ses corrélations avec l'hypothalamus, la thyroide et les gonades. Thesis, Marseille.
Leray, C. (1971). See Lajonchere: 324.
Leray, C. & Carlon, N. (1963). See Pillay: 0736.
Leray, C. & Stahl, A. (1961). See Pillay: 0738.

Liao, I-C. (1975). Experiments on induced breeding of the grey mullet in Taiwan from 1963 to 1973. *Aquaculture*, **6**, 31–58.
Libosvarsky, J. & Darrag, A. E. H. (1975*a*). Winter catches of grey mullet (*Mugil capito* Cuv.) in Lake Borullus, ARE. *Zool. Listy*, **24**, 81–95.
Libosvarsky, J. & Darrag, A. E. H. (1975*b*). Fishing success and weather conditions in winter at Lake Borullus, ARE. *Zool. Listy*, **24**, 175–84.
Linder, D. R., Strawn, K. & Luebke, R. W. (1975). The culture of striped mullet (*Mugil cephalus* L.) in ponds receiving heated effluent from a power plant. *Aquaculture*, **5**, 151–61.
Livni, L. (1971). See Lajonchere: 336.
Lo Bianco, S. (1888). See Lajonchere: 337.
Lo Bianco, S. (1909). Notizie biologische riguardanti specialmente el periodo di matura sessuale degli animali del Golfo di Napoli. *Mitth. Zool. Stat. Neapel*, **19**, 513–61.
Lozano-Rey, L. (1947). Peces ganoideos y fisostomos. *Mem. Real Acad. Ci. Madrid*, **11**, 839.
Lumare, F. & Villani, P. (1972). Contribute alla fecondazione artificiale di *Mugil cephalus* (L.). *Boll. Pesca Piscic. Idrobiol.* **27**(2), 255–61.
Marinaro, J-Y. (1971). Contribution à l'étude des oeufs et larves pélagiques de poissons méditerranéens. V. oeufs pélagiques de la baie d'Alger. *Pelagos*, **3**(1), 1–118.
Marion, A. F. (1890*a*). Oeufs flottants et alevins observés dans le Golfe de Marseille durant l'année 1890. *Ann. Mus. Hist. Nat. Marseille*, **4**, 112–21.
Marion, A. F. (1890*b*). Essai d'élevage de quelques alevins. *Ann. Mus. Hist. Nat. Marseille*, **4**, 121–4.
Marquez, R. M. (1974). Observaciones sobre mortalidad total y crescimento en longitud de la Lisa (*Mugil cephalus*) en lagune de Tamiahua, Mexico. *Inst. Nac. Pesca*, SC/3, 1–16.
Moe, M. A. (1966). See Pillay: 0824.
Moore, R. H. (1974). General ecology, distribution and relative abundance of *Mugil cephalus* and *Mugil curema* on the south Texas coast. *Contrib. Mar. Sci.* **18**, 241–55.
Morovic, D. (1954). See Pillay: 0840.
Morovic, D. (1957). See Pillay: 0842.
Morovic, D. (1963). See Pillay: 0845.
Nair, G. S. (1957). See Pillay: 0861.
Nash, C. E. & Kuo, C. M. (1975). Hypothesis for problems impeding the mass propagation of grey mullet and other finfish. *Aquaculture*, **5**, 119–33.
Nash, C. E., Kuo, C. M. & McConnel, S. C. (1974). Operational procedures for rearing larvae of the grey mullet (*Mugil cephalus* L.) *Aquaculture*, **3**, 15–24.
Navarro, F. de P. (1927). Observaciones sobre el Mar Menor (Murcia). *Inst. Esp. Oceanogr.* **2**, 1–63.
Nikolskii, G. V. (1961). See Pillay: 0871.
Ochiai, A. & Umeda, S. (1969). Spawning aspects of the Grey Mullet, *Mugil cephalus* living on the coastal region of Kochi Prefecture. *Jap. J. Ichthyol.* **16**, 50–4.
Olivereau, M. (1968). See Pillay: 0901.
Oren, O. H. (1971). See Lajonchere: 400.
Oren, O. H. (1974). IBP/PM. International symposium on the grey Mullets and their culture: aftermath and reflections. *Bamidgeh*, **26**, 23–7.
Oren, O. H. (1975). Opening address IBP/PM international symposium on the grey mullets and their culture. *Aquaculture*, **5**, 3–8.
Orlandi, S. (1902). See Lajonchere: 401.

Paolucci, C. (1917). I prodotti del mare sul mercato di Cagliari. *Riv. Pesca Idrobiol.* **11**, 81–109.
Perlmutter, A., Bograd, L. & Pruginin, J. (1957). See Pillay: 0943.
Peterson, G. L. & Shehadeh, Z. H. (1971). See Lajonchere: 423.
Popescu, L. (1961). Contribution à la connaissance de la prolificité de quelques espèces de poissons d'eau douce et marins des eaux roumaines. *Bull. Inst. Cercet. Piscicol.* **20**, 39–44.
Rafail, S. Z. (1968). See Pillay: 1008.
Rafail, S. Z. & Hamid, E. M. (1974). The abundance of mullet fry at the sides of Mex canal. *Bull. Inst. Ocean. Fish. ARE*, **4**, 97–129.
Rojas, J. C. (1972). Contribucion al conocimiento de la biologia de las lagunas y Rios de Campoma y buena vista (Venezuela) especialmente del Robalo, *Centropomus parallelus* R. *Cuadern. oceanogr.* **3**, 3–36.
Rossignol, M. (1951). See Pillay: 1048.
Sanzo, L. (1931). Uova e larve di Mugilidi. *Atti Soc. Ital. Prog. Sci.* **20**, 324–5.
Sanzo, L. (1932). See Pillay: 1074.
Sanzo, L. (1936). See Pillay: 1075.
Sanzo, L. (1938). Uova e larve di *Mugil labeo* Cuv. *Rapp. P.-V. Comm. Int. Explor. Mer Mediterr.* **11**, 73–6.
Sarig, S. (1966). Fisheries and fish culture in Israel in 1965. *Bamidgeh*, **18**, 37–50.
Sarig, S. (1969). See Lajonchere: 463.
Sarig, S. (1974). Fisheries and fish culture in Israel in 1973. *Bamidgeh*, **26**, 57–83.
Savchuk, M. H. (1967). See Pillay: 1088.
Savchuk, M. Y. (1968). See Pillay: 1089.
Savchuk, M. Y. (1973). Feed migrations of the Mugilidae fry at the Crimea and west Caucasius coast. *Gidrobiol. Zh. USSR*, **5**, 28–35.
Senkevich, N. K. & Kulikova, N. I. (1970). Intraspecific differentiation of serum proteins in the blood of Black Sea Mullets. *J. Ichthyol.* **10**, 538–44.
Serbetis, C. D. (1939). See Pillay: 1117.
Shehadeh, Z. H. & Ellis, J. N. (1970). See Pillay: 1127.
Shehadeh, Z. H., Kuo, C. M. & Milisen, K. K. (1973*a*). Induced spawning of grey mullet *M. cephalus* L. with fractionated salmon pituitary extract. *J. Fish Biol.* **5**, 471–8.
Shehadeh, Z. H., Kuo, C. M. & Milisen, K. K. (1973*b*). Validation of an *in vivo* method for monitoring ovarian development in the grey mullet (*M. cephalus*) *J. Fish Biol.* **5**, 489–96.
Shehadeh, Z. H., Kuo, C. M. & Nash, C. E. (1973*c*). Establishing brood stock of grey mullet (*Mugil cephalus*) in small ponds. *Aquaculture*, **2**, 379–84.
Shehadeh, Z. H., Madden, W. D. & Dohl, T. P. (1973*d*). The effect of exogenous hormone treatment on spermiation and vitellogenesis in the grey mullet, *M. cephalus. J. Fish. Biol.* **5**, 479–87.
Sidhu, G. S., Vale, G. L., Shipton, J. & Murray, K. E. (1972). Nature and effects of kerosene-like taint in mullet. In *FAO Technical Conference on Marine Pollution*, ed. M. Ruivo 546–50.
Stahl, A., Seite, R. & Leray, G. (1960). See Lajonchere: 485.
Stenger, A. H. (1959). See Pillay: 1192.
Sylvester, J. R. & Nash, C. E. (1975). Thermal tolerance of eggs and larvae of Hawaiian striped mullet, *Mugil cephalus* L. *Trans. Am. Fish. Soc.* **104**, 144–7.
Sylvester, J. R., Nash, C. E. & Emberson, C. R. (1975). Salinity and oxygen tolerances of eggs and larvae of Hawaiian striped mullet, *Mugil cephalus* L. *J. Fish. Biol.* **7**, 621–9.

Tamura, T. (1974). El cultivo de la Lisa. *Docum. Minist. Pesquera Lima, Peru*, **43**, 24–6.
Tang, Y. (1964). See Pillay: 1213.
Thierberger-Abraham, B. (1967). See Pillay: 1220.
Thomson, J. M. (1951). See Pillay: 1223.
Thomson, J. M. (1953). See Pillay: 1225.
Thomson, J. M. (1955). See Pillay: 1228.
Thomson, J. M. (1957*a*). See Pillay: 1230.
Thomson, J. M. (1957*b*). See Pillay: 1232.
Thomson, J. M. (1963). See Pillay: 1238.
Thomson, J. M. (1966). See Pillay: 1240.
Thong, L. H., (1969*a*). See Pillay: 1241.
Thong, L. H., (1969*b*). Contribution à l'étude de la biologie des Mugilidés (Poissons Téléostéens) des côtes du Massif Armoricain. *Bull. Soc. Sci. Bretagne*, **44**, 137–8.
Thong, L. H., (1971). See Lajonchere: 514.
Timoshek, N. G. & Shilenkova, A. K. (1974). The nature of the oogenesis and spawning of Black Sea mullet. *J. Ichthyol.* **14**, 727–34.
Tseng, W. Y. (1975). Mullet fisheries of Taiwan catch pattern. *Aquaculture*, **5**, 113.
Tung, I. H. (1968). Studies on the fishery biology of the grey mullet (*Mugil cephalus*) in Taiwan. *FAO Fish. Rep.* no. 63, 52–3.
Varagnolo, S. (1964*a*). See Lajonchere: 529.
Varagnolo, S. (1964*b*). Variazoni diurne della presenza degli stadi di sviluppo di alcuni Teleostei marini nel plancton di Chioggia. *Boll. zool.* **31**, 1037–47.
Varagnolo, S. (1964*c*). Calendario di comparse di uova pelagiche di Teleostei marini nel plancton di Chioggia. *Arch. Ocean. Limnol. Venezia*, **13**, 249–97.
Vialli, M. (1937). See Lajonchere: 531.
Vladimirov, V. I. (1973). The effect of the growth rate of spawners on the survival and abundance of the progeny. *J. Ichthyol.* **13**, 801–12.
Wallace, J. H. (1973). Aspects of the life histories of estuarine fish in Natal with emphasis on reproduction and recruitment. In *South African Natural Oceanography Symp. Cape Town, August 1973*, 31–2.
Wallace, J. H. (1975). Aspects of the biology of *Mugil cephalus* in a hypersaline estuarine lake on the east coast of South Africa. *Aquaculture*, **5**, 111.
Webb, B. F. (1973). Fish populations of the Avon-Heathcote estuary. II. Breeding and gonad maturity. *New Zeal. J. Mar. Fresh. Res.* **7**, 45–66.
Wheeler, A. (1969). The Fishes of the British Isles and NW. Europe. McMillan, London.
Wimpenny, R. S. (1934). An analysis of Arabian seine net haul on the sea coast near Ashtoun El Gameel, August 1928–May 1929. *Notes Mem. Fish. Res. Directorate, Egypt*, **2**, 1–113.
Wimpenny, R. S. (1936). See Pillay: 1313.
Wimpenny, R. S. & Faouzi, H. (1935). See Pillay: 1315.
Yang, W. T. & Kim, U. B. (1962). See Pillay: 1319.
Yashouv, A. (1967). *Researches on breeding of Mugilidae*. Fish Breeding Experimental Stat., Dor.
Yashouv, A. (1969). See Pillay: 1326.
Yashouv, A. & Berner-Samsonov, E. (1930). See Pillay: 1328.
Yashouv, A., Berner-Samsonov, E. & Hines, R. (1969). See Pillay: 1329.
Zhitenev, A. N., Kalinin, D. S. & Abayev, Y. I. (1974). The state of the gonads of the striped mullet (*M. cephalus*) and the sharpnose mullet (*M. saliens*) leaving estuaries to spawn, and their reaction to a pituitary injection. *J. Ichthyol.* **14**, 232–9.

5. Age and growth of grey mullet

J. P. QUIGNARD & H. FARRUGIO

Mugilidae are coastal marine and less frequently fresh-water fish; some of them seem capable of spending a great part of their life in fresh or brackish water and others only some months each year. These euryhaline and relatively warmth-loving fish are widely distributed in tropical and subtropical waters.

There is a paucity of local and critical information on their biology but the following pattern is generally acknowledged: young grey mullet live, before maturity, predominantly in the coastal systems of river and lakes (salt, brackish or fresh water lakes); they move to the sea when they mature (first spawning migration); after spawning they return to coastal lakes, estuarine waters and sometimes reach fresh water some miles from the coast. Such migrations are then undertaken each year but permanent sea living populations do exist. These migrations induce large variations in the growth rate that can change year after year, naturally with age and also with the biotope if the grey mullet do not return periodically to the same feeding area; this possibility is suggested by Kesteven (1942, 1953) who indicates that the populations of the rivers, lakes and sea of *Mugil cephalus*, in Australia, intermingle yearly. Into the puzzle of the growth rate enter also other variations due to the long spawning period and the sexual physiology. Grant, Spain & Jones (1977) find some degree of sexual dimorphism in nine Australian mullet; generally, other authors do not acknowledge morphological variations between sexes. Therefore growth has been often measured without regard to the sex of the fish. But we must notice that some authors have detected differences between males and females. Here also we have to mention the possible existence of difference racial stocks. Often this possibility has not been taken into account (Kesteven, 1942, 1953; Sarojini, 1957, 1958; De Sylva, Stearns & Tabb, 1956; Hongskul 1970) but now as we have the concrete evidence of the possible presence of separate races in the same area (Peterson & Shehadeh, 1971) it is of basic importance to determine if the fish studied belong to a homogeneous population or come from different racial stocks.

Age determination and back calculation of lengths

Growth of fish may be studied only if ageing is possible. The methods employed for determining the age of the grey mullet are numerous but the most frequently used is the interpretation and counting of growth zones or of 'stop growth annuli' (Berg & Grimaldi, 1967) which appear in the scales.

Age and growth

Other hard parts of the grey mullet: otoliths, operculum, fins rays and other methods: tagging and length frequency distribution of a population (Peterson method) have been used simultaneously with scale reading to get better data on age and growth.

Scales

It is necessary to give some information concerning the difficulties of scale reading. The structure and development of the scales of grey mullet have been described by several authors (Jacot, 1920; Hubbs, 1921; Schultz, 1946; Pillay, 1951; Devasundaram, 1952; Thomson, 1957; Libosvarsky, 1976). The annuli (winter rings, marks, breaks or checks) which are the boundaries between two successive growth zones often have been described as a hyaline line on the oral sector and breaks in the pattern of the circuli elsewhere (Pillary, 1954; Kennedy & Fitzmaurice, 1969; Libosvarsky, 1976). The annuli of the mullet cannot be compared with those of salmon, herring or mackerel (Jacot, 1920).

Generally the scales used for age determination are situated under the first dorsal fin or the middle part of the body at the extremity of the pectoral fins. Kennedy & Fitzmaurice (1969) write that 'comparison of sets of scales from different regions of the body. . . gave no definite indication that any one region of the body was likely to yield more consistently statisfactory scales than any other' for age reading but Brulhet (1975) writes that the scales of the caudal peduncle of *Mugil cephalus ashenteensis* are more readable than the others.

A lot of authors admit without criticism that the number of annuli on the scales are related to the age and size of fish and that only one annulus (break) is formed yearly. On a statistical basis this usual pattern is certainly true, results of tagging (Kesteven, 1942, 1953) and the study of the evolution of the anterior margin of the scales (Farrugio & Quignard, 1974; Farrugio, 1975; Cassifour, 1975) supports this point of view and validate the scale reading method for age determination. But El Maghraby, Hashem & El Sedfy (1973) point out that 'fry rings' are found in most of the scales of *M.* (*Liza*) *ramada* and Cech & Wohlschlag (1975), using scale analysis, find two periods of slow growth (winter and summer) for *M. cephalus* living in Texas coastal waters, so the difficulties in age reading are sometimes increased by the 'formation of more than one break quite close together on some scales' (Pillay, 1954). Most of authors (Arne, 1938; Pillay, 1954; Thong, 1969*a*, *b*; Kennedy & Fitzmaurice, 1969) reject the breaks which do not conform with the remainder of the sample; in his study Cassifour (1975) writes that 40% of the scales for *M.* (*Liza*) *ramada* and *M.* (*Chelon*) *labrosus* were discarded in working out the final results. But these researchers don't explain the criteria of their choice. What is the right break? We have no valid information about the formation of these 'additional rings', we do not know if they develop as a result of periodic or non-periodic variations in the growth of the fish while on the other hand we have no more information on the possible 'omission' of breaks. In conclusion to his study Devasundaram (1952) stated that breaks on the scales of *M. cephalus* are highly irregular in respect to time and consequently cannot be employed for age determination.

There is a considerable lack of useful information on the time and period of formation and appearance of the 'stop growth annuli'. There is some indication that the development of the annuli can be protracted (Kesteven, 1953) and that their time and period of formation and appearance vary with the age (Kennedy & Fitzmaurice, 1969). These problems are of basic importance for a good interpretation and comparison of the intra- and extra-specific growth rate. For instance the yearlings of *M.* (*Chelon*) *labrosus* from the Irish sea show clearly a 'winter annulus' in June and the adults only in late July (Kennedy & Fitzmaurice, 1969). *M. cephalus* spawns in North Queensland (Australia) from May to August (Grant & Spain, 1975) and the first scale mark is formed six months later. Also the age at succeeding scale mark formation is at 0.5–1.5–2.5 years; the same species from the Salton Sea (USA) spawns from late November to January (Hendricks, 1961) and breaks develop in winter, the grey mullet of this country are approximately one year old by the time the first scale mark is formed; in this case the 'real biological years' coincide with the annual scale mark at the beginning of the life. But afterwards there may be an alteration in the length of time of the ring formation (Le Gall, 1969). Few authors (Grant & Spain, 1975) take these facts into consideration when they are calculating growth curves.

In spite of what we have just written we can accept that, in temperate and subtropical waters, the usual pattern would be an increasing of the growth from spring to summer, with leveling off at the end of autumn and virtual cessation of growth in winter, with the result of the formation of an annulus which becomes visible in spring. Also breaks (stop growth annuli) can be considered as 'winter rings' (Thomson, 1951; Broadhead, 1953; Denizci, 1958; Erman, 1959; El Zarka *et al.*, 1970; El Sedfy, 1972; Brulhet, 1974, 1975; Farrugio & Quignard, 1974; Libosvarsky, 1976). In tropical waters marks are formed during the monsoon season (Pillay, 1954; Sarojini, 1957, 1958; Grant & Spain, 1975).

These are other problems in age determination with scales, and the authors agree that this method is rather more difficult to apply in the case of the mullet than is that of the other fish living in temperate water. The rings left by a period of cessation of growth are often not clearly marked (Erman, 1959) and when the grey mullet becomes old (at more than four years) the centre of the scale becomes unrecognizable, first and sometimes second breaks are obscured and can rarely be determined with certainty, also what is often taken for the first ring is really the second. Such a disappearance of the innermost rings is quoted for *M.* (*Liza*) *ramada* (Wimpenny, 1932; Elster & Jensen, 1960; Cassifour, 1975; Libosvarsky, 1976); *M.* (*Chelon*) *labrosus* (Rossignol, 1951;

Kennedy & Fitzmaurice, 1969; Cassifour, 1975); *M. cephalus* (Paget, 1922; Erman, 1959; Hendricks, 1961) and *M. (Valamugil) seheli* (Grant & Spain, 1975). Difficulties in age determination result also from the crowding of the peripheral annuli in the scales of old fish (Hendricks, 1961; Grant & Spain, 1975).

In making back-calculation of lengths at earlier ages there are two main sources of a lack of precision. Firstly there is little basis for predicting what type of relationship exists between length (and growth) of the scale and length (and growth) of the fish. Many researchers such as Wimpenny (1932), base the method of back-calculated length on the assumption that the growth in length of the grey mullet is directly proportional to the growth of the scale and apply Lea's formula (Lea, 1910) but this point of view is not always exact, there are different coefficients of regression of scale size against fish size (Kesteven, 1942; Farrugio & Quignard, 1974; Sancho, 1975; Cassifour, 1975).

Then Lee's phenomenon (Lee, 1920) of a progressive decrease in the calculated intermediate lengths is described for *M. (Liza) ramada* (Wimpenny, 1932; Zaky-Rafail, 1968), *M. (Liza) auratus* (Thong, 1969) and *M. (Liza) saliens* (Zaky-Rafail, 1968). In these cases the mean length 'l_1' calculated from scales is underestimated and this may be attributed to Lee's phenomenon. But the data of some authors give no evidence of it for *M. (Liza) saliens* (El Zarka & El Sedfy, 1970), *M. (Chelon) labrosus* (Rossignol, 1951; Morovic, 1960, 1964; Farrugio & Quignard, 1974; Farrugio, 1975) *M. cephalus* (Zaky-Rafail, 1968; Farrugio, 1975) and *M. tade* (Pillay, 1954). In other cases the estimates from scale and length distribution do not differ significantly from each other *M. cunnesius* and *M. parsia* (Sarojini, 1957, 1958), *M. cephalus* (Thakur, 1967; Zaky-Rafail, 1968).

Age and growth history determination from scales of grey mullet are not easy. Very clear and typical breaks which are not always noticeable, crowded sequence of markings and the possible presence of supernumerary (accessory) rings, uncertainty of the (one or two) innermost annual rings and occasionally Lee's phenomenon (Lee, 1920) are the cause of many difficulties in getting correct data and in comparing the results of the different researchers. We must add that there is a large amount of uncertainty concerning the biological age of fish at succeeding scale marks. In spite of this, researchers such as Serbetis (1939), Pillay (1954), Hendricks (1961) and Thomson (1966) and so on, think that of all 'the structures examined in an effort to assess the age of grey mullet, scales seemed the most feasible' (Hendricks, 1961).

Otoliths and scales

Otoliths are seldom used alone in age determination (Denizci, 1958; Ezzat, 1964: *M. cephalus*; Ezzat, 1964: *M. (Liza) auratus*) but generally simultaneously with scales (Serbetis, 1939; Cassifour, 1975; Zaouali, 1976: *M.*

(*Liza*) *ramada*; Erman, 1959, 1961; Karvounaris, 1963; Kennedy & Fitzmaurice, 1969; Cassifour, 1975: *M.* (*Chelon*) *labrosus*). Otolith reading is not always easy. Thong (1969*a*, *b*) indicates that only 67%, 66% and 64% of the sagitta of *M.* (*Liza*) *ramada*, *M.* (*Chelon*) *labrosus*, *M.* (*Liza*) *auratus* can be taken into account and this author finds differences in back calculated lengths between scales and otoliths. The sagitta of *M. cephalus* more than four years old (Erman, 1959) and of *M.* (*Chelon*) *labrosus* more than two years old (Kennedy & Fitzmaurice, 1969) is too opaque to be used and Brulhet (1974) cannot read otoliths of *M. cephalus* at all.

Opercular bones and scales

Hickling (1970) used these two structures but as with the scales and the otoliths, on the opercular bones the first ring (*M.* (*Chelon*) *labrosus*) and sometimes the first three rings (*M.* (*Liza*) *auratus*) are not distinct enough.

Spiny rays, otoliths and scales

Erman (1959, 1961) detects the first mark on cross sections of the ventral spiny ray.

Petersen's method and scales

Generally Petersen's method cannot be used for determining age and growth rate because of the difficulties in obtaining samples and because spawning takes place over several months, but length frequency is sometimes appropriate to determine 'L_1' as in the case of *M.* (*Liza*) *saliens* (Zaky-Rafail, 1968 – differences between scales and Petersen's method – El Zarka, 1964); *M. cephalus* (Thomson, 1951; Broadhead, 1958; Anderson, 1958; Zaky-Rafail, 1968); *M.* (*Chelon*) *labrosus* (Hickling, 1970; Cassifour, 1975); *M.* (*Liza*) *ramada* (Zaky-Rafail, 1968 – large difference between scale back calculated length and Petersen's method – Albertini-Berhaut, 1975; Cassifour, 1975); *M.* (*Liza*) *auratus* (Thong, 1969 – l_1 length frequency 106 mm, back calculated length 91 mm – Hickling, 1970) and *M. parsia* and *M. cunnesius* (Sarojini, 1957, 1958). We should take note that Thomson (1951) and Thakur (1967) succeeded in determining age and growth of *M. cephalus* with Petersen's method, they observed five modes in the size frequency distribution which coincide with the observed breaks of the scales.

Monthly growth of yearlings and juveniles based on length frequency is sometimes given: for *M. cephalus* (Coste, 1855 quoted by Brunelli, 1916; Jacot, 1920; Higgings, 1928; Faouzi, 1938; Gunter, 1945; Kilby, 1949; Broadhead, 1953; Cain & Dean, 1976); *A. forsteri* (Thomson, 1957); *M.* (*Liza*) *saliens* (Zaky-Rafail, 1968; El Zarka, 1964; El Zarka, El Maghraby

& Abdel-Hamid 1970; Albertini-Berhaut, 1975); *M. parsia* and *M. cunnesius* (Sarojini, 1957, 1958) and *M.* (*Chelon*) *labrosus* (Kennedy & Fitzmaurice, 1969).

Tagging and scales

Tagging operations are not numerous, but very important, because they confirm the validity of the scale reading technique for *M. cephalus* (Kesteven, 1942, 1953; Thomson, 1951) and show for *M.* (*Chelon*) *labrosus* and *M.* (*Liza*) *ramada* the presence of supernumerary breaks (Cassifour, 1975).

Growth in length

In Tables 5.2 to 5.11 we have separated marine and lake (estuarine) populations. When the researchers do not substitute growth data into the Von Bertalanffy growth equation (as in Tables 5.2 to 5.12), we have estimated the parameters from their results and for more homogeneity we have transformed fork length and standard length into total length (Table 5.1).

Even if researchers have sometimes refused to accept the data which are not in agreement with 'the reasonable estimate of age' (Zaky-Rafail, 1968) there is little agreement between the results except that the rate of growth is highest during the first and sometimes second years of life and noticeably slows down after the first sexual maturity; this phenomenon is not apparent in long-lived, slow growing and late maturing populations (Kennedy & Fitzmaurice, 1969; Hickling, 1970). Some researchers have found differences in the growth rate between males and females but others working with the same species and in the same area indicate no variations with sex. When differences are noticeable the females grow faster than the males but these differences appear only after the second year of life. Significant differences have been pointed out for *M. cephalus* (Ezzat, 1965; Brulhet, 1974, 1975; Farrugio, 1975), *M.* (*Liza*) *saliens* (Ezzat, 1965; El Zarka, 1968, El Zarka & El Sedfy, 1970), *M.* (*Liza*) *auratus* (Ezzat, 1965), *M.* (*Liza*) *ramada* (Elster & Jensen, 1960; Ezzat, 1965; El Maghraby *et al.*, 1973; Farrugio, 1975), *M.* (*Chelon*) *labrosus* (Morovic, 1964; Farrugio, 1975), *Aldrichetta forsteri* (Thomson, 1957). But no significant differences between sexes are mentioned for *M. cephalus* (Dannevig, 1902; Kesteven, 1942; Thomson, 1951; Morovic, 1957; Erman, 1959; Thakur, 1967; Cech & Wohlschlage, 1975; Grant & Spain, 1975), *M.* (*Chelon*) *labrosus* (Kennedy & Fitzmaurice, 1969; Hickling, 1970), *M. curema* (Angell, 1973), *M. parsia* and *M. cunnesius* (Sarojini, 1957, 1958), *M.* (*Valamugil*) *seheli* (Grant & Spain, 1975). Even in these cases, sometimes the females grow slightly faster (Thomson, 1951; Hickling, 1970; Cech & Wohlschlag, 1975), live longer than the males (Thomson, 1951) or at least are predominant among older fish (Hickling, 1970).

Table 5.1. *Total length, fork length and standard length relationships of various mullet*

Mugil cephalus	
Kesteven, 1942 (cm)	LS = 0.784 LT + 0.046
	LF = 0.866 LT + 0.735
Ezzat, 1965	LS = 0.847 LT − 4.130
	LF = 0.939 LT − 2.950
Farrugio, 1975 (cm)	LS = 1.080 LF + 0.030
	LT = 1.200 LS − 0.850
	LS = 0.990 LF − 0.610
Mugil (Chelon) labrosus	
Farrugio & Quignard, 1974 (cm)	LT = 1.240 LS − 0.470
	LS = 0.900 LF − 0.330
	LT = 1.110 LF − 0.680
Mugil (Liza) ramada	
Ezzat, 1965	LS = 0.809 LT + 5.700
Farrugio & Quignard, 1974 (cm)	LT = 1.200 LS + 0.780
	LS = 0.920 LF − 1.040
	LT = 1.100 LF + 0.130
Mugil (Liza) auratus	
Ezzat, 1965	LS = 0.842 LT − 3.940
Sancho, 1975	LF = 0.875 LT + 7.000
Farrugio, 1975 (cm)	LT = 1.230 LS + 0.210
	LS = 0.990 LF − 0.610
	LT = 1.230 LF − 2.090
Mugil (Liza) saliens	
Ezzat, 1965	LS = 0.840 LT + 1.660
Farrugio, 1975 (cm)	LT = 1.130 LS + 1.100
	LS = 0.930 LF − 1.050
	LT = 1.060 LF + 0.330
Mugil cunnesius	
Sarojini, 1958 (cm)	LS = 0.799 LT − 0.432

Abbreviations for Tables 5.1 to 5.14: a, intercept of the line with the Y axis (length-weight relationship $W = aLT^n$); Bon, ageing by opercular bones; F, female; FF, sexually mature female; juv, juvenile male or female; L_1, L_2..., length at the first, second...annulus; LF, fork length; Lfr, length frequency; LS, standard length; LT, total length; L_∞, k, t_o, l_o, parameters of the von Bertalanffy growth model (Ford–Walford); M, male; MM, sexually mature male; N, number of samples; n, length-weight regression coefficient; Oth, otolithometry; Ray, ageing by ray section; Scl, scalimetry; Tag, tagging. Measurements given in mm unless otherwise stated).

The values given in *italic* are estimated values for total length, calculated by the authors from original data on fork length and standard length (see p. 160).

The effect of the locality on growth rates has to be looked at carefully. There is little agreement between the data concerning the growth rate of *M. cephalus*. Devasundaram (1952), eastern Indian waters; Alexandrova (1957), Nikolskii (1954), Black sea; Serbetis (1939) and Alessio (1976), Italian brackish waters and Heldt (1948), lake of Tunis and Kelbia; give lengths at the formation of the first and second ring which are approximately those given by most of the researchers at the second and third ring and sometimes more. The same

Table 5.2. *Growth rate of* Mugil (Liza) auratus *caught in the sea*

Authority and area	Method	Length	Range of length	No. and sex	L_1	L_2	L_3	L_4	L_5	L_6	L_7	L_8	L_∞	k	t_0	l_0
Hickling, 1970	Scl +	LT	200–410	—	(50)	107	187	242	265	343	—	—	1164	0.061	0.242	−17.23
Great Britain (Devon, Wales, Scilly)	Bon + Lfr	—	—	—	—	—	—	258	290	300	—	—	—	—	—	—
Thong, 1969*a, b*	Scl	LF	68–338	86	91	162	223	257	293	326	359	—	—	—	—	—
France (Brittany)		*LT*	*81–414*	—	*109*	*197*	*272*	*314*	*358*	*399*	*439*	—	577.1	0.201	−0.083	9.57
Thong, 1971	Scl	LF	172–361	127	80	172	215	256	282	313	—	—	—	—	—	—
France (Morbihan)		*LT*	*209–442*	—	*96*	*209*	*262*	*313*	*345*	*383*	—	—	—	—	—	—
Arne, 1938	Scl	LT	—	—	112	173	271	311	349	—	—	—	511	0.224	−0.019	2.175
France (Bay of Biscay)																
Serbetis, 1939	Scl	LT	73–300	71	113	183	300	—	—	—	—	—	—	—	—	—
Italy (Adriatic)																
Nikolskii, 1954																
Black Sea	?	LT	—	—	120	210	260	300	360	—	—	—	515.6	0.229	−0.214	24.67
Caspian Sea		LT	—	—	210	280	330	360	410	430	460	470	599.8	0.177	−1.560	144.80
Alexandrova, 1964*a, b*	?	LT	—	—	130	240	327	375	415	450	480	—	541.2	0.311	0.116	−19.81
Black Sea																

Table 5.3. *Growth rate of* Mugil (Liza) auratus *caught in coastal lakes*

Authority and area	Method	Length	Range of length	No. and sex	L_1	L_2	L_3	L_4	L_5	L_6	L_7	L_8	L_∞	k	t_0	l_0
Sancho, 1975	Lfr	LF	—	144	170	240	—	—	—	—	—	—	—	—	—	—
Spain (Cadix)		*LT*	—	—	*207*	*293*	—	—	—	—	—	—	—	—	—	—
Salt water		LF	—	—	210	260	—	—	—	—	—	—	—	—	—	—
		LT	—	—	*256*	*350*	—	—	—	—	—	—	—	—	—	—
Coutelan, 1953	Scl	LT	—	—	125	176	216	319	—	—	—	—	—	—	—	—
France (Thau)																
Brackish water																
Ezzat, 1965																
France (Berre)	Scl	LT	90–420	1055 F	131	201	266	325	369	410	—	—	685.5	0.140	−0.499	46.26
Open salt lake				M	127	188	254	303	340	—	—	—	577.5	0.160	−0.501	44.59
Ezzat, 1964	Scl	LT	—	—	140	194	240	315	—	—	—	—	—	—	—	—
France (Berre, Port de Bouc)																
Heldt, 1948	Scl	LT	—	—	188	270	320	—	—	—	—	—	398	0.495	−0.292	53.52
Tunisia (Tunis)																
Salt lake																
Serbetis, 1939	Scl	LT	140–430	40	135	250	335	392	—	—	—	—	537	0.340	0.152	−28.44
Italy (near Rome)																

Table 5.4. *Growth rate of* Mugil cephalus *from sea*

Authority and area	Method	Length	Range of length	No. and sex	L_1	L_2	L_3	L_4	L_5	L_6	L_7	L_8	L_9	L_{10}	L_∞	k	t_o	l_o
Probatov &																		
Terechensko (*in*																		
Nikolskii, 1954)	?																	
Black Sea		LT	—	—	120	210	260	300	360	—	—	—	—	—	516	0.230	−0.210	24.68
Caspian Sea		LT	—	—	210	280	330	360	410	470	510	—	—	—	600	0.180	−1.560	144.80
Alexandrova, 1957	Scl	LF	—	—	220	330	480	560	600	—	—	—	—	—	—	—	—	—
(*in* Thomson, 1963)		*LT*	—	—	*237*	*356*	*518*	*605*	*648*	—	—	—	—	—	859	0.270	−0.100	22.30
Black Sea																		
Ilin, 1949		LF	—	—	108	163	208	248	281	312	343	373	404	430	—	—	—	—
(*in* Deniczi, 1958)		*LT*	—	—	*117*	*176*	*225*	*268*	*303*	*337*	*370*	*403*	*436*	*464*	1089	0.052	−1.620	87.54
Black Sea																		
Deniczi, 1958	Oth	LF	—	807	47	145	216	300	418	515	—	—	—	—	—	—	—	—
Istanbul and		*LT*	—	—	*51*	*157*	*233*	*324*	*451*	*556*	—	—	—	—	—	—	—	—
surroundings																		
Erman, 1959	Oth	LF	35–620	349	—	171	253	333	426	503	550	590	620	—	—	—	—	—
	+	*LT*	*38–670*	—	—	*185*	*273*	*360*	*460*	*543*	*594*	*637*	*670*	—	—	—	—	—
Bosphorus and	Ray	LF	—	293	65	161	241	329	380	—	—	—	—	—	—	—	—	—
Marmara Sea		*LT*	—	—	*70*	*174*	*260*	*355*	*410*	—	—	—	—	—	790	0.160	0.455	−59.55
Berg *et al.*, 1949	?	LF	—	—	180	290	380	420	440	450	—	—	—	—	—	—	—	—
(*in* Thomson, 1963)		*LT*	—	—	*194*	*313*	*410*	*454*	*475*	*486*	—	—	—	—	515	0.510	−0.041	10.65
Italy																		
Cech & Wohlschlag,	Scl	LS	122–286	M	139	191	229	262	—	—	—	—	—	—	—	—	—	—
1975																		
USA (Texas)		*LT*	*132–309*	*89*	*166*	*228*	*274*	*313*	—	—	—	—	—	—	450	0.240	−0.900	88.76
		LS	—	F	146	197	240	269	286	—	—	—	—	—	—	—	—	—
		LT	—	—	*174*	*235*	*287*	*322*	*342*	—	—	—	—	—	407	0.320	−0.710	82.96
Broadhead, 1958	Scl	LF	—	M	166	255	301	—	—	—	—	—	—	—	—	—	—	—
USA (N & NW	+	*LT*	—	—	*179*	*275*	*325*	—	—	—	—	—	—	—	379	0.652	0.021	−5.32
Florida)	Tag	LF	—	M	166	256	302	—	—	—	—	—	—	—	—	—	—	—
		LT	—	—	*179*	*276*	*326*	—	—	—	—	—	—	—	379	0.660	0.036	−9.18
		LF	—	F	173	271	313	—	—	—	—	—	—	—	—	—	—	—

		LF	—	M	150	213	—	—	—	—	—	—	—	—	—	—	—	—
		LT	—	—	*162*	*230*	—	—	—	—	—	—	—	—	—	—	—	—
		LF	—	F	144	214	—	—	—	—	—	—	—	—	—	—	—	—
		LT	—	—	*155*	*231*	—	—	—	—	—	—	—	—	—	—	—	—
Jacot, 1920	Scl	LF	—	—	182	222	—	—	—	—	—	—	—	—	—	—	—	—
USA (East Florida)		*LT*	—	—	*197*	*240*	—	—	—	—	—	—	—	—	—	—	—	—
Zaky-Rafail, 1968	Scl	LT	—	—	157	344	471	—	—	—	—	—	—	—	740	0.390	0.383	−118
Egypt																		
Brulhet, 1974, 1975	Scl	LS	90–510	M	—	321	425	493	540	—	—	—	—	—	—	—	—	—
Mauritania		*LT*	*107–611*	286	—	*384*	*509*	*591*	*647*	—	—	—	—	—	—	—	—	—
		LS	90–771	—F	—	355	501	589	645	—	—	—	—	—	—	—	—	—
		LT	*107–924*	—	—	*425*	*600*	*705*	*773*	—	—	—	—	—	—	—	—	—
Devasundaram,	Scl	LF	—	—	300	420	550	640	—	—	—	—	—	—	—	—	—	—
1952 (in Ezzat,		*LT*	—	—	*324*	*454*	*594*	*691*	—	—	—	—	—	—	1439	0.133	−0.870	157.28
1965)																		
India																		
Kesteven, 1942	Scl	LT	100–600	1659	144.5	229.5	317.5	387.6	473.0	537.3	—	—	—	—	1729	0.060	−0.510	49.11
Australia																		
Thomson, 1951	Scl	LF	—	113 M	130	238	326	394	432	—	—	—	—	—	—	—	—	—
West Australia		*LT*	—	—	*140*	*257*	*352*	*425*	*466*	—	—	—	—	—	609	0.300	0.143	−26.57
	Scl	LF	—	67 F	133	244	346	408	440	—	—	—	—	—	—	—	—	—
		LT	—	—	*144*	*263*	*374*	*441*	*475*	—	—	—	—	—	591.5	0.340	0.200	−41.42
	Scl	LF	—	1646 MF	140	241	331	405	467	507	555	—	—	—	—	—	—	—
		LT	—	—	*151*	*260*	*357*	*437*	*504*	*547*	*599*	—	—	—	828	0.180	−0.129	19.04
	Lfr	LF	—	MF	150	240	340	380	420	—	—	—	—	—	—	—	—	—
		LT	—	—	*162*	*259*	*367*	*410*	*433*	—	—	—	—	—	505	0.399	0.040	−8.07
	Scl	LF	—	1639 MF	149	231	317	357	477	537	—	—	—	—	—	—	—	—
		LT	—	—	*161*	*249*	*342*	*429*	*515*	*580*	—	—	—	—	1752	0.061	−0.515	54.30
	Scl	LF	—	670 MF	145	242	335	405	467	508	555	—	—	—	—	—	—	—
		LT	—	—	*157*	*249*	*362*	*437*	*504*	*549*	*599*	—	—	—	866	0.163	−0.212	29.48
Thomson, 1963	Scl	LF	—	—	140	245	336	405	467	505	537	—	—	—	—	—	—	—
Australia		*LT*	—	—	*151*	*265*	*363*	*437*	*504*	*545*	*580*	—	—	—	727	0.229	0.006	1.058
Grant & Spain, 1975,	Scl	LF	155–574	—	106	268	342	374	452	505	532	566	—	—	—	—	—	—
Australia		*LT*	*170–620*	323	*114*	*289*	*369*	*404*	*488*	*545*	*574*	*611*	—	—	604.6	0.344	0.140	−29.83
(N Queensland)			—	—	(L0.5)	(L1.5)	(L2.5)	(L3.5)	(L4.5)	(L5.5)	(L6.5)							

Table 5.5. *Growth rate of* Mugil cephalus *from coastal lakes*

Authority and area	Method	Length	Range of length	No. and sex	L_1	L_2	L_3	L_4	L_5	L_6	L_7	L_∞	k	t_o	l_o
Ezzat, 1964	Oth	LT	—	—	176	268	319	359	—	—	—	417.7	0.47	−0.169	31.76
France (Berre)															
Salt lake															
Ezzat, 1965	Oth	LT	130–490	2124 M	140	228	284	334	350	—	—	—	—	—	—
France (Berre)				F	140	232	298	352	386	—	—	—	—	—	—
Salt lake															
Coutelan, 1953	Scl	LT	—	—	160	240	280	350	400	—	—	—	—	—	—
France (Thau)															
Salt lake															
Serbetis, 1939	Scl	LT	19–580	—	220	352	475	490	527	—	—	563	0.561	0.083	−26.82
Italy (near Rome)		LT	—	—	215	366	460	504	529	—	—	565	0.580	0.143	−50.44
Morovic, 1954, 1957	Scl	LT	—	539	166	246	318	384	426	451	539	1041	0.090	−1.100	99.92
Italy (Venezia)		LT	—	—	163	243	314	374	420	451	—	611	0.210	−0.465	56.05
Italy (culture)		LT	—	—	244	319	370	422	—	—	—	619	0.214	−1.370	157.60
Alessio, 1976	Scl	LF	407–540	38	200	325	400	455	500	520	—	—	—	—	—
Italy (Orbetello)		*LT*	*439–583*	—	*216*	*351*	*432*	*491*	*540*	*562*	—	615.0	0.400	−0.044	10.73
Brackish water															
Morovic, 1957															
Yugoslavia (Vransko)															
Fresh water	Scl	LT	—	—	128	201	302	368	416	440	—	599	0.220	−0.005	0.72
(Pantan)															
Fresh water		LT	100–500	1723	131	216	304	361	408	436	—	566	0.244	−0.050	6.89
(Neretva)															
Estuary		LT	—	—	134	218	313	370	413	448	—	590	0.234	−0.083	11.32
Morovic, 1964															
Yugoslavia															
(Pantan)	Scl	LT	—	—	135	224	303	351	381	461	—	742	0.152	−0.450	48.62
(Vransko)		LT	—	—	159	249	314	363	412	464	—	683	0.173	−0.600	66.99
Heldt, 1948															
Tunisia															
[illegible]	Scl	LT			290	450	525	575	[illegible]	[illegible]	[illegible]	620.4	0.650	−0.048	−19.59

Tunisia (Tunis)		*LT*	—	—	*178*	*273*	*343*	*393*	*441*	*495*	—	693	0.190	−0.630	78.47
Salt lake		LS	—	51 M	124	226	277	317	(365)	—	—	—	—	—	—
		LT	—	—	*148*	*270*	*331*	*379*	*437*	—	—	519	0.360	0.081	−15.61
Marquez, 1974	Scl	LT?	260–455	166	161.8	156.7	332.8	366.5	375.6	408.3	—	510	0.340	−0.114	19.39
Mexico															
Hora & Pillary, 1962	Lfr	LT	—	—	140	240	330	390	—	—	—	636	0.230	−0.045	6.71
Indo-pacific															
Thakur, 1967	Lfr	LT	—	—	115	210	288	381	457	—	—	1940	0.050	−0.17	17.36
India (Bengal)		LT	—	—	148	257	366	430	483	—	—	896.2	0.150	−0.20	26.49
Estuary		LT	—	—	175	256	325	404	459	—	—	1063	0.096	−0.83	81.64
	Scl	LT	—	M	165	243	314	—	—	—	—	1034	0.094	−0.85	79.31
		LT	—	F	165	248	314	—	—	—	—	570	0.23	−0.49	60.62
		LT	—	MF	187	257	325	432	521	—	—	—	—	—	—

Table 5.6. *Growth rate of* Mugil (Chelon) labrosus *caught in the sea*

Authority and area	Method	Length	Range of length	No. and sex	L_1	L_2	L_3	L_4	L_5	L_6	L_7	L_8	L_9	L_∞	k	t_o	l_o
Hickling, 1970	Scl +	LT	230–600	453	—	—	151	251	277	342	381	380	413	—	—	—	—
Great Britain (Devon, Wales, Scilly)	Bon + Lfr	—	—	278	(50)	90	161	246	263	299	354	386	421	745.2	0.093	0.199	−13.92
Kennedy & Fitzmaurice, 1969	Scl +	LF	150–585	144	41	125	169	214	251	286	316	344	371	—	—	—	—
Ireland	Oth	*LT*	*166–649*	—	*45*	*138*	*187*	*237*	*278*	*317*	*350*	*381*	*411*	—	—	—	—
Thong, 1969*a*, *b*	Scl	LF	42–563	253	109	182	228	278	317	349	387	—	—	—	—	—	—
France (Brittany)		*LT*	*46–624*	—	*120*	*201*	*252*	*308*	*351*	*387*	*429*	—	—	624	0.157	−0.410	38.87
Arne, 1938	Scl	LT	—	—	132	196	237	319	—	—	—	—	—	—	—	—	—
France (Bay of Biscay)																	
Nikolskii, 1954	?	LT?	—	—	160	230	270	290	320	350	380	—	—	446	0.234	−0.943	88.29
Caspian Sea																	
Erman, 1961	Scl +	LF	120–592	310	151	236	305	366	422	475	518	552	561	—	—	—	—
Turkey (Dardanelles)	Oth +	*LT*	*132–656*	—	*167*	*261*	*338*	*405*	*468*	*526*	*574*	*612*	*622*	787	0.168	−0.289	37.28
Rossignol, 1951	Scl	LT?	170–790	—	120	179	234	316	351	538	572	550	616	1099.7	0.090	0.172	−17.19
Morocco (Atlantic)			—	—	102	210	260	359	386	473	519	550	610	1082	0.096	0.130	−13.56

Table 5.7. *Growth rate of* Mugil (Chelon) labrosus *caught in coastal lakes*

Authority and area	Method	Length	Range of length	No. and sex	L_1	L_2	L_3	L_4	L_5	L_6	L_7	L_8	L_9	L_∞	k	t_0	l_0
Cassifour, 1975	Scl	LF	—	—	84	167	237	277	311	339	363	394	422	—	—	—	—
France (Arcachon)		*LT*	—	—	*93*	*185*	*262*	*307*	*344*	*376*	*402*	*437*	*467*	538.5	0.219	0.115	−13.72
Fish ponds																	
Coutelan, 1953	Scl	LT	—	—	135	200	270	—	—	398	—	—	—	—	—	—	—
France (Thau)																	
Salt lake																	
Karvounaris, 1963	Scl +	LT	257–557	—	270	323	386	425	453	497	—	—	—	739	0.132	2.51	208.4
Italy (Paola)	Oth																
Brackish water																	
Morovic, 1960	Scl	LT	—	484	95	155	233	324	383	445	487	523	—	989	0.093	0.064	−5.94
Yugoslavia (Pantan)																	
Fresh water																	
Morovic, 1964	Scl	LT	—	F	102	176	253	315	358	399	428	—	—	596	0.180	−0.015	1.636
Yugoslavia (Pantan)		LT	—	M	93	165	242	307	350	387	415	—	—	575	0.185	0.075	−8.09
fresh water		LT	—	MF	94	166	237	312	361	399	425	—	—	613	0.171	0.092	−9.76
(Neretva) estuary	Scl	LT	—	F	115	209	269	320	360	400	—	—	—	510	0.252	−0.031	3.98
		LT	—	M	103	193	251	308	350	396	—	—	—	566.6	0.198	−0.042	4.66
		LT	—	MF	110	203	261	317	355	399	—	—	—	529	0.231	−0.033	4.06
(Vransko) fresh water	Scl	LT	—	F	133	234	307	353	379	390	—	—	—	423.6	0.447	0.123	−24.01
		LT	—	M	123	233	296	335	367	384	—	—	—	410	0.470	0.282	−58.09
		LT	—	MF	130	230	302	346	374	388	—	—	—	423.5	0.431	0.130	−24.44
Heldt, 1948	Scl	LT	—	—	258	391	468	500	—	—	—	—	—	542	0.643	−0.014	4.94
Tunisia (Tunis)																	
Salt lake																	
Farrugio & Quignard, 1974	Scl	LS	160–320	51	129	199	283	295	—	—	—	—	—	—	—	—	—
Tunisia (Tunis)		*LT*	*198–396*	—	*159*	*246*	*350*	*365*	—	—	—	—	—	426.7	0.498	0.135	−29.76
Salt lake																	
Farrugio, 1975	Scl	LS	218–430	77 M	124	190	257	294	—	—	—	—	—	—	—	—	—
Tunisia (Tunis)		*LT*	*270–533*	—	*153*	*235*	*318*	*364*	—	—	—	—	—	557	0.247	−0.26	34.53
Salt lake		LS	—	82 F	128	186	267	308	351	386	—	—	—	—	—	—	—
		LT	—	—	*158*	*230*	*330*	*381*	*435*	*478*	—	—	—	727	0.165	−0.435	50.47

Table 5.8. *Growth rate of* Mugil (Liza) ramada *from the sea*

Authority and area	Method	Length	Range of length	No. and sex	L_1	L_2	L_3	L_4	L_5	L_6	L_7	L_8	L_9	L_∞	k	t_0	l_0
Hickling, 1970	Scl	LT	300–500	—	(50)	118	202	264	294	331	—	—	—	492.9	0.202	0.494	−51.75
Great Britain																	
Thong, 1969*a*, *b*	Scl	LF	45–423	89	100	192	253	306	352	386	407	—	—	—	—	—	—
France (Brittany)		*LT*	*50–465*	—	*110*	*211*	*278*	*337*	*387*	*425*	*448*	—	—	540.9	0.256	0.140	−19.72
Arne, 1938	Scl	LT	—	—	131	181	210	234	272	316	—	—	—	1457	0.030	−2.43	102.37
France (Bay of Biscay & Fish Ponds)																	
Alexandrova, 1962	Scl	LT	—	—	148	287	367	447	—	—	—	—	—	615	0.338	0.117	−37.88
(Black Sea)																	
Wimpenny, 1932	Scl	LT	—	92	140	139	128	—	—	—	—	—	—	—	—	—	—
Egypt (Alexandria)			—	—	—	193	188	—	—	—	—	—	—	296	0.410	−0.543	59.74
			—	—	—	—	228	—	—	—	—	—	—	—	—	—	—
Zaky-Rafail, 1968	Scl	LT	—	338	147	238	314	377	427	462	—	—	—	623.9	0.217	−0.221	29.17
Egypt	Lfr	LT	—	—	180	240	314	377	427	462	—	—	—	797.0	0.122	−0.97	89.51

Table 5.9. *Growth rate of* Mugil (Liza) ramada *caught in coastal lakes*

Authority and area	Method	Length	Range of length	No. and sex	L_1	L_2	L_3	L_4	L_5	L_6	L_7	L_8	L_9	L_∞	k	t_0	l_0
Cassifour, 1975	Scl	LF	—	—	103	177	219	257	282	310	330	347	366	443	0.247	−0.132	14.24
France (Arcachon)		*LT*	—	—	*113*	*195*	*241*	*253*	*310*	*341*	*363*	*382*	*402*	—	—	—	—
Fish ponds																	
Ezzat, 1965	Scl	LT	115–425	1399 M	114	190	263	312	333	414	—	—	—	534.6	0.205	−0.229	24.55
France (Berre)		LT	—	F	111	194	280	325	356	420	—	—	—	494.5	0.266	−0.016	2.145
Open salt lake																	
Serbetis, 1939	Scl	LT	250–478	59	145	265	350	420	460	—	—	—	—	578	0.326	0.126	−24.27
Italy (near Rome)																	
Brackish & fresh water																	
Heldt, 1948	Scl	LT	—	—	305	355	400	—	—	—	—	—	—	805	0.105	−3.520	249.4
Tunisia (Kelbia)																	
Brackish water																	
Farrugio & Quignard, 1974	Scl	LS	208–285	—	132	201	243	270	—	—	—	—	—	377	0.475	−0.152	26.25
Tunisia (Tunis)		*LT*	*205–343*	—	*159*	*242*	*292*	*325*	—	—	—	—	—	—	—	—	—
Salt lake																	
Farrugio, 1975	Scl	LS	212–322	64 M	136	208	243	270	—	—	—	—	—	360.6	0.559	−0.070	13.97
Tunisia (Tunis)		*LT*	*256–387*	—	*164*	*250*	*292*	*325*	—	—	—	—	—	—	—	—	—
Salt lake		LS	—	90 F	148	201	276	307	—	—	—	—	—	565.4	0.226	−0.544	65.47
		LT	—	—	*178*	*242*	*332*	*369*	—	—	—	—	—	—	—	—	—
Zaouali, 1976	Scl + others	LT	—	—	250	335	380	430	—	—	—	—	—	524.8	0.350	−0.865	137.20
Tunisia (Kelbia)																	
Brackish water																	
Wimpenny, 1932	Scl (1928)	LT	125–245	372	159	169	134	—	—	—	—	—	—	—	—	—	—
Egypt (Mariout)			—	—	—	187	168	—	—	—	—	—	—	—	—	—	—
Fresh water				—	—	—	183	—	—	—	—	—	—	—	—	—	—
Wimpenny, 1932	Scl (1929)	LT	—	657	140	139	127	138	—	—	—	—	—	—	—	—	—
Egypt (Bardawil)			—	—	—	191	156	165	—	—	—	—	—	—	—	—	—
			—	—	—	—	193	195	—	—	—	—	—	—	—	—	—
			—	—	—	—	—	208	—	—	—	—	—	—	—	—	—

Brackish water	Scl	LT	—	—	132	190	228	—	—	—	—	—	—	300	0.420	−0.370	43.47
Paget, 1922																	
Egypt (Mariout)	?	LT	—	—	100	—	—	—	—	—	—	—	—	—	—	—	—
Fresh water			—	—	230	—	—	—	—	—	—	—	—	—	—	—	—
Faouzi, 1938	?	LT	—	—	210	—	—	—	—	—	—	—	—	—	—	—	—
Egypt (Mariout)																	
Elster & Jensen, 1960	Scl	LT	360–640	40 M	—	—	—	—	463	497	516	510	540				
Egypt (Mariout, Nhoza		LT	—	54 F	—	—	—	—	510	548	553	613	605	—	—	—	—
hydrodrome)		LT	—	153 MF	—	—	385	416	486	523	534	561	572	—	—	—	—
Fresh water																	
El Maghraby *et al.*, 1973	Scl	LT	—	M	123	226	299	—	—	—	—	—	—	476.6	0.344	0.133	−22.33
Egypt (Borullus)		LT	—	F	129	238	313	—	—	—	—	—	—	478.4	0.374	0.159	−29.41
Libovarsky & Darrag, 1975	?	LT	—	—	149	190	—	—	—	—	—	—	—	—	—	—	—
Egypt (Borullus)																	

Table 5.10. *Growth rate of* Mugil (Liza) saliens *from the sea*

Authority and area	Method	Length	Range of length	No. and sex	L_1	L_2	L_3	L_4	L_5	L_6	L_∞	k	t_o	l_o
Zaky-Rafail, 1968	Scl+ Lfr	LT	—	209	76	187	236	280	320	—	364	0.447	0.523	−95.56
Egypt														
Aleev, 1956 (cited by Thomson, 1966)		?	—	—	87	145	179	209	230	258	320.2	0.258	−0.263	21.08
Black Sea			—	—	69	163	232	268	291	318	353.2	0.408	0.461	−73.06
Alexandrova, 1964*a*, *b*	Scl	LT	—	271	161	257	328	382	—	—	541	0.290	−0.217	33.16
Black Sea														

Table 5.11. *Growth of* Mugil (Liza) saliens *from coastal lakes*

Authority and area	Method	Length	Range of length	No. and sex	L_1	L_2	L_3	L_4	L_5	L_6	L_∞	k	t_0	l_0
Ezzat, 1965	Scl	LT	120–130	F	81	138	189	235	277	—	662.5	0.102	−0.272	18.24
France (Berre)			—	M	81	138	188	228	263	—	452.9	0.168	−0.172	12.92
Open salt lake														
El Zarka, 1968	Scl	LT	120–350	F	120	165	220	270	310	—	1322	0.043	−1.126	62.49
Egypt (Qarun)			—	M	120	165	200	240	275	—	765.3	0.069	−1.504	75.08
Salt lake														
El Zarka *et al.*, 1970	Scl	LT	130–150	F	125	184	224	271	307	—	536.6	0.146	−0.826	60.95
Egypt (Qarun)			—	M	123	182	205	242	275	—	369.6	0.235	−0.752	59.83
(Edku)	Scl	LT	—	F	90	145	176	—	—	—	216	0.573	0.060	−7.58
Fresh water			—	M	82	135	167	—	—	—	215.8	0.504	0.052	−5.78

Table 5.12. *Growth rate of various mullet*

Authority, species, area	Method	Length	Range of length	No. and sex	L_1	L_2	L_3	L_4	L_5	L_6	L_7	L_8	L_9	L_{10}	L_∞	k	t_o	l_o
Mugil vaigiensis					(L0.5)	(L1.5)	(L2.5)	(L3.5)	(L4.5)									
Grant & Spain, 1975	Scl	LF	?–651	290	38	194	296	340	404	—	—	—	—	—	677.2	0.240	0.450	—
Australia																		
Mugil seheli					(L0.5)	(L1.5)	(L2.5)	(L3.5)	(L4.5)	(L5.5)	(L6.5)							
Grant & Spain, 1975	Scl	LF	?–627	165	219	334	380	433	480	537	—	—	—	—	609.6	0.311	0.170	—
Australia																		
Aldrichetta forsteri																		
Thomson, 1957	Scl	LF	50–315	M	111	190	247	292	328	—	—	—	—	—	438	0.270	−0.082	9.63
W. Australia			—	F	111	197	264	317	352	379	392	—	—	—	447	0.310	−0.076	−10.52
E. Australia	Scl	LF	—	M	50	129	195	249	—	—	—	—	—	—	510	0.190	0.450	−45.75
			—	F	52	136	210	267	306	—	—	—	—	—	472	0.233	0.510	−59.85
Mugil tade																		
Pillay, 1954	Scl	LT	40–679	68	165	246	290	340	393	438	474	525	578	613	1167	0.065	−1.420	103.54
India (Bengal)																		
Mugil parsia																		
Sarojini, 1957	Scl	LS	—	—	61	112	206	239	—	—	—	—	—	—	439.5	0.210	0.430	−41.28
India (Bengal)			—	—	87	140	188	222	—	—	—	—	—	—	380	0.210	−0.240	18.69
Mugil cunnesius																		
Sarojini, 1958	Scl +	LS	?–233	—	84	131	175	205	233	—	—	—	—	—	362	0.192	−0.37	24.97
India (Bengal)	Lfr																	
Mugil curema																		
Richards & Castagna,																		
1976	?	LF	25–142	480	203	288	327	345	353	—	—	—	—	—	360	0.78	−0.06	—
USA (Virginia)																		
Rhinomugil corsula																		
Gopalalzrishnan, 1974	—	—	—	—	—	350	—	—	—	—	—	—	—	—	—	—	—	—
India (Bengal)	—	—	—	—	—	450	—	—	—	—	—	—	—	—	—	—	—	—

problem is present in *M.* (*Chelon*) *labrosus* (Heldt, 1948: Lake of Tunis; Karvounaris, 1963: Paola Italia); *M.* (*Liza*) *saliens* (Alexandrova, 1964*b*: Black Sea); *M.* (*Liza*) *ramada* (Heldt, 1948; Zaouali, 1976; Lake of Kelbia, Tunisia; Elster & Jensen, 1960: Lake Maruit Egypt) and *M.* (*Liza*) *auratus* (Nikolskii, 1961: Caspian Sea). These data suggests the possibility of error in the interpretation of the annuli rather than large variations according to the environment.

On the other hand Deniczi (1958) who uses the otolith-reading technique and Erman (1959) the otolith and spiny fin ray techniques underestimate l_1 and l_2 of *M. cephalus*. The data given by Thong (1969*a*, *b*) suggests that back-calculated lengths from otolith-reading are lower than those obtained from scale-reading in *M.* (*Liza*) *auratus*, *M.* (*Liza*) *ramada*, *M.* (*Chelon*) *labrosus*. But Ezzat (1964) using otoliths of *M. cephalus* and of *M.* (*Liza*) *auratus* and Erman (1961) those of *M.* (*Chelon*) *labrosus* do not calculate low $l_1 . l_2 . l_3 \ldots$

Berg, Bogdarov, Kojia & Pass (1949) and Hendricks (1961) describe an abnormally slow growth of *M. cephalus* living in the Black Sea and in the Salton Sea (USA) and report that this mullet can reach about thirteen and sixteen years of age (exactly thirteen and sixteen annual breaks). Thomson (1966) writes that 'there is fair agreement between American (Broadhead, 1958) and Australian (Kesteven, 1942; Thomson, 1951) results' for *M. cephalus* but researchers have given pertinent data concerning local differences of the growth rate for this same species and others. Thomson (1951) and Broadhead (1953) find variations in populations of *M. cephalus* of adjacent estuaries; El Zarka *et al.* (1970) between *M.* (*Liza*) *saliens* living in Lake Quarun and Lake Edku (Egypt). These variations may be due, mainly, to the density of the populations and to the quantity of available food.

Some authors emphasize the differences of growth rate between lake dwelling and sea dwelling grey mullet (Serbetis, 1939; Bougis, 1952; Cain & Dean, 1976: *M. cephalus*, *M.* (*Liza*) *auratus*, *M.* (*Liza*) *ramada*). If we do not take into consideration any questionable data (Devasundaram, 1952; Alexandrova, 1957, 1964*a*, *b*; Nikolskii, 1954; Serbetis, 1939; Alessio, 1976; Heldt, 1948) we find there are seldom consistent, and never striking, differences between these two biotypes as far as growth in length is concerned. If on the Mediterranean coasts the growth of grey mullet in the 'lagoons' is slightly higher than in the sea, on the Atlantic coasts (Bay of Biscay, Brittany) the researchers have observed that it is the contrary during the first three years of life of *M.* (*Chelon*) *labrosus* and *M.* (*Liza*) *ramada* (Arne, 1938; Thong, 1969*a*, *b*; Cassifour, 1975). Except for some which are land-locked (Kelbia, Tunisia; Valli, Italy; Karoum, Egypt and the fish ponds of Arcachon) the lakes are mainly accessible to the sea and the grey mullet can migrate at least once a year to the sea where they intermingle with other neighbouring populations. As we have no evidence of homing, we may suppose that these migrations permit some lessening in the variations between local growth rates. But for some populations spending a large part of their life in eutrophic lakes such as Vranko (Yugoslavia) and the Lake of Tunis, we can agree with Morovic (1964), and Farrugio (1975) who describe an higher growth rate in those lakes than in the others. We must also notice that *M.* (*Chelon*) *labrosus*, *M.* (*Liza*) *ramada*, *M.* (*Liza*) *auratus* grow faster in the Mediterranean Sea, along the coasts of Morocco and in the Bay of Biscay than on the English and Irish coasts (Kennedy & Fitzmaurice, 1969; Hickling, 1970). The differences between these two latter localities and the others are so important that we are tempted to incriminate the method of age determination, however we must take into consideration that these grey mullets are here at the northern part of their distribution area.

Growth in weight

No researchers back-calculate the weight at each stop growth ring. They give the length–weight relationship which may be expressed by the power equation $W = aLT^n$ (Tables 5.13, 5.14) from which we can obtain data on age-weight relations.

In the early life of the grey mullet there is a fairly low ratio of weight to length (Thomson, 1951; Broadhead, 1953; Erman, 1959; Ezzat, 1965; Farrugio, 1975). But the weight increases rapidly when the fish become mature for the first time and it continues to do so as the fish get older.

Generally weight-length data show very uniform values in a given species up to the length of the first or second maturation, then above that the variations become more and more noticeable (Kesteven, 1942; Pillay, 1954; Erman, 1961; Cassifour, 1975). There is often a large disparity between spawning and non-spawning females (Erman, 1959, 1961, Farrugio & Quignard, 1974; Farrugio, 1975) and no important differences between non-spawning males and females (Erman, 1961; Farrugio & Quignard, 1974; Farrugio, 1975). But the regression coefficient is often higher for the females than for the males (Morovic, 1957; Erman, 1959; Hendricks, 1961; Kennedy & Fitzmaurice, 1969), however Thomson (1951) and Denizci (1958) for *M. cephalus* and Sarojini (1957, 1958) for *M. parsia* and *M. cunnesius*, write that there is no marked difference in the length-weight relationship between males and females.

Generally the weight of grey mullet of the same age, living in lakes (lagoons) and estuaries is higher (Morovic, 1957; Denizci, 1958) than that of the fish living in the sea (except for mature females).

Table 5.13. *Length–weight relationship of various mullet (total length, mm; weight, g; W* = aLT^n)

Authority and area	Sex	a	n
	M. cephalus (sea)		
Broadhead, 1953	M+F (juv.)	0.0108	1.7407
USA (Florida)	FF	$5.9707.10^{-6}$	3.1065
	MM	$3.5470.10^{-6}$	3.2067
Erman 1959			
Marmara Sea	M+F	$6.5375.10^{-6}$	3.0946
Bosphorus	M+F	$4.4183.10^{-5}$	2.7329
Deniczi, 1958	M+F	$1.1680.10^{-5}$	2.9999
Turkey (Istanbul)			
Kesteven, 1942	M+F	0.01	3.0000
Australia			
Zaky-Rafail, 1968	M+F	$1.7198.10^{-5}$	2.8962
Egypt			
	M. cephalus (lakes)		
Morovic, 1954	M+F	$9.2299.10^{-6}$	3.0138
Italy (Venice)			
Marquez, 1974	M+F	$5.2600.10^{-5}$	2.8670
Mexico			
Hora & Pillay, 1962	M+F	0.0002	2.8779
India			
Serbetis, 1939	M+F	$1.6675.10^{-5}$	2.9151
Italy (near Rome)			
Morovic, 1957	M+F (juv.)	0.0007	2.2279
Yugoslavia (Pantan)	F	$1.7528.10^{-5}$	2.8937
	M	$1.2736.10^{-5}$	2.9495
Chow, 1958	M+F	$0.3080.10^{-5}$	3.0250
China			
Ezzat, 1965	M	$2.9716.10^{-6}$	3.2000
France (Berre)	F	$4.8305.10^{-6}$	3.1200
	M+F	$9.2299.10^{-6}$	3.0138
Farrugio, 1975	M	$2.8565.10^{-6}$	3.2223
Tunisia (Tunis)	F	$2.3432.10^{-5}$	2.8613
	M. labrosus (sea)		
Hickling, 1970	M+F	$9.7668.10^{-7}$	3.3105
		$7.9410.10^{-7}$	3.4005
Great Britain	(Gutted fish)	$3.1195.10^{-7}$	3.4800
		$2.4466.10^{-7}$	3.6106
		$1.4420.10^{-5}$	2.9201
Thong, 1969*a*, *b*	M+F	$1.2802.10^{-5}$	2.9035
France (Brittany)		$3.6730.10^{-6}$	3.1719
		$3.4947.10^{-6}$	3.1716
		$2.0496.10^{-5}$	2.8349
Erman, 1961		$3.3356.10^{-5}$	2.8514
Turkey (Dardanelles)		$2.6092.10^{-5}$	2.8976
	M. labrosus (lakes)		
Farrugio & Quignard, 1974	M (Juv)	$2.4903.10^{-6}$	3.2404
Tunisia (Tunis)	M (Adult)	$1.3435.10^{-6}$	3.3468
	F (Juv)	$7.9899.10^{-6}$	3.0387
Cassifour, 1975	M+F	$5.8640.10^{-6}$	3.0968
France (Ponds of Arcachon)			

Table 5.13. (*cont.*)

Authority and area	Sex	a	n
Karvounaris, 1963	M+F	$30.9500.10^{-6}$	2.825
Italy (Paola)			
Farrugio, 1975	M	$7.0095.10^{-6}$	3.0511
Tunisia (Tunis)	F	$8.2869.10^{-6}$	3.0338
	M. ramada (sea)		
Hickling, 1970	M+F	$5.4691.10^{-6}$	3.0401
Great Britain			
Thong, 1969*a*, *b*	M+F	$1.2050.10^{-5}$	2.9679
France (Mont St Michel Rance estuary)	M+F	$3.6183.10^{-6}$	3.1476
Zaky-Rafail, 1968	M+F	$1.3465.10^{-5}$	2.9141
Egypt			
Albertini-Berhaut, 1975	From 15 to 67 mm	$1.0500.10^{-2}$	3.4300
France (Marseille)	More than 67 mm (M+F)	$8.0400.10^{-2}$	2.5200
	M. ramada (lakes)		
Cassifour, 1975	M+F	$1.9204.10^{-7}$	3.6695
France (Arcachon, fish ponds)			
Ezzat, 1965	M	$2.0730.10^{-3}$	2.0200
France (Etang de Berre)	F	$1.0146.10^{-3}$	2.1700
	M+F	$0.1663.10^{-4}$	2.8710
Serbetis, 1939	M+F	$1.0540.10^{-5}$	2.9528
Italy (near Rome)			
El Maghraby *et al.*, 1973	M	$2.7227.10^{-5}$	2.7610
Egypt (Borullus)	F	$1.2502.10^{-5}$	2.9390
	FF	$5.9979.10^{-6}$	3.0460
	M+F	$3.2809.10^{-5}$	2.8070
Libovarski *et al.*, 1975	LT 80 to 195 mm	0.0300	2.5010
Egypt (Borullus)	(M+F)	$3.3342.10^{-3}$	3.2650
Farrugio & Quignard 1974	M	$2.9324.10^{-5}$	2.7908
Tunisia (Tunis)	F	$3.0147.10^{-5}$	2.7816
	MM	$3.9032.10^{-5}$	2.7455
	FF	$1.1109.10^{-5}$	2.9962
Farrugio, 1975	M	$6.7030.10^{-6}$	3.0514
Tunisia (Tunis)	F	$2.2422.10^{-5}$	2.8624
	M. auratus (sea)		
Hickling, 1970	M+F	0.00019	4.0500
Great Britain			
Thong, 1969*a*, *b*	M+F	$1.0023.10^{-5}$	2.8767
France (Mont St Michel)			
Thong, 1971	M+F June	$4.6866.10^{-5}$	3.0446
France (Morbihan)	July	0.0001	2.6108
	Aug.	0.0001	2.4931
	M. auratus (lakes)		
Thong, 1969*a*, *b*	M+F	$4.2950.10^{-6}$	3.0540
France Rance estuary			
Morlaix		$2.4776.10^{-6}$	3.1528
Morbihan		$7.2277.10^{-6}$	2.9557

Table 5.13. (*cont.*)

Authority and area	Sex	a	n
Ezzat, 1965	F	$1.5620.10^{-4}$	2.4900
	M	$6.4968.10^{-5}$	2.6200
	M+F	$0.6312.10^{-4}$	2.6700
Ezzat, 1964	M+F	$4.3752.10^{-5}$	2.7280
France (Berre and Port de Bouc)			
Serbetis, 1939	M+F	$4.9162.10^{-5}$	2.6955
Italy (near Rome)			
	M. saliens (sea)		
Zaky-Rafail, 1968	M+F	$1.4992.10^{-5}$	2.8991
Egypt			
	M. saliens (lakes)		
Ezzat, 1965	F	$2.9369.10^{-5}$	2.7600
France (Berre)	M	$5.0326.10^{-5}$	2.6600
El Zarka & El Sedfy, 1970	M+F	$2.6920.10^{-5}$	2.7709
Egypt (Qarun)			
El Zarka, 1968	F	$1.3588.10^{-5}$	2.9028
Egypt (Qarun)	M	0.0002	2.3782
	M+F	$3.4702.10^{-5}$	2.7179

Table 5.14. *Length-weight relationships of other mullet from India and America*

Species	Authority and area	Length	a	n
Mugil tade	Pillay, 1954 India	LT(cm)	0.3337	2.6198
Mugil parsia	Sarojini, 1957 India	LS (cm)	0.04182	2.860
Mugil cunnesius	Sarojini, 1958 India	LS (cm)	0.19570	2.982
Mugil curema	Angell, 1973 Venezuela	LF (mm)	$9.2.10^{-5}$	2.660
	Richards & Castagna, 1976 Virginia	LF (cm)	0.0063	3.208

Growth under experimental conditions

For sometime now, we have had information about the growth of grey mullet (*M. cephalus*, *M.* (*Liza*) *ramada*, *M.* (*Liza*) *auratus*, *M. curema* and *M. tade*) in mono and polyculture systems (Hora & Pillay, 1962; Yashouv, 1966, 1972; Yashouv & Ben-Shachar, 1970; Vallet, Berhaut, Leray, Bonnet & Pic, 1970; Kuo, Shehadeh & Milisen, 1973; Ros & Vicent, 1973; Linder, Strawn & Luebke, 1975; Pruginin, Shilo & Mires, 1975; Raunich, 1975; Reich, 1975; Linder *et al.*, 1975; Liao, 1975; Chervinski, 1976; Houde, Berkeley &

Klinousky, 1976; De Silva & Perera, 1976). The interpretation of the results of these experiments is not very easy but three points must hold the attention.

Generally in fish ponds fluctuations in the growth (length and weight) are important (intra-and extra-specific competition, population density, effect of natural and additional feeding, improvement of the environmental factors, light and temperature, this last factor being very important). Secondly the possibility, reported by some authors, that the females grow faster than the males, is corroborated experimentally (Chervinski, 1976), thirdly if, in the natural environment there is no appreciable relation between the rate of increase of length and weight and the salinity, under experimental conditions such a relationship arises (De Silva & Perera, 1976).

References

Albertini-Berhaut, J. (1975). Biologie des stades juvéniles de Téléostéens Mugilidae *Mugil auratus* Risso 1810, *Mugil capito* Cuvier 1829 et *Mugil saliens* Risso 1810. III – Croissance linéaire et pondérale de *Mugil capito* dans le Golfe de Marseille. *Aquaculture*, **5**, 179–97.

Alessio, G. (1976). Riproduzione artificiale e piscicoltura intensiva di specie ittiche marine come possibilità di sfruttamento della laguna di Orbetello. *Ateneo Parmense, Acta Nat.* **12**, 315–32.

Alexandrova, K. (1957). The growth of the grey mullet (*Mugil cephalus* L.) on the Bulgarian Black Sea coast. *Tr. Nauchno-Issled. Inst. Ribar. Ribna Prom. Stn.*, **1**, 61–76.

Aleksandrova, K. (1962). Quelques données sur la biologie du *M. ramada* Risso. *Izv. Tsentr. Nauch. Izsl. Inst. Rib. Varna.* **1**, 117–31 (in Russian, French summary).

Alexandrova, K. (1964*a*). Peculiarities in the growth of *Mugil auratus*, R. along the Bulgarian coast of the Black Sea. *Izv. Tsentr. Nauch. Izsl. Rib. Varna*, **4**, 80–7.

Alexandrova, K. (1964*b*). The growth of *Mugil saliens* Risso along the Bulgarian coast of the Black Sea. *Izv. Nauch-voizol Rs. st. Okeanogr. Varna*, **5**, 117–28.

Anderson, W. W. (1958). Larval development, growth and spawning of striped mullet (*Mugil cephalus*) along the South Atlantic coast of the United States. *Fish. Bull. U.S. Fish. Wildl. Serv.* **58**(144), 500–19.

Angell, C. L. (1973). Algunos aspectos de la biologia de la Lisa *Mugil curema* Valenciennes en aguas hipersalinas del nororiente de Venezuela. *Mem. Soc. Cienc. Nat.* **96**, 223–37.

Arne, P. (1938). Contribution à l'étude de la biologie des muges du Golfe de Gascogne. *Rapp. P.-V. Comm. Int. Explor. Méditerr.* **11**, 77–115.

Berg, A. & Grimaldi, E. (1967). A critical interpretation of the scale structure used for the determination of annuli in fish growth studies. *Mem. Ist. Ital. Ictiobiol.* **21**, 225–39.

Berg, L. S., Bogdarov, A. S., Kojia, N. E. & Pass, T. A. (1949). *Commercial fishes of the USSR food industry*. Moscow.

Bougis, P. (1952). La croissance des poissons méditerranéens. Océanographie méditerranéenne, journées d'études du Laboratoire Arago. *Act. Sc. Indust.* **1187**, 118–46.

Broadhead, G. C. (1953). Investigations of the black mullet *Mugil cephalus* L. in northwest Florida. *Florida St. Board Conserv., Mar. Lab., Tech. Sèr.* **7**, 1–33.

Broadhead, G. C. (1958). Growth of the black mullet (*Mugil cephalus* L.) in the W. and NW Florida. *Florida St. Board Conserv., Mar. Lab., Tech. Sèr.* **25**, 1–29.
Brulhet, J. (1974). Situation et perspectives des pêcheries du mulet jaune de Mauritanie. *Pêche Marit.* (1159), 702–6.
Brulhet, J. (1975). Observations on the biology of *Mugil cephalus ashenteensis* and the possibility of its aquaculture on the Mauritanian coast. *Aquaculture*, **5**, 271–81.
Brunelli, G. (1916). Ricerche sul novellame dei muggini con osservazioni e considerazioni sulla mugginicoltura. *Mem. R. Com. Talassogr. Ital.* (54), 1–45.
Cain, R. L. & Dean, J. M. (1976). Annual occurrence, abundance and diversity of fish in a South Carolina intertidal creek. *Mar. Biol.* **36**, 369–79.
Cassifour, P. (1975). Contribution à l'étude de la biologie des Mugilidés des réservoirs à poissons du bassin d'Arcachon. Thesis, Bordeaux.
Cech, J. J. & Wohlschlag, D. E. (1975). Summer growth depression in the striped mullet, *Mugil cephalus. Contrib. Mar. Sci.* **19**, 92–100.
Chervinski, J. (1976). Growth of the golden grey mullet (*Liza aurata* (Risso)) in salt water ponds during 1974. *Aquaculture*, **7**, 51–7.
Chow, B. F. (1958). Growth characteristic of four species of pondfish in Hong Kong. *Hong Kong Univ. Fish. J.* **2**, 30–6.
Coutelan, R. (1953). *Contribution à l'étude des muges de l'Etang de Thau.* DES Université des Sciences et Techniques du Languedoc, Montpellier.
Dannevig, H. C. (1902). The sea mullet, *Mugil dobula* Gunther. *Rep. Fish. NSW*, (2), 26–33.
Denizci, R. (1958). Some thoughts about the biology of common grey mullet (*Mugil cephalus* L.) in the waters of Istanbul and its surroundings. *Rapp. P.-V. Comm. Int. Explor. Mediterr.* **14**, 359–68.
De Silva, S. S. & Perera, P. A. B. (1976). Studies on the young grey mullet, *Mugil cephalus* L. I. Effects of salinity on food intake, growth and food conversion. *Aquaculture*, **7**, 327–38.
De Sylva, D. P., Stearns, H. B. & Tabb, D. C. (1956). Populations of the black mullet (*Mugil cephalus* L.) in Florida. *Flo. State. Board Conserv. Tech. Ser.* **19**, 1–45.
Devasundaram, M. P. (1952). Scale study of *Mugil cephalus* L. of Chilka lake. *J. Madras Univ.* (*B*), **22**(1), 147–63.
El Maghraby, A. M., Hashem, M. T. & El Sedfy, H. M. (1973). Some biological characters of *Mugil capito* (Cuv.) in Lake Borullus. *Bull. Inst. Oceanogr. Fish.* **3**, 55–88.
El Sedfy, H. M. (1972). A biological study of the mullet fishery in lake Borullus MSc. Thesis, University of Alexandria.
Elster, H. J. & Jensen, K. W. (1960). *Limnological and Fishery Investigations of the Nozha Hydrodrome near Alexandria, Egypt 1954–1956. Notes and Mem. no. 43.* Hydrobiological Department, Alexandria Institute of Hydrobiology.
El Zarka, S. El-D. (1964). Acclimatization of *Mugil saliens* (Risso) in Lake Quarun UAR *Proc. gen. Fish. Comm. Mediterr.* **7**(34), 337–46.
El Zarka, S. (1968). Rehabilitation of the fisheries of an inland saline lake in the United Arab Republic. *Stud. Rev. Gen. Fish. Counc. Mediterr.* **35**, 21–44.
El Zarka, S., El Maghraby, A. M. & Abdel-Hamid, Kh. (1970). Studies on the distribution, growth and abundance of migratory fry and juveniles of mullet in a brackish coastal lake (Edku) in the United Arab Republic. *Etud. Rev. FAO-GPM*, no. 46, 1–19.
El Zarka, S. and El Sedfy, H. M. (1970). The biology and fishery of *Mugil saliens* (Risso) in Lake Quarun, United Arab Republic, *UAR. J. Oceanogr. Fish.* **1**, 26 pp.

Erman, F. (1959). Observations on the biology of the common grey mullet (*Mugil cephalus*). *Proc. Tech. Pap. Gen. Fish. Counc. Mediterr.* **5**, 157–69.

Erman, F. (1961). On the biology of thicklipped grey mullet (*Mugil chelo* Cuv.) *Rap. P.-V. Comm. Int. Explor. Sci. Mer. Mediterr.* **16**, 277–85.

Ezzat, A. (1964). Contribution à l'étude de la biologie des Mugilidae de la région de l'étang de Berre et de Port de Bouc. *Rec. Trav. Stn. Mar. Endoume*, **47**(31), 187–202.

Ezzat, A. (1965). Contribution à l'étude de la biologie de quelques Mugilidae de la région de l'étang de Berre et de Port de Bouc. Thesis, Marseille.

Faouzi, H. (1938). Quelques aspects de la biologie des muges en Egypte. *Rap. P.-V. Comm. Int. Mer Méditerr.* **11**, 63–8.

Farrugio, H. (1975). Les muges (Poissons, Téléostéens) de Tunisie. Répartition et pêche. Contribution à leur étude systématique et biologique. Thesis, Université des Sciences et Techniques du Languedoc Montpellier.

Farrugio, H. & Quignard, J. P. (1974). Biologie de *Mugil* (*Liza*) *ramada* Risso 1826 et de *Mugil* (*Chelon*) *labrosus* Risso, 1826 (Poissons, Téléostéens, Mugilidés) du Lac de Tunis: âge et croissance. *Bull. Inst. Océanogr. Pêche Salammbô*, **3**(1–4), 139–52.

Gopalakrishnan, V. (1974). Taxonomy and biology of tropical finfish for coastal aquaculture in the Indo-Pacific Region. In *Coastal aquaculture in the Indo-Pacific Region*, ed. T. V. R. Pillay, Fishing News (Books) Ltd., London, pp. 120–49.

Grant, C. J. & Spain, A. V. (1975). Reproduction, growth and size allometry of *Mugil cephalus* L. (Pisces: Mugilidae) from North Queensland inshore waters. *Aust. J. Zool.* **23**, 181–201.

Grant, C. J., Spain, A. V. & Jones, P. N. (1977). Studies of sexual dimorphism and other variation in nine species of Australia mullets (Pisces, Mugilidae). *Aust. J. Zool.* **25**, 615–30.

Gunter, G. (1945). Studies on marine fishes of Texas. *Publ. Inst. Mar. Sci. Univ. Texas*, **1**(1), 1–90.

Heldt, H. (1948). Contribution à l'étude de la biologie des muges des lacs Tunisiens. *Bull. Stn. Océanogr. Salammbô*, **41**, 1–35.

Hendricks, L. J. (1961). The striped mullet *Mugil cephalus* L. *Fish. Bull. Calif.* **113**, 95–103.

Hickling, C. F. (1970). A contribution to the natural history of the English grey mullet (Pisces, Mugilidae). *J. Mar. Biol. Ass. UK*, **50**, 609–33.

Higgings, E. (1928). Progress in biological enquiries 1926. *Rep. US Comm. Fish.* **8**, 515–59.

Hongskul, V. (1970). Serum protein polymorphism in grey mullet (*Mugil cephalus* L.). *Proc. Indo-Pac. Fish. Counc.* **13**(2), 107–2.

Hora, S. L. & Pillay, T. V. R. (1962). Handbook on fish culture in the Indo-Pacific region. *FAO Fish. Biol. Tech. Pap.* **14**, 1–203.

Houde, E. D., Berkeley, S. A. & Klinousky, J. J. (1976). Culture of larvae of the white mullet, *Mugil curema. Aquaculture*, **8**, 365–70.

Hubbs, C. L. (1921). Remarks on the life history and scale characters of American mullets. *Trans. Am. Microsc. Soc.* **40**, 26–7.

Ilin, B. C. (1949). *M. cephalus* L. *M. auratus*, *M. saliens* Risso. In *The industrial Fishes of SSSR*, Moscow.

Jacot, A. P. (1920). Age, growth and scale characters of the mullet *Mugil cephalus* and *Mugil curema. Trans. Am. Microsc. Soc.* **39**(3), 199–230.

Karvounaris, D. (1963). Alcune osservazioni su *Mugil chelo* Cuv. nel lago di Paola. *Boll. Pesc. Piscic.* **18**, 71–89.

Kennedy, M. & Fitzmaurice, P. (1969). Age and growth of thicklipped grey mullet *Crenimugil labrosus* in Irish waters. *J. Mar. Biol. Ass. US* **49**, 683–99.

Kesteven, G. L. (1942). Studies in the biology of Australian mullet. I. Account of the fishery and preliminary statement of the biology of *Mugil dobula* Gunter. *Bull. Aust. CSIRO Melb.* **157**, 1–99.

Kesteven, G. L. (1953). Further results of tagging sea mullet, *Mugil cephalus* L. on the eastern Australian coast. *Aust. J. Mar. Freshwater Res.* **4**, 251–306.

Kilby, J. D. (1949). A preliminary report on the young striped mullet (*Mugil cephalus* L.) in two gulf coast areas of Florida. *Q. J. Florida Acad. Sci.* **2**(1), 7–23.

Kuo, C. M., Shehadeh, Z. H. & Milisen, K. K. (1973). A preliminary report on the development, growth and survival of laboratory reared larvae of the grey mullet, *Mugil cephalus* L. *J. Fish. Biol.* **5**, 459–70.

Lea, E. (1910). On the methods used in the herring investigations. *Publ. Circ. Cons. Int. Explor. Mer. Copenhague*, **53**, 7–175.

Lee, R. M. (1920). A review of age and growth determination in fishes by means of scales. *Fisheries Fish. Invest. London*, **2**(4), 1–32.

Le Gall, S. (1969). Croissance d'un poisson Téléostéen *l'Agonus cataphratus* (L.) rapports avec le cycle sexuel et le cycle de l'activité thyroidienne. *Vie Milieu*, (*A*), **20**, 153–234.

Liao, I. C. (1975). Experiments on induced breeding of the grey mullet in Taiwan from 1963 to 1973. *Aquaculture*, **6**(1), 31–58.

Libosvarsky, J. (1976). Lepidological note on grey mullet (*Mugil capito*) from Egypt. *Zool. Listy*, **25**(1), 73–9.

Libovarsky, J. & Darrag, A. H. A. (1975). Winter catches of grey mullet (*Mugil capito* Cuvier, 1829) in Lake Borullus, ARE. *Zool. Listy*, **24**(1), 81–95.

Linder, R. D., Strawn, K. & Luebke, R. (1975). The culture of striped mullet (*Mugil cephalus* L.) in ponds receiving heated effluent from a power plant. *Aquaculture*, **5**, 151–61.

Marquez, R. (1974). Observaciones sobre mortalidad total y crecimiento en longitud de la lisa (*Mugil cephalus*) en la laguna de Tamiahua, ver., Mexico. *Inst. Nac. Pesq. Subsecr. Pesq. Secr. Indust. Com. Mexico, INP/SC*, **3**, 1–16.

Morovic, D. (1954). Contribution à la connaissance de la croissance annuelle de *Mugil cephalus* (L.) dans quelques 'Valli de pesca' du littoral Venetien. *Rap. P.-V. Comm. Int. Explor. Sci. Mer. Mediterr.* **12**, 203–17.

Morovic, D. (1957). *Les muges de l'Adriatique avec la Bibliographie des muges*. Inst. Zaslatkovodno Ribar. Zagreb.

Morovic, D. (1960). Contribution à la connaissance de la croissance annuelle du muge *Mugil chelo* Cuv. dans l'étang 'Pantan' (Dalmatie). Rapport préliminaire. *Rap. P.-V. Comm. Int. Explor. Sci. Mer. Mediterr.* **15**(3), 115–8.

Morovic, D. (1964). Contribution à la connaissance de la croissance annuelle de *Mugil cephalus* (L.) et de *M. chelo* (C.) dans l'Adriatique. *Acta Adriatica*, **11**, 195–204.

Nikolskii, G. V. (1954). *Special Ichthyology*. Translated from Russian by J. I. Lengy & Z. Krauthamer. Published by the Israel Programme for Scientific Translation, IPST cat. no. 233.

Paget, G. (1922). The determination of the state of growth of bouri, *Mugil cephalus* in Lake Mariyut. *Rep. Fish. Serv. Egypt*, 143–9.

Peterson, G. L. & Shehadeh, Z. H. (1971). Subpopulations of the Hawaiian grey mullet *Mugil cephalus*: analysis of variations of nuclear eye-lens protein electropherograms and nuclear eye-lens weights in the grey mullet, induced breeding, and larval rearing. *Mar. Biol.* **11**(1), 52–60.

Pillay, S. R. (1954). The biology of the grey mullet *Mugil tade* Forsk with observation on its fishery in Bengal. *Proc. Natl. Inst. Sci. India*, **20**, 187–217.

Pillay, T. V. R. (1951). The structure and development of the scales of five species of grey mullet of Bengal. *Proc. Natl. Inst. Sci. India*, **17**, 413–24.

Pruginin, Y., Shilo, S. & Mires, D. (1975). Grey mullet: a component in polyculture in Israel. *Aquaculture*, **5**, 291–8.

Raunich, A. (1975). Analisi biometrica dell'accrescimento stagionale in una popolazione di cefali (*Mugil cephalus* L.) del primo anno di età, in condizioni ambientali controllate. *Statistica*, **35**(2), 107–29.

Reich, K. (1975). Multispecies fish culture (polyculture) in Israel. *Bamidgeh Bull. Fish. Cult. Israel*, **27**(4), 85–100.

Richards, C. E. & Castagna, M. (1976). Distribution, growth and predation of juvenile white mullet (*Mugil curema*) in ocean-side waters of Virginia's eastern shore. *Chesapeake Sci.* **17**(4), 308–9.

Ros, A. O. & Vicent, J. R. (1973). Primeras experiencias sobre cultivos de peces en el Mar Menor. *Bol. inst. Esp. oceanogr.* **163**, 1–20.

Rossignol, M. (1951). *Mugil chelo* Cuvier. Notes sur les muges des côtes marocaines. *Ann. biol. Cons. int. Expl. Mer. Copenhague*, **8**, 89–90.

Sancho, J. M. (1975). Estudio comparativo del crecimiento del Albur, *Mugil auratus*, en tres esteros de la provincia de Cadiz y parametros mas importantes obtenidos. *Publ. Tec. Junta Est. Pesca*, **11**, 231–43.

Sarojini, K. K. (1957). Biology and fisheries of the grey mullets of Bengal. I. Biology of *Mugil parsia* Hamilton with notes on its fishery in Bengal. *Ind. J. Fish.* **4**, 160–207.

Sarojini, K. K. (1958). Biology and fisheries of the grey mullets of Bengal. II. Biology of *Mugil cunnesius* Valenciennes. *Ind. J. Fish.* **5**, 56–76.

Schultz, L. P. (1946). A revision of the genera of mullets fishes of the family Mugilidae, with description of three new genera. *Proc. US Natl. Mus.* **96**, 377–95.

Serbetis, C. D. (1939). L'età e l'accrescimento dei Mugilidi. *Boll. Pesca Piscic. Idrobiol.* **15**(6), 628–707.

Thakur, N. K. (1967). Studies on the age and growth of *Mugil cephalus* L. from the Mahanadi estuarine system. *Proc. Natl. Inst. Sci. India.* (*B*), **33**, 128–43.

Thomson, J. M. (1951). Growth and habits of the sea mullet, *Mugil dobula* Günther in Western Australia. *Aust. J. Mar. Freshwater Res.* **2**(2), 193–225.

Thomson, J. M. (1957). Interpretation of the scales of the yellow-eye mullet *Aldrichetta forsteri* (Cuvier & Valenciennes) (Mugilidae). *Aust. J. Mar. Freshwater Res.* **8**, 14–28.

Thomson, J. M. (1963). Synopsis of biological data on the grey mullet *Mugil cephalus* L. 1758. *Fish. Synop. Div. Fish. Oceanogr. CSIRO*, *Aust.* (1), 1–75.

Thomson, J. M. (1966). The grey mullet. *Oceanogr. Mar. Biol.* **4**, 301–35.

Thong, L. H. (1969*a*). Contribution à l'étude de la biologie des Mugilidés (Poissons, Téléostéens) des côtes du Massif Armoricain. *Bull. Soc. Sci. Bretagne*, **44**, 137–8.

Thong, L. H. (1969*b*). Contribution à l'étude de la biologie des Mugilidés (Poissons, Téléostéens) des côtes du Massif Armoricain. *Trav. Fac. Sci. Rennes, Série Océanogr. Biol.* **2**, 55–136.

Thong, L. H. (1971). Notes sur la croissance du mulet *Mugil* (*Liza*) *auratus* Risso, 1810 (Téléostéen, Mugilidé) dans la partie orientale du Golfe du Morbihan. *Trav. Lab. Biol. Halieutique. Univ. Rennes*, **5**, 3–27.

Vallet, F., Berhaut, J., Leray, C., Bonnet, B. & Pic, C. (1970). Preliminary experiments on the artificial feeding of Mugilidae. *Helgoländer Wiss. Meeresunters.* **20**, 610–9.

Age and growth

Wimpenny, R. S. (1932). *Observations on the age and growth* of two Egyptian mullets, *Mugil cephalus* and *Mugil capito.* Rep. Fish. Serv. Egypt, Coastguards and Fisheries Service, Ministry of Finance, Cairo.

Yashouv, A. (1966). Breeding and growth of grey mullet (*Mugil cephalus* L.) *Bamidgeh,* **18**(1), 3–13.

Yashouv, A. (1972). Efficiency of mullet growth in fishponds. *Bamidgeh,* **24**(1), 12–25.

Yashouv, A. & Ben-Shachar, A. (1970). Breeding and growth of Mugilidae. II. Feeding experiments under laboratory conditions with *Mugil cephalus* and *Mugil capito* C. *Bamidgeh,* **19**(2/3), 50–66.

Zaky-Rafail, S. (1968). Investigations of mullet fisheries by beach seine on the UAR Mediterranean coast. *Stud. Rev. Gen. Fish. Counc. Mediterr.* **35**, 1–19.

Zambriborshch, F. S. (1962). Data on the biology of grey mullets *Mugil auratus, Mugil saliens* and *Mugil cephalus. Tr. Odessk. Derzh. Univ* (*Biol.*) **152**(11), 11–39.

Zaouali, J. (1976). Données biologiques sur les Mugilidae, Anguillidae et Cyprinidae du Lac Kelbia. In *25è Congrès Assemblée Pleinière CIESM*, Split, October 1976.

6. Food and feeding in grey mullet*

J. BRUSLE

Introduction

The mullet are widespread estuarine and coastal Teleosts and important members of the brackish water communities. A great deal has been published about the food of the Mugilidae and is best summarized by Pillay (1953), Thomson (1963, 1966), Odum (1968*a*, *b*, 1970), Hickling (1970), Albertini-Berhaut (1973) and Zismann, Berdugo & Kimor (1975).

Mullet have been described variously, as vegetarian, omnivorous, planktophagous, or as devouring small crustaceans; they have been named 'mud-eaters' by Zenkevich (1963) 'iliophagous' by Pillay (1953), 'detritus feeders' by Rajan (1964), 'algal-feeders' by Hiatt (1944), 'feeders on micro- and meio-benthos' by Hickling (1970), 'interface feeders' by Odum (1970), 'deposit feeders' by Fagade & Olaniyan (1973) and 'soft bottom feeders' by Blaber (1976).

The most noticeable feature is that mullet are a major component in the flow of energy through the ecosystem because they feed on the lowest trophic level (Odum, 1970). These fish are able to utilize either the direct grazing or plant detritus food chains as an energy source and obtain their energy directly from the first trophic level.

Studies of aspects of food and feeding on certain mugilids and *M. cephalus* in particular have been carried out recently (Blaber, 1976, 1977; Blaber & Whitfield, 1977; Ching, 1977; Chervinski, 1976; De Silva & Wijeyaratne, 1977; Farrugio, 1976; Homen, 1976; Lasserre, Renaud-Mornant & Castel, 1975; Marais, 1976; Masson & Marais, 1976; Moriarty, 1975, 1976; Pauly, 1975; Payne, 1975, 1976; Perera & De Silva, 1978), dealing with feeding ecology, ethology or physiology. Moreover, the artificial feeding in aquaculture trials and in mullet farming (Ghosh, Berkeley, Das & Das, 1974; Houde, Berkeley, Klinovsky & Schekter, 1976; Kuo, Shehadeh & Milisen, 1973; Kurian, 1975; Nash & Kuo, 1975; Pillay, 1975; Pozuelo, 1975; Ros & Vicent, 1973; Rothbard, 1976; Vallet, Berhaut, Leray, Bonnet & Pic, 1970; Yashouv, 1972) was in progress.

Thus, it seemed of interest to give a general account of new research and to summarize recent information relative to food and feeding in mullet.

* The fish are named as in 'Mugilidae' by Trewavas (1973), in *Catalogue des poissons de l'Atlantique du NE et de la Méditerranée* (CLOFNAM, ed. Hureau & Monod). *Mugil cephalus* (181.1.1, p. 567); *Chelon labrosus* (181.2.1, p. 569); *Liza ramada* (181.3.1, p. 570); *Liza aurata* (181.3.2, p. 571); *Liza* (*Protomugil*) *saliens* (181.3.4, p. 572).

Larval feeding

In natural conditions

A wide variety of ingested organisms, of animal and plant foods, is found in the stomachs of juvenile mullet, in *Liza aurata*, *L. ramada* and *L. saliens* from Marseilles (Albertini-Berhaut, 1973: p. 257, Table 2) in *M. parsia* from India (Ghosh *et al.*, 1974), in *Mugil cephalus* and *L. saliens* from Bardawil lagoon (Kimor, 1974; Zismann, Berdugo & Kimor, 1975: p. 70, Table 3) in *Chelon labrosus* from the Bassin d'Arcachon (Lasserre *et al.*, 1975), in *M. cephalus*, *L. ramada* and *Chelon labrosus* from Tunisian lakes (Farrugio, 1976: p. 46, Table 3), in *M. cephalus* from Negombo lagoon in Sri Lanka (De Silva & Wijeyaratne, 1977) and in eleven species of Mugilidae from south-east African estuaries (Blaber & Whitfield, 1977). In this last study food items from stomachs (Blaber & Whitefield, 1977: p. 288, Table 2) were classified as planktonic and undergoing vertical migration (Calanoida, Cyclopoida, Ostracoda, Amphipoda, Isopoda and zoea larvae) microbenthic (<0.1 mm) or meiobenthic (0.5–1 mm). The fry (15–50 mm SL) of *C. labrosus*, in the semi-enclosed fish reservoirs of the Bassin d'Arcachon (Lasserre *et al.*, 1975, p. 410, Table VI) was mostly carnivorous with a preference for pelagic organisms including the calanoid copepod *Acartia*, various harpacticoid and cyclopoid copepods associated with algae (*Cladophora* and *Enteromorpha*).

However, differences in diet are quite evident according to age and size. It is widely known that juvenile mullet are initially carnivorous and planktonic feeders (Hickling, 1970; Odum, 1970). The carnivorous habits of the mugilid postlarvae were mentioned by Odum (1970) who indicates that they feed on mosquito larvae, copepods and other zooplankton until they reach a length (SL) of about 30 mm. Egusa (1950) pointed out that *M. cephalus* starts feeding on benthic amphipods and harpacticoid copepods on reaching a body length of 2.1–3.95 mm.

The food supply of juvenile mullet of *Liza aurata*, *Liza ramada* and *Liza saliens* (measuring 15–65 mm SL) from the gulf of Marseilles, consists mainly of benthic crustaceans (harpacticoid copepods), as shown by Albertini-Berhaut (1973, 1975). The 'fry' (15–50 mm SL) are carnivorous microphages with crustaceans as principal food, whereas the 'fingerlings' (51–65 mm SL) are herbivorous microphages with diatoms and other microalgae as food (Albertini-Berhaut, 1973: p. 259, Fig. 2). Three successive diets have been demonstrated (Albertini-Berhaut, 1974: p. 19, Fig. 4): the first (15–25 mm SL) is exclusively animal (mainly crustaceans), the second (25–55 mm) is mixed (animal and plant) and the third (>55 mm) is almost exclusively compound of plants (algae, benthic diatoms and sediment). There are fluctuations, according to size, of the vacuity coefficient ($V = 32.77\%$ in fry, 10.20% in fingerlings), the fullness of the stomach ($R = 74\%$ in fry, then 41% and after 64%, Albertini-Berhaut, 1974: p. 17, Fig. 2) and the repletion index (which decreases with age, from 2.32% in the fry to 1.22% in the fingerlings, Albertini-Berhaut, 1974: p. 18, Fig. 3). Thus, feeding is more frequent with animal than with plant diets. The diet of young Mugilidae changes when the species reach about 50 mm length: the diatoms become the primary source of food (35000/stomach in *L. saliens* and 200000/stomach in *L. ramada*; Albertini-Berhaut, 1973).

A similar study was carried out on the gut contents of mugilids of 4.27–22.88 mm (TL) from the Haifa Bay region (Zismann, Berdugo & Kimor, 1974, 1975). Copepods were found to be the most important food item in the gut contents, at all times of the year. It was of interest to note that dinoflagellates, Rotifera and Protozoa were not found in any of the gut contents, although these organisms occurred in the sea, at times in swarms. The smaller-sized fish feed on the smaller copepods while the larger-sized fish feed on a wider range of size and species of Copepoda. A number of common planktonic organisms, of a size that might be swallowed by the young mullet were not found in the gut contents (seven species of copepods out of the twenty-five found in the local marine environment), indicating that these fish are selective in their feeding in the natural environment, selected species of copepods belonging to the orders Calanoidea, Cyclopoida and Harpacticoida (Zismann *et al.*, 1974, 1975; p. 74, Fig. 3). Stomach contents of *M. cephalus* and *L. saliens* post-larvae consisted of the same species of Copepoda (*Paracalanus*, *Enterpina*, *Oithona*) with the same pattern of selectivity in Haifa (Zismann *et al.*, 1974, 1975) but, in Marseilles, attraction to benthic Crustacea was greater for *L. saliens* than for other species (Albertini-Berhaut, 1973). A pattern of selectivity can be seen in young mullet, depending the season of availability of the various copepods and the selection is not only for size, fish picking out their preferred species (Zismann *et al.*, 1974, 1975).

Liza ramada and *Mugil cephalus* fry apparently discover their foods (Cladocera, copepods) visually; the wide spectrum of kinds of food found in mullet intestine reported in the literature indicated that the mullet feed mainly on all available food. However, a fundamental difference in the feeding habit of the two mullet species was shown (Yashouv & Ben Schachar, 1967): *L. ramada* showed a preference for chironomid larvae. In the same way, certain differences are quite evident between species from Tunisian lakes (Farrugio, 1976): in *Chelon labrosus*, harpacticoid copepods are exclusively present in stomachs, while copepods, Nematoda and various species are found in *L. ramada* and benthic diatoms, similar to the adult diet, are observed in *M. cephalus*.

In natural estuarine conditions, mullet (*Mugil parsia*) fry, although exhibiting a preference for certain planktonic organisms (*Daphnia*, *Mysis*, *Brachionus*, *Cypris*) feed mostly on diatoms. It was only after the consumption of the preferred items which were completely consumed within the first hour that the fry fed on other available organisms which may be considered as 'occasional' or 'emergency' food (Ghosh *et al.*, 1974).

Feeding behaviour varies with length and *M. cephalus* from Japan is successively a zooplankton feeder, then a mixed food feeder, finally a plant feeder (Suzuki, 1965: p. 300, Fig. 1). However, the change of food from animal to plant type is not abrupt but gradual and takes place when the fish grow to a length of 23–33 mm. According to Odum (1970), juvenile mullet up to 30 mm (SL) are primarily carnivorous. In Sri Lanka lagoons, it was shown by De Silva & Wijeyratne (1977) that the carnivorous habit of *M. cephalus* is given up long before the fish reach a size of 30 mm in length. In grey mullet of 20–55 mm (TL), animal material constituted only a very minor proportion. The diet consisted of eighteen genera of diatoms (Bacillariophyceae) which are the most predominant food items, accounting for 55% of the total diet (De Silva & Wijeyaratne, 1977: p. 163, Fig. 5), eight genera of green algae (which accounted for 22.3%), six genera of blue-green algae (which accounted for 6.1%) and eight other food groups but the animal material together accounted for less than 1%. The great majority of the diatoms and almost all the green and blue-green algae are planktonic and no sand and detritus was found to occur in fish smaller than 25 mm. However, the percentage occurrence of detritus and sand particles increased with increase in body length, which undoubtedly indicated a bottom feeding habit and was an indication of the gradual transition from a planktonic to a benthic feeding habit (almost completely benthic grazing habits were found in fish over 55 mm in length).

In Israel, *Acartia* generally appears to be a 'favourite' food. As the fish increase in size and approach the beaches, estuaries and lagoons, their feeding pattern changes basically and they begin to feed on a variety of phytoplankton and zooplankton organisms (Zismann *et al.*, 1975: p. 74, Fig. 3). Thus *M. cephalus* of 22 mm TL and over and *L. saliens* over 14 mm TL begin to change their diet when they come close to the shore and enter lagoons (Zismann *et al.*, 1974). Copepoda and Cladocera appear to be the most important components in the diet of mullet up to 30 mm (with insect larvae, algae and detritus); over 60 mm TL, the stomach contents included Polychaeta, Cladocera, Insecta, Sagitta, detritus, blue-green algae, diatoms and a small amount of sand (Anon. 1976).

A similar observation has been made in South African estuaries where changes in the feeding ecology of juveniles (10–59 mm SL) of eleven species of Mugilidae were investigated by Blaber & Whitfield (1977). When juveniles enter estuaries at a length of about 10 mm, they change their feeding habits in the following sequence: from zooplankton in the neuston to zooplankton in the benthos (at 10–15 mm) from zooplankton in the benthos to meiobenthos (at 10–20 mm) and meiobenthos to sand particles and associated microbenthos (at 15–25 mm). However, the percentage frequency of sand grains varies according to different species: fish of 10–15 mm of *M. cephalus*, *Liza alata* and *Liza dumerili* consume no sand, while it may be present in up to 10% of *Mugil capensis* and *Valamugil buchanani*. All species show a similar pattern of two changes (Blaber & Whitfield, 1977: p. 282, Fig. 3) but interspecific competition is probably prevented by the rapid switch to the adult diet of microbenthos and sand particles. It is suggested that the change from planktonic macrophagous carnivore to benthic microphagous omnivore can only take place in the estuaries when they enter at different times according to spawning periods.

In Malaysia, in the stomach contents (Ching, 1977: p. 302, Table IV) of juvenile *Liza melinoptera* (1.8–4.4 cm), decaying detritus predominated but were of finer particles than those eaten by adults and some unidentified organic masses were present. Consumption of diatoms (*Pleurosigma*, *Coscinodiscus*) was limited to those with diameters of less than 40 μm. The selective ability of the mullet was reflected in the approximate mean particle size of 14 μm in the stomach. A high percentage of particles in the sediment (66%) was less than 12 μm but a much higher percentage of these particles was found in the stomach (87%) and particles greater than 220 μm occurred in sediments while few greater than 120 μm occurred in the stomach. Studies of late juveniles indicated a general decrease of plankton algal food and a corresponding increase of organic particles; at this time, sand particles of size greater than 20 μm became more prominent (Ching, 1977).

The adaptative variation in ethology of juvenile feeding was observed by Odum (1968*b*) in estuarine waters of Georgia: striped mullet (*M. cephalus*) and silver mullet (*M. curema*) deviated from their 'normal diet' to a diet consisting of 95% of the dinoflagellate (*Kryptoperidinium*) near the surface of the waters, during a 'bloom period', which resulted in an increase in inorganic material of stomach contents from 20 to 50% (dry weight). When the dinoflagellate bloom ends, there is a return to normal benthic feeding such as benthic diatoms, detritus particles and sediment. Feeding uptake is doubled during such a period but dinoflagellates are poorly assimilated. Young specimen (< 10 cm) of *L. richardsoni* in the Swartkops estuary (Marais, 1976) also deviated from their diet, consisting mainly of diatoms and detritus obtained from the sediment, to utilize a bloom of mysid (*Mesopodopsis*) and their gut contents expressed as a percentages of body mass reached 13.3% at this time, whereas it does not exceeded 10% on a normal diet. In the present case, protein content was more than doubled (from 15% to 36%) while ash content decreased (from 65% to 43%, dry weight).

Food and feeding

In artificial breeding

According to Kuo *et al.*, (1973), in *Mugil cephalus*, no food was taken during the first four days despite the fact that the mouth opened on the second day at 24 °C and the third day at 22 °C. Larvae presumably survived on endogenous food reserves (yolk sac) during the first four days of larval life. These mullet larvae, which begin feeding on the third or fourth day, require large quantities of a single larval food smaller than the nauplii of *Artemia salina* (Nash & Kuo, 1975). They are entirely carnivorous during early development (Nash & Kuo, 1975: p. 123, Fig. 1), becoming omnivorous and capable of ingesting plant material sometime before metamorphosis.

Successful larval rearing techniques practised on a laboratory scale, then later in farm fish, were presented by Nash, Kuo & McConnell (1974): they used zooplanktonic organisms such as oyster veligers $< 70\ \mu m$ (Liao *et al.*, 1971, cited by Nash *et al.*, 1974), both oyster and sea-urchin larvae followed by day old nauplii of *Artemia salina* and wild zooplankton $< 150\ \mu m$ (Kuo *et al.*, 1973). In addition, many common phytoplanktonic organisms such as *Gymnodinium*, *Chlorella*, *Dunaliella* and *Isochrysis* have been tried. Rotifers (*Brachionus sp.*) and *Isochrysis* were the most successful during the early days of development, followed by *Artemia* nauplii on the seventh day after hatching (Nash *et al.*, 1974: p. 22, Fig. 6). Larvae up to four weeks old are vulnerable to problems resulting from ingestion of the empty cysts or shells of *Artemia* (Nash *et al.*, 1974). The food was only partially digested and some larvae suffered from intestinal blockage and subsequently died from rupture of the abdomen as shown by Kuo *et al.*, (1973).

Many experimental foods were used as larval diet: cooked egg yolk, soybean milk, pig liver, flour, milk powder, egg albumin, fish meal, yeast, artificially fertilized oyster eggs, trochophore larvae, Rotifers, Copepods and *Artemia* nauplii in *Mugil cephalus* of Taiwan (Liao, Lu, Huang & Lin, 1974) and of Hawaii (Nash & Kuo, 1975; p. 122, Table I). Quality, size, motility and density of the food items were found to be important factors. In *Liza hematochaela*, the following diet was offered to larvae: at 5–20 days, rotifers; at 12–30 days, *Artemia* and copepod *Tigriopus* and from 25 days, minced fish meat (Rothbard, 1976). In Florida, larvae of *Mugil curema* were reared and fed with nauplii and copepodites of copepods (Houde *et al.*, 1976). In Cadiz, Rotifera (*Brachionus plicatilis*) was used for feeding juveniles (28 mm TL) *Liza ramada* when breeding in tanks (Pozuela, 1975). In India, *Mugil parsia* fry exhibited a preference for *Daphnia*, *Brachionus* and *Cypris* in this order; subsequently the sequence of preference was changed to *Brachionus*, *Daphnia* and *Cypris*, whereas phytoplankton were added later, the fish fed with mixed populations of zoo- and phyto-plankton exhibited maximum growth (Ghosh *et al.*, 1974). Finally, it has been suggested by Nash & Kuo (1975) that too much food can be as harmful for larvae as too little.

Adult feeding

In natural conditions

The wide spectrum of food kinds found in mullet stomach and intestines reported in the literature indicate that the mullet feed on all available food. They are generally herbivorous, feeding on algae and diatoms but also consume animal prey (zooplankton and zoobenthos).

The composition of the diet of *M. cephalus* 'the universally successful species' (Marais, 1976), was studied in particular and divided by Thomson (1966) into the following categories: (1) microalgae including epiphytic and benthic forms, (2) decaying plant detritus and (3) inorganic sediment particles, but other mullet include a proportion of animal material (Crustacea, Polychaeta, molluscs, Nematoda in their diet. Suzuki (1966) observed five groups of elements in Japanese Mullet: (1) bottom living microcrustaceans, (2) planktonic microcrustaceans, (3) diatoms and other microalgae, (4) organic particles and (5) sand grains and muds. Hiatt (1944) stated that Hawaiian mullet feed mainly on diatoms and blue-green algae. Thomson (1959) found that *M. cephalus* at Lake Maquarie (Australia) feeds on diatoms, filamentous algae and blue-green algae; often, it feeds on detritus including mud and sand (Thomson, 1963). Luther (1962) reported that the main foods of *M. cephalus* in eastern India are decayed organic matter, algae and Foraminifera; Odum (1970) showed that for the Floridan striped mullet, *M. cephalus*, very little of the food ingested originates from the water column and that the diet is comprised of ephiphytic and benthic diatoms, Dinoflagellates, blue-green algae and decaying macroplant detritus (*Thalassia*, *Spartina*) associated with inorganic sediment particles. Microalgae are the primary source of nutrition and fish exhibits a preference for living microalgae and plant detritus are available (Odum, 1970: p. 228, Fig. 1). *M. cephalus* from St Lucia (South Africa) appears to prefer feeding on the gasteropod *Assiminea*, foraminifera and diatoms rather than on plant material (Blaber, 1976: p. 270, Table 2). Indeed, it is likely that *M. cephalus* from St Lucia Lake prefers feeding on substrates rich in animals such as foraminifera, in areas where plants, diatoms and blue-green algae together with particulate organic matter form the bulk of the available food (Blaber, 1976). However, in estuaries of Natal and Pondoland, pennate diatoms were the most important and most frequently occurring food item in *M. cephalus*, but particulate organic matter and terrestrial plant debris were also ingested in large quantities (Blaber, 1977). In the stomach contents of *M. cephalus* and *L. dumerili* from the Swartkops estuary, diatoms and algae dominated the food material and, in addition, there was animal material including ciliates, Rotifera, nematodes, Foraminifera, fragments of Crustacea, larval Mollusca and dinoflagellates (Masson & Marais, 1975).

In other mullet species, Sarojini (1954, in Ghosh *et al.*, 1974) observed that the bulk of the stomach content of *M. parsia* in India consisted of decayed organic matter followed by algae (*Oscillatoria*). Edwards (1973*a*) pointed out that *M. curema* from Venezuela was the dominant herbivore consuming benthic diatoms together with quantities of sand and macroalgae such as *Enteromorpha* with attached epifauna, and Fagade & Olaniyan, (1973) concluded that the main food items in the stomachs of seven species from a Lagos lagoon are diatoms, organic detritus and sand grains. In the fish reservoirs of the Bassin d'Arcachon, France, the food preferences of *Chelon labrosus* were presented by Lasserre *et al.* (1975: p. 410, Table VI). Notable quantities of plant food were found, including green and blue-green algae coated with epiphytic diatoms but also prey belonging to the macrofauna (*Corophium*, *Abra*, *Hydrobia*, *Nereis*, chironomid larvae) and chiefly to the meiofauna (dominant prey being Nematodes and Harpacticoid Copepods, associated with ostracods, oligochaetes, polychaetes, ciliates and turbellarians). The predominance, in the diet of five species of mullet from St Lucia Lake (Blaber, 1976: p. 273, Table 3) of relatively large food organisms (mean diameter greater than 100 μm), the high numbers of *Assimea* and *Aptinoptychus* and the absence of significant numbers of very abundant animals such as ostracods and harpacticoid copepods, suggest that the fish are capable of selecting particular food items or substrates rich in these animals. Moreover, no apparent changes in the diet and food preference according to age were observed in any of the species and no seasonal variation was recorded in the diet by Blaber (1976).

Inorganic particles or sediment

Almost all species of mullet were found to ingest large quantities of inorganic particles together with food items. It has been suggested that the function of these particles is to act as a grinding paste in the degradation of plant cell walls in the pyloric portion of the stomach (Thomson, 1966). The mineral matter in the muscular pyloric stomach, which functions as a gizzard, is considered as an essential grinding paste able to aid the trituration of food. Thus, in Taiwan, sand grains make up 65% of the stomach contents when *M. cephalus* is fed on diatoms, and only 35% when fed on copepods (Tung, 1971). It is probable that the fish are only able to digest plant material due to breaking up of the plant cells by the grinding action of the sand grains (Blaber, 1976). The large amount of inert sand particles ingested by *L. dumerili* from the Swartkops estuary (Marais, 1976) probably assisted in mechanically breaking up the diatoms in the gizzard-like stomach, to enable maximum assimilation efficiency. Blaber (1976: p. 273, Table 3) has shown that the substrate particle size ingested by each species may be different and that grey mullet from St Lucia Lake exhibit a marked sand grain size preference, each species being remarkably constant in the size of particles in their stomachs. Individual species of mullet exhibit a marked preference for particles of a certain diameter (four species of *Valamugil*, particles 100–200 μm mean diameter; *M. cephalus*, *L. macrolepis* and *L. dumerili*, 200–300 μm; *L. alata*, coarser particles with a mean diameter of 500 μm). The occurrence of a particular food item in the diet may depend upon whether it occurs on substrates of the preferred particle size. Blaber (1976) suggests that, in St Lucia Lake, the size of particles present in the stomach may be a combination of the choice of a suitable substrate and a filter mechanism, the distribution of mullet being related to their particle size preferences. Odum (1970) in Florida found that *M. cephalus* prefers very fine particles of less than about 10 μm in diameter, these being richer in adsorbed organic material. Similarly, in the Tiber river, the diameter of the fine particles contained in the stomach of *M. cephalus* varied between 100 μm and 10 μm but was almost always smaller than 25 μm (Capanna, Cataudella & Monaco, 1974). A similar estimation was demonstrated by Moriarty (1976) in Australia: the seagrass sediments were sandy with particles up to 400 μm in diameter but in the mullet contents, particles were rarely > 200 μm and mostly < 100 μm, with a much higher proportion of small sand particles (10–50 μm). These fish were therefore selecting small particles which were richer in organic matter, especially algae and bacteria. The dry weight of sediment in the full gut of mullet (130–300 g) was measured $Y = 0.157 \times -15.5$ (Y, dry weight of gut contents; X, fresh weight of fish and the standard error of the slope was ± 0.021). *M. cephalus* in St Lucia Lake are markedly different since their mean preferred particle diameter is 225 μm despite finer grained substrates being available (Blaber, 1978).

According to Odum (1968) there is evidence for the tendency of *M. cephalus* to select and ingest fine sediment particles, while expelling coarser material, by means of a pharyngeal filter apparatus (Thomson, 1966), both a pharyngeal filter and gill rakers being used in the selection of food (Pillay, 1953). Thus, Capanna *et al.*, 1974, from anatomical and histological studies observed that Mugilidae have a highly complex pharyngobranchial structure, provided with parts specialized for different functions, and stated that the complex morphology of the pharyngobranchial organ explains the selectivity of the alimentary regime as shown by Odum (1968). The pharyngobranchial organ functions as follows: the particles ingested with the intake flow are detained in the oral cavity and channelled towards the pharyngeal region where selection of the food is accomplished by the mechanical (denticulated portion in contact with branchispinal combs) and gustatory elements. Only fine particles, smaller than the distance between two teeth can pass through this barrier while larger alimentary particles are hindered. The mass of non-ingested particles is removed from the filter palissade by means of vibratory movements imparted by the intrinsic muscles to the organ's denticulated valve; this mass of non-ingested particles is then flushed out of the mouth. The gustatory factor is due to the presence of numerous taste buds and chemical analysis is of primary importance in selecting the suitable source of food. According to Capanna *et al.* (1974), the filtering and gustatory functions of the pharyngobranchial organ are closely integrated so as to justify the particular alimentary habits of Mugilidae.

It was suggested by Odum (1968) that these small inorganic sediment particles are much richer both in adsorbed organic material and in absorbed bacteria, protozoa and other microorganisms than the coarser material that the mullet rejects. The selectivity results in higher organic values of the stomachs contents than of the sediments. It is of great interest to demonstrate the ability of mullet to collect organic material both through absorption and adsorption. Thus, Payne (1976: p. 283, Table 4), in a study of the reliance of mullet on organic matter for food, has observed that, in an estuary in Sierra Leone, the time when mullet (*L. falcipinnis*, *L. dumerili*, *L. grandisquamis*, *M. cephalus ashenteensis* and *M. curema*) are most abundant, is closely correlated with the period of high organic content of the sediment. The concentration of organic matter in the stomachs was consistently higher in *M. cephalus ashenteensis* and *L. falcipinnis* than that in the stomach of *L. dumerili hoefleri*. It would seem to be due to differences between the feeding methods of these mullet: the stomach anatomy of *L. dumerili hoefleri* (in which the general organic content of ingested matter is very low, the grain size of the sand is larger and the frequency of blue-green algae is lower) is somewhat different from that of the majority of mullet and had led to grouping this species in the subgenus *Protomugil*. There was always a higher proportion of organic matter in the stomach contents than in the sediment (the concentration in the stomach ranged from 3.8 to 30.3 times that in the sediments for *L. falcipinnis*, 3.7–22.3 and 1.4–6.2 times for *M. cephalus ashenteensis* and *L. dumerili hoefleri* respectively). This appears to be due to the efficiency of the filtering apparatus or the selection by the fish or a combination of both of these. According to Odum (1970), it is necessary for *M. cephalus* to filter 100 g (dry weight) of sediment to obtain 1 g dry weight of digestible material and, for 24 h, one sample fish of 200 mm length is able to filter 1500 g of sediment (dry), i.e., more than 450 kg/year, corresponding to an area of 45 m^2 and a depth of $\frac{1}{2}$ cm as evaluated by Alvarez-Lajonchere (1976).

Thomson (1954) suggested that much of the unrecognizable detritus could be bacteria and protozoa, this rich protozoa–bacteria detritus system being of great potential food value. Bacteria and diatoms seemed of equal importance in the food of the mullet and the biomass of microorganisms was probably sufficient for the energy needs of the fish. Further, Odum (1970) found that the fine particles eaten by mullet were covered with bacteria and protozoa. Webb (1973) observed in the gut contents of *Aldrichetta forsteri* from Avon-Heathcote estuary (New-Zealand) heavy quantities of mud of three distinct types, which contained substantial quantities of soluble organic and inorganic nutrients which support a population of microorganisms. The importance of bacteria in the diet of *M. cephalus* has been demonstrated quantitatively by Moriarty (1976) in Australia where bacteria comprise about 15 to 30% and diatoms 20 to 30% of the organic carbon in the stomach. The biomass of bacteria ingested by *M. cephalus* was estimated by measuring the muramic acid in their gut contents (Moriarty, 1976). Thus, algae (mainly diatoms and blue-green algae – 1×10^4 cells/g sediment in the gut) as well as bacteria are the main food items in mullet (Moriarty, 1976: p. 136, Table IV; p. 137, Table V). The large amount of muramic acid in the gut contents of the animals investigated by Moriarty (1975) shows that bacteria are important in their diet. Total organic carbon was 10 to 20 times that in the *in situ* sediment and this author suggests that the lowest stomach pH (3.5–4.0) is probably insufficient to act as the primary mechanism of bacterial cell lysis and that the process is therefore more likely to be enzymatic.

Thus, although being able to utilize either the direct grazing or plant detritus food chains as an energy source, mullet exhibit a preference for living microalgae when both food sources are present in abundance. Such an algal diet has a higher caloric content than a macroplant detritus diet. Bacteria and protozoa, absorbed with fine particles of decaying plant detritus may be important in the diet of the mullet, either as a source of essential nutrients or by providing assistance in the breakdown of plant material (Oren, 1971).

Finally, it is interesting to note that vitamin B_{12} is abundant in sediments (Starr *et al.*, cited by Vallet, 1970) and the concentration of B_{12} in the liver of mullet is 3.795 ng/gr fresh weight according to Valencia (cited by Vallet, 1970).

Variations according to species, age and localities

Variations in the diet of mullet occur according to species, age, localities and seasons. This last point will be discussed later under 'feeding periodicity' (p. 206).

Species variations were shown in the differences in the feeding of the four species of mullet living in the Swartkops estuary and the life forms of the diatoms which had been eaten gave some indication of the type of substratum on which the fish fed. In *M. cephalus*, *L. dumerili* and *L. richardsoni*, which fed on the sediment, motile benthic diatoms and some attached diatoms occurred most commonly whereas in *L. tricuspidens*, which fed mainly on algal mats, the attached diatoms occurred almost exclusively. Moreover, *M. cephalus* and *L. richardsoni* ingested other sediment particles and finer material than *L. dumerili* (Masson & Marais, 1975). In this South African estuary, Marais (1976) demonstrated species differences. Thus, *M. cephalus* feeds mostly on diatoms and on very fine material (inorganic sediment particles rich in adsorbed organic material and in absorbed bacteria and protozoa). *L. dumerili* consumed larger diatoms (motile free living forms on sandy substrate) than those of *M. cephalus* at the same station and selected sand particles smaller than 0.5 mm but not smaller than 0.125 mm, therefore it fed on much coarser material (mean diameter, 0.17 mm) than *M. cephalus* (mean diameter, 0.06 mm) of the same size range caught in the same area. *L. richardsoni*, which preferred sandy areas, had an extremely efficient selective ability to obtain a large amount of very fine material from an area with a relatively coarse substrate. *L. tricuspidens*, which feeds on a rocky substratum, ingested fibrous matter and bits of *Zostera* plants in addition to filamentous algae and attached diatoms, but the stomach contained practically no sand particles.

Species differences have been also recorded by Farrugio (1976: p. 45, Table 1) between mullet from Tunisian lakes: *C. labrosus* and *L. ramada* were mainly carnivorous (benthic crustaceans: amphipod *Corophium*) whereas *M. cephalus* had a herbivorous diet (benthic diatoms and Foraminifera). However, the diet of *C. labrosus* is generally of mixed type, herbivorous (diatoms and Cyanophycea) and carnivorous (Nematoda and harpacticoid copepods) according to Kreft & Michaelis (1976). Observations of the feeding habits of *M. cephalus* and *L. ramada* in aquaria have shown that these two species differ in the choice of food as well as the way and rate of collecting foods (Yashouv & Ben Schachar, 1967). These differences are most pronounced when the food is abundant and, under conditions of scarcity, fish are less meticulous in their choice. *M. cephalus* shows a preference for either plant detritus or live plant material when both are present in abundance. Faced with a choice between these two resources, *M. cephalus* feeds almost exclusively on plant material (Odum, 1970). In South Africa (Blaber, 1976), five species of mullet ingest inorganic particles mixed with food items but individual species of mullet differ in the size of inorganic particles ingested and exhibit marked sand grain size preferences (Blaber, 1976: p. 270, Table 2). Similarly in Natal and Pondoland estuaries (Blaber, 1977) there was a marked difference in the size of particles ingested by each species: *Valamugil robustus* and *V. buchanani* consumed fine particles, 200 μm, while *M. cephalus* showed a strong peak at 250 μm, *L. macrolepis* overlapping with *M. cephalus* and *L. dumerili* ingested mainly coarse particles of 400–500 μm in diameter (Blaber, 1976: p. 269, Fig. 2). Finally, in an estuary of Sierra Leone, it was shown by Payne (1976) that *L. falcipinnis* differs from *L. dumerili hoefleri* in its much higher frequency of intake of blue-green algae (Payne, 1976: p. 283, Table 3), and in a notably smaller sand grain size in the former than in the latter (Payne, 1976; p. 283, Table 4).

It has been noticed that the intestine length varies according to the species diet. Thus, Adriatic lagoons, the ratio of intestine length to body length of mullet was 2.41–4.90 (Homen, 1976). Mullet that feed mainly on living plants have a shorter intestine in relation to body length. The annual increase in body length of these fish exceeds that of fish that feed mainly on detritus. Hiatt

(1944) found that Hawaiian *M. cephalus* have an intestine to fish length ratio of 3.2/1 and Hickling (1970) confirms that in *C. labrosus* the gut may be five times the length of the body, correlated with the large quantities of indigestible organic matter in their diet. Odum (1970) showed that in the same species in Florida the ratio between gut and body length increases with age according to a change in diet from one predominantly of diatoms to one of plant detritus. He also indicated that mullet populations adjust morphologically to the type of food available in the environment by lengthening the intestine for coarser diets and he suggested that an intestine to fish ratio of 3.2 in *M. cephalus* is adequate to assimilate a diatom diet, while a longer intestine is needed to extract nourishment from plant detritus and blue-green algae. Hickling (1970) has suggested that the longer gut of larger mullet may be an adaptation to food which contains a high proportion of indigestible matter. Evidence of a very significant correlation for the linear regression of body length on body length to gut ratio in *M. cephalus* from St Lucia Lake ($r = 0.554$) was shown by Blaber (1976). Measurements of the total gut length indicate that the ratio of gut length to body length increases with age, but here the diet remains the same; thus, the ratio of total gut length to standard length ranged from 2.0/1 for fish of 120 mm to 6.5/1 for fish of 500 mm. The ratios of intestine length to standard body length (IL/SL) were also determined in mullet from the Swartkops estuary (Marais, 1976). It was clear that *M. cephalus* had a ratio (4.47 to 4.65) nearly twice that found in the other species, indicating that it was morphologically well adapted to digest either a predominantly microalgal diet or a macroplant detrital diet. Low ratios of intestinal to fish length obtained for *L. tricuspidens* (2.06) and *L. richardsoni* (2.45) are in accordance with Odum's general theory. A large stomach and a short intestinal length, as in *L. dumerili*, could be advantageous to ingest large amounts of inert material together with relatively small quantities of nutritious microalgae. A large stomach permits ingestion of a large amount of sediment whereas a short intestinal length would suffice for the digestion and assimilation of easily digestible food items. Thus, the smallest stomachs were observed in the two species (*M. cephalus* and *L. richardsoni*) which ingested the finest material. Moreover, in the case of *M. cephalus* and especially of *L. tricuspidens*, well developed and sharp pointed teeth would be essential to dislodge and reduce in size attached diatoms and filamentous algae as well as pieces of macroplant detritus, prior to ingestion (Marais, 1976); in contrast, the species (*L. dumerili* and *L. richardsoni*) which consumed the largest amount of inorganic material with food items have very weakly developed teeth.

Finally, Marais (1976) revealed marked interspecific differences in *chemical composition* of alimentary canal contents. *L. tricuspidens* had the highest moisture (65.6%), protein (6.4%) and carbohydrate (7.03%) contents and the lowest ash content (20.8%); it therefore consumed the most nutritious diet, fairly large (> 1 mm in diameter) fibrous macroplant material which was probably of greatest nutritional significance). *L. dumerili* in contrast, had a lower percentage moisture (43.1%), protein (3.6%) and carbohydrate (3.40%) content and higher values for ash content (49.8%); it therefore ingested food material (diatoms) with the lowest nutritional values. Nutritional value of food material ingested by *M. cephalus* and *L. richardsoni* fell between values recorded for *L. dumerili* and *L. tricuspidens*.

Differences in diet occur according to age. A study of stomach contents (*L. aurata* and *L. ramada*) revealed (Vallet, Berhaut, Leray, Bonnet & Pic, 1970) that a change occurs in the diet of young individuals (15–40 mm SL, 4–5 months old): Copepoda, Amphipoda and Nematoda, indicating an essentially *carnivorous* diet, whereas older fish (> 50 mm SL) feed on diatoms, plant fragments, sand and mud, indicating that they were mainly *herbivorous* and *detritivorous*. These basic modifications in the digestive capacities indicate that a food rich in proteins is necessary in the first months of life of the mullet. Careful examination of stomach contents of fish of larger size than 50 mm SL showed that plant substance is hardly digested, a similar observation was reported by Yashouv & Ben Schachar (1967) and tends to confirm Odum's opinion (1968) according to which the intestinal flora (bacteria, flagellates) acts as a secondary source of metabolites. Evolution is quite evident from the pelagic life of the young, feeding on plankton, to the adult feeding on plant and detritus.

In South Carolina, *M. cephalus* and *M. curema* also shift their feeding behaviour as they grow, young feeding on planktonic and benthic crustaceans and larger fish consuming detritus; thus they shift their trophic level with growth (Cain & Dean, 1976). In *M. cephalus* from Negombo lagoon, there was an increase in the occurrence of sand detritus in the food with increasing length (De Silva & Wijeyaratne, 1977). Finally, the stomachs of the smaller size range (6–10 cm) of *L. dumerili* from the Swartkops estuary (Marais, 1976) contained much finer material (46% silt and clay in the stomach contents) than larger specimens (31% and 20% respectively); these largest specimens, not only contained the largest sand-particles but also consumed the largest variety of diatoms, algal plants and, in addition, more animal material (ciliates, Rotifera, nematodes, Foraminifera, larval Mollusca and dinoflagellates).

No change in diet according to the length and age of the fish (*L. macrolepis* and *M. cephalus*) was noted in Natal (Blaber, 1977).

Variation in the diet according to the locality may be observed. The diet of the various species of Mugilidae from Natal and Pondoland (Blaber, 1977) is similar, but varies from one estuary to another and is probably determined largely by the occurrence of particular food items on substrates of the preferred particle size of each species. Stomach contents of several species of mullet consist of variable food material ingested according to different stations of the Swartkops estuary system (Marais, 1976). Thus, *M. cephalus* consumed mostly diatoms (*Navicula, Nitschia, Cyclotella*) in some areas and selected fine particles (nearly 60% of the dry content) of silt and clay and very fine material, with the smaller diatom species typically associated with such material, in others.

The nature of the component of diet varies with the substratum, thus, on Australian sea-grass flats (*Zostera*) epiphytic and epibiontic diatoms are important (Thomson, 1959). In Adriatic lagoons, seventy-nine algal species (Chrysophyta: 79% and Cyanophyta 5%) were identified in the gut contents of mullet from the Sas lake and only forty-two species in guts of mullet from Diujiska lake (Homen, 1976).

In South Carolina (Cain & Dean, 1976) formation of detritus from *Spartina* (extensive cover of Spartina marsh in the intertidal area) provides a direct food source for estuarine fish and plentiful food supply for mullet which constitute 42% of the biomass. Along the Mauritanian coast, the stomachs of *M. cephalus ashentensis* are full of a mixture of sand and slime with planktonic Dinoflagellata, planktonic and benthic diatoms and with much detritus (Brulhet, 1975).

In artificial conditions

Artificial diets for mullet are known in experimental conditions, in aquaria with the purpose of precisely defining physiological requirements and also in semi-extensive or intensive breeding in lakes and ponds (aquaculture trials).

Experiments on artificial feeding have been carried out on fry and young individuals in Israel where fish (*L. auratus*) were fed with pellets containing 25% crude protein (Chervinski, 1976) or with pelleted food composed of wheat, fish meal and soya flour containing 21% protein. The addition of plankton noticeably increasing the growth (Yashouv & Schachar, 1967). In India, *M. parsia* shows an optimal growth in 180 days when fed with 21.6% of protein of vegetable origin in the supplementary food (Ghosh, Das & Das, 1976).

Many artificial foodstuffs were used in mullet breeding such as: fish meal (protein: 52.9%, lipids 21.6%, salts 30%, ash 16.6%) in the Adriatic *M. cephalus*, *L. aurata* and *L. ramada* (Morovic, 1976); rice bran and peanuts and/or soybean meal in *M. cephalus* of Taiwan (Chen, 1974); bread, fish meal and a mixture of potatoes, decayed vegetables, eggs, Cephalopoda, *Ulva*, *Enteromorpha* and *Chaetomorpha*, in five species in Tunisia (Pillai, 1975); fish meal (*Boops boops*), soya and maize with vitamins and antibiotics (proteins: 27–28%) in *L. aurata* at Mar Menor (Ros & Ortega, 1973); ground nut cake in mullet of Lagos Lake (Sivalingam, 1973); a mixture of rice bran and fish meal (3/1 proportion by weight) containing protein: 66.2%, lipids 5.3% and ash 23.4% as percentage values of the dry weight in *M. cephalus* of Sri Lanka (De Silva & Pereira, 1976). This food intake was salinity-dependent, for any given body weight, intake per unit weight is lower at 10‰ salinity than at higher salinities and the mean monthly food intake per day is highest at 30‰ salinity (De Silva & Pereira, 1976: p. 330, Fig. 2) *L. aurata* and *L. ramada* from Marseilles were offered food of varied composition with a vitamin mixture (Vallet *et al.* 1970: p. 610, Table 1; p. 611, Table 2). When kept without food in urea solutions, it was observed by Leray (1970) and Vallet *et al.* (1970) that the urea moderates the weight loss caused by fasting, a concentration of 4 g urea/l being proved most efficient.

Mullet are capable of assimilating nitrogen in the form of urea as well as in the form of proteins and the assimilation of urea seems to be due to bacterial activity (Albertini-Berhaut & Vallet, 1971). Thus, urea may be considered as an economic source of nitrogen to be incorporated in the diet of mullet and its use may be very advantageous in aquaculture, either as an economical source of nitrogen or as a fertilizer in breeding tanks. In India, urea has been used as a general fertilizer for fishponds and artificial manuring, using organic and inorganic manures. It has been found to have a remarkable effect on the growth of mullet in ponds (Kurian, 1975). In Hong-Kong, pig, duck and human excreta are used according to Pillai (1962) and, in the utilization of human manure, toilets are constructed over the fish ponds and the people of the areas are invited to use them.

Fertilization of the ponds is known in the Mex fish farm in Egypt (Wahby, 1974); in Lagos Lake (Sivalingam, 1973) using phosphate fertilizers as in Taiwan (Chen, 1974); superphosphate, ammonium sulfate and poultry manure as in Dor, Israel (Yashouv & Halevy, 1973). Fertilization by sewage effluents in a system of 'bheris' or brackish water ponds of the delta area of West Bengal has been described by Pakrasi, Das & Thakurta (1964). Diluted effluents from the Calcutta sewage works are admitted to increase fertility and develop fish culture of *M. tade* and *M. parsia*. Another sewage system for mullet culture was reported in Auckland, New Zealand, by Slack (1975) and the possibilities of development of industrial mariculture by using the organic waste waters for the fertilization of Tunis Lake were presented by Stirn, Kerzan & Kubik (1974).

Sometimes, natural foods are used for mullet culture. Thus, when reared in Indian ponds, *M. cephalus* is reported to consume considerable quantities of decayed plants. Enormous masses of plant material washed ashore each year after floods could be used as a supplementary food in mullet farms of Chilka Lake fisheries (Jhingram & Natarajan, 1974).

Owing to their dependence upon the availability of physical surfaces to concentrate food available for grazing microflora and microfauna, it was suggested by Odum (1970) that production in mullet ponds could be raised by increasing surface areas of the sediment particles (less than 200 μm), this being easily achieved by putting strips of synthetic (polypropylene) films and periodical addition of very finely chopped macroplant detritus.

Daily food ration

From the point of view of aquaculture, it was shown that in grey mullet culture, a ration of 8% of the body weight is sufficient to obtain almost maximum growth (Vallet *et al.*, 1970). The daily food was equivalent to 5% of the body weight in young *L. aurata* in Marseilles (Albertini-Berhaut & Vallet, 1971), 5–8% in *M. cephalus* in Sri Lanka (De Silva & Pereira, 1976). It was salinity-dependent and the highest food intake was at a salinity of 30‰ and and the lowest at 10‰ according to De Silva and Pereira (1976: p. 330, Fig. 2); in Marseilles, a salinity of 20‰ affords maximum growth and highest efficiency of food turnover in *L. ramada* (Leray, 1970).

The maintenance ration (nitrogen balance) appears to be (at 15 °C) about 3% of the wet weight, corresponding to a daily food energy intake of 0.92 cal for a fish (*L. aurata*) of 10 g and maximum efficiency was observed at intake rates of 6 g wet weight for 100 g weight (Vallet *et al.*, 1970). At 15 °C, young *L. aurata* (10 g) require a daily ration of 1.8 cal for high growth efficiency, a diet rich in proteins being necessary for the growth of young fish (SL = 21 mm). Observations indicate that the fish become most excited when fed a ration below maintenance requirements and that this agitation resulted in general mortality.

Digestive physiology

As assumed by Pereira & De Silva (1978), there is no information available on any aspects of digestion of grey mullet and few papers deal with digestive physiology.

Retention time

Retention time of the ingested material is highly variable. Depending upon the intensity of feeding, it ranges from 2 to 6 h. From a study of a mud detritus mixture containing fluorescent particles, Odum (1970) determined the retention time of food material and concluded that, for *M. cephalus*, it ranged from 2 to 6 h, indicating a turnover rate of five times in 24 h. In young *C. labrosus* from English waters, at 16–17 °C, food pellets of mixed mud and cod roe passed through the gut in 8–9 h. Marais (1976) noticed that the retention time appears to be shorter (3–4 h) in *L. richardsoni*, and in *L. dumerili* (probably between 1 and 2 h).

Assuming a turnover rate of 6 times and 12 times respectively the maximum stomach content, expressed as a per cent of bodymass, amounted to 3.5% in *L. richardsoni* and 5% in *L. dumerili*. Thus, a 100 g specimen of *L. richardsoni* would consume 21 g wet material per day and one of *L. dumerili* about three times more (Marais, 1976). Average retention time for sediment in the mullet was found by Odum (1970) to be 4 or 5 h. For a feeding period of 12 h and a retention time of 4 h, a fish of 200 g fresh weight would eat about 50 g dry sediment/day which corresponds to 1.5 g carbon/day for a sediment in the stomach containing 3% of carbon.

Digestibility and assimilation

Tung (1971) observed that 2% diatoms and 2% copepods are intact in the posterior part of the intestine but diatoms were shown (Payne, 1976) to be readily utilized as food by all species of mullet, their empty frustules being apparent in the rectum contents of the fish. Digestion is not always complete in *C. labrosus* from English waters and, though the animal material is completely used, many of the algal cells still retain their green contents in the rectum (Hickling, 1970).

The digestive tract of *M. cephalus* is efficient at breaking up diatoms but other particles such as plant detritus and the green algae are only partially destroyed and from 20 to 30% of the dinoflagellate (*Kryptoperidinium*) was found to pass through the digestive tract in a viable condition (Odum, 1970). In *L. tricuspidens* from the Swartkops estuary (Marais, 1976) trituration of the larger plant material in the pyloric stomach was very incomplete and digestion of contents of broken cells was not very complete, either in stomach or in the pyloric caecae. In *L. falcipinnis*, the mean organic content in the stomach being 21.94 mg carbon/g dry sediment and 10.60 mg carbon/g dry weight in the faeces, it is thought by Payne (1976) that 51.7% (11.34 mg carbon/g dry weight) of the weight of the original organic matter in the stomach has been assimilated. Pereira & De Silva (1978) presented results on gastric digestion of young (10–50 mm TL) *M. cephalus* in Negombo Lagoon (Sri Lanka). The rate of digestion was found to be salinity-dependent, being slower at lower salinities (< 1‰) than at higher salinities. This is correlated with higher food intake at lower salinities. They have clearly shown (Pereira & De Silva, 1978: p. 385, Table 2) that salinity is an important factor determining gastric digestion, food assimilation and therefore growth; digestion tends to proceed faster with increasing salinity (fastest at 30‰ and slowest at 1‰). This is correlated with higher food intake at the lower salinities; the higher intake at < 1‰ probably leads to a lowering of the rate of digestion, processes which are interrelated with the metabolic cost of osmoregulation. The rate of digestion was also found to be dependent on body size, increasing with increasing body weight.

Hickling (1970) found that assimilation efficiency of *C. labrosus* to be about 45% while Payne (1976) showed that the assimilation efficiency of *L. falcipinnis* was 52%. However, it would be advantageous for omnivorous Mugilidae to consume as much animal material as possible and thereby increase their assimilation efficiency. *L. tricuspidens* ingested food material with significantly lower ash percentage (22.09%) and more protein (6.22%)

and carbohydrates (7.59%) than other species. *L. dumerili* ingested food with the highest ash (49.6%) and lowest protein (3.6%) and carbohydrate (3.3%) percentages (Marais, 1976). However, irrespective of species, the youngest and smallest (< 15 cm) consumed food of higher feeding values (highest protein and lowest ash content) correlated with their more efficient selection of smaller particles (the smaller the particle, the higher its relative food content as indicated by Odum, 1970).

The conversion efficiency (CE) which was defined as the percent of dry food material that is converted into fish flesh, decreases with time and growth according to De Silva & Pereira (1976); in *M. cephalus*, high food intake results in a low CE and vice-versa. The nature of the protein incorporated in the food does not seem to have any influence on the conversion ratio (Vallet *et al.*, 1970).

The most efficient transfer of energy from primary production to fish occurs through the fish herbivores (Odum, 1970) and when efficiency of assimilation was assumed to be 80% for carnivores and 60% for herbivores, it was 70% for detritus feeders (Miller *et al.*, 1971 & Tsikan-Lukana, 1970, in Edwards, 1973*b*) which is a conspicuous advantage for mullet.

In summary and according to Odum (1970), the maximum power output for *M. cephalus* results from a relatively large and continuous ingestion rate and short retention time coupled with an assimilation efficiency which is probably higher for components of the diet such as diatoms, but rather low when the entire bulk of ingested matter is considered.

Ecology of the feeding grounds

Apart from temperature and salinity, another factor which might limit the distribution of the species of mullet could be the availability of suitable food. Odum (1970) is of the opinion that the most important primary producers used by mullet are marsh grasses (*Spartina*), sea grasses (*Zostera*, *Thallassia*), mangrove (*Rhizophora*) and attached micro-algae and benthic microflora (benthic diatoms, dinoflogellates and filamentous green and blue-green algae) of various ecosystems. According to Capanna *et al.* (1974), a filtering system such as the pharyngobranchial organ constitutes a considerable advantage for colonization of new habitats by mullet. They particularly colonize productive waters such as those of lakes, saline areas and estuaries. Thus, suitable feeding grounds are known in the following locations.

The sea

Along the shores of South Texas (Moore, 1974), Australia (Moriarty, 1976), Mauritania and all around the islands of Banc d'Arguin (Brulhet, 1975), on intertidal flats with *Spartina* in Carolina (Cain & Dean, 1976), in bays and shallow waters of the Caspian and Aral seas (Zenkevich, 1963) and of the Azov sea (Savchuk, 1973, 1976), on coastal marine ecosystems of Venezuela (Edwards, 1973) and in the mangroves of Madagascar (Kiener, 1976).

Littoral lagoons

The rich plant and animal microorganisms living on the bottom of many lakes which are connected with the sea have been described as providing a rich source of suitable food for the mullet, in the southern Mediterranean sea: Tunis and Ichkeul lakes in Tunisia (Pillai, 1975; Farrugio, 1976), Lake Borullus and Egyptian delta lakes (El Maghraby, Hashem & El Sedfy, 1974), in tropical Africa: Lagos lagoon in Nigeria (Fagade & Olaniyan, 1973, 1974) Sakumo lagoon in Ghana (Pauly, 1975) and South Africa St Lucia Lake (Blaber, 1976) and in India: Lake Chilka (Rajan, 1964).

Estuaries

The most studied are the Black Johnson estuary of the Sierra Leone River (Payne, 1975, 1976) and the estuaries in Natal (Wallace, 1973), in Natal and Pondoland (Blaber, 1977) and the Swartkops estuary near Port Elizabeth, South Africa (Masson & Marais, 1975; Marais, 1976). Thus, in the Swartkops estuary, different species of mullet were encountered at or near areas where substrates containing food items of their distinct choice were abundantly present (Marais, 1976).

Some of these areas function as 'nursery habitats', 'nursing grounds' or 'mullet nurseries', with extensive utilization of the shallower waters by fish during the juvenile phase of their life cycles, these rich feeding grounds or 'fattening grounds' (Wallace, 1973; Savchuk, 1968) being therefore completely dominated by mullet (larvae, postlarvae, fingerlings, under-yearlings...).

The fry of the mugilid species generally migrate from the sea where they are born, towards the coast, reaching some bays and moving into the shallow waters among the eel grass and algae in which they are hidden from enemies and in which they find abundant microzoobenthos (harpacticoid copepods, amphipods...) and later micro-phytobenthos (algae, diatoms...) very suitable for early development and growth. They generally feed on a particularly fine filtrate and sedimental ground which is a determining factor in colonizing special alimentary substrata (Capanna *et al.*, 1974).

Feeding migrations

These are known in Florida (Tabb & Manning, 1961), on the shores of the northwestern part of the Black sea from the Dnieper to the Danube estuaries

(Savchuk, 1968, 1973) and in other countries. As asserted by Thomson (1955), mullet are 'trophically brackish water fish' and fry movements are 'trophic migrations'; thus they generally enter coastal lagoons and brackish ponds for a 'trophic life' and fast growth during the first years of life (0 group) in the Venetian 'valli' (De Angelis, 1959, 1960, 1967), in lakes (Edku, Borullus...) of the Nile delta (El Maghraby *et al.*, 1974); in the 'reservoirs à poissons' of Arcachon, in the Bay of Biscay (Stequert, 1972; Lasserre Renaud-Mornant & Castel, 1975), in the Berre lagoon near Marseilles (Ezzat, 1965) and in lagoons of the Gulf of Lion (Herve, 1978).

Such a feeding movement occasionally spreads out toward fresh water. Thus, Johnson & McClendon (1970) indicated an upstream migration of striped mullet (*M. cephalus*) in the Colorado River.

Ethology of feeding and feeding periodicity

The first to describe the behaviour of mullet during their feeding were Tabb & Manning (1960). They observed that *M. cephalus* and *M. curema* from Florida bays create dense 'muds' when feeding on muddy bottoms and that they 'stand on their heads' to graze the bottom and rotate slowly in a clockwise direction. As they feed in this manner, the expulsion of inedible material is jetted out through their gill chambers and deposited around the perimeters of the depressions. Feeding on soft, bare mud bottoms, they create conical depressions 18 to 20 in in diameter and 2 to 4 in in depth which serve to concentrate flocculent materia. However, such an interpretation was thought to be probably incorrect by Thomson (1966).

Mullet have three ways of obtaining food (Hickling, 1970). In the first, large edible items are ingested by a sudden extension of the protrusible mouth into a circular sucker and simultaneously a powerful pumping of water creates a suction current. Another feeding process involves scraping to skim off fine algal bottom and browsing on microalgae attached to submerged surfaces. Finally, they can stir up a cloud of mud and filter and concentrate the finer fraction of the bottom substrate on which they feed. For sediment feeding, the mullet inclines its body at an angle of 15 to 30 degrees to the surface of the sediment and extends the premaxillaries so that particles are picked up in one of two ways: either by taking up small mouth-fulls at random or by skimming along the bottom with the lips barely in contact with the sediment and sucking up the uppermost layer (Odum, 1970). Any material found unsuitable (usually large particles) is expelled in a small cloud; consequently the presence of mullet is often characterized by a trail of these clouds behind feeding fish. According to Odum, (1968), *M. cephalus* feeds either by sucking up the surface layer of the mud or by grazing on submerged surfaces such as rocks, and *Zostera* leaves. In one African lake, mullet fed mainly on bottom deposits (deposit feeding and detritus feeding) by stirring up the bottom and filtering particles brought into suspension with their gill-rakers, as shown by Fagade & Olaniyan (1973). The role of visual stimuli in feeding behaviour was experimentally studied by Olla & Jamet (1974). As the fish began to graze, they would tilt their heads down towards the sand, ingesting particles of sand and food. The fish tended to remain grouped, feeding together in the same general area. It was apparent that there was a strong social interaction between fish during feeding and that social stimuli (visual type) play a role in facilitating feeding responses. When isolated *M. cephalus* viewed a feeding group the initiation of feeding was greatly facilitated and when viewing a non-feeding group, they show inhibition of feeding. The adaptative and selective advantage of schooling for feeding was thus discussed.

Owing to the micro-benthic way of life, mullet spend a lot of energy in search of food. This behaviour is reflected in a basal rate higher than that in other teleost fish (David, 1968, David & Warren, 1968 in Vallet *et al.*, 1970).

Feeding periodicity

Two types of feeding rhythms have been determined, feeding varying according to a seasonal pattern and to a daily pattern.

Seasonal variations

M. cephalus do not feed at the same intensity throughout the year, as suggested by Hiatt (1944) and Luther (1962). Thus, the feeding intensity of *M. cephalus* from Negombo Lagoon (Sri Lanka) followed a seasonal pattern with the intensity increasing from April and reaching a peak during June–August and thereafter showed a downward trend (De Silva & Wijeyaratne, 1977). In Lake Vrana, Homen (1976) states that Adriatic mullet break their feeding under unfavourable conditions and in Florida Bay, it was shown by Tabb & Manning (1961) that *M. cephalus* and *M. curema* feed most actively in these areas throughout the summer and autumn, then move out with the first cold weather. Mullet with empty digestive tracts were also found during extreme weather conditions, in Georgia when there was a great deal of turbidity and turbulence (Odum, 1970), in the Bay of Biscay with the winter cold (Le Dantec, 1955) and in the Sea of Marmara during the winter (Erman, 1959).

In South Carolina, Cain & Dean (1976) have shown that species (*M. cephalus*, *M. curema*) occupying the intertidal area of *Spartina* marsh shift their energy sources from season to season. In an estuary of Sierra Leone a discernable change in the diet of mullet was recorded by Payne (1976), between the wet and dry seasons: during the rains, the gut contents were composed almost entirely of detritus and benthic diatoms with sand, whilst in the dry season the proportion of detritus declined and was replaced by a corresponding increase in the quantity of blue-green algae (filamentous species). In this estuary, mullet tend to be most abundant as the rains are finishing, when the organic matter in the sediment is at a maximum (Payne, 1975). In the Avon-Heathcote estuary of New Zealand, it was shown by Webb (1973) that the yellow-eyed mullet *Aldrichetta forsteri* were found to be filled with mud during the winter months and that algae (*Ulva* and *Enteromorpha*) increased in abundance from mid-October, with a marked increase in the utilization of these algae during summer months. Similarly, the proportion of 'full' stomachs of *C. labrosus*, from English waters, appears to be highest in the late summer and autumn months (Hickling, 1970) blue-green algae were most abundant in the gut contents of *M. cephalus* from Australia during the spring (Oct.–Nov.) according to Moriarty (1976) but at all times diatoms were the dominant algae ingested.

No seasonal changes in diet were noted in *M. cephalus* and *L. macrolepis* from Natal (Blaber, 1977) although terrestrial plant debris may be more available after heavy floods during the rainy season. Such an absence of significant seasonal variation in the diet of any of the mullet from St Lucia Lake with the exception of Cyanophyta and green flagellates (Blaber, 1976), corresponds with the situation reported for *M. cephalus* in Hawaii (Hiatt, 1944) and *M. cephalus* and *L. macrolepis* in India (Luther, 1962).

Finally, little feeding was observed during spawning migration and it is generally reported that feeding stops with sexual maturity, stomachs being empty during reproduction (Ghazzawi, 1933; Kesteven, 1942; Odum, 1970; Tung, 1971; Brulhet, 1975).

Daily rhythms

Some rhythms were observed according to the tidal cycle, others according to the photoperiodic cycle and others according to a combination of these. Under normal conditions, mullet appear to feed almost continuously. However, the intensity of feeding is not always the same. In areas which are influenced by tidal exchange, Odum (1970) showed that there is a definite relationship between the rate at which food is ingested and the state of the tide, with a marked increase in the amount of food ingested as the tide rises, when optimal feeding areas become accessible to the mullet (Odum, 1970: p. 232, Fig. 2).

It was noticed by Marais (1976) that, for *L. dumerili* and *L. richardsoni* from the Swartkops estuary, feeding intensity was more pronounced during the day, the period of lowest feeding intensity being at low tide during the night; *L. richardsoni* was considerably influenced by *tidal movement* and showed a larger decrease than *L. dumerili* in ingestion rate as low tide was approached (especially when low tide occurred during the night).

Moriarty (1976: p. 138, Table VI) has shown that mullet feed only during the day and not at night on the Australian sea-grass flats: no mullet caught at night were feeding, indeed all had empty guts whereas most mullet caught during the day had full guts, although occasionally one or two were found with empty or half empty guts at midday. Those caught in the morning had sediment in the stomach and anterior part of the intestine, indicating that they had begun feeding a little before capture but had not fed over night. Fish caught in the evening had sediment in the posterior part, indicating that they had stopped feeding some time before capture. A further daily rhythm of feeding was recorded in Adriatic mullet (Homen, 1976): in Lake Vrana, different species fed intensively by daylight, particularly in the morning, whereas at night they break their feeding. A similar point of view was adopted by Blaber (1976: p. 272, Fig. 2) who showed that *L. macrolepis* and *L. dumerili* feed predominantly during daylight hours at St Lucia Lake and that in *M. cephalus* the peak feeding time for about 18 h. However, *M. cephalus* and *L. dumerili* continue feeding at night but at a lower intensity and *L. macrolepis* ceases feeding at night. *M. cephalus* from the coastal Negombo lagoon (Sri Lanka) was also found to show diurnal periodicity in feeding activity (De Silva & Wijeyaratne, 1977). Fish do not appear to feed continuously and peaks of activity occurred at dawn and around midday, when the light intensity is highest; they were unrelated to the state of the tide. A diurnal feeding rhythm was still evident in artificial breeding in 29–32 day-old larvae of *M. cephalus* reared in Taiwan and fed with *Artemia* (Liao *et al.*, 1974 in Nash & Kuo, 1974).

An opposing point of view was presented by Tabb & Manning (1962). From ecological observations, they concluded that feeding activity of *M. cephalus* and *M. curema* in Florida Bay was during the night: fish would move to shallow waters in the evening and visit muddy depressions many times each night, retreating to deeper channels during the daylight hours. Such a pattern was, however, never confirmed.

Feeding and pollution

It is well known that fish concentrate various elements from their surrounding water and thus concentrate them into the food chain. Mullet being detritus and sediment feeders and this sediment sometimes being mixed with waste industrial by-products as shown by Webb (1973), contamination by pollutants has been observed.

The ingestion of sediment which could be enriched with heavy metals as a result of industrial activity would probably lead to the ingestion of greater quantities of heavy metals. Thus, in Tasmania, the zinc concentration in *M. cephalus*, expressed on a dry matter basis, was 13.3 ppm, with the highest concentration at 38.9 ppm in some individuals, higher than the concentration in fish with pelagic feeding habits (Eustace, 1974), the position of the fish in the food chain appeared to be an important factor determining its mercury content. Establier (1975) found that the levels of mercury in sediment from

Cadiz Bay was 0.12–0.18 ppm and the concentration in mullet (*M. cephalus*) was below 0.12 ppm but he was not able to establish a direct relationship according to trophic activity of mullet. In contrast, in mullet from the Tyrrhenian coast, Renzoni, Bacci & Falciai (1973) found that total mercury concentration was (mg/kg wet weight) 0.39 ppm in muscle, 5.36 in liver and 4.71 in kidney, whilst in the sediment, mercury concentrations were 0.10 to 2.72 mg/kg wet weight. In mullet from Tasmania, the results of Ratkowsky, Dix & Wilson (1975) do not support any hypothesis which suggests that bottom feeders may have higher mercury contents than free swimmers.

Sidhu, Vale, Shipton & Murray (1972) indicated that Australian mullet (*M. cephalus*), being a detritus feeder with relatively high body fat is likely to take up petroleum hydrocarbons more readily than other species of fish living in the same environment, while bottom mud obtained from waters close to oil refineries impart a kerosene-like taint. The kerosene-like taint of these mullet is due to the presence of hydrocarbons in their flesh and originates from polluted sediments (Connell, 1971).

Accumulation of a pollutant (caesium) and its concentration in foods (algae) used by mullet according to trophic relationships was shown by Isaacs (1975) and that of another pollutant (Sr^{85}) from polluted sediment by Amiard (1975).

The problem of pollution by pesticids seems particularly interesting. This, Fokakiv & Rand (in Kleerekoper, 1974) indicates that *M. cephalus* which normally displays positive rheotaxis become negatively rheotactic after short-term exposure to 0.016 ppm parathion. The consequence of such a change in rheotaxis for the capability to localize food may be important from an ecological point of view.

Feeding and parasitism

Mullet are often infected with parasites by a passive means when eating larvae generally in the encysted form. This is particularly evident with helminths, such as Trematoda. Thus, Maillard (1976), Fares & Maillard (1974, 1975) described in French Mediterranean lagoons the infection of *M. cephalus*, *L. ramada* and *Chelon labrosus* with Trematoda, Haploporidae and Haplosplanchnidae by eating cysts (encysted metacercariae).

In a study of the parasites of *L. aurata*, Reshetnikova (in Skinner, 1975) pointed out the difference in the types of parasites inhabiting juvenile and adult hosts: pelagic young were infected by parasites picked up with the food, planktonic Copepoda which serve as intermediate hosts. These parasites become scarcer in bottom feeding adults and eventually disappear in hosts that have been feeding on detritus for a long time. In Florida, young *M. cephalus* generally had different parasites than old ones. Unlike the young, the bottom-feeding adult comes into contact with parasites (Strigeoid and

Heterophyid metacercariae) which use shallow-water invertebrates as intermediate hosts (Skinner, 1975). Another group of parasites infecting adult mullet by way of their plant diet is the Digenea (Haplosplanchnidae) whose larvae encyst on vegetation.

Trophic relationships and interspecific competition

The main features of the feeding relationships in various ecosystems have been presented in the Black Sea (Zenkevich, 1963: p. 461, Table 188) in Lagos Lagoon (Fagade & Olaniyan, 1973: p. 224, Fig. 5), in Sakumo Lagoon, Ghana (Pauly, 1975: p. 58, Fig. 12) and in the Malgachian mangrove (Kiener, 1976: p. 633, Table 2). It is, indeed, of great interest to evaluate the contribution of the plankton, the benthos and the relation to the transport of energy and nutrients through the estuarine and lagoonal ecosystems. The Sakumo lagoon ecosystem is characterized by an energy storage in the form of organic sediment that is the food of many animals which represent most of the biomass and among them, the mullet (Pauly, 1975). Odum (1970) has suggested for mullet, two pathways, the grazing food chain, in which living plants are the primary energy source and the detritus food chain in which dead and decaying organic material are the energy source. Thus *M. cephalus* occupies a position between the two food chains and is able to utilize whichever is the easiest to exploit in a given situation. It was also shown by Odum (1970) that the change in the diet of mullet, from carnivorous to herbivorous when diatoms become the main primary source of the diet, corresponds with a 'telescoping of the food chain' (Hiatt, 1944). Odum (1968) believes that, for a 'broad spectrum' herbivore such as *M. cephalus*, food does not seem to be a limiting factor and that grey mullet seem to have evolved a diet in which they suffer no competition from other fish. The ability to ingest and to utilize benthonic and ephiphytic diatoms, dinoflagellates, protozoans and green as well as blue-green algae with macroplant detritus, ensures a constant energy source, without interferring in the growth of other fish in the same area (Oren, 1971). Interspecific competition seems, however, to be evident because the diets are the same, either between mullet of different species (*M. cephalus* and *L. macrolepis* in India, Luther, 1962), or between mullet and other fish (*Tilapia*) in Lagos Lagoon, Fagade & Olaniyan, 1973: p. 224, Fig. 5) which compete for the same food items. According to Moore (1974), in addition to detritus and inorganic substances which occupied about 90% of their stomach volume, *M. cephalus* and *M. curema* feed on the same type of diatoms, unicellular green algae and filamentous blue-green algae. Studies by Odum (1970) indicate that both species are opportunistic feeders with little difference in either their mode of eating or their choice of food. It is possible for the two species to feed on the same materials at the same time but differ markedly in assimilation efficiency (Moore, 1973). This point is important in determining the degree of adaptation of each species to any environment.

The interspecific relationships among sympatric Mugilidae have been studied by Blaber (1976, 1977). It was suggested (Blaber, 1976) that the competition for food is reduced by substrate particle size selection and perhaps by differences in feeding periodicity as shown for ten species of mullet in the St Lucia Lakes system of South Africa. Thus, interspecific competition may be reduced because peak feeding hours of the three dominant species are different at St Lucia and also by the particle size preference of each species. *L. alata* only feeds in areas when the sand grains are about 500 μm in diameter and thus does not compete with the other nine species of the lake. However, competition for food may occur between *M. cephalus*, *L. macrolepis* and *L. dumerili* and between four species of *Valamugil* since groupings co-exist in the same areas and share food resources. Co-existence may be possible in St Lucia, due to a superabundance of certain foods, *Assimea* and diatoms being present in excess (Blaber, 1976). In Natal and Pondoland estuaries (Blaber, 1977), fish all fed in the same area and the food items ingested by individuals of the four species were similar (Blaber, 1977: p. 267, Table 4) although *Valamugil* species consumed more particulate organic matter than *L. macrolepis* and *L. dumerili*, while *L. dumerili* was the only species that ingested very few flagellate Protozoa, Cyanophyta and Ostracoda. Moreover, each species ingests a wide range of particles sizes (Blaber, 1977: p. 268, Table 5) and, consequently, there is a minimum of overlap between the species. Within any one ecosystem most species of mullet avoid interspecific competition by selecting particles of different sizes (Blaber, 1977: p. 272, Table 6). Competitive interactions may occur among the *Valamugil* species which feed on substrates of a similar size but the food items ingested by this genus, mainly pennate diatoms and particulate organic matter, are nearly always present in excess, thereby ensuring that competion is reduced to a minimum.

Most species of Mugilidae are sufficiently flexible in their feeding ecology to cope with a variety of substrates and still avoid serious interspecific competition by partitioning the resources according to substrate particle size. Marais (1976) showed that, in the Swartkops estuary, *L. tricuspidens*, the least abundant mullet species consumed the most nutritious diet (fibrous macroplant material) whereas *L. dumerili*, the most abundant species, ingested food material with the lowest nutritional value (diatoms), this 'lower quality food' material being available in large quantities, whereas the amount of higher quality food material was limited.

All eleven species of mullet from southeast African estuaries are initially planktonic feeders. In spite of the similarity of diet of the various species of juveniles (10–59 mm SL), it was shown by Blaber & Whitfield (1977) that interspecific competition is unlikely because of three factors: (*a*) shoals consist of a single species of one length class, (*b*) species enter estuaries at different times according to the period of adult spawning, and (*c*) as initial growth is rapid, the changeover to the adult mode of feeding takes place within a few weeks.

Food and feeding

In the mixohaline and shallow waters of the reservoirs in the Bassin d'Arcachon, France, which constitutes a rich lagoonal ecosystem, it was shown by Lasserre *et al.* (1975) that the rich input of detritus, organic matter, bacteria and benthic microflora plays a prominent role in the food chain and that meiofauna and grey mullet compete for food, both feeding on detritus. The most abundant species of meiofauna were nematodes and harpacticoid copepods but ciliates, oligochaetes, tubellarians, ostracods, polychaetes and cnidarians are also fairly abundant, the number of all together (< 0–5 mm) being 12.6×10^6 individuals/m^2 in July when the meiofauna biomass reaches 13.9 g dry weight/m^2. Results of Lasserre *et al.* (1975) indicate that benthic invertebrates are important consumers of detritus which is accumulated from late autumn through winter and which provides an abundant food source on which luxurious development of meiofauna will occur. They also show that spring and summer growth retardation of *C. labrosus* can be explained in terms of a trophic competition between meiofauna and grey mullet, both communities being sustained primarily by feeding on detritus. The very high densities of meiofauna indeed consume much more energy in the form of organic matter (detritus + microflora + bacteria) than does the macrofauna. The energy store formed by detritus is dissipated by the meiofauna at relatively high rates and these chief bottom consumers in coastal lagoons compete with the grey mullet for the benthic food available. Significant amounts of detritus can therefore be lost by way of the meiofauna community through its high metabolic activity. Yashouv (1972) assumed that, in Israeli fish ponds, polyculture (grey mullet, *M. cephalus*, *L. ramada* + carp, *Cyprinus carpio* + tilapia, *Tilapia aurea*) give the best results; a beneficial effect, named '*synergism*', on the growth of mullet was observed. The fish seem to be influenced by mutual stimulation in their competition for food with the other fish groups in the ponds.

Finally, it was pointed out by Capanna *et al.* (1974) that mullet are capable of colonizing waters where polluted organic deposits have been registered and which are no longer suitable for other species of fish. Moreover, specific structural microvariations in oral morphology (teeth, papillae), in pharyngobranchial structures and in pyloric appendix anatomy offer a major possibility of exploitation of the same habitat by optimal utilization of very specialized niches.

Conclusion

Mullet feeding is of very great interest because these widespread and cosmopolitan fish are able to find their energy directly from the first and lowest trophic level. They are highly successful fish which utilize both benthic microplant material and macroplant detritus, being dependent on food chains derived from two main energy sources: detritus and direct consumption of diatoms, flagellates and other microorganisms and bacteria of the sand and mud according to Odum (1970). In general, mullet ingested whatever organic food of suitable size is common in the area in which feeding takes place (Marais, 1976) and few other fish can live on such poor material (Hickling, 1970).

Prospects in mullet rearing and breeding and in mullet farming are widespread, in temperate, subtropical and tropical areas, where these fish are naturally abundant. They constitute an available source of proteins and a highly rated food. The importance of grey mullet for artificial rearing was indicated by the first international symposium on aquaculture of grey mullet, in Haifa, June 1974. Oren expressed the opinion that *M. cephalus*, the universally successful species, was probably the first marine fish to be cultured. This species indeed appeared to have the highest potential and to be the most suitable fish; its high growth rate coupled with its high energy content and calorific value as well as its morphological adaptability to change in diet, make it an ideal fish of very suitable economic value for artificial rearing. The interest of studies on food and feeding habits for aquaculture, particularly for polycultural practices, is well known, but as pointed out by Oren (1975), its food and feeding habits at various stages of growth are still little understood. Much recent research, however, provides useful information which should be of great value to aquaculturists.

References

Less recent references are quoted from:

Pillay, S. R. (1972). A bibliography of the grey mullets, family Mugilidae. *FAO Fisheries Technical paper* no. **109**, Rome.

Alvarez-Lajonchere, L. (1974). Adiciones a la bibliografia de la familia Mugilidae (Teleostei) publicada por FAO en 1972. *Ciencias, Inv. marinas, Univ. La Habana*, **8**(13), 1–116.

Alvarez-Lajonchere, L. (1976). Segunda adicion a la bibliografia de la familia Mugilidae (Teleostei) publicada por FAO en 1972. *Ciencias Inv. marinas, Univ. La Habana*, **8**(23), 1–80.

Albertini-Berhaut, J. (1973). See Alvarez-Lajonchere, 1976: 008.

Albertini-Berhaut, J. (1974). Biologie des stades juvéniles de Téléostéens Mugilidae *Mugil auratus* R. 1810, *Mugil capito* C. 1829 et *Mugil saliens* R. 1810. II. Modifications du régime alimentaire en relation avec la taille. *Aquaculture*, **4**, 13–27.

Albertini-Berhaut, J. (1975). Allométrie chimique et changement de régime alimentaire chez *Mugil capito* (Téléostéen, Mugilidae). *C.R. Acad. Sci. Paris*, **280**, 1297–300.

Albertini-Berhaut, J. & Vallet, F. (1971). See Alvarez-Lajonchere, 1974: 009.

Alvarez-Lajonchere, L. (1976). *Contribucion al estudio del ciclo de vida de Mugil curema (Pisces, Mugilidae). 5.* Ciencias, Series 8, Investigaciones Marinas, no. 28, Univ. Habana.

Anon. (1976). Marine biological studies. In *Israel Oceanogr. and Limnol. Res., Ann. Rep. 1973–75.*

Blaber, S. J. M. (1976). The food and feeding ecology of Mugilidae in the St. Lucia lake system. *Biol. J. Linn. Soc.* **8**, 267–77.

Blaber, S. J. M. (1977). The feeding ecology and relative abundance of mullet (Mugilidae) in Natal and Pondoland estuaries. *Biol. J. Linn. Soc.* **9**, 259–75.

Blaber, S. J. M. & Whitfield, A. K. (1977). The feeding ecology of juvenile mullet (Mugilidae) in south-east African estuaries. *Biol. J. Linn. Soc.* **9**, 277–84.

Bruhlet, J. (1975). Observations on the biology of *Mugil cephalus ashentensis* and the possibility of its aquaculture on the Mauritanian coast. *Aquaculture*, **5**, 271–81.

Cain, R. L. & Dean, J. M. (1976). Annual occurrence, abundance and diversity of fish in a South Carolina intertidal creek. *Mar. Biol.* **36**, 369–79.

Capanna, E., Cataudella, S. & Monaco, G. (1974). The pharyngeal structure of Mediterranean Mugilidae. *Monit. Zool. Ital.* **8**, 29–46.

Chen, T. P. (1974). See Alvarez-Lajonchere, 1976: 147.

Chervinski, J. (1976). Growth of golden grey mullet (*Liza aurata* R.) in saltwater ponds during 1974. *Aquaculture*, **7**, 51–7.

Ching, C. V. (1977). Studies on the small *Liza malinoptera* V. *J. Fish. Biol.* **11**, 293–308.

Connel, D. W. (1971). See Alvarez-Lajonchere, 1974: 140.

De Angelis, C. M. (1959). See Pillay, 1972: 0290.

De Angelis, C. M. (1960). See Pillay, 1972: 0291.

De Angelis, C. M. (1967). See Pillay, 1972: 0292.

De Silva, S. S. & Wijeyaratyne, M. J. S. (1977). Studies on the biology of young grey mullet, *Mugil cephalus* L. II. Food and feeding. *Aquaculture*, **12**, 157–67.

Edwards, R. R. C. (1973*a*). Production ecology of two Caribbean marine ecosystems. I. Physical environment and fauna. *Estuarine and Coastal Mar. Sci.* **1**, 303–18.

Edwards, R. R. C. (1973*b*). Id. II. Metabolism and energy flow. *Estuarine and Coastal Mar. Sci.* **1**, 319–33.

Egusa, S. (1950). See Alvarez-Lajonchere, 1974: 175.

El Maghraby, A. M., Hashem, M. T. & El Sedfy, H. M. (1974). Species composition and distribution of the mullet fry in Lake Borullus. *Bull. Inst. Oceanogr. Fish. ARE*, **4**, 3–31.

Erman, F. (1959). See Pillay, 1972: 0352.

Establier, R. (1975). Contenido en mercurio de las anguilas (*Anguilla anguilla*) de la desembocadura del rio Guadalquivir y esteros de las salinas de la zona de Cadiz. *Invest. Pesq.* **39**, 249–55.

Eustace, I. J. (1974). Zinc, cadmium, copper and manganese in species of finfish and shellfish caught in the Derwent estuary, Tasmania. *Aust. J. mar. freshwater Res.* **25**, 209–20.

Ezzat, A. (1965). See Pillay, 1972: 0369.

Fagade, S. O. & Olaniyan, C. I. O. (1973). The food and feeding interrelationship of the fishes in the Lagos lagoon. *J. Fish. Biol.* **5**, 205–25.

Fagade, S. O. & Olaniyan, C. I. O. (1974). Seasonal distribution of the fish fauna of the Lagos lagoon. *Bull. Inst. Francais d'Afrique Noire*, **36**, 244–52.

Fares, A. & Maillard, C. (1974). Recherches sur quelques Haploporidae (Trematoda) parasites des muges de Méditerranée occidentale: systématique et cycles évolutifs. *Z. Parasitenkd.* **45**, 11–43.

Fares, A. & Maillard, C. (1975). Cycle évolutif de *Haplosplanchnus pachysomus* (Trematoda, Haplosplanchnidae) parasite de Mugilidés (Teleostei) *Bull. Mus. Nat. Hist.* **312**, 837–44.

Farrugio, H. (1976). Premières observations sur le régime alimentaire des muges des lacs Tunisiens. *Rapp. P.-V. Comm. Int. Explor. Mer. Méditerr.* **23**, 45–6.

Ghazzawi, F. M. (1933). See Pillay, 1972: 0449.

Ghosh, A. N., Das, P. R. & Das, L. K. (1974). Experimental observations on the food requirement of fry of *Mugil parsia* H. In: *Coastal Aquaculture in the Indo-Pacif. Region*, ed. T. V. R. Pillay, 429–37.

Hervé, P. (1978). Ichthyofaunes comparées de deux étangs littoraux du Roussillon: Canet-Saint-Nazaire et Salses-Leucate. Ecologie générale et Biologie de diverses espèces de Poissons. Thesis *Univ. Pierre et Marie Curie*, Paris.

Hiatt, R. W. (1944). See Pillay, 1972: 0543.

Hickling, C. F. (1970). See Pillay, 1972: 0545.

Homen, Z. (1976). A contribution to the knowledge of the Adriatic mullet nutrititon. In *Proceedings of the Twenty-fifth Congress CIESM*, Split, Oct. 1976.

Houde, D., Berkeley, S., Klinovsky, J. & Schekter, R. (1976). Culture of larvae of the white mullet, *Mugil curema* V. *Aquaculture*, **8**, 365–70.

Isaacs, J. D. (1975). Assessment of man's impact on marine biological resources. In *Marine Pollution and Marine Waste Disposal*, (ed. Perrson, E. A. & De Fraja-Frangpane, E.), *Pergamon Press*, Oxford.

Jhingram, V. G. & Natarajan, A. V. (1974). See Alvarez-Lajonchere, 1976: 265.

Johnson, D. W. & McClendon, E. L. (1970). See Pillay, 1972: 0613.

Kesteven, G. L. (1942). See Pillay, 1972: 0691.

Kiener, A. (1976). Spectre des niches écologiques des principaux groupes de poissons de la mangrove malgache. *Rev. Trav. Inst. Pêch. Marit.* **40**, 632–5.

Kimor, B. (1974). Euryhaline elements in the plankton of the Bardawil lagoon (North Sinai). In *Proceedings of the Twenty-fourth Congress CIESM*, *Monaco*.

Kleerekoper, H. (1974). Effect of exposure to a subacute concentration of parathion on the interaction between chemoreception and water flow in fish. In *Pollution and Physiology of Marine Organisms*, ed. Vernberg and Vernberg, Academic Press, New York.

Kreft, G. & Michaelis, H. (1976). Die Meeräche im niedersächsischen wattenmeer. *Natur. Mus.* **106**, 23–9.

Kuo, C. M., Shehadeh, Z. H. & Milisen, K. K. (1973). A preliminary report on the development, growth and survival of laboratory reared larvae of the grey mullet, *Mugil cephalus*. *J. Fish. Biol.* **5**, 459–70.

Kurian, C. V. (1975). See Alvarez-Lajonchere, 1976: 289

Lasserre, P., Renaud-Mornant, J. & Castel, J. (1975). Metabolic activities of meiofaunal communities in a semi-enclosed lagoon. Possibilities of trophic competition between meiofauna and mugilid fish. In *Proceedings of the Tenth European Symposium on Marine Biology*, *Ostend*, part 2, 393–414.

Le Dantec, J. (1955). See Pillay, 1972: 0728.

Leray, C. (1970). Experimental approaches to artificial feeding of some sea fish. In *Report on the* 1970 *Workshop on Fish Feeding Technology and nutrition*, FAO EIFAC, US Department Interior, no. 102, 169–71.

Liao, I. C., Lu, Y. J., Huang, T. L. & Lin, M. C. (1974). Experiments on induced breeding of the grey mullet, *Mugil cephalus* L. In *Coastal Aquaculture, in the Indo-Pacific region*, ed. T. V. R. Pillay, 213–43.

Luther, G. (1962). See Pillay, 1972: 0762.

Maillard, Cl. (1976). Distomatoses de poissons en milieu lagunaire. Thesis, Montpellier, AO 12304.

Marais, J. F. K. (1976). Comparative studies on the nutritional ecology of mullet in the Swartkops estuary. *Thesis, Univ. Port-Elizabeth.*

Masson, H. & Marais, J. F. K. (1975). Stomach content analyses of mullet from the Swartkops estuary. *Zool. Africana*, **10**, 193–207.

Moore, R. H. (1973). Age, growth, respiration and general ecology of the mullet: *Mugil cephalus* and *Mugil curema* on the South Texas coast. *Thesis, Univ. Austin, Texas.*
Moore, R. H. (1974). General ecology, distribution and relative abundance of *Mugil cephalus* and *Mugil curema* on the South Texas coast. *Contrib. Mar. Sci.* **18**, 241–55.
Moriarty, D. J. W. (1975). A method for estimating the biomass of bacteria in aquatic sediments and its application to trophic studies. *Oecologia*, **20**, 219–29.
Moriarty, D. J. W. (1976). Quantitative studies on bacteria and algae in the food of the mullet *Mugil cephalus* L. and the prawn *Metapeneus bennettae*. *J. exp. Mar. Biol. Ecol.* **22**, 131–43.
Morovic, D. (1976). On certain theoretical and practical problems in raising fish of the Mugilidae family. *Poljopr. Znan. Smotra*, **37**, 47–56 (in Serbo-Croate, Italian summary).
Nash, C. E. & Kuo, C. M. (1975). Hypothesis for problems impeding the mass propagation of grey mullet and other finfish. *Aquaculture*, **5**, 119–33.
Nash, C. E., Kuo, C. M. & McConnell, S. C. (1974). Operational procedures for rearing larvae of the grey mullet, *Mugil cephalus* L. *Aquaculture*, **3**, 15–24.
Odum, W. E. (1968*a*). See Pillay, 1972: 0884.
Odum, W. E. (1968*b*). See Pillay, 1972: 0885.
Odum, W. E. (1970). See Pillay, 1972: 0886.
Olla, B. L. & Samet, C. (1974). Fish-to-fish attraction and the facilitation of feeding behaviour by visual stimuli in striped mullet, *Mugil cephalus*. *J. Fish. Res. Board Can.* **31**, 1621–30.
Oren, O. H. (1971). See Alvarez-Lajonchere, 1974: 400.
Oren, O. H. (1975). See Alvarez-Lajonchere, 1976: 363.
Pakrasi, B. B., Das, P. R. & Thakurta, S. C. (1964). See Pillay: 0917.
Pauly, D. (1975). On the ecology of a small West African lagoon. *Ber. deutsch. wiss. Komm. Meeresforsch.* **24**, 46–62.
Payne, A. I. (1975). See Alvarez-Lajonchere, 1976: 377.
Payne, A. I. (1976). The relative abundance and feeding habits of the grey mullet species occurring in an estuary in Sierra Leone, West Africa. *Mar. Biol.* **35**, 277–86.
Perera, P. A. B. & De Silva, S. S. (1978). Studies on the biology of young grey mullet (*Mugil cephalus*). Digestion. *Mar. Biol.* **44**, 383–87.
Pillai, T. G. (1962). See Pillay, 1972: 0964.
Pillai, T. G. (1975). Possibilités d'aquaculture et développement de la pêche en eau douce et saumâtre en Tunisie. *Bull. Pêches Tunisie*, **2**, 69–130.
Pillay, T. V. R. (1953). See Pillay, 1972: 0975.
Pozuelo, M. M. (1975). Cultivo del rotifero *Brachionus plicatilis* obtencion de un zooplancton artificial. *Publ. Tec. Dir. Gen. Pesca Marit.* **11**, 401–16.
Rajan, S. (1964). Environmental studies of the Chilka lake. I. Feeding spectrum of fishes. *Indian J. Fish.* **11**(2), 521–32.
Ratkowsky, D. A., Dix, T. G. & Wilson, K. C. (1975). Mercury in fish in the Derwent estuary, Tasmania, and its relation to the position of the fish in the food chain. *Austr. J. Mar. Freshwater Res.* **26**, 223–31.
Renzoni, A., Bacci, E. & Falciai, L. (1973). Mercury concentration in the water, sediments and fauna of an area of the Tyrrhenian coast. *Rev. Intern. Oceanogr. Med.* (31–32), 17–45.
Ros, A. O. & Vincent, J. R. (1973). Primeras experiencias sobre cultivos de peces en el Mar Menor. *Bol. Inst. Esp. Oceanogr.* **163**, 3–20.
Savchuk, M. Y. (1968). See Pillay, 1973: 1089.

Savchuk, M. Y. (1973). Les migrations alimentaires des alevins de Mugilidae sur le littoral de Crimée et du Caucase. *Gidrobiol. USSR*, **2**(5), 28–35.
Sidhu, G. S., Vale, G. L., Shipton, J. & Murray, K. E. (1972). Nature and effect of kerosene-like taint in mullet. In *Proceedings of FAO Technical Conference on Marine Pollution*, Rome, December 1970, ed. M. Ruivo, FAO, pp. 546–50.
Sivalingam, S. (1973). Brackish water fish culture project, Lagos: selection of species for culture. In *Annual Report Federal Department Fisheries, Nigeria*, 62–9.
Skinner, R. (1975). Parasites of the striped mullet, *Mugil cephalus*, from Biscayne bay, Florida, with descriptions of a new genus and three new species of Trematodes. *Bull. Mar. Sci.* **25**(3), 318–45.
Slack, E. B. (1974). Sewage and aquacultural production. Coastal aquaculture and environment. *IPFC Proc.* **15**(2), 70–5.
Stequert, B. (1972). Contribution à l'étude du bar (*Dicentrarchus labrax*) des réservoirs à poissons de la région d'Arcachon. *Thesis, Univ. Bordeaux.*
Stirn, J., Kerzan, I. & Kubik, L. (1974). The possibilities of development of industrial maricultures by using the organic waste waters for the fertilization of primary producers. *Acta Adriat.* **16**(25), 423–34.
Suzuki, K. (1965). See Pillay, 1972: 1204.
Tabb, D. B. & Manning, R. B. (1961). A checklist of the flora and fauna of Northern Florida Bay and adjacent brackish waters of the Florida mainland collected during the period July 1957 through September 1960. *Bull. Mar. Sci. Golf. Caribb.* **11**, 552–649.
Thomson, J. M. (1954). See Pillay, 1972: 1227.
Thomson, J. M. (1959). See Pillay, 1955: 1228.
Thomson, J. M. (1963). See Pillay, 1972: 1238.
Thomson, J. M. (1966). See Pillay, 1972: 1240.
Tung, I. H. (1971). See Alvarez-Lajonchere, 1974: 524.
Vallet, F. (1970). See Pillay, 1972: 1261.
Vallet, F., Berhaut, J., Leray, C., Bonnet, B. & Pic, P. (1970). See Pillay, 1972: 1262.
Wahby, S. D. (1974). Fertilizing fish ponds. I. Chemistry of the waters. *Aquaculture*, **3**, 245–59.
Wallace, J. H. (1973). Aspects of the life histories of estuarine fish in Natal with emphasis on reproduction and recruitment. In *Proceedings of the South Afr. Nat. Oceanogr. Sympos.* Cape Town, 31–2.
Webb, B. F. (1973). Fish populations of the Avon-Heathcote estuary. *New Zeal. J. Mar. freshwater Res.* **7**, 223–34.
Yashouv, A. (1972). See Alvarez-Lajonchere, 1974: 546.
Yashouv, A. & Ben Schachar, A. (1967). Breeding and growth of Mugilidae. II. Feeding experiments under laboratory conditions with *Mugil cephalus* L. and *Mugil capito* C. *Bamidgeh*, **19**, 50–66.
Yashouv, A. & Halevy, A. (1973). Observations on the growth of hybrid fingerlings of Tilapia crosses (*Tilapia aurea* × *T. volcani*) *Bamidgeh*, **25**, 85–8.
Zenkevich, L. (1963). See Alvarez-Lajonchere, 1974: 562.
Zismann, L., Berdugo, V. & Kimor, B. (1974). The food and feeding habits of early stages of grey mullets in the Haifa Bay region. In *Israel Oceanographic Limnological Research, Haifa Laboratory, Annual Report* 72–73.
Zismann, L., Berdugo, V. & Kimor, B. (1975). See Alvarez-Lajonchere, 1976: 508.

7. Energy metabolism of mullets

M. N. KUTTY

Introduction

Energy utilization for maintenance, activity, growth and reproduction in the life of an individual fish has to be accounted for precisely, in preparing an energy budget which is meaningful for commercial fish production. No such complete study exists for any single species of fish. However, several studies on energetics of fish have been made at different levels, which have been brought together and subjected to analysis yielding, in some cases, most useful information. One might refer especially to the two most notable among these, one is Fry's study on the influence of environment on animal activity, which provides the basic philosophy for studies on energetics and survival of fish (Fry, 1947) and the other is that of Winberg (1956), which critically reviews and collates vast materials of study on fish energetics. Further results of studies on fish energetics have been discussed in several later publications (e.g., Brett, 1970; Hastings & Dickie, 1972; Niimi & Beamish, 1974).

Just as in early agriculture, a vast amount of information on growth and culture of fish has accumulated empirically. This is specially true of culture of fish in China and the Indo-Pacific Region (Hora & Pillay, 1962; Bardach, Ryther & McLarney, 1972). Many of these culture practices have been subjected to scientific scrutiny and improvements are being made (Pillay, 1973; Jhingran, 1975). New information on growth and yield of fish adopting the bioenergetics approach might soon form the basis of attempts for optimizing the yield of selected species as for the 'factory-fish' of Brett (1974).

In studying energy relations of fish, a complete estimation of energy flow through the fish body is obtained from the energy input as food, that stored as fish meat (growth), that dissipated as heat (for maintenance, specific dynamic action and activity) and the energy thrown out of the body unused. It is known that heat dissipated for maintenance and activity is a large portion of the energy intake of the organism. This aspect of energy metabolism of mullets forms the major portion of this discussion, as there is more information available on this aspect of the energetics of grey mullets than any other. Also it is partly because of the involvement of the author in these studies on grey mullets, especially the species, *Rhinomugil corsula*. The available information on energy budgeting of mullets is limited, as indeed is that on several other aspects of the physiology of such a euryplastic group as the mullets. The present discussion therefore will be restricted, but will thus point out areas which need to be explored.

Some aspects of energy metabolism of three species of mullet, *Rhinomugil corsula*, *Mugil cephalus* and *Liza macrolepis* are known. Much of the available information is on routine metabolism, but some aspects of metabolism of actively swimming fish are also known. The influence of salinity and temperature, perhaps the two most important factors, on metabolism of mullets is little explored with the exception of the recent contribution of Wohlschlag & Moore (unpublished) on the oxygen consumption of *M. cephalus* under field conditions of temperature and salinity. Hypoxic stress on the physiology and metabolism appear to be better documented. Sources and biochemical mechanisms of substrate utilization are least known, but some guesses at substrate utilization have been made from total metabolism studies, involving oxygen consumption, carbon dioxide production and nitrogen excretion in *R. corsula*. How these studies contribute to some understanding of the energy metabolism of grey mullets in general is discussed herein.

Levels of metabolism

'Standard' and 'active' metabolism are two most recognized levels of metabolism, the former associated with the quiescent fish maintaining its minimum energy requirements for survival and the latter associated with the maximally respiring animal. The concepts of these two terms and routine metabolism accepted herein are according to the exposition of Fry (1947, 1957). 'Routine metabolism' is defined as that related to randomly active fish which has been encouraged to be quiescent, i.e., in a darkened chamber, without any undue external disturbance.

As explained 'active metabolism' is that related to the maximal activity of which the fish is capable, even though loosely it has been used to mean metabolism associated with any level of forced activity. The difference between the standard and active metabolism, i.e., the energy available for any activity above the standard, was termed by Fry (1947) as the scope for activity or metabolic scope. Fry (1957) demonstrated that there is a linear relation between cruising speed and the square root of metabolic scope. With more refined measurements Brett (1964) showed that metabolic rate and activity (swimming speed) of fish have an exponential relation, in common with other power-performance relations. Brett (1964) also showed that depending on the level of activity imposed and the temperature, the active metabolism can be from ten to twenty times the standard metabolism in young sockeye salmon. This capacity for increase of metabolism was termed by him as the 'metabolic expansibility', which is equal to that which the mammals are capable of.

Measurement of energy metabolism

Measurement of oxygen consumption is the easiest, most satisfactory and well recognized method of studying energy metabolism (Winberg, 1956). As Winberg showed, disregarding the nitrogen excretion portion would not cause any significant error in energy estimates based on an average oxycalorific value for the oxygen consumed. In this reasoning it is implicit that adequate ambient oxygen concentrations are maintained. But under certain conditions (intense activity, hypoxia) an anaerobic component may be involved (Brett, 1964; Kutty, 1968*a*) and then separate measures for anaerobic metabolism such as oxygen debt (Brett, 1964), respiratory quotient (RQ, volume of carbon dioxide produced/volume of oxygen consumed; Kutty, 1968*a*; 1972) or lactic acid accumulation (Black, Robertson & Parker, 1961) and excretion (Karuppannan, 1972) become necessary. This may be specially so for fishes adapted to the tropical environment which tends to be more often hypoxic than the temperate. Simultaneous measurement of nitrogen excretion can certainly add accuracy to metabolism estimates.

The nitrogen values can be used to obtain an estimate of the amount of protein utilized (Kutty, 1972; Kutty & Mohamed, 1975). Further, protein being the dearest of the dietary components, studies of nitrogen excreted and retained are of utmost importance in aquaculture. Measurements of depletion of substrates can also be used to estimate energy utilization (e.g., Idler & Clemens, 1959; Brett, 1973). Perhaps the relative amounts of energy trapped from protein, carbohydrate and fat, and proportional exhaustion, can be estimated from measurements of oxygen consumption, carbon dioxide production and nitrogen excretion, as was attempted for *Rhinomugil corsula* by Kutty & Mohamed (1975). Since the predominant nitrogenous excretory product in fish is ammonia, a measure of nitrogen excreted as ammonia might give a reasonable measure of protein degraded, but for exact measurement of nitrogen excretion of fish, total nitrogen excretion has to be measured (Fromm, 1963; Fromm & Gillette, 1968), as was done for *R. corsula* by N. Sukumaran (unpublished observations) in the author's laboratory.

Measurement of heat changes of fish body can also be used for estimates of energy utilization as theoretically recognized and tested for fish (Kutty & Kulandaivelu, 1973), but for the desired accuracy the techniques have to be improved.

For metabolism measurements certain precautions are usually observed prior to the tests (Fry, 1957). The usual method is to starve the fish for at least 24 h (Beamish, 1964*a*) prior to transfer of the fish to the respirometer, but when influence of feeding itself is studied (Pierce & Wissing, 1974) the method has to be suitably modified. The fish is otherwise left overnight in the respiration chamber before experiment. Thus the fish is in the post-absorptive state and free from handling effect (Black *et al.*, 1961) when the

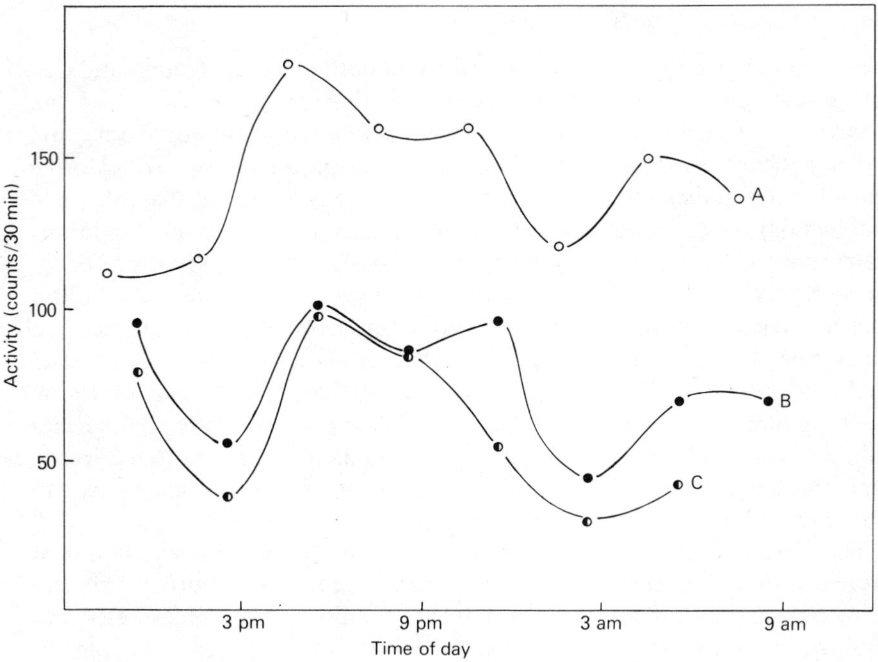

Fig. 7.1. Random activity of *Rhinomugil corsula* (45.9 g, 19.1 cm) in relation to time of day. A, activity of the fish 9 h after handling (transfer to apparatus) and 36 h after feeding; water was flushing through the respirometer throughout the 24 h experiment; B & C, data obtained on the same fish 33 h after handling and 50 h after feeding, while the water supply was alternatively open (B) and shut off (C) for 30 min periods. Each point on the graph is a mean of three values – consecutive 30 min periods. After Kutty, Mohamed, Thiagarajan & Leonard, 1971*b*).

estimates of metabolism are made. The influence of handling and feeding on the routine metabolism of *R. corsula* can be found from the data presented in Table 7.1. Handling usually causes elevation of metabolic rate and also an increase in anaerobic energy utilization as is evident from the RQ which is over unity in handled fish. In the post-absorptive state the routine RQ is less than unity but not as low as in mullet under prolonged starvation. The latter suggests increased utilization of fat. The experimental variables might explain the ammonia quotients (AQ, volume of ammonia excreted/volume of oxygen consumed, which can be used to interpret the intensity of protein degradation (Kutty, 1972). The implications of changes in ammonia excretion and protein utilization are discussed separately. The study of RQ and AQ has special significance because of the influence of composition of diet on routine metabolism. Job (1976) observed that *R. corsula* are non-selective feeders taking either plankton or benthic food depending on their availability. In the laboratory too the mullet feed on any of the diets given.

The random activity of the mullet (Table 7.1) decreased, the longer it was

Table 7.1. *Influence of handling and feeding on routine metabolism of* Rhinomugil corsula *(approximately 50 g) tested in fresh water at 30 °C*

Series	Experimental details	Oxygen consumption (ml/kg/h)	RQ	AQ	Random activity (counts/hour)
1	Starved for 20 days (after capture) in glass aquarium at 30 °C				
	Immediately after handling	(1) 180	1.63	0.14	438
	48 h after handling	(4) 113	0.65	0.13	4
2	Fed with groundnut cake and rice bran (2:1 ratio) in culture pond for over a year and transported 30 miles to laboratory				
	12 h after feed and immediately after handling (transfer to respirometer)	(1) 149	1.64	0.34[a]	152
	48 h after feed and 36 h after handling	(4) 151	0.91	0.35	17
3	Fed with wheat hearts, goat liver and fish muscle (2:1:1 ratio) in glass aquaria at 30 °C, for over 10 days				
	36 h after feed and 12 h after handling	(19) 118	0.91	0.09	19

Data for series 1 and 2 from Narayanan (1974), and for series 3 from Kutty & Mohamed (1975).
Numbers in parentheses indicate the number of determinations.
[a] Nitrogen excretion can be high in injured animals (Bell, Davidson & Scarborough, 1970) but no injury was evident in the fish tested.

in the respiration chamber, as was also observed by Kutty, Karuppannan, Narayanan & Mohamed (1971*a*, and Fig. 7.1). This is due, partly, to starvation (Beamish, 1964*a*). However, as is well known, activity patterns can change markedly with the time of day (Fig. 7.1), and for correlating random activity and metabolism a full range of random activity can be obtained if the fish is observed for over 24 h.

Standard metabolism is often obtained by finding the metabolism at zero activity by extrapolation from a plot of routine oxygen consumption against corresponding random activity (Spoor, 1946; Beamish & Mookherjii, 1964). Such measurements of standard metabolism was obtained for *R. corsula* by Kutty, Mohamed, Thiagarajan & Leonard (1971*b*) and Kutty & Mohamed (1975), using Fry's respirometer (Fry & Hart, 1948). The apparatus used for mullet was a modified version (Kutty *et al.*, 1971*b*, and Fig. 7.2) of the Fry apparatus described by Smit (1965). This is an annular chamber made of transparent acrylic plastic ('Perspex') with a photocell system for counting activity. The photocells were fixed in the inner hollow of the annular chamber and beams of light were focused from outside onto the photocells. When the

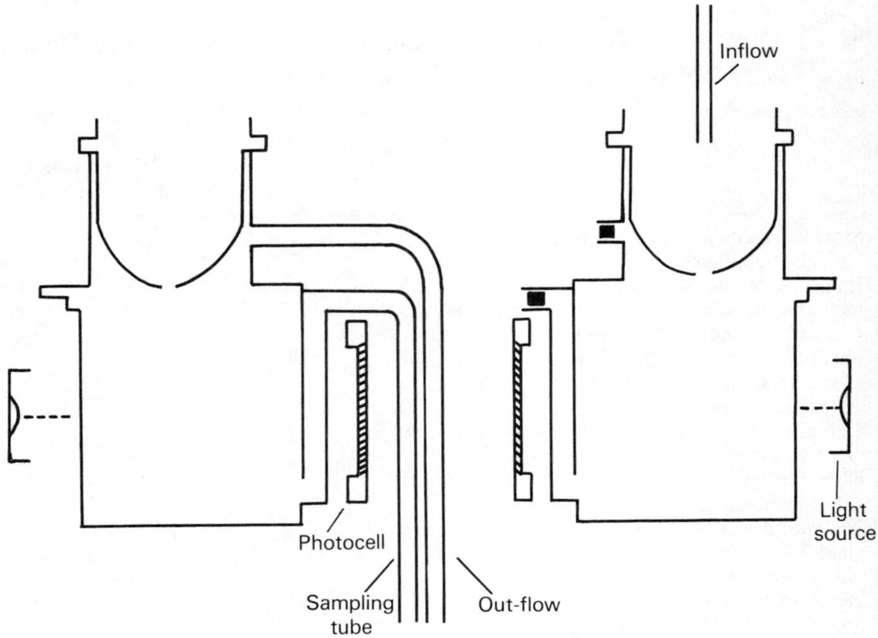

Fig. 7.2. Section through Fry's respirometer (annular chamber) described by Kutty, Mohamed, Thiagarajan & Leonard (1971*b*).

fish moved, it cut the light beam and the event was registered electronically on a chart recorder.

An improved fish activity counter developed in the author's laboratory in collaboration with K. Thiagarajan and A. N. Leonard (Electric Control Equipment Co., Madurai) makes use of a photocell system with light source fixed at the centre (Fig. 7.3). The photocell rack moves outside the annular chamber shadowing the fish. The unit can be adjusted to measure correctly linear movements of fish. A section through the respiration chamber in the new apparatus is shown in Fig. 7.4. The construction is similar to that described by Kutty *et al.* (1971*b*) (Fig. 7.2) but for the double wall serving as a water jacket and slight changes in plumbing. The two-well adaptation on top of the respirometer effectively reduces the diffusion of gases into and from the chamber.

There are several other techniques which can be used for estimating the standard metabolism of fishes (Beamish & Mookherjee, 1964; Fry, 1957). With proper design of apparatus and accurate regulation of temperature, the heat flow meter (Beamish & Mookherjii, 1964; Mathur & Srivastava, 1970) can be used effectively. Here use is made of the principle of heat dissipation due to the turbulence caused by the activity of the fish. This method is perhaps

Fig. 7.3. Photographs of respirometer and random activity counter used for mullet studies in the author's laboratory. Here the light source (L) is in the centre of the annular chamber and the photocells (PR) fixed at the periphery would move around the chamber seeking the shadow of the fish (mullet, *Rhinomugil corsula*, in position in lower photograph only). The two-well modification on the top of the respirometer (Figs. 7.2 & 7.4) minimises diffusion of gases into the chamber. (Photographs, Dr D. Padmanabhan).

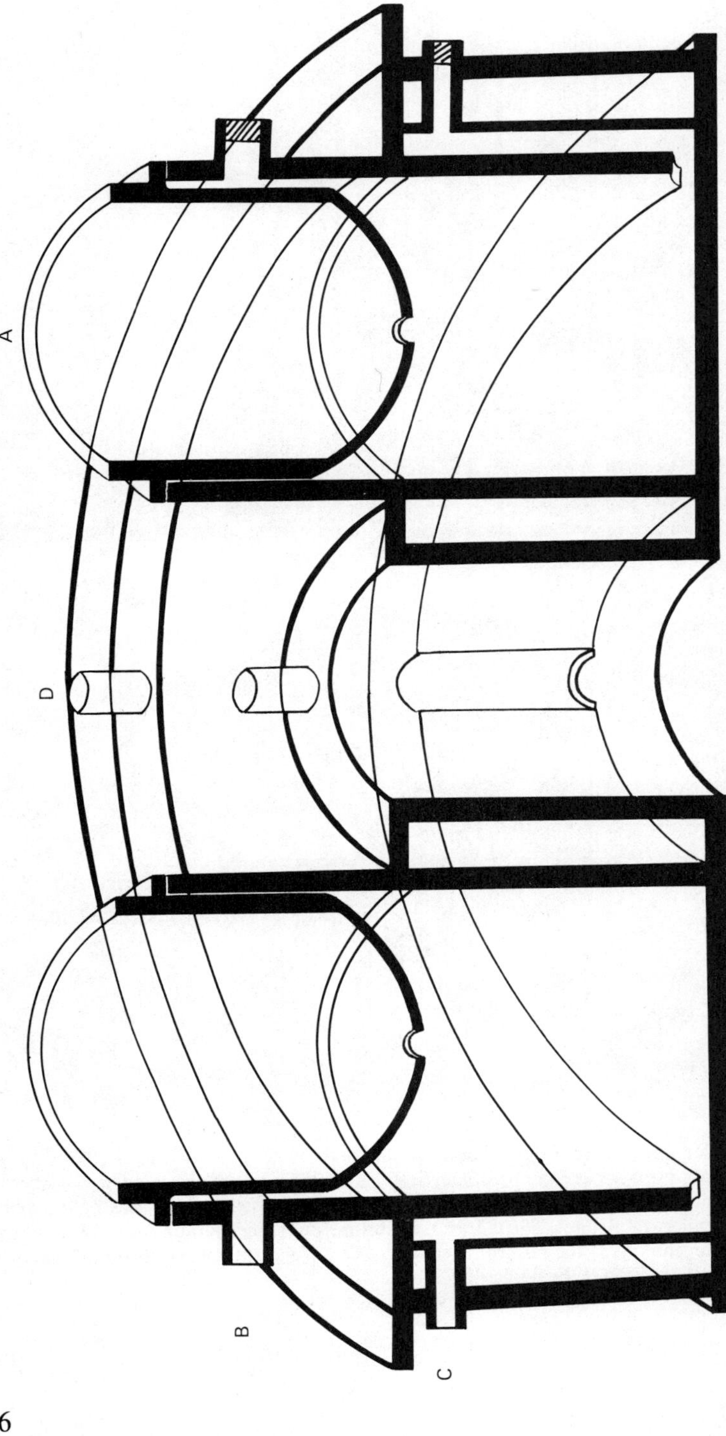

Fig. 7.4. Diagramatic cross section through the double walles acrylic plastic ('Perspex') respiration chamber (see Fig. 3). A three dimensional view of one half of the chamber is suggested. A, B & C denote inflow, outflow and sampling arrangements. D indicates provision for circulation of water through the jacket, which avoids the heating of enclosed water by the central bulb (L, Fig 3) when the chamber is closed.

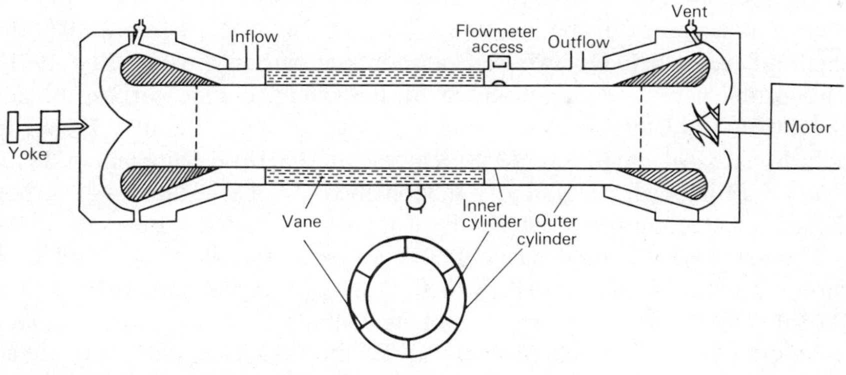

Fig. 7.5. Blazka's respirometer (after Fry, 1971).

more accurate than any other method described so far for determination of standard metabolism of fish, but has not been used for mullet.

Both the rotating annular chamber (Fry & Hart, 1948) and the tunnel-type of swimming apparatus (Blazka's apparatus, Blazka, Wolf & Cepela, 1960) have been used for studying metabolism of mullets under forced activity. Kutty (1969) used a rotating chamber for studying the metabolism of *Liza macrolepis*. Richmonds, Arthur, Thangapandian & Loyola (1971) used a modified version of the original Blazka apparatus for their studies of *R. corsula* under forced activity. The apparatus had a 7 l 'Perspex' double cylinder chamber, with less refinements than the one figured (Fig. 7.5). A propeller fixed at one end can be rotated and water currents, at desired speeds, can be created in the inner chamber where the fish swims. The water pushed by the propeller, returns through the outer cylinder and a steady water current is established. The grid closing the hind end of the inner cylinder is electrified and can be used to prod a 'lazy' fish to swim or to ascertain if the fish is really fatigued (Kutty, 1968*b*) but the electrical stimulus is hardly ever required for *R. corsula* to swim.

Analytical techniques

The Winkler estimation of water samples gives sufficient accuracy for determination of dissolved oxygen, as is universally recognized. Electrometric measurements can also be relied upon, provided the method is constantly counter-checked with Winkler estimates. For expression of units of dissolved gases, see Fry (1957).

Carbon dioxide estimation in water, accurate enough to yield carbon dioxide output values for fish, is beset with many difficulties (Kutty, 1968*a*). The major difficulty is the large amount of bound carbon dioxide normally present in natural water. Since the carbon dioxide added by the fish is very small when compared with the already existing carbon dioxide in natural water, the accuracy of the measure depends on determining accurately the small difference between two large amounts of carbon dioxide (Fry, 1971). This problem can be circumvented by reducing the total carbon dioxide content in ambient water, by using a phosphate buffer (Kutty, 1968*a*) or decarbonated water (Blazka, 1958; Kutty *et al.*, 1971*a*; Karuppannan, 1972; Kutty & Mohamed, 1975). Kutty & Mohamed obtained estimates of carbon dioxide production for *R. corsula* by using decarbonated water.

The van Slyke method has sufficient accuracy for the purpose of total carbon dioxide estimations (Bosworth, O'Brien & Amberson, 1936; Kutty, 1968*a*). Direct titrimetry was found inadequate, but the Maros-Schulek technique (Maros, Schulek, Molnár-Perl & Pinter-Szakács, 1961) as modified by Kutty *et al.* (1971*a*) (Fig. 7.6), where the distillation and titration are done without the solutions coming into contact with atmospheric carbon dioxide, can be used for measurement of carbon dioxide production in fish, as was done for *R. corsula* (Kutty & Mohamed, 1975).

Ammonia in water can be measured using the method of Stroganov (1962) or the Conway diffusion technique. Kutty & Mohamed (1975) followed the former method, as described by Kutty (1972) for their mullet studies. Total nitrogen excreted by fish can be measured using the Kjeldahl method. Sukumaran, N. (unpublished) has obtained total nitrogen excretion in *R. corsula* using the Kjeldahl method.

Routine and standard metabolism

Kutty & Mohamed (1975) provide information on routine oxygen consumption, carbon dioxide output, ammonia excretion, RQ, AQ and random activity of *Rhinomugil corsula* from a transplanted stock (Vaigai Reservoir, Madurai). Their observations on fish acclimated to 30 and 35 °C in air-saturated fresh water and tested under the same conditions are shown in Table 7.2. These *R. corsula* appear to be aerobic in air-saturated water, the RQ values at both temperatures being just below unity. These routine values are discussed below from the point of view of energy derived from different sources. In Figs. 7.7 and 7.8 are given plots of logarithms of the routine metabolic rates and quotients against corresponding values of random activity for *R. corsula* at 30 and 35 °C respectively. Standard rates and quotients obtained from the plots are also indicated in Table 7.2. The standard oxygen consumption and carbon dioxide production at 35 °C are higher than the corresponding values at 30 °C. These rates in randomly active fish increase with increasing activity. This change is not so clear for ammonia excretion, for the trends are opposite for values at 30 and 35 °C, but the slopes concerned are very low (Figs. 7.7 & 7.8).

The slopes of regression lines of RQ on activity are close to zero at both

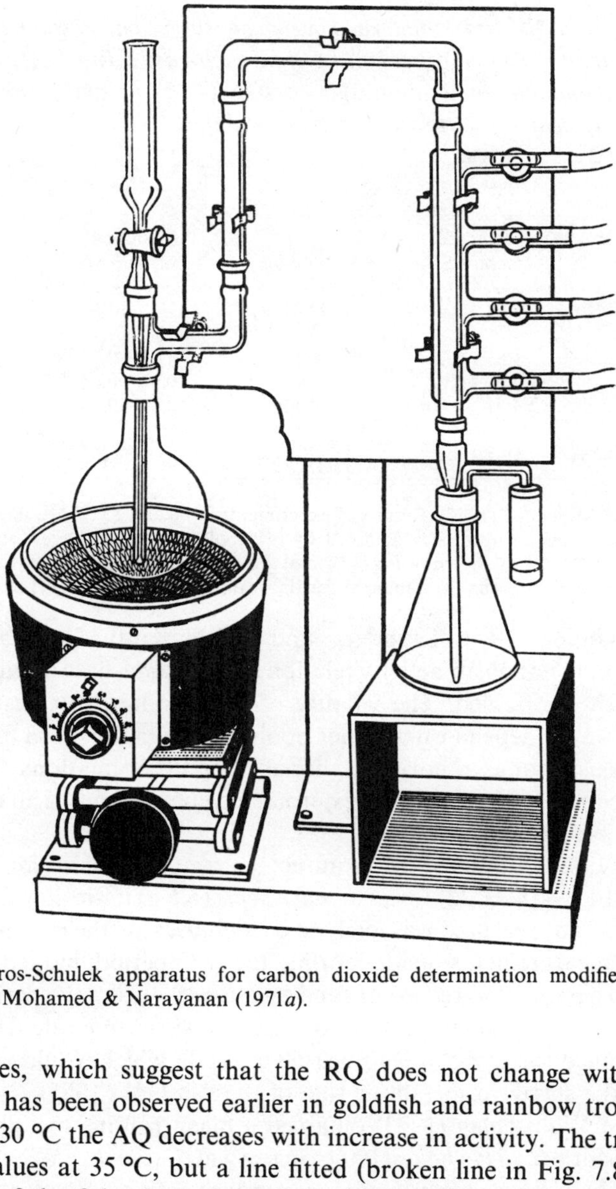

Fig. 7.6. Maros-Schulek apparatus for carbon dioxide determination modified by Kutty, Karuppanan, Mohamed & Narayanan (1971*a*).

temperatures, which suggest that the RQ does not change with random activity, as has been observed earlier in goldfish and rainbow trout (Kutty, 1968*a*). At 30 °C the AQ decreases with increase in activity. The trend is less clear for values at 35 °C, but a line fitted (broken line in Fig. 7.8) through the values of the fish which showed the lowest activity, has a pronounced negative slope. Since the AQ can be taken as a measure of relative protein utilization (Kutty, 1972), the protein catabolism may be increasing with decrease in random activity. Also it can be seen that there is a greater component of protein energy in standard rather than in routine metabolism of the fish. A similar conclusion was also reached from the studies of smolting

Table 7.2. *Routine and standard oxygen consumption, carbon dioxide production, ammonia excretion (ml/kg/h in each case), RQ, AQ, and random activity (counts/h) in* Rhinomugil corsula *(42 g) at ambient oxygen concentrations near air saturation in fresh water*

	30 °C		35 °C	
	Routine (mean ± SE)	Standard	Routine (mean ± SE)	Standard
O_2	117.7 ± 5.0	92.8	117.3 ± 2.8	110.6
CO_2	106.1 ± 2.8	86.0	112.1 ± 3.7	98.0
NH_3	10.5 ± 0.7	11.1	11.7 ± 0.9	10.8
RQ	0.91 ± 0.01	0.93	0.95 ± 0.02	0.91
AQ	0.091 ± 0.006	0.129	0.99 ± 0.007	0.097 (0.174)[a]
Random activity	19.1 ± 2.1	—	19.0 ± 2.7	—

After Kutty & Mohamed (1975). Results of experiments at 30 and 35 °C are shown separately. The standard values are estimates obtained by extrapolation of the regression lines to zero activity through the plots shown in Figs. 7.7 and 7.8.
[a] The two values denote the 'Y' intercept for the two regressions for AQ (Fig. 7.8).

Atlantic salmon (R. L. Saunders and M. N. Kutty, unpublished). In estimation of metabolism/activity relation of *R. corsula*, data obtained for two or three fish are pooled. The validity of the standard rates estimated for *R. corsula* would depend on whether pooling of data for more than one fish in the plots used for extrapolations, has affected the estimations. Since fish of similar size are used in the same respirometer, the method of analysis would appear acceptable.

Kutty (1969) obtained a routine oxygen consumption value of 183 ± 20 ml/kg/h ($N = 22$) for *Liza macrolepis* (5.6 g) in fresh water at 29 °C. The metabolism measured was likely to be high because the routine value was based on measurements made during the 7 h period immediately after handling. There was no record of random activity during the period, but the fish tested did not remain motionless. The lowest routine rate (single value) Kutty obtained for *L. macrolepis* was 60 ml/kg/h, which would appear to be closest to the standard metabolic rate of the fish (Job, 1955).

Cech and Wohlschlag (1973) obtained a mean routine metabolic rate of 84 ml/kg/h for *Mugil cephalus* (100 g) in sea water at an average temperature of 18 °C. The fish was restrained from random activity, unanaesthetized, in a van Dam respirometer (van Dam, 1938). The experiment was designed specifically for studying the influence of hypoxia on ventilation parameters along with metabolic rate and will be referred to again. The lowest metabolic rate measured for *M. cephalus* was 44 ml/kg/h at 15 °C, which appears to be the closest value to the standard metabolism of the fish. In a more recent study

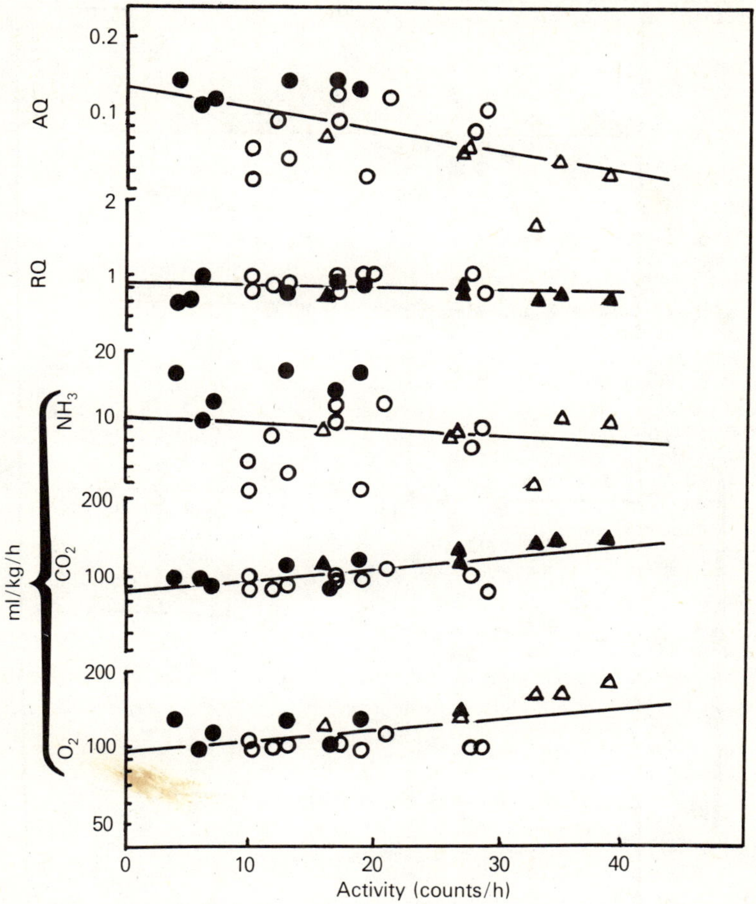

Fig. 7.7. Oxygen consumption, carbon dioxide production, ammonia excretion, RQ and AQ in relation to random activity in the mullet, *Rhinomugil corsula* acclimated to and tested in air-saturated water at 30 °C. Data from individual fish are indicated by separate symbols. The lines fitted through the plots are according to the regression equations calculated for the pooled data for all fish in each case (After Kutty & Mohamed, 1975).

(J. J. Cech in Wohlschlag & Moore, unpublished) the standard metabolism of *M. cephalus*, apparently close to 100 g, is shown to be ranging between 70 and 140 ml O_2/kg/h, the lowest and the highest values corresponding to winter and summer conditions respectively. Wohlschlag, Longley, Moore and Parker (unpublished) studied the routine oxygen consumption of *M. cephalus* exposed to the polluted waters of Galveston Bay (Texas coast) and found that the oxygen consumption of stressed fish was 6–14% less than that of the control, in several tests stretched over the year. The evidence would be more convincing had the standard or another activity-defined rate, instead of the

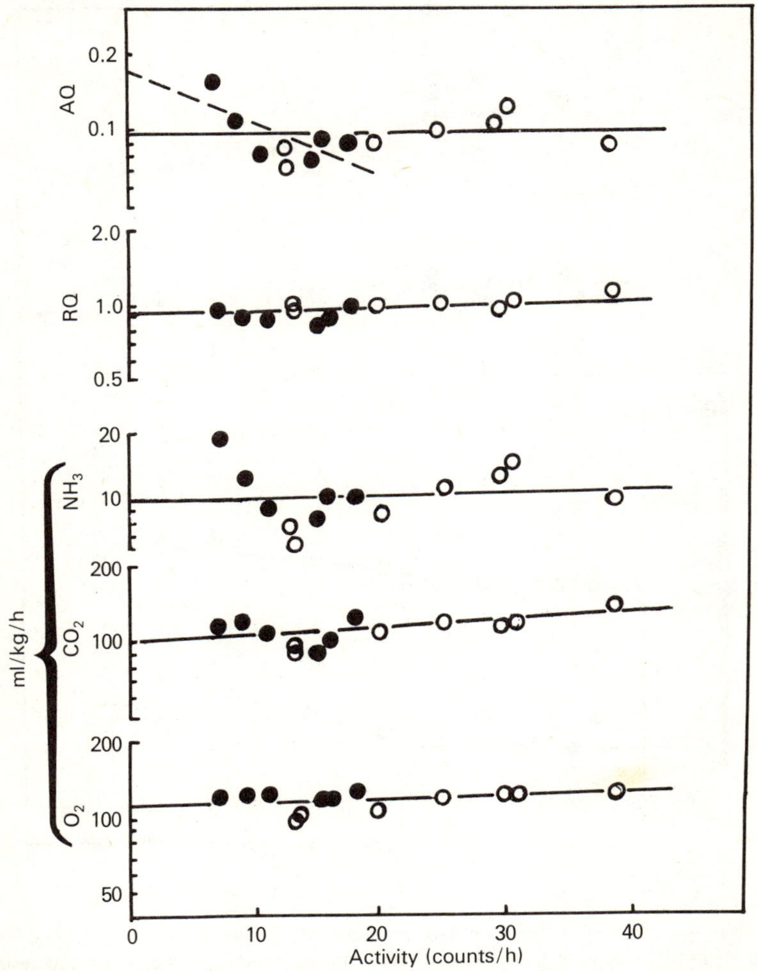

Fig. 7.8. Oxygen consumption, carbon dioxide production, ammonia excretion, RQ and AQ in relation to random activity in *Rhinomugil corsula* acclimated to and tested in air-saturated water at 35 °C. See legend for Fig. 6 for further explanation (After Kutty & Mohamed, 1975).

routine rate, been used for comparison, for there is no indication as to how random activity itself was affected by the polluted waters.

Energy sources

The metabolic aspects discussed are concerned specifically with the energy dissipated by the body as heat, for maintenance, activity and possibly also specific dynamic action (SDA) (see General considerations, p. 253). As an example for estimating energy sources, the routine oxygen consumption, carbon dioxide production and ammonia excretion for *R. corsula* at 30 °C

Table 7.3. *Estimates of energy utilization (heat dissipated) in* Rhinomugil corsula *(42 g) during random activity in fresh water at 30 °C*

	Amino acid mixture		Whole protein (beef) (Lowey's elemental composition)	
Energy source (1)	Glycine and histidine, mole ratio of 3:2 (Creac'h, 1966, carp muscle, free amino acids) (2)	Lysine, glycine, taurine and histidine, mole ratio of 10:17:30:37 (Konosu, Ozay & Hashimoto 1964, *Mugil cephalus* muscle, free amino acids) (3)	Fish system (4)	Mammalian system (5)
Protein RQ	1.24	0.99	1.00	0.80
Protein AQ[a]	0.621	0.478	0.335[b]	0.268[c]
	gcal/kg fish/h (%)			
rotein	79 (13.9)	102 (17.9)	146 (25.7)	176 (31.1)
arbohydrate	248 (43.7)	300 (52.5)	256 (45.1)	340 (60.2)
at	241 (42.4)	168 (29.5)	165 (29.1)	50 (8.8)
Total	568 (100.0)	570 (99.9)	567 (99.9)	566 (100.1)

our separate estimates are indicated assuming that different substrates have been used, in addition the RQ ıd AQ values are given for proteins. The calculations are based on routine metabolism data at 30 °C (routine xygen consumption, 117.7 ml/kg/h; routine RQ, 0.91; routine AQ, 0.091). The calorific value for protein r the first three estimates was taken as 4.57 gcal/ml oxygen; the oxycalorific values for carbohydrate and t for all estimates are according to conventional calculations for mammals.
ammonia quotient – mole to mole relation of ammonia excreted to oxygen consumed.
AQ value obtained assuming that all 'nitrogen' degraded results in ammonia production and all the elemental 'hydrogen' in protein is used for ammonia formation.
AQ value obtained assuming that all 'nitrogen' excreted is in the form of ammonia.

can be taken (Table 7.3). In the table (columns 2–5) results of four separate treatments of data are shown. Most of the earlier energy estimates for fish assumed that protein degradation for fish was almost similar to that of mammals. The treatments of data for the four estimations in Table 7.3, differ in the assumptions made about protein metabolism. In the first two treatments (columns 2 & 3) the protein substrates degraded are taken as mixtures of amino acids and in the last two treatments (columns 4 & 5), the protein substrate is assumed to be the same as 'beef' protein. It is known that the amino acid composition of flesh of several fishes is almost identical to that of beef (Borgstrom, 1962). From this point of view the treatments of columns 4 & 5, might not be out of place, but there are other differences to consider. Gerking (1955) points out from his studies on endogenous nitrogen excretion in bluegill sunfish that 'there may be fundamental difference in the protein metabolism of homoiothermous and poikilothermous animals'. He also cites the studies of Phillips (1948), Tunison, Brockway, Shaffer, Maxwell, McCary, Palm & Webster (1943) and Tunison, Phillips, Shaffer, Maxwell, Brockway & McCary (1944) on brook trout, to support his conclusion. There are also other studies on protein degradation in fish (Braffield & Solomon, 1972; Kutty, 1972) to suggest that protein metabolism in fish is quite different from that of mammals, and that the energy estimations of fish have to be re-examined. It is under these considerations that the energy estimates based on the degradation of amino acid mixtures (Table 7.3, columns 2 & 3) are made. It is also evident from the studies on *Tilapia mossambica* (Kutty, 1972; Karuppannan, 1972) and on smolting Atlantic salmon (R. L. Saunders & M. N. Kutty, unpublished) that at times fish excrete ammonia far in excess of that which could be accounted for by the direct oxidation of whole proteins. Some of the AQs (caused by changes in ambient oxygen, activity or season) suggested degradation of certain predominant amino acids. Oxidation of certain amino acids would indicate a much different RQ (and also AQ) than that conventionally accepted for proteins, as pointed out by Braffield & Solomon (1972) and Kutty & Mohamed (1975). Thus for glycine, an RQ of 1.33 and an AQ of 0.667 would be indicated:

$$\text{Glycine } (C_2H_5O_2N) + 1.5\ O_2 \longrightarrow 2\ CO_2 + H_2O + NH_3,$$
$$\text{therefore the RQ} = CO_2/O_2 = 2/1.5 = 1.33$$
$$\text{and the AQ} = NH_3/O_2 = 1/1.5 = 0.667.$$

Among the amino acids, arginine, which seems to have a dominant role in smolting salmon (Malikova, 1957), displays the highest AQ value (0.9). In column 2 (of Table 7.3) a mixture of glycine and histidine, in 3:2 mole ratio, which makes up 80% of the free amino acids in the muscle of carp (Creac'h, 1966), is taken as the substrate. In column 3 a mixture of four amino acids, which also make up over 80% of the free amino acids of muscle of *Mugil cephalus* (Konosu, Ozay & Hashimoto, 1964) is taken as substrate. RQs and AQs for the protein substrates concerned (as indicated in the table) are used to apportion energy, by the usual methods, i.e., by separating the protein and non-protein portion of RQ and using appropriate oxycalorific values. Except for the last column, the oxycalorific value for protein is taken as 4.57 gcal/ml oxygen (Baffield & Solomon, 1972).

It is evident from Table 7.3 that depending on the protein substrate degraded and the method of analysis, wide ranges in the energy derived from the three sources, can be shown. Assuming the same oxygen consumption, for the amino acid mixtures as substrates the percentage energy derived from protein is about half as that for the beef protein as substrate; and in the case of fat, the energy derived is estimated to be five times lower (column 5) than for the amino acid mixture (column 2). Gerking (1955) points out that the nitrogen excretion per kilocalorie of energy utilized in fish (bluegill) is three and one-half times greater than in mammals. This suggests fundamental differences in protein metabolism in fish, as already referred to. From the data in Table 7.3, the routine ammonia excretion of *R. corsula* can be made out as 11.8 mg/kcal, which suggests even a higher disparity than that suggested by the values based on the bluegill. This confirms that the fish need not be following the metabolic pattern of mammals for breakdown of proteins. It is perhaps not one of direct breakdown of predominant amino acid mixtures as assumed for estimations shown in columns 2 & 3 in Table 7.3, but these estimates certainly indicate lesser energy derived from proteins. It is possible that fish oxidize free amino acids more readily than mammals, even though it is realized that the free amino acids are in dynamic equilibrium with the structural proteins. The terminal deamination alone of amino acids and excretion of ammonia can also be a pathway of nitrogen elimination (Forster & Goldstein, 1969), which would indicate a low energy turnover of proteins in fish. Anaerobic breakdown of amino acids along with carbohydrates (Hochachka & Somero, 1973; Hochachka, Owen, Allen & Whittow, 1975) can lead to excess ammonia release (also see the next section, p. 252). An aspect which cannot be ignored in this context, but is not taken into account here, is the change in nitrogen excretion pattern caused by the interconversion of substrates. The present analysis is made from the point of view of complete oxidation of substrates.

From the discussion on the analyses presented in Table 7.3 it would appear that the estimate based on mammalian system may be less applicable in fish. Possibly a picture nearer the actual is exemplified by the estimate shown in column 4 in Table 7.3, i.e., breaking down of whole proteins (and perhaps also some free amino acids) resulting predominantly in the production of ammonia. It is noteworthy that the total energy derived for the observed oxygen consumption, RQ and AQ, assuming different proteinaceous substrates and their breakdown, are almost the same (final column, Table 7.3), a fact clearly brought out by Winberg (1956) in a different context. However, in energy metabolism studies it is important to know the relative amounts of substrates consumed for energy liberation especially from the point of view of dietary requirements.

Factors affecting energy metabolism

Factors influencing metabolism, and indeed life in general, are usually separated into two categories, the intrinsic and the extrinsic. A discussion of these factors as affecting fish metabolism is given by Brett (1970). Among intrinsic factors can be listed activity, size, social interaction (excitement, schooling), sex and maturity and among extrinsic factors may be listed temperature, salinity, dissolved oxygen, dissolved carbon dioxide, pH, light, pressure, water currents and the season of the year (see Fig. 7.16, discussed later).

Energy metabolism of mullets

In the present discussion some of the salient factors influencing metabolism only will be taken up. The most important of these factors is activity, which as pointed out already, when combined with a factor like temperature can cause about a twenty-fold change in metabolism of fish (Brett, 1964). Activity itself can be divided into two categories, the random activity and forced activity. The former has already formed part of the discussion on routine metabolism. The forced activity in response to swimming in water currents, a rheotactic response, can be taken as an intrinsic response because it is the option of the fish to swim or lag. It is this aspect which will be discussed in the following section under activity.

Activity (exercise)

Richmonds *et al.* (1971) studied the oxygen consumption of *R. corsula* (16 cm) and found a wide scatter of values, ranging between 160 and 440 ml/kg/h, at the swimming speeds tested, apparently caused by the different levels of excitement of the fish (Brett, 1964; Smit, 1965). The lowest values obtained for the levels of activity appear to be too high and a reasonable extrapolation to zero activity could not be made from their data. They, however, estimate a standard oxygen consumption of 140 ml/kg/h for the mullet at 35 °C.

Oxygen consumption, carbon dioxide production and nitrogen excretion (ammonia nitrogen and total nitrogen) obtained by N. Sukumaran (unpublished) for *R. corsula* forced to swim at 20 and 75 cm/sec for 5 h are displayed in Fig. 7.9. It is seen that both short-term and long-term oxygen consumption, carbon dioxide production and nitrogen excretion are higher at 75 cm/sec than at 20 cm/sec. With increase in the duration of exercise the carbon dioxide production, which was always higher than the oxygen consumption in the first hour of swimming, decreased much more steeply than the oxygen consumption, in agreement with the earlier observations on goldfish and rainbow trout (Kutty, 1968*a*) and *Tilapia mossambica* (Karuppannan, 1972). The first hour RQs for *R. corsula* were 1.17 and 1.54 and for the fifth hour the RQs were 0.82 and 0.82, respectively for the low and high swimming speeds, suggesting a burst of anaerobic energy utilization in the first hour, declining subsequently to reach an aerobic state, again agreeing with the observations on other teleosts.

The trend of nitrogen (both ammonia nitrogen and total nitrogen) excretion in relation to the duration of exercise is reversed when compared with the other two metabolic parameters, the nitrogen excretion increasing markedly with the duration of exercise. The ammonia excretion pattern is similar to that of *T. mossambica* (Kutty, 1972; Karuppannan, 1972). Both total nitrogen and ammonia nitrogen excreted were higher in fish swimming at the higher speed. The ammonia portion excreted accounted for 70–80% and 60–65% of total nitrogen excretion of *R. corsula* swimming at 20 and 75 cm/sec,

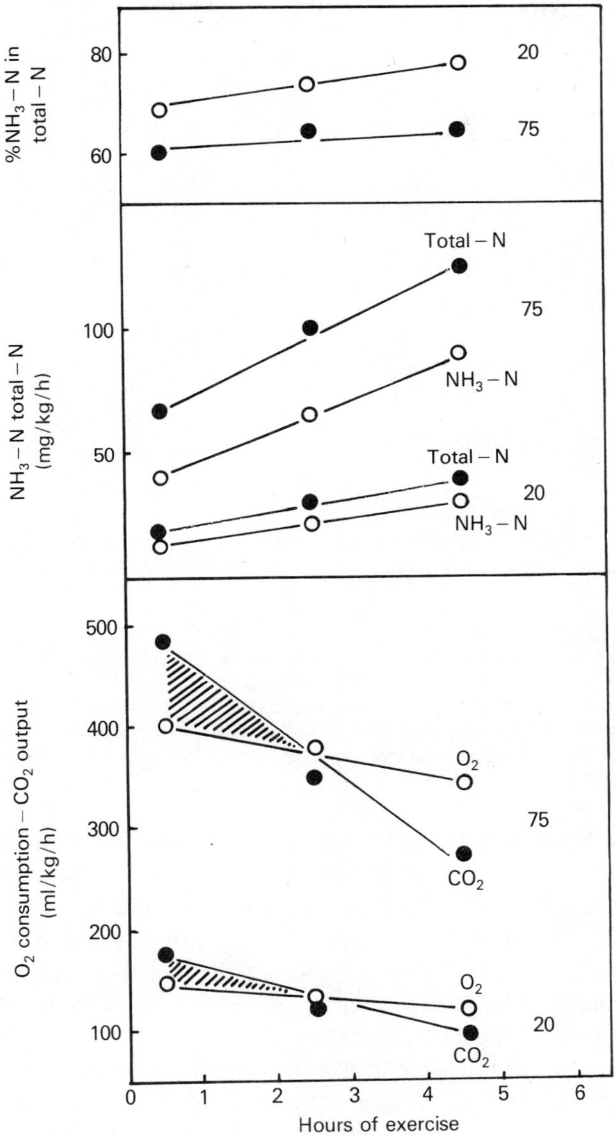

Fig. 7.9. Influence of swimming speed and duration of swimming on the oxygen consumption, carbon dioxide production, nitrogen (total-N and NH_3^-N) excretion in *Rhinomugil corsula* (mean TL, 16.8 cm; weight, 43.1 g) acclimated and tested in fresh water at 30 °C. The top panel shows percentage of ammonia-nitrogen in total nitrogen excreted. The figures 20 and 75 denote the swimming speeds (cm/sec). The values plotted are means of three determinations. The hatched areas between oxygen consumption and carbon dioxide production curves indicate the extent of anaerobic metabolism.

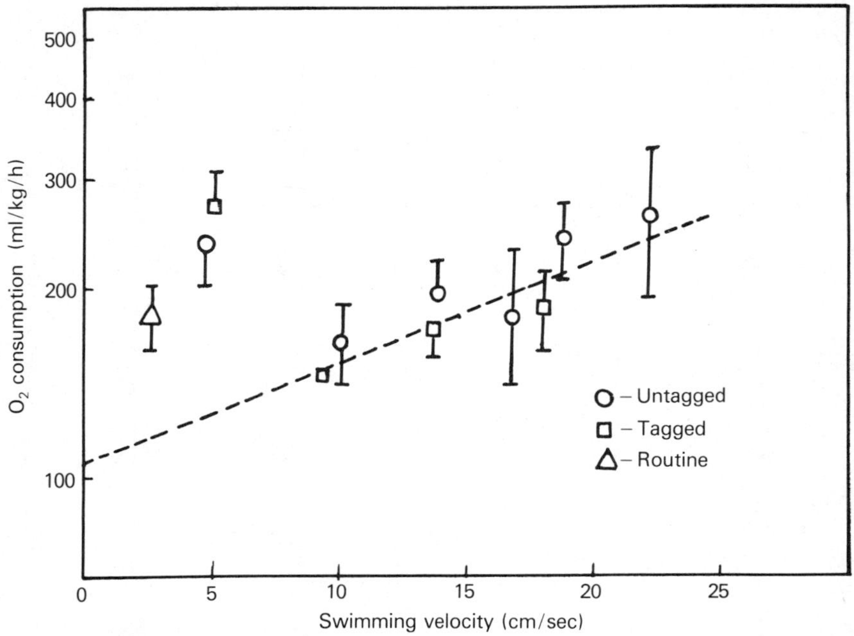

Fig. 7.10. Oxygen consumption in relation to swimming velocity in tagged (plastic opercular tag) and untagged *Liza macrolepis* (total length of fish used, 10.3–11.2 cm) acclimated to, and tested in, fresh water at 29 °C. In all cases, metabolic rates indicate steady rates determined $1\frac{1}{2}$ h after the initiation of exercise in Fry's respirometer. The activity for routine rate is arbitrary. Vertical bars through points indicate standard error of the mean. The broken line shown in the figure is fitted by eye (After Kutty, 1969).

respectively. These patterns suggest high use of proteins when the intensity and duration of exercise are increased, as observed by Karuppannan in *T. mossambica*, but the Tilapia study was based on ammonia excretion only. The significance of exchange of ammonium ions for sodium ions at the gills for ionic regulation and acid-base balance (Maetz & Garcia Romeu, 1964; Garcia Romeu & Motais, 1966; Forster & Goldstein, 1969; Kutty, 1972) in fish under stress of anaerobic metabolism, is referred to later. This aspect cannot be without significance in fish under exercise as well.

Kutty (1969) studied the oxygen consumption of *Liza macrolepis* (10 cm) forced to swim at swimming speeds up to 22 cm/sec. He did not find any difference in the oxygen consumption of a tagged mullet from that of untagged (Fig. 7.10). The routine metabolic rate of the mullet is also included in Fig. 7.10. A significant feature in the figure is that the metabolic rate at the lowest swimming speed tested (5 cm/sec) and the mean routine metabolic rate of the fish (arbitrarily shown at 2.5 cm/sec) are higher than several metabolic rates obtained at higher speeds. This is in agreement with the observations of Brett (1964) and Smit (1965), who stated that fish swim much less efficiently at lower speeds. The line drawn through the plot of values at speeds higher than 5 cm/sec indicates the average requirement of energy for *L. macrolepis* swimming at the concerned levels of activity. Extrapolation of this line to zero activity suggested a value of about 105 ml/kg/h, which is obviously higher than the standard oxygen consumption of the fish. Extrapolation of a line drawn through the lowest values of oxygen consumption (not shown in the figure) might more appropriately suggest the actual standard metabolic rate. This should be near the lowest routine metabolic rate (60 ml/kg/h), obtained for the mullet at 29 °C as already observed.

Wohlschlag & Moore (unpublished) studied the oxygen consumption of *Mugil cephalus* induced to swim in a rotatable chamber (Wohlschlag, 1964). They obtained multiple regression equations of oxygen consumption (grouped for different seasons in brackish and marine waters) for fish of different sizes, forced to swim up to 32 m/min (53 cm/sec) at field conditions of temperature and salinity. The fish were fed as under natural conditions and their metabolic rates were measured in the first hour after transfer to the respirometer. Under the conditions described the fish might have derived some amount of energy anaerobically (Kutty, 1968a; 1972; Karuppannan, 1972), but since the conditions for all fish were about equal the comparisons appear to be valid. We shall refer here to the relationship of oxygen consumption to swimming speed alone, as the metabolic relations to size, temperature and salinity are covered separately hereunder. The regression coefficient for swimming speed was about 0.01 to 0.02 (log units of oxygen consumed 1 m/min increase in speed), which is slightly lower than those obtained for sockeye salmon (Brett, 1964; Brett & Glass, 1973), but compares well for fish in general.

Size

The relation of size and metabolism is usually expressed by the equation, $y = ax^b$, where x is the body weight and y the rate of oxygen consumption. The slope value (b) is generally accepted as 0.8 for routine metabolism of fish (Winberg, 1956), but several deviations from this value have been reported and this value cannot be accepted as universal (see Fry, 1971). Wohlschlag & Moore (unpublished) in their study on seasonal aspects of oxygen consumption of *M. cephalus* forced to swim in a rotatable chamber (as mentioned previously) obtained weight coefficients ranging between 0.3 and 0.9. Studies made on the active metabolism of salmonids have indicated a weight proportionality (Brett, 1965; Rao, 1968), but a decrease in slope with increase in temperature has been reported by Job (1955). Wohlschlag & Moore explain that the very low slope values obtained for *M. cephalus* might have resulted from the depression of metabolism of the larger fish by any stress effect. They also suspect that in their fish samples tested under field conditions a mixing of 'sub populations' could have taken place. Size effect on the metabolism of mullets seems to have been insufficiently explored, also in view of the fact that there is no information on the effect of fish size on standard metabolism for any mullet.

Temperature

The range of distribution of *Mugil cephalus* is known to be 12–25 °C, but the lower limit of temperature controlling the natural distribution of the species seem to be between 16 and 18 °C except for the Mediterranean and Black seas (Thomson, 1966). Zambriborshel (1962) studying *M. cephalus* in wintering pens observed that temperatures below 5 °C were injurious to the fish. There is possibly a large difference between the avoidance temperature and the lethal temperature of the mullet. It is obvious that *M. cephalus* of Indian coasts inhabit waters warmer than 25 °C. Kuthalingam (1959) acclimated *M. cephalus* fry to 32–34 °C, which showed preference for water at 29–32 °C. *M. cephalus* are certainly the most widely distributed grey mullet and their temperature tolerance appears stretchable, depending on the natural temperatures at the geographical locations wherein they occur.

There does not appear to be any planned study on the temperature tolerance and resistance (Fry, 1947) of *M. cephalus* or other mullets such as those available for goldfish (Fry, Brett & Clawson, 1942) and for certain salmonids (Brett, 1952). Kutty & Murugapoopathy (1969) observed that in comparison with *Tilapia mossambica* the low temperature tolerance of *Liza macrolepis* is less. *L. macrolepis* acclimated to 30 °C died at 16 °C (median lethal time of 2000 min). With acclimation to lower temperatures the low lethal temperature can be shifted downward, but it appears that the ultimate lower lethal temperature of this species would not be lower than 10 °C.

Rhinomugil corsula have been acclimated to 35 °C (Kutty & Mohamed, 1975); acclimations at higher temperatures have not been tried. It is likely that the summer kills observed in *R. corsula* in Vaigai Reservoir (Madurai) are due to high temperature exposure (Mohamed, M. P. & A. Basil, unpublished observations). From the natural temperatures of distribution of *R. corsula* the 20 °C range, between 15 and 35 °C, appears to be the meaningful range within which energy metabolism of the fish has to be studied; work on this is in progress in the author's laboratory. The information available on *R. corsula* at 30 and 35 °C has already been dealt with (see the section on routine and standard metabolism. The routine metabolic rates and quotients at these two temperatures are not significantly different from each other (Table 7.2). Random activity at the two temperatures are also similar and therefore it would appear that the temperature/metabolism relation has reached a plateau at 30–35 °C. However, it would be hazardous to guess at the temperature relation outside this narrow range (Fry, 1957), especially when it concerns routine metabolism (Fry & Hochachka, 1970).

Cech & Wohlschlag (1973) observed that the routine metabolic rate of *M. cephalus* of the Texas coast was significantly influenced by temperature. Between 13 and 22 °C the oxygen consumption of *M. cephalus* (100 g) ranged between 44 and 129 ml/kg/h in air-saturated sea water. There is no information on the specific temperature history of the fish in these experiments, excepting that the test temperatures were similar to the natural temperatures off the Texas coast. In the van Dam respirometer used by these authors the fish were apparently unable to make any locomotory movements but excitement could not be ruled out.

In *M. cephalus* exercised in a rotating chamber Wohlschlag & Moore (1975) obtained temperature co-efficients in the common range of 0.02–0.03 (partial regression co-efficients for oxygen consumption measured under field conditions of temperature – see the section on activity, p. 236. However, the coefficient obtained during summer was high (0.05). R. H. Moore, (in Wohlschlag & Moore, unpublished) found high coefficients of 0.05 in spring and 0.04 in summer for fasted and thermally acclimated *M. cephalus*, while for *Mugil curema* the values were between 0.02 and 0.04 through the year studied. In general the temperature coefficients for the mullets studied were within the range of those obtained for other fishes.

Salinity

Mullets in general tolerate a wide range of salinity. Simmons (1957) observed that *M. cephalus* can tolerate salinity ranges between a trace and 75‰. Greater tolerance, from trace to 113‰, has been reported for *M. cephalus* (Zenkevitch, 1963; Odum, 1970). The fish have also been known to live hundreds of miles up fresh-water rivers, such as the Colorado (Dill, 1944). In spite of their wide tolerance to salinity *M. cephalus* naturally spawn only in the sea. But there are other grey mullets such as those of the genera *Rhinomugil*, *Trachystoma* and *Agonostomus* which live in fresh water, but it is not known if they spend their life completely in fresh water (Thomson, 1966). However *Rhinomugil corsula* adapt completely to fresh water and breed in fresh water (Hora, 1938; Pakrasi & Alikunhi, 1952; Narayanan, 1974). This is significant because the species can be cultured thus without complications of inducing spawning. It is possible to judge the salinity preferences of mullets from their natural distributions (Bograd, 1961; Thomson, 1966), but there appears to be no experimental study of this.

Acclimation of mullet to fresh water has been done regularly as an aquacultural practice. Devanesan & Chacko (1943) successfully acclimated *Mugil troschelii* and *Mugil waigiensis* by gradual replacement of salt water with an equal volume of fresh water. Venkataraman (1944) suggested prolonged conditioning of twelve days for fresh water acclimation of *Mugil seheli*. Job & Chacko (1947) reduced the conditioning time of *M. seheli* to

18 h for acclimating the mullet successfully to fresh water. Pillay (1947) and Spurgeon (1947) suggest that mullet fry become naturally acclimated to fresh water while ascending the fresh water course from estuaries. Systematic experiments in acclimating *Mugil parsia* fry to fresh water by stages of 6–12 h was conducted by Ghosh (1967), who obtained as high as 99% survival in his tests. Young *Liza aurata* die on direct transfer from sea water to fresh water, but 100% survival is reported when the salinity is decreased gradually from 10% sea water to 100% fresh water (Chervinsky, 1974).

The influence of salinity on the oxygen consumption of *M. cephalus* in both marine and brackish waters under field conditions has also been studied by Wohlschlag & Moore (unpublished) (see pp. 231–232). They did not observe any significant influence of salinity on metabolism within the brackish or marine test groups, but in all cases *M. cephalus* in marine water had a higher metabolic rate than in brackish water. At the same time considering weights, temperatures and swimming velocities, the marine rates were higher by about 40% in summer, 24% in autumn, 15% in winter and 4% in spring, without adjustment for seasonal changes in salinity of the brackish water. This observation clearly suggests that the mullet spend more energy while in marine water, which as recognized, is owing to the greater osmotic work required in marine water than in 'iso-osmotic' brackish water (Rao, 1968). In support of this G. H. Collins (in Wohlschlag & Moore, unpublished) found that the blood iso-osmotic level of *M. cephalus* was 8‰ salinity and that their standard metabolism in brackish water was below that in sea water.

Certain other useful aspects of osmoregulation in mullets have also been investigated. Lasserre (1971) studied the Na^+–K^+ dependent ATPase activity in *Crenimugil labrosus* (Risso). The ATPase activity in both gills and kidneys were found to be higher in *C. labrosus* acclimated to fresh water than in the sea-water acclimated forms. In general it is seen that the N^+-K^+ ATPase activity, i.e., the salt regulatory activity is greater in the gills and kidneys of fishes of sea water origin, when living in fresh water and vice versa.

McFarland (1965) studied the effect of hypersalinity on serum and muscle ion concentration in *M. cephalus*. Hyper-saline adaptations are of special interest to mullet culturists and attempts have already been made for hyper-saline cultivation of mullets (e.g. Zismann & Ben Tuvia, 1975; Wallace, 1974). A point in question is as to what happens to the extra energy which might be required to pump the salt out of the fish body. Would the fish be taking in more food for this, or would it be at the sacrifice of growth? It is possible that energy loss for ion-osmoregulation in hypersaline conditions is not proportional to salinity gradient, possibly due to changes in permeability characteristics of the exchange epithelium (Gordon, 1963; Rao, 1968; Ananthakrishnan & Kutty, 1974). From this it would appear that the influence of salinity on energy metabolism and growth of *M. cephalus* needs immediate attention.

Oxygen

The capacity of mullets to tolerate a low level of dissolved oxygen has not been explored, except for brief indications from general biology studies. Nair (1944) showed that the mullets, *Mugil parsia* and *Mugil tade* are intolerant of sewage pollution, death being caused due to deficiency of oxygen. McFarland & Moss (1967) observed that *M. cephalus* often encounter hypoxic waters while schooling and such schools disperse temporarily into smaller groups in areas around low-oxygenated water. *M. cephalus* displayed symptoms of dyspnoea at pO_2 concentrations near 20 mm Hg at 12–19 °C (Itazawa, 1959: Fig. 1). Kutty & Mohamed (1975) showed that *Rhinomugil corsula* lose equilibrium at 0.8 mg O_2/l in freshwater at 30 °C. J. M. Thomson (personal communication) observed that while oxygen was indicated as zero in stream water into which discharges from a health food factory were made, *M. cephalus* crowded around the discharge pipe from the factory. It is not known, however, whether these fish were managing without oxygen for long periods. Mullets can make use of the thin oxygenated surface layer of water by skimming it (Odum, 1970; Cech & Wohlschlag, 1973; Narayanan, 1974). Further the positive rheotaxis of the fish might attract them to waters which would otherwise be avoided (Kutty, M. N. & N. Sukumaran, unpublished).

Wohlschlag & Moore (unpublished) have observed a depression in the oxygen consumption of *M. cephalus* during summer and ascribed it to partial anaerobiosis, apparently caused by a reduction in dissolved oxygen in the warm waters of the summer and consequent changes in the biochemical mechanisms. While this metabolic depression was evident in fish exercised, apparently at levels close to their natural activity, under field conditions, it was not observed by J. J. Cech, (in Wohlschlag & Moore, unpublished) in their estimates of standard metabolism of thermally acclimated mullet. These authors also found a reduction in growth of the mullet during summer, as judged from the terminal growth zone of the fish scales.

It has been shown that the mullet, *Rhinomugil corsula* can supplement energy anaerobically under hypoxic conditions. Kutty & Mohamed (1975) studied the influence of dissolved oxygen on the routine oxygen consumption, carbon dioxide production and ammonia excretion in *R. corsula* in fresh water. They allowed the fish to reduce the oxygen content of the water in a closed respirometer (see the section on energy metabolism, p. 232) until it became asphyxiated. Their observations on *R. corsula* at 30 and 35 °C are shown graphically in Figs. 7.11 and 7.12 respectively. The trends of results at the two temperatures are strikingly similar. The lowest curve in each figure indicates random activity of the mullet. The changes in metabolism of *R. corsula* in relation to ambient oxygen is in general agreement with similar studies on goldfish (Kutty, 1968*a*) and *Tilapia mossambica* (Kutty, 1972). It is evident that the routine oxygen consumption at 30 °C increased with

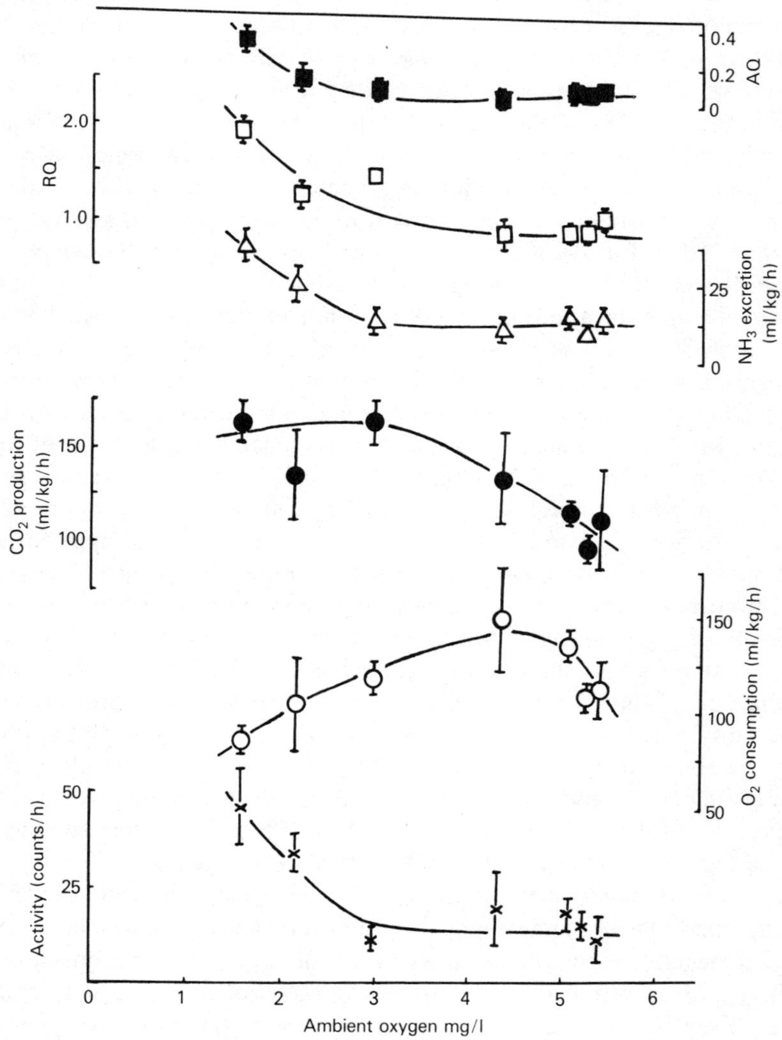

Fig. 7.11. Oxygen consumption, carbon dioxide production, ammonia excretion, RQ, AQ and random activity in relation to ambient oxygen in *Rhinomugil corsula* acclimated to and tested at 30 °C. Ambient oxygen was reduced by the respiration of the fish itself. The experiment was terminated when the test fish lost its equilibrium. Each value plotted is a mean ($\pm$SE) of ten determinations. Standard error is not indicated if it falls within the area of the symbol shown in the figure (After Kutty & Mohamed, 1975).

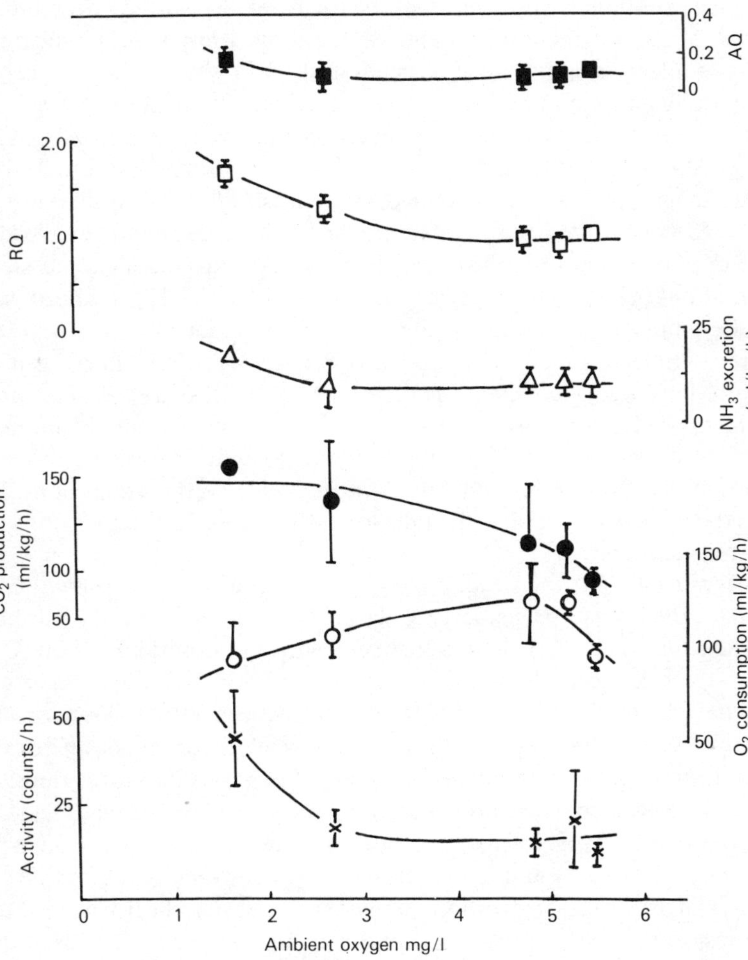

Fig. 7.12. Oxygen consumption, carbon dioxide production, ammonia excretion, RQ, AQ and random activity in relation to ambient oxygen in *Rhinomugil corsula* acclimated to and tested at 35 °C. See legend for Fig. 7.11 for further explanations (After Kutty & Mohamed, 1975).

reduction in ambient oxygen down to about 4 mg/l and subsequently decreased with further reduction in ambient oxygen. The carbon dioxide output of the fish increased with the decrease in ambient oxygen, but at the lowest oxygen concentrations tested it reached a plateau, indicating that the carbon dioxide production, though still limited, is relatively less dependent on ambient oxygen. The routine ammonia excretion remained almost unchanged at the higher oxygen concentrations, but when the ambient oxygen decreased lower than about 3 mg/l, the excretion rate rose up gradually to reach a maximum value at the lowest oxygen concentration tested.

As in earlier studies with fish it can be assumed that an RQ near unity is indicative of aerobic metabolism and values above unity would indicate the intensity of anaerobiosis. Thus it is seen in Figs. 7.11 and 7.12 that *R. corsula* are essentially aerobic at ambient oxygen concentrations above 3 mg/l, and below this level the fish become more anaerobic the lower the ambient oxygen. The RQ sustained at the lowest oxygen level (1.6 mg/l) tested is about 2, about the same magnitude as that sustained by goldfish at 20 °C. But unlike the goldfish, *R. corsula* were not tested for chronic exposure to low oxygen. Almost in parallel with the change in RQ, and AQ increases from a steady value of about 0.1 at high oxygen levels (Figs. 7.11 and 7.12) to about 0.3 at 1.6 mg O_2/l indicating increased protein utilization under hypoxic conditions. A similarity in trends of RQ and AQ with reference to ambient oxygen was also noticed in *Tilapia mossambica* (Kutty, 1972); goldfish and *Barbus sarana* (Mohamed, 1974) and it was suggested that the increased amount of ammonia produced and excreted by fish might prevent acidosis during hypoxia and also help in conservation of sodium ions (Kutty, 1972) as the ammonium ions produced can be exchanged at the gills for sodium ions in the ambient water as referred to earlier.

It is significant that the routine oxygen consumption of *R. corsula* under hypoxia (at the lowest oxygen level tested) is not different from the normoxic oxygen consumption. This is in agreement with the observations on *Mugil cephalus* (Cech & Wohlschlag, 1973) and also on goldfish (Kutty, 1968*a*) even though deviations from this pattern have also been noted (Doudoroff & Shumway, 1970; Mohamed, 1974). However, when data on routine oxygen consumption at decreasing oxygen concentrations over the total range are considered, oxygen consumption at the lowest oxygen level tested is less than that at the middle range of oxygen, as exemplified by the curves described for *R. corsula* (Figs. 7.11 and 7.12). The decrease in oxygen consumption near the asphyxial oxygen concentration was evidently because the fish were at their zone of respiratory dependence and not due to change in activity, for at these oxygen levels the random activity was actually increasing. There is little chance that oxygen was supplemented from the swim bladder in the mullet, for this contribution is shown to be meagre in the case of *M. cephalus* (Moore, 1970). However, as explained, there was anaerobic energy supplement in all the cases known, as was obvious from the hypoxic RQs, which are over unity. In spite of this fact, if oxygen consumption corrected for the anaerobic energy supplement at the low oxygen is compared with oxygen consumption at the high level, the efficiency of work (energy utilized per unit work, in this case random activity) is much higher in fish in low oxygen concentrations (Beamish, 1964*b*; Kutty, 1968*a*). It may be because these fish spend less energy on 'avoidable' oxygen consumption such as excitement. This would suggest that the efficiency of energy utilization, rated as performance of fish at different oxygen levels, is subject to behavioural regulation (Kutty, 1968*b*; Fry, 1971).

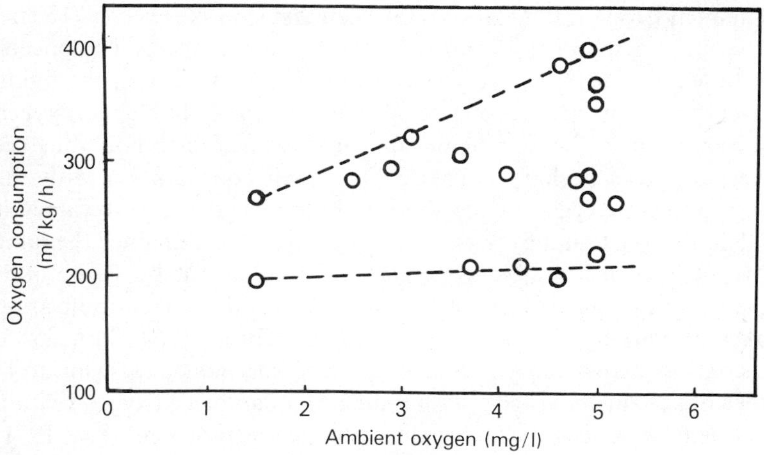

Fig. 7.13. Oxygen consumption in mullet *Rhinomugil corsula* (average TL, 14 cm), forced to swim at a speed of 43 cm/sec at 35 °C at various ambient oxygen concentrations. Each point represents a single determination. See text for explanation (After Richmonds *et al.*, 1971).

The fact that the routine oxygen consumption near air saturation and that in low oxygen are about equal in *M. cephalus* has been explained as due to ventilatory adjustments serving a constant oxygen supply to the gills under conditions of oxygen depletion. The adjustments in *M. cephalus* were found to be threefold increase in ventilation volume, twofold increase in opercular stroke volume and about singlefold increase in ventilation frequency (Cech & Wohlschlag, 1973). These observations are in agreement with those made on other teleosts. Cech & Wohlschlag (1973) found also a decrease in percentage utilization of oxygen, with the increase in ventilation volume in *M. cephalus* in low oxygen as was again observed in other species (van Dam, 1938; Saunders, 1962 and others). The energy cost of the increase in ventilation to maintain adequate oxygen supply in *M. cephalus* is probably small (Cameron & Cech, 1970). Except for the study of Beamish (1964*b*) none of those discussed has studied the influence of ambient oxygen change on standard metabolism in fish. Beamish showed in brook trout, carp and goldfish that standard oxygen consumption remains fairly constant over a wide range of ambient oxygen down to a critical level where it rises to a peak, indicating the increased cost of ventilation.

The exact level of anaerobic metabolism sustained by *R. corsula* in the tests described cannot be known, but it might be fairly high. It was suggested by Kutty & Mohamed (1975) that this could be close to 50% of the total energy derived by the hypoxic mullet (at 1.6 mg O_2/l), but this conclusion was based on an indirect estimation and has to be confirmed.

The only study on the influence of ambient oxygen on the metabolism of mullet swimming actively is that of Richmonds *et al.* (1971) on *R. corsula*. A reference to their experimental set-up has already been made (in the section on measurement of energy metabolism). Their data on *R. corsula* (16 cm) forced to swim at a speed of 43 cm/sec at various ambient oxygen concentrations below air saturation, are given in Fig. 7.13. It is seen in the figure that the scatter of oxygen consumption rates is wider at higher oxygen concentrations than at low. It is significant that the line drawn bounding the highest values shows a sharp decrease in oxygen consumption with the decrease in ambient oxygen, while the line bounding the lowest values is horizontal, indicating that the lowest oxygen consumption remained the same irrespective of change in ambient oxygen. As pointed out by Richmonds *et al.* if the range of values of oxygen consumption at a given concentration can be taken as a measure of excitement of the fish (Brett, 1964; Smit, 1965) it can be concluded that oxygen consumption of the mullet to swim at 43 cm/sec does not change markedly with reduction in ambient oxygen (within the levels tested). It can also be said that 'excitement metabolism' (Fry, 1971) decreases with an increase in stress due to low oxygen. A similar reduction in oxygen consumption and total metabolism (taking the RQ into consideration) with decrease in ambient oxygen was observed in the case of goldfish (Kutty, 1968*a*), possibly due again to the central regulation of metabolism enabling the animal under stress to direct the energy available for actual work alone (Fry, 1971). Richmonds *et al.* determined only a few RQ values (between 3 and 4 mg oxygen/l), but these did not show that *R. corsula* were anaerobic under the conditions of their experiments, but observations of Kutty & Mohamed (1975) suggest that the mullet forced to activity might have derived some energy anaerobically at lower ambient oxygen concentrations. Perhaps the increase in cost of ventilation with decrease in ambient oxygen in *R. corsula* swimming steadily was small under the test conditions.

Oxygen debt

Kutty & Mohamed (1975) subjected *R. corsula* to a hypoxic phase until the fish became asphyxiated and subsequently studied its recovery in air-saturated water. The metabolic rates and quotients obtained during the course of their experiments are shown in Figs. 7.14 and 7.15, the trends of which during the hypoxic phase are broadly similar to that described in Figs. 7.11 and 7.12. While the oxygen was being depleted the decrease in oxygen consumption was accompanied by an increase in carbon dioxide production. This clearly suggests an anaerobic involvement, but on exposure to air-saturated water (recovery) the fish became aerobic again (RQ near unity). The oxygen consumption and carbon dioxide production almost immediately reached the pre-hypoxic levels, which were maintained during the recovery period studied. A significant feature here is that oxygen consumption does not show a post-hypoxic increase, as normally expected in cases where an oxygen debt is incurred (e.g., Blazka, 1958, for brown trout). There are, however, instances in fish where an oxygen debt is not incurred,

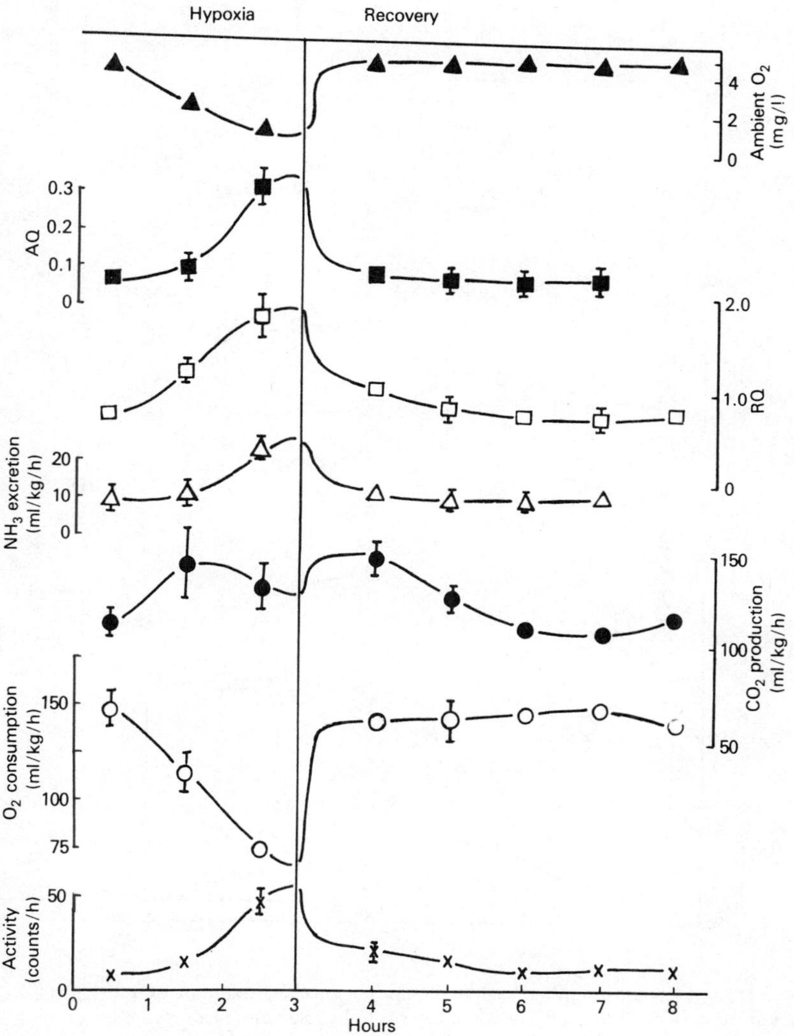

Fig. 7.14. Oxygen consumption, carbon dioxide production, ammonia excretion, RQ, AQ and random activity in *Rhinomugil corsula* subjected to hypoxia and subsequent recovery in air-saturated water. Acclimation and test at 30 °C. The end of hypoxic phase is indicated by the vertical line cutting across the curves fitted through the mean values. Each point shown is a mean (±SE) of five determinations on separate fish – see text for further details (After Kutty & Mohamed, 1975).

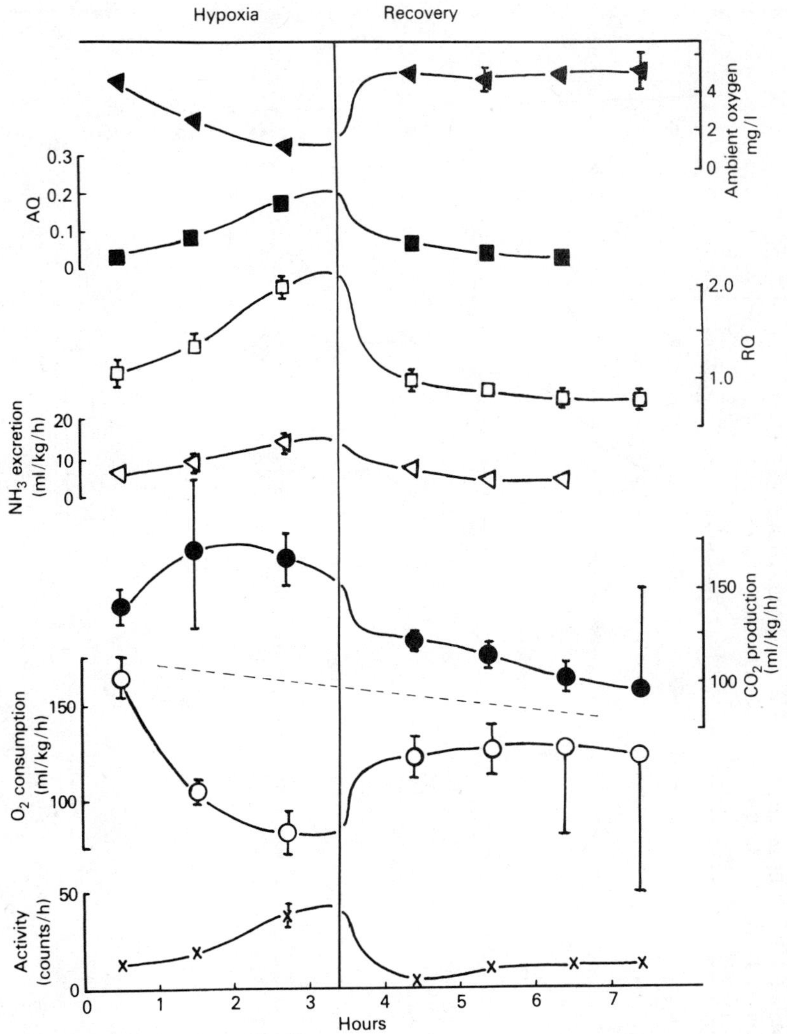

Fig. 7.15. Oxygen consumption, carbon dioxide production, ammonia excretion, RQ, AQ and random activity in *Rhinomugil corsula* subjected to hypoxia and recovery. Acclimation and test at 35 °C. See legend for Fig. 7.14 for further details.

as was observed in the crucian carp (Blazka, 1958). A measure of lactic acid accumulated in blood and muscle of the fish subjected to hypoxia would confirm this aspect for *R. corsula.* Indeed the available information on the vascular physiology of mullets is scanty in spite of those given by Itazawa (1959), McFarland (1965) and Cameron (1970). In the absence of the information on lactic acid accumulation a fact which supports that *R. corsula* do not accumulate an oxygen debt is that at both 30 and 35 °C the fish exposed to hypoxia did not exhibit an increase in oxygen consumption during recovery after hypoxia.

Biochemical mechanisms

The most well known anaerobic energy route in fish is glycolysis, resulting in the production of lactic acid (Black, 1955; 1957*a*, *b*, *c*; Black, Chin, Forbes & Hanship, 1959; Black, Robertson, Hanship & Chin, 1960; Hayashi, Green & Black, 1964; Heath & Pritchard, 1962; Caillouet, 1964; Karuppannan, 1972), but certain fishes such as the crucian carp which do not incur an oxygen debt, do not accumulate lactic acid (Blazka, 1958). This observation, however, has to be viewed in the context of the temperature relations of the enzyme, lactic dehydrogenase (Hochachka & Somero, 1973). The lactic acid produced would cause acidosis as well as the release of extra carbon dioxide mostly by displacement from the alkaline or bicarbonate reserve (Black, 1958; Black *et al.*, 1959; Kutty, 1968*a*). Carbon dioxide of this origin is referred to as 'acidic' or 'non-metabolic', as opposed to the 'organic' or 'metabolic' carbon dioxide which arises from the carbon atoms in the substrate degraded (see Peters and van Slyke, 1946). The extra carbon dioxide released during anaerobiosis, such as under hypoxia, can be of either or both kinds described. In the case of crucian carp it has been suggested that the anaerobic carbon dioxide released is organic (Blazka, 1958). Ekberg (1962) showed that excised gills of crucian carp when kept under nitrogen produced both acidic and organic carbon dioxide. Kutty (1968*a*) showed that goldfish sustained hypoxia for weeks, supplementing energy anaerobically by sustaining an RQ of 2, producing large amounts of extra carbon dioxide which was probably organic in origin (Hochachka, 1961). This is also supported by the observation that goldfish are not known to produce or excrete lactic acid (Prosser, Barr, Pinc & Lauer, 1957), even though this point has not been established definitely for goldfish. *Tilapia mossambica* sustain an RQ over unity (up to 8) (Kutty *et al.*, 1971*a*; Kutty, 1972) for short periods under hypoxia; the fish do accumulate lactic acid in muscle subsequent to exercise (Karuppannan, 1972) and thus it is likely that the extra carbon dioxide produced by *T. mossambica* under hypoxia is 'acidic'. While the magnitude of the RQ would indicate the intensity of anaerobiosis, total anaerobic capacity can be judged from the period for which the high RQ is sustained.

As pointed out earlier, *Rhinomugil corsula* are not known to accumulate lactic acid during hypoxia. However, since the fish do not incur an oxygen debt, it is likely that a good portion of the extra carbon dioxide produced (as signified by RQs over unity) during hypoxia is organic. The fish, however, sustain partial anaerobiosis only, since they are asphyxiated at an oxygen concentration of 0.8 mg/l. It is significant that the curve from oxygen consumption during hypoxia and recovery in *R. corsula* at 35 °C (Fig. 7.15) is almost mirror-imaged in the curve for carbon dioxide production. (This is not so evident in *R. corsula* at 30 °C, Fig. 7.14). The broken line drawn through the middle of the two curves indicates this. This would suggest a 1:1 compensation for the lack in oxygen consumption by carbon dioxide production. A similar ratio has been suggested in estimates of hypoxia and recovery in goldfish as well (M. N. Kutty, unpublished). Assuming that the substrate degraded is glycogen, it can be estimated that a mole of carbon dioxide ('acidic') resulting from glycolysis is indicative of 4.3 times less energy release than a mole of carbon dioxide produced ('organic') aerobically i.e. through complete degradation of mole of glucose.*

If energy is estimated thus, the extra carbon dioxide released by *R. corsula* during hypoxia would be very small. As pointed out earlier an indirect estimate, however, suggests that the hypoxic *R. corsula* (at 1.6 mg O_2/l at 30 °C) obtain about 50% of their energy needs from anaerobic sources. This again suggests that *R. corsula*, as do some other fishes such as the crucian carp and goldfish, might be obtaining anaerobic energy at least partially through pathways other than glycolysis, even though Hochachka's (1969) estimate of the energy derived thus was small. As a mechanism of anaerobic metabolism in crucian carp under anoxia, Blazka (1958) suggested conversion of carbohydrate into fat and Blazka and Kopecky (1961) suggested decarboxylation of pyruvate and condensation of resulting dicarbon units giving rise to adequate energy anaerobically. However, Blazka's observation that lower fatty acids accumulate in anoxic carp muscle could not be confirmed (Burton & Spehar, 1971).

Simultaneous breakdown of carbohydrate (glucose) and protein (amino acids) anaerobically, to yield appreciable quantities of energy, almost as efficiently as in aerobic systems, has been suggested by Hochachka & Somero (1973). This might result in the accumulation of multiple end products, such as alanine, succinate and lactic acid as observed in some diving vertebrates (Hochachka *et al.*, 1975) and release of extra carbon dioxide and probably also ammonia, but W. R. Driedzic & P. W. Hochachka (personal communication) found little evidence in the carp muscle for substantiating this fully. These workers observed that in the exercised carp muscle the lactic acid accumulation was not high and that most of the ammonia excreted came from adehylates. Whether this is true of anoxic carp muscle also is not known, but the muscle was also shown to accumulate a small amount of lactic acid, but none of the other known end products of anaerobic metabolism.

This aspect has not been checked in *R. corsula*, but it is known that

* Six moles of carbon dioxide are produced from one mole glucose (from glycogen) aerobically resulting in 39 moles of ATP, i.e., 39/6 = 6.5 moles of ATP per mole of CO_2; and 2 moles of CO_2 (displaced by lactic acid from 'alkaline reserve') are indicative of 3 moles of ATP, produced anaerobically from a mole of glucose (source: glycogen), i.e., 3/2 = 1.5 moles ATP per mole of CO_2. Therefore, the ratio of efficiency of energy release of aerobic and anaerobic CO_2 is 6.5/1.5 = 4.33.

anaerobic carbon dioxide production (RQ over unity) is correlated well with increased excretion of ammonia (AQ over 0.1), as was shown also in *Tilapia mossambica* under hypoxia (Kutty, 1972). It is likely that such unknown pathways, coupling anaerobic breakdown of carbohydrate and protein so resulting in more efficient energy release than in conventional glycolytic pathway, do exist in *R. corsula* (see also the section on routine and standard metabolism). Further investigations on this line would be most worthwhile. It must, however, be stated that unlike crucian carp, which are known to survive under complete lack of oxygen at low temperatures (Blazka, 1958) and goldfish which survive in 15% air saturation for months sustaining an RQ greater than 2 (Kutty, 1968*a*) the influence of chronic exposure to low oxygen in *R. corsula* has not been investigated.

General considerations

From the preceding discussion it is obvious that information on the total energy relations of any single species of mullet is far from complete. However, it may be worthwhile spelling out the different aspects of energy utilization in fish in general and their relevance to the available information on mullets.

The fate of food energy when it passes through the fish body is shown graphically in Fig. 7.16. The general scheme given has bearing on the studies on the bioenergetics of farm animals (Brody, 1945; Kleiber, 1961; Blaxter, 1965), but in the case of fish the scheme has been reinforcd notably from the studies of Winberg (1956), Paloheimo & Dickie (1965), Warren & Davis (1967), Brett (1970), Hastings & Dickie (1972) and Niimi & Beamish (1974). The Figure is largely a revised version of Brett (1970: Figs. 1 and 2). The estimates shown are presented on the basis that the fish is randomly active and fed with a ration less than satiation level, but above maintenance. The 'hatched' blocks represent the end levels of energy utilization, which are displaced upwards stepwise, passing through different energy stages beginning from the base, which represents food energy or intake energy. The apportioning of the food energy is suggested by the size of the blocks. From the food energy, energy losses take place at different steps. Thus of the energy contained in the food, 95% is assimilated (5% lost as faeces) and at the next step 10% of the food energy is lost in urine and branchial ammonia excreted (non-faecal loss) (Niimi & Beamish, 1974). Out of the remaining 85% (metabolizable energy), 10% is again shown as a loss as 'heat increment' or 'specific dynamic action' (SDA). The SDA is still not clearly defined, but it is considered as the energy needed for absorption, digestion, transport and deposition of energy in the tissues (as proteins, fats and carbohydrates). It is difficult to distinguish, experimentally, the energy spent for the actual act of feeding from SDA (see Leroy, 1965). The information on the cost of feeding and its separation from total metabolism appears inadequate (Muir & Niimi, 1972;

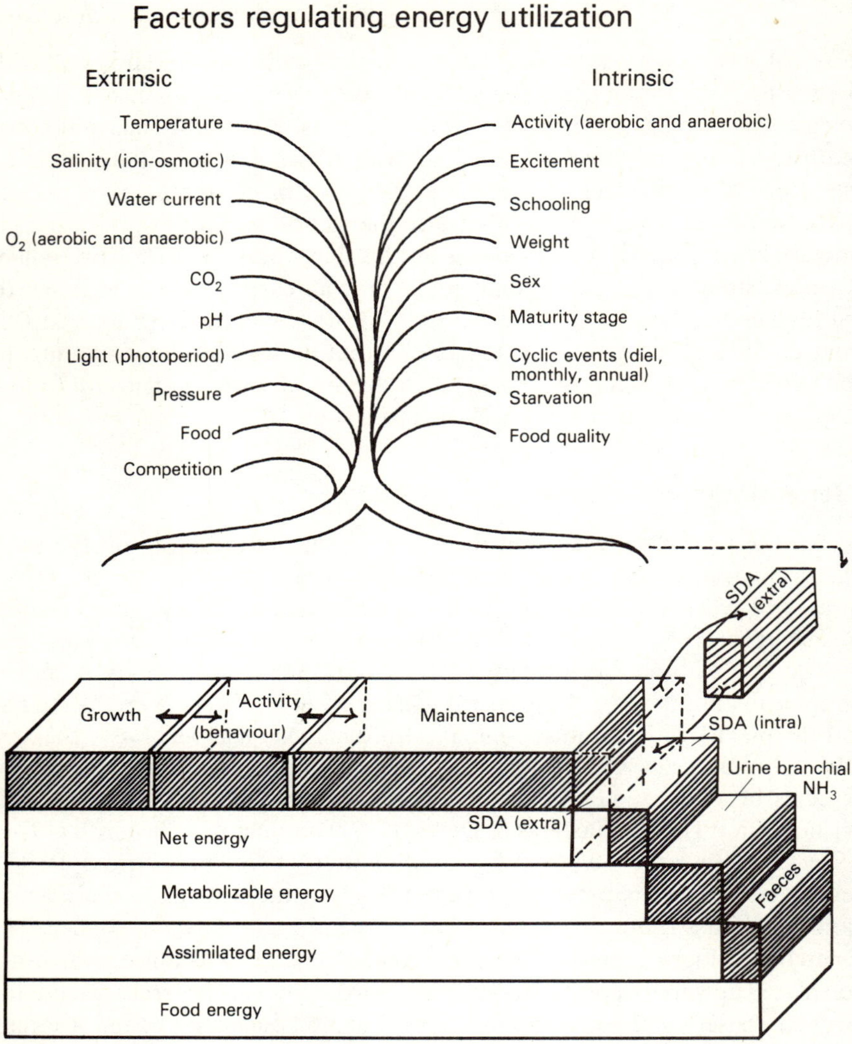

Fig. 7.16. Utilization of food energy. The hatched blocks represent the end levels of energy utilization which are shown displaced upwards through different energy stages from the base (food energy). The influence of the various factors (extrinsic and intrinsic) on metabolism, growth and overall energy utilization is also indicated. The apportioning of food energy is suggested by the size of the blocks. Specific dynamic action (SDA) is shown as of two portions: from source outside (extra, i.e., body store) food and from food itself (intra). The net energy available is shown as being utilized for maintenance (standard metabolism), growth and activity, in this case at routine level; with increase in activity (up to full 'scope') the portion of growth will be impinged, indicated by the thick arrow on the partition between 'growth' and 'activity'. The various factors are shown as influencing the total metabolism, but these might influence maintenance, growth and activity, and perhaps also the other losses of energy, each separately. The double-headed arrows on the partitions of compartments of net energy use, suggest the plasticity of quantities of energy for the different functions, depending on the specific extrinsic and intrinsic conditions prevailing. The food intake is assumed to be at some level of ration between maintenance and satiation.

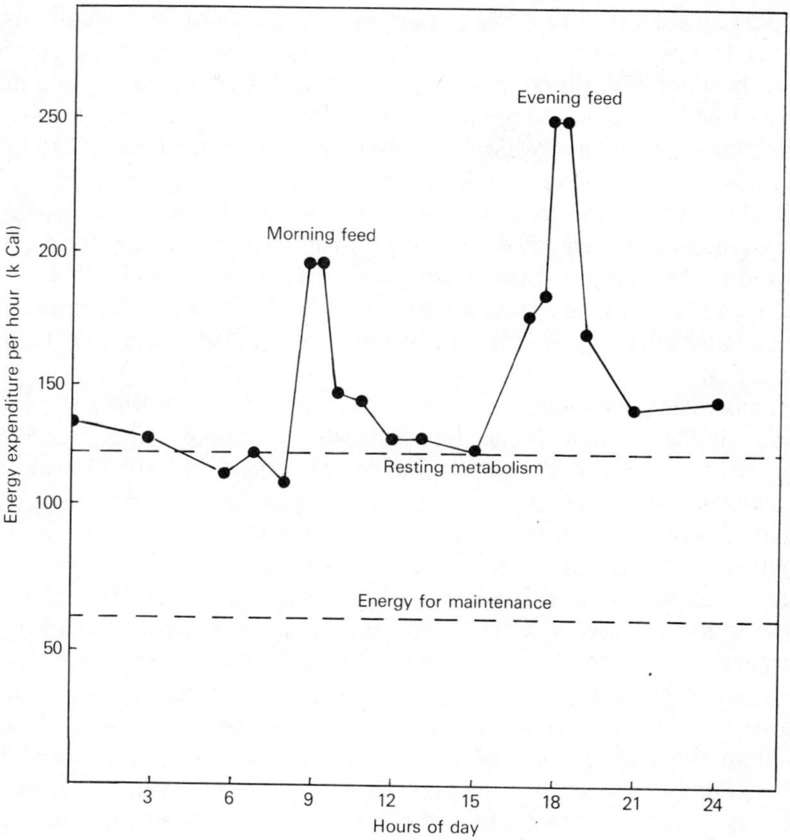

Fig. 7.17. Energy expenditure of a 48 kg pig measured at different times during the day (observations repeated during a period of 1 week). Each point corresponds to several measurements taken at the time of day indicated. (After Leroy, 1965).

Pierce & Wissing, 1974). The deficiency in experimental techniques will have to be overcome, so as to estimate energy utilization connected with feeding as that obtained for mammals, e.g., data on the pig (Fig. 7.17). Such a routine level of metabolism of fish can now be understood better with the knowledge of the other aspects of total metabolism.

Another aspect of SDA, brought to our notice by Niimi and Beamish (1974) is also worth mentioning here. How much of the heat increment due to SDA is accounted for from the food intake itself? The food energy could not have been involved at all in the physical transport of the food from the mouth to the site of digestion and assimilation. The energy for the secretion of digestive enzymes is also taken from the animal and not from food. This may be so for the reconstituion of energy deposit also. Thus the origin of the energy for SDA is not wholly the food itself, but the animal body. Thus it may not be proper to indicate the 10% loss of energy entirely as SDA from metabolizable energy. If it came from the animal can the SDA be taken as an 'energy debt' from the animal for digestion of food, as far as the energetics of food intake is concerned, since the energy can certainly be adjusted in the course of time and from the food energy itself? The SDA is then shown in Fig. 7.16 in two portions, from sources outside food ('extra' and from food itself ('intra'). The 'extra' SDA is shown as a 'block' of energy outside the 'food steps'. The arrows indicate the 'energy debt' being drawn from animal store for digestion of food and being replenished to the store from food energy itself subsequently. Again this aspect is elaborated here less with the view to adding precision to estimates of energy dissipation than as a topic of significance in studies of energetics.

The net energy available to the fish under routine conditions, is shown as that available for maintenance (less than 40% of intake energy, but 50% of net), activity and growth. The maintenance energy is indicated by standard metabolism. A measure of energy available for growth and activity could be obtained from the measurement of 'metabolic scope' (see Kerr, 1971). A 'competition' between the two forms of energy (for activity and growth) has been recognized (Brett, 1970). Standard metabolism would also shift according to the factors imposed. For this reason, double-headed arrows are shown over the partitions within the net energy block, to indicate movement of the partition walls suggesting change in relative energy use for maintenance, growth and activity. An activity increase (up to the full scope level) might result in the impingement of the growth compartment. This could be a behaviourally induced or regulatory situation caused by the environment (Fry, 1971), almost always taking place in one direction; the thick arrow pointing towards the growth compartment is suggestive of this. It is also recognized that almost all the factors extrinsic and intrinsic, which act on the organism will influence the size of one or the other or all the compartments of net energy shown in Fig. 7.16. The influence of factors such as temperature, water current, dissolved oxygen, etc., seem to be obvious, but the precise influence of several other factors is not well known.

As far as the present discussion on mullet is concerned, there appears to be virtually no information on food energy turnover, except for the studies of Odum (1970) and Job (1976). Job measured the caloric value of food in the fore and hind gut of freshly caught *Rhinomugil corsula* (150 g) and observed that the energy absorbed from food as it passed down the gut was about 40%. This is higher than the values he obtained for *Tilapia mossambica* and *Cyprinus carpio* from the same impoundment, but lower than in experimentally fed laboratory fish. *R. corsula* being an omnivore (see the Introductory section) could also, like *Mugil cephalus* (Odum, 1970), achieve a greater power output from a high rate of ingestion with only a moderate assimilation efficiency.

The information available on metabolizable energy in mullet data cannot be used to compose such a figure as shown in Fig. 7.16. The nearest usable data are on the standard metabolism (maintenance) and the metabolic scope of *R. corsula*. The standard metabolism of the fish at 30 °C is about 90 ml O_2/kg/h and metabolic scope about 350 ml O_2/kg/h, on the assumption that fish swimming at 75 cm/sec (*c*. 4.5 × body length/sec) sustains a near-active metabolism. The metabolic scope obtained experimentally appears to be different in concept, at least temporarily from the scope for growth (Warren & Davis, 1967). The former suggests the immediate capacity of the fish for activity, which if sustained long will result in no growth, whereas the latter (the long-term scope) would include scope for growth and activity limited totally by food intake. Thus the former or the short-term scope appears to be partially independent of food intake, but can be sustained for long periods only at maximum level of feeding, subject to physiological and behavioural restrictions.

The observations on *R. corsula* also suggest that in the short-term scope there would be an anaerobic component, probably about 20% of the active oxygen consumption when the activity is sustained for about one hour, but much higher for shorter bursts of activity such as the leaping of mullets. This may be significant especially when the energy requirement for foraging (see Kerr, 1971) and escape from predation is considered. The long-term metabolic scope may however, be fully aerobic, except when oxygen supply is restricted. It is not known if *R. corsula* are capable of obtaining energy anaerobically for sustained periods, but as has already been pointed out, in the case of goldfish, the anaerobic energy supplement under low oxygen was significant (RQ of 2) (Kutty, 1968*a*) and the crucian carp are reported to sustain anoxia for months in low temperatures (Blazka, 1958). *R. corsula* can also sustain an RQ of 2 for about an hour in a low oxygen tension. The efficiency of energy transfer from the intake would be different under these circumstances.

How temperature and other factors would affect this picture is also vitally important. The energy utilized in activity for example a fish swimming in a water current, can make a vast change in the energy made available for growth. As is well recognized, in the energy budgeting of migratory fish this aspect is of prime significance. Many mullets do make migratory movements between rivers and seas. *R. corsula*, however, remain and breed in fresh water under landlocked conditions, but the fish are mainly riverine. In this context the observations on metabolism of *R. corsula* at various swimming speeds are again significant. A point of interest here is the increased utilization of protein in *R. corsula* swimming for long periods (with a 4–10 times higher AQ than in resting fish).

Salinity also has an important role on the energy metabolism of both resting and active mullets (cf. Rao, 1968), but mullets such as *R. corsula* which breed in fresh water have an added advantage in culture work. Significant advances in induced breeding of mullets (Tang, 1964; Yashouv, 1967; Liao, 1974; Kuo, Nash & Shehadeh, 1974; Sebastian & Nair, 1974) have given hope to intensive culture of the fish. This hope can be materialized if mullet production involving profitable energy and economic relations, can be worked out in a controlled system such as that schemed for Brett's sockeye salmon.

Acknowledgements

I am grateful to Dr S. V. Job for critical reading of the manuscript and to Drs V. G. Jhingran and J. M. Thomson for providing useful information. Mr Mohamad Kasim assisted in the drawings and Mr N. Sukumaran in the preparation of the manuscript.

References

Ananthakrishnan, K. R. & Kutty, M. N. (1974). Mortality and breathing rate at high ambient temperatures in the cichlid fish, *Tilapia mossambica* Peters. *Indian J. Exper. Biol.* **12**, 55–9.

Bardach, J. E., Ryther, J. H. & McLarney, M. O. (1972). *Aquaculture – The Farming and Husbandry of Freshwater and Marine Organisms*, Wiley-Interscience, New York.

Beamish, F. W. H. (1964*a*). Influence of starvation on standard and routine oxygen consumption. *Trans. Am. Fish. Soc.* **93**, 103–7.

Beamish, F. W. H. (1964*b*). Respiration of fish with special emphasis on standard oxygen consumption. III. Influence of oxygen. *Can. J. Zool.* **42**, 355–66.

Beamish, F. W. H. & Mookherjii, P. S. (1964). Respiration of fishes with special emphasis on standard oxygen consumption. I. Influence of weight and temperature on respiration of goldfish, *Carassius auratus* L. *Can. J. Zool.* **42**, 161–75.

Bell, G. H., Davidson, J. N. & Scarborough, H. (1970). *Text Book of Physiology and Biochemistry*, 7th edn. The English Language Book Society, London.

Black, E. C. (1955). Blood levels of haemoglobin and lactic acid in some freshwater fishes following exercise. *J. Fish. Res. Board of Canada*, **12**, 917–28.

Black, E. C. (1957*a*). Alteration in the blood level of lactic acid in certain salmonid fishes following muscular activity. I. Kamloops trout, *Salmo gairdneri*. *J. Fish. Res. Board Can.* **14**, 117–34.

Black, E. C. (1957*b*). Alteration in the blood level of lactic acid in certain salmonid fishes following muscular activity. II. Lake trout, *Salvelinus namaycush*. *J. Fish. Res. Board Can.* **14**, 645–9.

Black, E. C. (1957*c*). Alteration in the blood level of lactic acid in certain salmonid fishes following muscular activity. III. Sockeye salmon, *Oncorhyncus nerka*. *J. Fish. Res. Board Can.* **14**, 807–14.

Black, E. C. (1958). Hyperactivity as a lethal factor in fish. *J. Fish. Res. Board Can.* **15**, 573–86.

Black, E. C., Chin, W. G., Forbes, F. D. & Hanship, A. (1959). Changes in pH, carbonate and lactate of the blood of yearling Kamloops trout, *Salmo gairdneri*, during and following muscular activity. *J. Fish. Res. Board Can.* **16**, 391–402.

Black, E. C., Manning, G. T. & Hayashi, K. (1966). Changes in levels of haemoglobin, oxygen, carbon dioxide, pyruvate and lactate in venous blood of rainbow trout

(*Salmo gairdneri*) during and following severe muscular activity. *J. Fish. Res. Board Can.* **23**, 783–95.

Black, E. C., Robertson, A. C., Hanslip, A. R. & Chin, W. G. (1960). Alterations in glycogen, glucose and lactate in rainbow and kamloops trout, *Salmo gairdneri*, following muscular activity. *J. Fish. Res. Board Can.* **17**, 487–500.

Black, E. C., Robertson, A. C. & Parker, R. R. (1961). Some aspects of carbohydrate metabolism in fish. In *Comparative Physiology of Carbohydrate Metabolism in Heterothermic animals*, ed. A. W. Martin, University of Washington Press, Seattle, Washington.

Blaxter, K. L. (1965). *Energy metabolism* (European Association for Animal Production Publication No. 11), Academic Press, London and New York.

Blazka, P. (1958). The anaerobic metabolism of fish. *Physiol. Zool.* **31**, 117–28.

Blazka, P. & Kopecky, M. (1961). Basic mechanisms of adaptations to anoxia and hypoxia. III. The physiology of adaptive phenomena. In *Proceeding of the Fifth National Congress of the Physiological Society*, Karlovy Vary, Czeckoslovakia, 13–16 June 1961.

Blazka, P., Wolf, M. & Cepala, M. (1960). A new type of respirometer for the determination of the metabolism of fish in the active state. *Physiol. Bohemoslov.* **9**, 553–59.

Bograd, L. (1961). Occurrence of *Mugil* in the rivers of Israel. *Bull. Res. Counc. Israel*, (*B*) **9**, 169–90.

Borgstrom, G. (1962). *Fish as Food. 2. Nutrition, sanitation and utilization*, Academic Press, London and New York.

Bosworth, M. W., O'Brien, H. & Amberson, W. R. (1936). Determination of the RQ in marine animals. *J. Cell. Comp. Physiol.* **9**, 77–87.

Braffield, A. E. & Solomon, D. J. (1972). Oxy-calorific coefficients for animals respiring nitrogenous substrates. *Comp. Biochem. Physiol.* **43**, 837–41.

Brett, J. R. (1952). Temperature tolerance in young Pacific salmon, genus *Oncorhynchus*. *J. Fish. Res. Board Can.* **9**, 265–323.

Brett, J. R. (1964). The respiratory metabolism and swimming performance of young sockeye salmon (*Oncorhynchus nerka*). *J. Fish. Res. Board Can.* **21**, 1183–226.

Brett, J. R. (1965). The relation of size to rate of oxygen consumption and sustained swimming speed of sockeye salmon (*Oncorhynchus nerka*). *J. Fish. Res. Board Can.* **22**, 1491–501.

Brett, J. R. (1970). Fish – the energy cost of living. In *Marine Aquaculture*, ed. W. J. McNeil, pp. 37–52. Oregon State University Press.

Brett, J. R. (1973). Energy expenditure of sockeye salmon, *Onchorhyncus nerka*, during sustained performance. *J. Fish. Res. Board Can.* **30**, 1799–809.

Brett, J. R. (1974). Marine fish aquaculture in Canada. In *Aquaculture in Canada* (Bulletin of the Fisheries Research Board of Canada No. 188) ed. H. R. MacCrimmon, J. E. Stewart & J. R. Brett, pp. 55–84, Ottawa. Fisheries and Marine Service, Department of the Environment.

Brett, J. R. & Glass, N. R. (1973). Metabolic rates and critical swimming speeds of sockeye salmon (*Oncorhyncus nerka*) in relation to size and temperatures. *J. Fish. Res. Board Can.* **30**, 379–89.

Brody, S. (1945). *Bioenergetics and Growth.* Reinhold Publishing Corporation, New York.

Burton, D. T. & Spehar, A. M. (1971). A Re-evaluation of the anaerobic end products of fresh-water fish exposed to environmental hypoxia. *Comp. Biochem. Physiol.* **40**, 945–54.

Caillouet, C. W. Jr. (1964). Blood lactic acid concentration of unexercised and exercised mature carp in winter and summer. *Iowa State J. Sci.* **38**, 309–22.

Cameron, J. B. (1970). The influence of environmental variables on the haematology of pinfish (*Lagodon rhomboides*) and striped mullet (*Mugil cephalus*). *Comp. Biochem. Physiol.* **32**, 175–92.

Cameron, J. N. & Cech, J. J. Jr. (1970). Notes on the energy cost of gill ventilation in teleosts. *Comp. Biochem. Physiol.* **34**, 447–55.

Cech, J. J. Jr., & Wohlshlag, D. E. (1973). Respiratory responses of the striped mullet, *Mugil cephalus* (L.) to hypoxic conditions. *J. Fish Biol.* **5**, 421–8.

Chervinsky, J. (1974). *Liza aurata* (Risso), (Pisces, Mugilidae) and its adaptability to various saline conditions. In *Proceedings of the IBP/PM International Symposium on the Grey Mullets and their Culture, Haifa*, 2–8 June 1974.

Creac'h, Y. (1966). Protein thiols and free amino acids of carp tissues during prolonged starvation. *Arch. Sci. Physiol.* **20**, 115–21.

Devanesan, D. N. & Chacko, P. (1943). Possibility of culture of certain marine mullets in freshwater tanks. *Proc. Natl. Inst. Sci. India*, **9**, 249–50.

Doudoroff, P. & Shumway, D. L. (1970). Dissolved oxygen requirements of freshwater fishes. In *FAO Fisheries Technical Paper No. 86 FIRI/T86, Water Pollution*, FAO, Rome.

Dill, W. A. (1944). The fishery of the lower Colorado River. *California Fish Game*, **30**, 109–214.

Ekberg, D. E. (1962). Anaerobic and aerobic metabolism in the gills of the crucian carp. *Comp. Biochem. Physiol.* **5**, 123–8.

Forster, P. R. & Goldstein, L. (1969). Formation of excretory products. Excretion, ionic regulation and metabolism. In *Fish Physiology*, vol. I, ed. W. S. Hoar & D. J. Randall, pp. 318–50. Academic Press, London and New York.

Fromm, P. O. (1963). Studies on renal and extra-renal excretion in a freshwater teleost, *Salmo gairdneri. Comp. Biochem. Physiol.* **10**, 121–8.

Fromm, P. O. & Gillette, J. R. (1968). Effect of ambient ammonia on blood ammonia and nitrogen excretion of rainbow trout (*Salmo gairdneri*). *Comp. Biochem. Physiol.* **26**, 887–96.

Fry, F. E. J. (1947). Effects of environment on animal activity. *Univ. Toronto Stud. Biol. Ser.* **55**, 1–62.

Fry, F. E. J. (1957). The aquatic respiration of fish. In *Physiology of Fishes*, vol. 1, ed. M. E. Brown, pp. 1–63. Academic Press, New York.

Fry, F. E. J. (1971). The effect of environmental factors on the physiology of fish. In *Fish Physiology*, vol. 6, ed. W. S. Hoar & D. J. Randall, pp. 1–98. Academic Press, London and New York.

Fry, F. E. J., Brett, J. R. & Clawson, G. H. (1946). Lethal limits of temperature for young goldfish. *Rev. Can. Biol.* **1**, 50–6.

Fry, F. E. J. & Hart, J. S. (1948). The relation of temperature to oxygen consumption in goldfish. *Biol. Bull. Woodshole*, **94**, 66–77.

Fry, F. E. J. & Hochachka, P. W. (1970). Fish. In *Comparative Physiology of Thermoregulation*, ed. G. C. Whittow, pp. 79–134. Academic Press, London and New York.

Garcia Romeu, F. & Motais, R. (1966). Mise en evidence d'echanges Na^+/NH_4^+ chez l'anguille d'eau douce. *Comp. Biochem. Physiol.* **17**, 1201–4.

Gerking, S. D. (1955). Endogenous nitrogen excretion of bluegill sunfish. *Physiol. Zool.* **28**, 283–9.

Ghosh, A. (1967). Acclimatization of the mullet (*Mugil parsia* Hamilton) fry in fresh water. *Proc. Zool. Soc., Calcutta*, **20**, 49–55.

Gordon, M. S. (1963). Chloride exchange in rainbow trout (*Salmo gairdneri*) adapted to different salinities. *Biological Bulletin*, Woodshole, **124**, 45–54.

Hastings, W. H. & Dickie, L. M. (1972). Feed formulation and evaluation. In *Fish Nutrition*, ed. J. E. Halver, pp. 327–74. Academic Press, London and New York.

Hayashi, K., Green, N. W. & Black, E. C. (1964). Carbohydrate metabolism during transportation of live rainbow trout, *Salmo gairdneri*. *Rep. Fac. Fish., Prefect. Univ., Mie*, **5**, 51–125.

Heath, A. G. & Pritchard, A. W. (1962). Changes in the metabolic rate and blood lactic acid of the bluegill sunfish, *Lepomis machrochinus* Raf. following severe muscular activity. *Physiol. Zool.* **35**, 323–9.

Hochachka, P. W. (1961). Glucose and acetate metabolism in fish. *Can. J. Biochem. Physiol.* **39**, 1930–41.

Hochachka, P. W. (1969). Intermediary metabolism in fishes. In *Fish Physiology*, vol. 1, ed. W. S. Hoar & D. J. Randall, pp. 351–89. Academic Press, London and New York.

Hochachka, P. W., Owen, T.G., Allen, J. F., & Whittow, G. C. (1975). Multiple end products of anaerobiosis in diving vertebrates. *Compar. Biochem. Physiol.*, **50B**, 17–22.

Hochachka, P. W. & Somero, G. N. (1971). Biochemical adaptation to the environment. In *Fish Physiology*, vol. 6, ed. W. S. Hoar & D. J. Randall, pp. 100–56. Academic Press, London and New York.

Hochachka, P. W. & Somero, G. N. (1973). *Strategies of Biochemical Adaptation*, W. B. Saunders, Philadelphia & London.

Hora, S. L. (1938). Notes on the biology of freshwater grey mullet, *M. corsula* Ham. with observations on the probable mode of origin of aerial vision in fishes. *J. Bombay Nat. Hist. Soc.* **40**, 62–8.

Hora, S. L. & Pillay, T. V. R. (1962). *Handbook of Fish Culture in the Indo-Pacific Region*. FAO Fisheries Biology Technical Paper No. 14.

Idler, D. R. & Clemens, W. A. (1959). *The Energy Expenditures of Fraser River Sockeye Salmon during Spawning Migration to Chilko and Stuart Lakes*. (International Pacific Salmon Fishery Commission, Progress Report 6).

Itazawa, Y. (1959). Gas content of the blood in response to that of medium water in fish. II. Comparison of responses in several species. *Bull. Jap. Soc. Sci. Fish.* **25**, 301–6.

Jhingran, V. G. (1975). *Fish and Fisheries of India*. Hindustan Publishing Corporation, New Delhi.

Job, S. V. (1955). The oxygen consumption of *Salvelinus fontinalis*. *Univ. Toronto Biol. Ser.* **61**, 1–39.

Job, S. V. (1976). Energy turnover of fish in freshwater impoundments. *J. Madurai Univ.* **5**, 79–81.

Job, T. J. & Chacko, P. I. (1947). Rearing of salt water mullet in fresh water of Madras. *Indian Ecol.* **2**, 12–20.

Karuppannan, N. V. (1972). Studies on the locomotory metabolism of *T. mossambica* (Peters), Thesis, Madurai University, India.

Kerr, S. R. (1971). Analysis of laboratory experiments on growth efficiency of fishes. *J. Fish. Res. Board Can.* **28**, 801–8.

Kleiber, M. (1961). *The fire of life. An introduction to Animal Energetics*. John Wiley & Sons, New York.

Konosu, S., Ozay, M. & Hashimoto, Y. (1964). Free amino acids in the muscle of a few species of fish. *Bull. Jap. Soc. Sci. Fish.* **30**, 930–4.

Kuo, C. M., Nash, C. E. & Shehadeh, Z. H. (1974). A procedural guide to induce spawning in grey mullet (*Mugil cephalus* L.). *Aquaculture*, **3**, 1–14.

Kuthalingam, M. D. K. (1959). Temperature tolerance of larvae of ten species of marine fishes. *Cur. Sci.* **28**, 75–6.

Kutty, M. N. (1968*a*). Respiratory quotients in goldfish and rainbow trout. *J. Fish. Res. Board Can.* **25**, 1689–728.

Kutty, M. N. (1968*b*). Influence of ambient oxygen on the swimming performance of goldfish and rainbow trout. *Can. J. Zool.* **46**, 647–53.

Kutty, M. N. (1969). Oxygen consumption in the mullet *Liza macrolepis* with special reference to swimming velocity. *Mar. Biol.* **4**, 239–42.

Kutty, M. N. (1972). Respiratory quotient and ammonia excretion in *Tilapia mossambica. Mar. Biol.* **16**, 126–33.

Kutty, M. N., Karuppannan, N. V., Narayanan, M. & Mohamed, M. P. (1971*a*). Maros-Schulek technique for measurement of carbon dioxide production in fish and respiratory quotient in *Tilapia mossambica. J. Fish. Res. Board Can.* **28**, 1342–4.

Kutty, M. N. & Kulandaivelu, G. (1973). Recording body temperature of fish with thermister thermometer. *Biol. News, Madurai University*, **2**, 26–9.

Kutty, M. N., Mohamed, M. P., Thiagarajan, K. & Leonard, A. N. (1971*b*). Modification of Fry's fish activity counter and respirometer. *Indian J. Exp. Biol.* **9**, 218–22.

Kutty, M. N. & Mohamed, M. P. (1975). Metabolic adaptations of mullet *Rhinomugil corsula* (Hamilton) with special reference to energy utilization. *Aquaculture*, **5**, 253–70.

Kutty, M. N. & Murugapoopathy, G. (1969). Low lethal temperatures of the fishes *Liza macrolepis* (Smith) and *Tilapia mossambica* (Peters). *Cur. Sci.* **38**, 171–2.

Lasserre, P. (1971). Increase in ($Na^+ + K^+$) dependent ATPase activity in gills and kidneys of two euryhaline marine teleosts, *Crenimugil labrosus* (Risso, 1826) and *Dicentrarchus labrax* (L. 1758), during adaptation to fresh water. *Life Sci.* **10**(2), 113–9.

Lercy, A. M. (1965). Analysis of the origin of the heat loss of pigs fed normally. In *Energy metabolism*, ed. K. L. Blaxter, Academic Press, London and New York. pp. 37–47.

Liao, I. C. (1974). Experiments on the induced breeding of the grey mullet in Taiwan from 1963–73. In the *Proceedings of the IBP/PM International Symposium on the Grey Mullets and their Culture*, Haifa, 2–8 July 1974. Abstracts of Papers, Section B, III, 5.

Maetz, J. & Garcia Romeu, F. (1964). The mechanism of sodium and chloride uptake by the gills of a freshwater fish, *Carassius auratus.* III. Evidence for NH_4^+/Na^+ and HCO_3^-/Cl^- exchanges. *J. Gen. Physiol.* **47**, 1209–27.

Malikova, E. M. (1957). Biochemical analysis of young salmon at the time of their transformation to a condition close to the smolt stage, and during retention of smolts in fresh water. *Tr. Latv. Otdel.* VNIRO, **2**, 241–55 (Translated from Russian by Fisheries Research Board of Canada, Translation Series No, 232, 1959).

Maros, L., Schulek, E., Molnár-Perl, I. & Pintèr-Szakács, M. (1961). Einfaches destillationverfahren zur titrimetrischen bestimmung von Kohlendioxyd. *Anal. Chim. Acta*, **25**, 390–9 (Fisheries Research Board of Canada Translation Series No. 596, 1965).

Mathur, G. B. & Shrivastava, B. D. (1970). An improved activity meter for the determination of standard metabolism in fish. *Trans. Am. Fish. Soc.* **99**, 602–3.

McFarland, W. N. (1965). The effect of hypersalinity on serum and muscle ion concentrations in the striped mullet, *Mugil cephalus. Publ. Inst. of Mar. Sci.* **10**, 173–86.

McFarland, W. N. & Moss, S. A. (1967). Internal behaviour in fish schools. *Science, New York*, **156**, 260–62.

Mohamed, M. P. (1974). Studies on the influence of hypoxia on fish metabolism and activity. Thesis, Madurai University, Tamil Nadu, India.

Moore, R. H. (1970). Changes in the composition of the swim bladder gas of the striped mullet, *Mugil cephalus*, during hypoxia. *Comp. Biochem. Physiol.* **34**, 895–9.

Muir, B. S. & Niimi, A. J. (1972). Oxygen consumption of the euryhaline fish aholeahole (*Kuhlia sandivicensis*) with reference to salinity, swimming and food consumption. *J. Fish. Res. Board Can.* **29**, 67–77.

Nair, K. K. (1944). The effect of Calcutta sewage on fish life. *Proc. Natl. Inst. Sci., India*, **10**, 459–62.

Narayanan, M. (1974). Studies on the biology of the mullet, *Rhinomugil corsula.* Thesis, Madurai University, Tamil Nadu, India.

Niimi, A. J. & Beamish, F. W. H. (1974). Bioenergetics and growth of large mouth bass (*Micropterus salmoides*) in relation to body weight and temperature. *Can. J. Zool.* **52**, 447–56.

Odum, W. E. (1970). Utilization of the direct grazing and plant detritus food chains by the striped mullet, *Mugil cephalus.* In *Marine Food Chains*, ed. J. H. Steele, pp. 222–40. Oliver & Boyd, Edinburgh.

Pakrasi, B. B. & Alikunhi, K. H. (1952). On the development of the grey mullet *Mugil corsula* (Hamilton). *J. Zool. Soc. India*, **4**, 123–40.

Paloheimo, J. E. & Dickie, L. M. (1965). Food and growth of fishes. I. A growth curve derived from experimental data. *J. Fish. Res. Board Can.* **22**, 521–42.

Peters, J. P. & Van Slyke, D. D. (1946). *Quantitative Clinical Chemistry*, 2nd edition, vol. I. William & Wilkins, Baltimore.

Philips, A. M. Jr. (1948). The utilization of carbohydrates by trout. *Fish. Res. Bull.* NY St. Conserv. Dept. **11**, 6–44.

Pierce, J. & Wissing, T. E. (1974). Energy cost of food utilization in bluegill (*Lepomis macrochirus*). *Trans. Am. Fish. Soc.* **103**, 38–45.

Pillay, T. V. R. (1947). Possibilities of mullet culture in India. *Indian Farming*, **8**, 544–8.

Pillay, T. V. R. (ed.) (1973). Coastal aquaculture. In *Proceedings of the IPFC Symposium on Coastal Aquaculture, 1970*, Fishing News (Books) Ltd., London.

Prosser, C. L., Barr, L. M., Pinc, R. A. & Lauer, C. Y. (1957). Acclimation of goldfish to low concentration of oxygen. *Physiol. Zool.* **30**, 137–41.

Rao, G. M. M. (1968). Oxygen consumption of rainbow trout (*Salmo gairdneri*) in relation to activity and salinity. *Can. J. Zool.* **46**, 781–6.

Richmonds, C. R., Arthur, M. T., Thangapandian, G. & Loyola, S. I. (1971). Oxygen consumption in the mullet *Rhinomugil corsula* (Hamilton) in relation to swimming speed. *Biol. News, Madurai University*, **1**, 5–8.

Saunders, R. L. (1962). The irrigation of the gills in fishes. I. Studies on the mechanism of branchial irrigation. *Can. J. Zool.* **39**, 637–53.

Sebastian, M. J. & Nair, V. A. (1974). The induced spawning of the grey mullet, *Mugil macrolepis* (Agans) Smith and the large scale rearing of its larvae. In *Proceedings of the IBP/PM International Symposium on the grey mullets and their culture*, Haifa, 2–8 June 1974. Abstracts of papers, Section B, III, 4.

Simmons, E. G. (1957). An ecological survey of the upper Laguna Madre of Texas. *Publ. Inst. Mar. Sci., Univ. Texas*, **4**, 156–200.

Smit, H. (1965). Some experiments on the oxygen consumption of goldfish (*Carassius auratus* L.) in relation to swimming speed. *Can. J. Zool.* **43**, 623–33.

Spoor, W. A. (1946). A quantitative study of the relationship between the activity and oxygen consumption of the goldfish. *Biol. Bull., Woodshole*, **91**, 312–25.

Spurgeon, V. D. (1947). Natural acclimatization of two species of mullets, *Mugil dussumieri* (CV) and *Mugil oeur* (Forsk) to freshwater conditions in the Nellore coast. *Cur. Sci., India*, **16**, 123–4.

Stroganov, N. S. (1962). Methods of study of respiration of fish and methods for ammonia determination, used in studies on fish metabolism. In *Techniques for the investigation of Fish Physiology*, ed. E. N. Pavlovskii, pp. 106–11, Izdatel'stvo Akademii Nauk SSSR, (Translated by National Science Foundation Washington, DC, PST, Cat. no. 1130 by the Israel Programme for Scientific Translation, Jerusalem, (1964)).

Tang, Y. (1964). Induced spawning of striped mullet by hormone injection. *Jap. J. Ichthyol.* **12**, 23–8.

Thomson, J. M. (1966). The grey mullets. *Oceanogr. Mar. Biol., Ann. Rev.* **4**, 301–35.

Tunison, A. V., Brockway, D. R., Shaffer, H. B., Maxwell, J. M., McCary, C. M., Palm, C. E. & Webster, D. E. (1943). Protein utilization by brook trout. *Cortland Hatchery Report, Albany*, NY, State Dep. **12**, 10–12.

Tunison, A. V., Phillips, A. M., Shaffer, H. B., Maxwell, J. M., Brockway, D. R. & McCary, C. M. (1944). Protein utilization by trout. *Cortland Hatchery Report, Albany*, NY State Dep. **13**, 13–14.

van Dam, L. (1938). On the utilization of oxygen and regulation of breathing in some aquatic animals. In Dissertation, Drukkerij 'Vonharding', Groningen.

Venkataraman, R. S. (1944). Acclimatization of salt water mullet *Mugil scheli* to freshwater. *Cur. Sci.* **13**, 239.

Wallace, J. H. (1974). Aspects of the biology of *Mugil cephalus* in a hypersaline estuarine lake on the east coast of South Africa. In *Proceedings of the IBP/PM International Symposium on the Grey Mullet and their Culture*, Haifa, 2–8 June 1974, Abstracts of papers, Section B, I, 4.

Warren, C. E. & Davis, G. E. (1967). Laboratory studies on the feeding, bioenergetics and growth of fish. In *The Biological Basis of Freshwater Fish Production*, ed. S. D. Gerking, pp. 175–214. Blackwell Scientific Publications, Oxford and Edinburgh.

Winberg, G. G. (1956). *Rate of metabolism and food requirements of fishes*. Belorussian State University. Minsk (Fisheries Research Board of Canada Translation Series No. 194).

Wohlschlag, D. E. (1964). Respiratory metabolism and ecological characteristics of some fishes in McMurdo Sound, Antartica. In *Antartic Research Series*, vol. 1, ed. H. W. Well, pp. 33–62. American Geophysical Union, Washington, DC.

Yashouv, A. (1967). Preliminary report on induced breeding of *M. cephalus* (L.) reared in captivity in fresh water ponds. *Bamidgeh*, **21**, 19–24.

Zambriborshel, F. S. (1962). The wintering biology of young mullets (*Mugil saliens* and *M. cephalus*), *Vopr. Ikhtiol*, **2**, 615–25.

Zenkevitch, L. (1963). *Biology of the Seas of USSR*, trans. S. Botcharskaya, Allen & Unwin, London.

Zismann, L. & Ben-Tuvia, A. (1975). Distribution of young mullets in the hypersaline Bardawil Lagoon on the Mediterranean coast of Sinai. *Aquaculture*, **6**, 59–75.

Note added in proof: Literature subsequent to 1975 has not been included for review or discussion in this chapter. Editor

8. Artificial propagation

C. E. NASH & R. M. KONINGSBERGER

History of mullet culture

Farming of the Mugilidae has been practised for centuries but generally the cultivation of this potentially invaluable source of animal protein for man has been small and non-intensive. Subsistence farming has been a tradition in the Mediterranean region, south-east Asia, Taiwan, Japan and Hawaii within fenced lagoons, creeks and swamps, and in man-made ponds.

Experimental work in intensive mullet culture, aimed at finding ways to optimize production which is a necessary forerunner of any large scale operation, is more recent. This increased effort has developed in approaches differing from region to region because of traditional needs and practices.

In the Mediterranean region, especially in Italy, the traditional valliculture methods employed in raising mullet are now advanced (D'Ancona, 1955; De Angelis, 1969). In Egypt a successful stocking experiment was carried out at Lake Qarun in 1921 (Faouzi, 1936) using *Mugil cephalus* and *Mugil capito*. No effort was made to optimize growth of the transplanted juveniles by supplemental feeding. A second transplanatation was made of *M. cephalus*, *M. capito* and *Mugil saliens*. The latter was reported to spawn successfully in the lake, but the stock of the other two species had to be replenished with juveniles from the estuaries (El-Zarka, 1968).

In the Soviet Union, experiments with mullet were carried out in the regions of the Black Sea and the Caspian Sea. In 1930 and 1934 *M. cephalus*, *Liza saliens*, and *L. auratus* were introduced into the Caspian Sea (Thomson, 1966). The transplantation of *L. saliens* and *L. auratus* was successful and their acclimatization, development and reproduction were studied in later years (Perceva-Ostroumova, 1951; Babaian, 1958). The work was followed by suggestions for further improvements in the region (Babaian, 1960; Chepurnov & Dmitriyev, 1962) and Apyekin & Tronina (1972) subsequently published the tentative results of experiments on the stimulation of the maturing and spawning of *M. cephalus*, *M. saliens* and *Mugil auratus*.

The mullet were introduced for a seconary crop in carp ponds in Israel in the 1950s (Perlmutter, Bograd & Pruginin, 1957; Pruginin & Kitai, 1957). Considerable research work was then performed on artificial feeding in ponds (Erman, 1958), seasonal and regional variations in the spawning season (Abraham, Blanc & Yashouv, 1966), histological studies of ovarian development in captivity (Abraham, Blanc & Yashouv, 1968), culture at different stock densities (Yashouv, 1966), and breeding and growth in captivity (Yashouv & Ben-Shachar, 1967).

Artificial propagation

In south-east Asia and the Far East efforts to intensify mullet culture were centred in Taiwan, India and Hong Kong. Farming of mullet was also practised in Japan, the Philippines, Indonesia and mainland China, but only as a seconary crop in the culture of other species. Research in these countries has been sporadic. There are two incidences of mullet rearing experiments in Japan. Between 1906 and 1909 at the Aichi Prefecture Fisheries Experiment Station feeding experiments with mullet juveniles were conducted and these were repeated forty years later (Tamura, 1966).

In the Philippines the possibility of raising mullet juveniles together with milkfish in the brackish-water ponds was first suggested by Adams, Montalban & Martin (1931). Although mullet were raised successfully with the milkfish for some time, a scientifically supervised rearing experiment was not conducted until 1953 (Blanco & Acosta, 1958). Polyculture of mullet with milkfish and carp was again proposed but has yet to be intensified. In 1971 preliminary experiments on induced spawning of *Mugil dussumieri* were carried out at the Marjan Sabalo Hatchery Experimental Station (Angelos, 1971) indicating that research on intensive mullet culture in the Philippines had not been abandoned.

In Hong Kong intensive culture of mullet was imposed successfully on the traditional practice of carp culture (Bromhall, 1954). The culturing method developed empirically used two pond systems stocked with *M. cephalus* as both the primary and the secondary crop (Lin, 1940). The ponds were fertilized with organic matter and the fish given artificial food. Culling started when the fish were about seven months old. Chow (1958) conducted growth studies on pond reared mullet and found that their growth rate compared favourably with that of natural stocks.

In India, with its extensive estuarine waters in Kerala, Bengal and Madras, mullet have been farmed from very ancient times. Research in intensive culture began in the 1920s when rearing experiments with young mullet were conducted in Madras at the Fisheries Department farm at Ippur (Campbell, 1921; Hornell, 1922) and at the Chingulpet Fort moat fish farm (Gravely, 1929). In the 1940s two further aspects were emphasized, namely studying the feasibility of acclimating mullet juveniles to fresh water, and developing practices in polyculture. The first acclimation experiments were undertaken in Madras with *Mugil troschelli* and *Mugil wagiensis* (Devanesan & Chacko, 1943; Job & Chacko, 1947). This was followed by acclimatization studies of *Mugil parsia* in Bengal (Mookerjee, Ganguly & Sircar, 1946) and of *M. cephalus* and *Mugil seheli* in Madras (Ganapati & Alikunhi, 1949).

Several other reports on existing farming practices with their problems were published at the time. Improvements for the brackish-water farming of mullet in the Ganges delta were suggested (Hora & Nair, 1944). Basu (1946) advanced the adaptation of Chinese and Philippine practices for Bengal farms, and Pillay (1947) published his extensive examination of culture in Bengal,

Madras and Kerala. The acclimation experiments led Panikkar (1951) to suggest that temperature and salinity tolerances of the mullet species be studied in detail, but little appeared in the literature until the report of Mohanty (1973). He recorded that juveniles of *M. cephalus* acclimated to fresh water were more tolerant of changes in conditions and that tolerance varied with length.

The most significant development in India has been the undertaking recently of the artificial propagation studies at Kerala. Ovulation but not fertilization was achieved with *M. cephalus* (Sebastian & Nair, 1973). However successful spawning and larval rearing were achieved by Sebastian and Nair (1974) with *Mugil macrolepis*.

Although the pond culture of Mugilidae is not practised on a large scale in Korea, the grey mullet is one of the important food fishes in its south-west region. One instance of research deserves mention. Yang & Kim (1962) described an experiment to obtain and hatch eggs of the grey mullet and to rear the larvae, but the work does not seem to have been continued.

In Taiwan, at about the same time, the knowledge gained from centuries of pond culture was utilized to begin an intensive programme on mullet culture. About 39 % of the commercial catch of *M. cephalus* consisted of pond reared fish, cultured in combination with carp in fresh-water ponds and with milkfish in brackish-water ponds. Tang (1964) estimated that ten million juveniles were required to support the pond culture of mullet each year. He pioneered work on the induced spawning of *M. cephalus* as he knew that these resources were not infinite.

In 1966 a comprehensive research programme was established at the Taiwan Fishery Research Institute for intensive culture and mass propagation of juveniles. In subsequent years the techniques of induced spawning of mature fish caught during their seasonal migration were refined and improved, and survival of the larvae extended to thirty days. In 1973 the larval survival had increased to 19.35% and the production of juveniles had been established. A summary of a decade of work at the Institute has been made by Liao (1974).

In the United States the earliest mention of mullet farming was made by the US Fish and Wildlife Service (Anon., 1940) and attributed to Prytherch. A general article by Sharpe (1945) about the same farm quoted a yield of 5000 lb of fish per acre. Hiatt (1944) obtained information on the role of mullet in the food cycle of Hawaiian fish ponds. Lunz (1951) described preliminary experiments of mullet culture in brackish-water ponds in South Carolina. In Florida an experiment to raise the pompano, *Trachinotus carolinus*, inadvertently resulted in a yield of which *Mugil curema* and *M. cephalus* constituted the majority of fish (Johnson, 1954).

At present, experimental work with Mugilidae is centred in Hawaii, Texas and Louisiana. In Hawaii, the Oceanic Foundation successfully achieved artificial spawning of *M. cephalus* following induced breeding of captive broodstock (Shehadeh & Ellis, 1970; Shehadeh, Kuo & Milisen, 1973*b*). Larval rearing efforts met with mixed success and individual survival figures (up to 25.5%) could not be guaranteed (Kuo, Shehadeh & Milisen, 1973*a*; Nash, Kuo & McConnell, 1974). In Texas mullet were raised successfully in ponds receiving the heated effluent from a power plant (Linder, Strawn & Luebke, 1974), and in Louisiana acclimatization experiments were conducted with mullet juveniles (Shireman, 1974).

From this brief historical survey it is evident that the direction of the experimental research and development for the culture of the Mugilidae has, of necessity, been toward induced spawning and larval rearing. For centuries the farming of mullet, as with other species, has depended on natural resources for the annual replenishment of stock. However the annual spawning migrations of the adults and the strength of the year-classes are all subject to environmental biological factors about which there is little agreement on importance. Present evidence indicates that these factors differ with locality, region and even between populations within the species. The natural availability of juveniles is therefore subject to unpredictable variations in occurrence and abundance. More importantly the juveniles are collected in those coastal and estuarine waters where pollution is threatening or in some cases has already changed the ecosystem. The future availability of resources in such waters is becoming increasingly uncertain. Finally, improvement in fishing techniques, particularly for schooling fish on their spawning migration, is affecting the natural population and hence their progeny.

The literature is full of reasons for the depletion of natural populations and weakening year-classes. Yet there is an anomaly for the proponents of culture. Commercial fishermen consider that too many juveniles are removed by pond fishermen thus depleting the resources of the natural fishery. Consequently limits have been imposed on the number of juveniles of certain species which can be removed for intensive culture.

The future of aquaculture is entirely dependent on the success of artificial propagation and larval rearing. Juvenile resources have to be independent of the natural population. Once this is achieved, then the benefits of genetics and selective breeding will make aquaculture the parallel of intensive agriculture with its fast growing strains and specialized breeds. When this occurs, the Mugilidae will really have an advantage over all other species because of the number of related types which can be used to develop particular farming strains for almost any region or specific location.

The need for an intensive effort in mass propagation has been summarized by Oren (1971). He said 'In order to achieve maximum production of mullets under controlled conditions, one of the axioms is a steady supply of fry independent of conditions in the natural environment, natural spawning and migration of fry inshore. Only when such a supply is available will the most economic and most efficient way of utilization of the mullet, as a consumer of the first trophic level, be channelled for the benefit of man.'

Adult broodstock

The success of an artificial propagation system for any plant or animal species depends primarily on the quality and quantity of broodstock resources. Large numbers of sexually mature individuals must be contained in good health under acceptable environmental conditions in order to reproduce and yield viable offspring. Some essential external conditions for broodstock fish are suitable water quality, a nutritious diet, a high standard of hygiene and limited physical disturbance. But of most importance the broodstock must be exposed to the correct environmental parameters which influence the physiological changes in the pituitary gland and stimulate the gonads to seasonal maturity.

The endocrine system of all vertebrates forms the main link between the reproductive organs and the environmental regulators. Rhythmic regulators such as temperature and photoperiod, mediated through the central nervous system, initiate neurosecretions which in turn regulate the activities of the pituitary gland. As one of many target organs the gonads are influenced accordingly. The reproductive cycles are thus regulated intimately by the trophic hormones of the pituitary.

Many experiments with fish have used gonad development to interpret the effects of certain environmental regulators on the reproductive cycle. Among the factors concerned temperature and photoperiod are the two most important which initiate pituitary activity for fish in temperate and subtemperate regions (Hoar, 1959). The relative importance of each varies with different species of teleosts. Photoperiod has been reported as the dominant factor influencing the reproductive cycle of *Enneacanthus obesus*, *Notropis bifrenatus* and *Fundulus confluentus* (Harrington, 1959), *Gasterosteus aculeatus* (Baggerman, 1957), *Salvelinus fontinalis* (Henderson, 1963) and *Oryzias latipes* (Yoshioka, 1962). Temperature has been shown as dominant for *Phoxinus laevis* (Bullough, 1940), *Apeltes quadracus* (Merriman & Schedl, 1941), *Gambusia affinis* (Medlen, 1951) and *Couesius plumbeus* (Ahsan, 1966).

Henderson (1963) concluded for *Salvelinus fontinalis* that the influence of an environmental regulator varied with the stage of gonad maturation. The period most responsive to a regulator could also vary between males and females of the same species, and the gametogenetic process may be independent of environmental regulators at certain stages of maturity.

The reproductive cycle of most vertebrates is under the dual control of an internal physiological rhythm and an external seasonal rhythm. The refractory period or resting stage of the reproductive cycle is considered to be the time during which the two rhythms coincide and reinforce each other. As the fish are exposed to changing environmental conditions such as photoperiod and temperature the external rhythm begins to dominate. Its influence on the reproductive processes is transmitted by changes in the quantity of gonadotropin released from the pituitary gland.

The physical containment and maintenance of adult grey mullet in the most suitable conditions for survival are not difficult problems. The fish readily become domesticated as testified by centuries of pond farming. They grow and are technically mature, although the final stages of gametogenesis are not completed and no incidence of total development and unassisted natural spawning in captivity has been recorded. At present these need the artificial stimulus of hormone injection to complete the breeding cycle.

In order to simulate the desirable environmental conditions for gametogenesis of captive broodstock, it is necessary to examine the conditions at locations where the grey mullet spawn in nature.

Natural spawning locations

The natural spawning locations of grey mullet species, which have been described or deduced by many workers throughout the world, do not indicate specific and similar environmental patterns. The overall picture is confused both by misidentification of generic types observed and the variety of facts on which the spawning record is being made. For example, many reports are based on the collection of adult fish with ripe gonads and usually loose eggs; others describe small numbers of eggs or larvae in plankton tows and make predictions about the spawning location based on tidal movement and stage of development of the samples. Some reports describe inshore schooling and then a migration offshore, presumed to be for spawning; others describe schooling inshore for spawning. The location of spawning grounds for the mullet can only be described as controversial.

Anderson (1958) reviewed the records of many American workers who suggested the time and place of spawning of *Mugil cephalus* along the South Atlantic and Gulf coasts of the United States. From the evidence of collection of larvae and the occurrence of juveniles on the coast from lower Florida to North Carolina, he believed that the striped or grey mullet spawned offshore over a broad area extending from about the 20 fathom line into the Gulf Stream. In contrast with the spawning of the silver mullet, *Mugil curema*, which began in early spring when water temperatures were rising over the continental shelf (Anderson, 1957), *M. cephalus* spawned during late fall or winter when water temperatures were falling. Breder (1940) observed and described an aggregation of adult fish in the shallow creeks of the Florida coast and believed that spawning was taking place. Although he collected no eggs, his detailed description of the movements of the males around the females compares exactly with the observed spawning behaviour of mullet in aquaria, and it must be assumed that the fish were attempting to spawn.

Broadhead (1953), Dekhnik (1953) and Arnold & Thompson (1958) provided authenticated instances of *M. cephalus* spawning at sea in surface waters, but over deep water (50 fathoms in the Black Sea, and 750 fathoms in the Gulk of Mexico). Fitch (1972) described the capture of a ripe female over 40 miles off the Baja California coast. Other more unusual spawning places have been suggested. Roughley (1916) believed spawning to occur in fresh water. Smith (1935), Breder (1940), Jacob & Krishnamurthi (1948) believed it took place in estuaries and tidal creeks, and Kesteven (1953) in the coastal surf zone of Australian waters.

Demir (1971) reviewed the information available on spawning of grey mullets in the North Atlantic. The spawning of *Mugil labrosus* was assumed to be during April in British waters, as schools of adults were observed offshore and the juveniles entered tidal pools on the Channel coast and in Southern Ireland in July and August. Kennedy & Fitzmaurice (1969) believed the spawning period to last several weeks in Irish waters with May as the peak spawning month. Hickling (1970), in his contribution to the natural history of the English grey mullets, noted the spawning period of *Crenimugil labrosus* from January to April and that one of its spawning locations was near the Isles of Scilly. Two ripening *Liza ramada* he found in autumn indicated a late spawning season. He did not find any *Liza aurata* with active gonads. Le Dantec (1955) noted that genital activity for *C. labrosus* lasted from January through until April for fish in the Biscay region.

In the Mediterranean, Caspian and the Black Seas, the spawning periods again varied from region to region. Yashouv & Berner-Samsonov (1970) produced an extensive review of the spawning seasons of five species of Mugilidae in the Black Sea and the Mediterranean. Hamis (1972) stated that the spawning areas of certain species of mullet were those places where the optimum conditions for the young were available. In the Black Sea, *Mugil saliens* spawned from late June to October, *M. cephalus* spawned from late May to late October and *M. auratus* spawned from early June to early November. In the Mediterranean the spawning of *M. saliens* took place between May and October, *M. cephalus* from early May to September, *M. auratus* from early September to late December, *L. ramada* from early October to late December, and *M. labrosus* from early December to early April.

Avanesov (1972) established that spawning of the grey mullet in the Caspian Sea occurred in a wide area beginning in the southern part at the end of May and extending northward. He recorded the most intensive spawning in the Turkmenian waters in the south in August, at a temperature of 25–29 °C and about seven miles offshore where the depth was 5–40 m. Spawning of the long-finned mullet he noted occurred later at a considerable distance offshore where depths were over 400 m.

Wimpenny & Faouzi (1935) first recorded the shoaling migrations of *M. cephalus* and *M. capito* during the spawning period. The schools moving from the Delta Lakes of Egypt to the sea were entirely composed of spawning fish. Thong (1969) observed migrations of *Mugil auratus*, *M. labrosus* and *M. ramada* during egg laying from the coastal regions of NW. France into the open sea.

In the Pacific and Indian Oceans little information on spawning of Mugilidae is available except for *M. cephalus* which spawns in late winter in the tropics and subtropics. Some summer spawners were recorded by Sarojini (1958). In tropical Bengal waters *Liza parsia* spawned between January and March, and *Liza cunnesius* from May onward and in the time of the monsoon. Kurian (1974) recorded the maturity of mullet in the ponds and lagoons of the Indian coast. He noted the different types of spawning migrations of *M. cephalus*, *M. tade* and *Liza macrolepis*, but all were seaward or near to the estuaries.

Wallace (1974) recorded the distribution of *M. cephalus* in the high saline lake system of St Lucia on the east coast of South Africa. He observed two seaward migrations in response to the changing hydrological and topographic features of the water system. No spawning was recorded in the lakes. Cervigón & Padrón (1974) observed *M. curema* in the high saline lagoons along the coast of Venezuela. They concluded that no spawning occurred in the lagoons but the adults migrated to the open sea and possibly for two spawning periods a year. Table 8.1 summarizes the spawning seasons of Mugilidae.

The records deducing the spawning locations from the numbers of eggs and larvae in plankton samples make no reference to the time of day at which spawning might occur. The observations of Deknik (1953) and Arnold & Thompson (1958) indicated that *M. cephalus* spawned at night at the water surface but over considerable depth. Anderson (1957) reported night spawning of *M. curema*.

Although much reference is made to water depth and location of spawning sites offshore, no workers report that the fish themselves descend deep to spawn. Hotta (1955) recorded very small larvae of *M. cephalus* in plankton tows from 100 fathoms deep near Japan, and Zviagina (1961) found eggs of *Liza haematocheila* deep in Peter the Great Bay.

The incidences of mullet eggs and larvae being taken in plankton tows are few. Demir (1971) recorded the occurrence of postlarvae of *M. auratus* and *M. labrosus* in British waters for the first time. Previously the eggs and larval stages of *Mugil labrosus* had only been recorded once by Sanzo (1936), and the eggs only of *Mugil auratus* by Sanzo (1931) and Vodyanitskii & Kazanova (1954). Although depth may not be important for the adults at spawning it may be important for incubation and larval development (pp. 281 and 289).

From histological examination of adults Stenger (1959) believed that *M. cephalus* spawned more than once a year. Bromhall (1954), on the evidence of the size and distribution of the juveniles, concluded that there were two periods of spawning in the vicinity of Hong Kong with a lunar period apart.

Table 8.1. *Spawning seasons of Mugilidae according to the literature*

		Apr.	May	June	July	Aug.	Sept.	Oct.	Nov.	Dec.	Jan.	Feb.	Mar.
Mugil saliens	Israel		—	—	—	—	—						
	Tunisia			—	—	—							
	Black Sea			—	—	—	—						
	Venetian lagoons			—	—	—							
Mugil chelo	Israel									—	—	—	—
	Tunisia									—	—		
Crenimugil labrosus	British waters	—									—	—	—
	Irish waters		—										
	Biscay region	—									—	—	—
Mugil labrosus	NW France	—										—	—
	N Adriatic											—	—
Mugil auratus	Israel								—	—	—		
	Tunisia						—	—					
	Black Sea			—	—	—	—	—	—				
	Messina									—	—	—	
	Castellón		—	—	—								
	NW France				—	—	—						
Mugil capito	Israel								—	—			
	Tunisia							—	—				
	NW France				—	—	—	—					
Mugil cephalus	Gulf of Mexico								—	—			
	Formosa								—	—	—	—	
	W Florida							—	—	—	—	—	
	E Florida, N Carolina							—	—	—	—	—	
	Israel							—	—	—			
	Corsica					—	—						
	Tunisia					—	—						
	Egypt				—	—	—	—	—				
	Black Sea			—	—	—	—						
	Australia (winter)	—	—	—	—								
	N Adriatic						—	—					
	SW India		—	—	—	—							
	Hawaii									—	—	—	

The references consulted were: Australia, Thomson (1963); Biscay region, Le Dantec (1955); Black Sea, Hamis (1972); British waters, Hickling (1970); Castellon, Belloc (1938); Corsica, Belloc (1938); Egypt, Paget (1923); Florida east coast North Carolina, Anderson (1958); Florida west coast, Broadhead (1953); Formosa, Tang (1964); Gulf of Mexico, Arnold & Thompson (1958); Hawaii, Kuo & Nash (1975); Irish waters, Kennedy & Fitzmaurice (1969); Israel, Yashouv & Berner-Samsonov (1970); Messina, Sanzo (1931); N Adriatic, Sanzo (1936); NW France, Thong (1969); SW India, Kurian (1974); Tunisia, Heldt (1948); Venetian lagoons, Gandolfi & Orsini (1970).

However it is generally assumed under normal circumstances that most grey mullet species produce only one brood of eggs each year, and that some females only spawn in alternate years after their first maturity (Thomson, 1955).

In summary, from the extensive and conflicting data and the possible misidentification by the observers, it is not possible to define a pattern in the environmental conditions necessary for successful natural spawning. Hamis (1972) was probably close to the truth when he stated that the spawning areas of certain species of mullet were those places where the optimum conditions for the young were available. More importantly, he should have included the optimum conditions for the incubation of eggs.

Photoperiod and temperature rhythms time the onset of gametogenesis and these rhythms, which differ from region to region throughout the world, influence the indigenous population of mullet species. Existing data on the influence of salinity, temperature and dissolved oxygen on the viability of the eggs of *M. cephalus* (see pp. 282–3) show that incubation and development occur to some degree within a wide range of these parameters. The results also show that there are optimum conditions for both.

Environmental data on the temperature of the water during spawning of *M. cephalus* indicates some adaptation from region to region, with temperatures recorded between 12 and 24 °C. All records show a strong preference by the fish for oceanic water as the medium for incubation, with salinities of 32–35 ‰. It is interesting to note the observations of Wallace (1974) and Cervigón & Padrón (1974) for fish which lived in hypersaline conditions but which migrated to normal oceanic salinities to spawn.

Kuo, Nash & Shehadeh, (1974*b*) demonstrated that environmental manipulation of photoperiod and temperature increased the individual spawning frequency of adults, and that the spawning season could be prolonged throughout the year. The manipulations were all made in sea water of salinity 32‰. Their evidence indicated that the spawning behaviour of the grey mullet species was not a strictly controlled and regulated act. In fact the evidence pointed to a loosely controlled behavioral response. Therefore the many controversial and apparently misleading observations on the spawning period and location will be authentic for that respective population at that specific time of year and in that particular region.

It is concluded that the adults move to spawn in the nearest location which will provide the eggs and larvae with the highest chances for survival, specifically to oceanic water of salinity of 32–35‰ but up to 39.5‰ for the eastern Mediterranean and 41‰ for the northern Red Sea have been recorded. The distance of the migration may be short or long, depending on the local conditions of topography and tidal movements. For example, in Hawaii where there is no strong tidal movement but a coastal gyre, the adults move offshore beyond the coastal reef a distance of one or two miles; in

Australia the fish move upcurrent many miles on both the west and east coasts before spawning, so that the long natural drift will return the juveniles back to the home estuaries several weeks later. Depth does not appear to be a vital factor for survival. Although the eggs have a natural buoyancy and are capable of surviving great pressures and can be found in deep water, the records do show that eggs and larvae develop predominantly in the upper ocean layers.

Artificial spawning conditions

No record has been made of the Mugilidae spawning unaided in captivity or in artificial conditions. Successful fertilization of hand stripped eggs and milt from adults caught at sea has been claimed by Sanzo (1936) and Belloc (1938). Other successes have been reported with fish matured in large ponds. The chances for this practice to become the base of an artificial propagation unit are as yet too small to be considered further. Reliable results have been obtained from fish matured in captivity or captured at sea but induced to complete gametogenesis and spawn by the injection of hormones.

Suitable artificial conditions for holding the broodstock of grey mullet can only be described as those enclosures which permit the female fish to develop their oocytes beyond the tertiary yolk globule stage (stage III as described by Kuo *et al.*, 1974*b*) and the males to complete spermiogenesis. Without hormone stimulation females will not mature to the ripe stage (stage IV) prior to ovulation, but will undergo atresia (stage V) and degenerate.

The first report on the induced spawning of *Mugil cephalus* reared in captivity in fresh water ponds was made by Yashouv (1969). Liao (1974) summarized the work in Taiwan between 1963 and 1973 on the propagation of *M. cephalus*, using spawning fish collected from the sea and from fish contained in large salt-water ponds. For the broodstock the workers in Taiwan relied on strong uninjured male and female specimens selected from the catches of commercial fishermen during the spawning run along the southwest coast of Taiwan. The selected fish were placed in strong plastic bags filled with sea water and inflated with oxygen. Most of the mullet caught were of the IV-year class and measured about 32–50 cm in length and weighed 1.0–2.1 kg each. The specimens were all sexually mature with well developed gonads, but the eggs of the females were never fully ripe for natural spawning.

Pond stock were maintained in Taiwan in fresh water at first, with sea water added slowly over a three-month period prior to spawning. The fish were fed a special diet and also injected periodically with mullet pituitary glands and Synahorin. Pond reared fish were easier to spawn than wild stock as they were more docile and free from injury. Large tanks specially constructed for holding broodstock were made of concrete and measured 5 × 7 × 1.5 m deep. Sea water was circulated through each and all were aerated continuously.

Artificial propagation

Shehadeh, Kuo & Nash (1973*c*) established broodstocks of *M. cephalus* in small, rubber-lined dirt ponds supplied with circulating sea water. In addition, the ponds contained a substrate of weighted polyethylene strips which increased the internal surface area and provided a stable supply of benthic diatoms, blue-green and filamentous algae. Three-year old fish survived readily in such conditions and matured the following year. They were then successfully induced to spawn by hormone injection. The small ponds were excavated and lined with butyl rubber sheet (1 mm thick) and had a volume of 26 m^3. A food supplement was provided in addition to the available natural growth.

Sebastian & Nair (1974) reported on the collection and holding of *Mugil macrolepis* in preparation for spawning. Mature fish were taken from local brackish waters in Chinese dip-nets and transported to small concrete tanks, 170 × 95 × 70 cm deep. The gravid female fish were 13–23 cm in length and 40–130 g in weight. Water in the tanks was changed intermittently.

Kuo *et al.* (1974*b*) described at length the holding of *M. cephalus* subjected to environmental manipulations of temperature and photoperiod. The time of onset of vitellogenesis was determined to be about eight weeks after exposure to a short photoperiod regime (6 L/18 D) at temperatures ranging from 17 to 26 °C. The response of oocyte development to the retarded photoperiod regime was consistent and unrelated to any other preconditioning of photoperiod changes, including a simulated natural light cycle. The data also indicated that development of the oocytes was accelerated by constant exposure to a temperature of 17 °C and a 6 L/18 D photoperiod, but that it was not completed as only limited yolk deposition occurred in the tertiary stage.

For *M. cephalus*, it is essential to maintain broodstock prior to spawning in full saline conditions (32–35‰, and up to 39‰ for the eastern Mediterranean). There is also evidence that the fish mature readily even after prolonged periods in captivity. Yashouv (1969) reported some success working with *M. cephalus* in fresh water, but the work of Hines & Yashouv (1971) on the increased activity of the spermatozoa of *M. capito* in sea water together with the practical experience of others working with *M. cephalus* indicate a preferential use of sea water for holding broodstock.

Ambient temperature and photoperiod conditions regulate the normal seasonal maturation of captive stocks. But the feasibility of breeding throughout the year or extending the breeding season by manipulation of the photoperiod and temperature regimes has been proved and will produce greater use and increased efficiency of propagation facilities.

Spawning behaviour and fertilization

Breder (1940) and Arnold & Thompson (1958) made some of the first observations on the spawning behaviour of *Mugil cephalus* in nature. They observed large numbers of fish schooling but scattered into small groups, generally made up of one large female and a varying number of smaller and more active males. The groups remained close together as if attached.

The induced spawning techniques developed by Shehadeh & Ellis (1970) and Shehadeh *et al.* (1973*b*) permitted natural spawning behaviour following injection of regulated doses of purified salmon pituitary gonadotropin. The final courtship and spawning behaviour described by them was similar to that observed by the others many years before.

As a consequence, for the final stages of the present induced breeding techniques with natural spawning, two or three males are placed in an aquarium with each recipient female about 2 h before spawning. A female will usually spawn some 12 h after receiving the second and last injection. The males become more active as hydration in the female progresses, indicated by distension of the belly and frequent excretion of calcium deposits.

Spawning is heralded by a violent quivering of the males which are then lying parallel to and facing the same way as the female and touching. The first release of a small number of ripe eggs stimulates the males to liberate spermatozoa. The female then responds with an explosive and continuous release of eggs.

Although male fish were once given exogenous hormone treatment to finalize maturity the practice was found unnecessary for spawning in the natural breeding season. If males are needed for spawning out of season then spermiation can be induced readily by the injection of 17-alpha methyltestosterone (Shehadeh, Madden & Dohl, 1972).

The effectiveness of the hypophysation technique for spawning depends ultimately on the selection of suitable recipient fish at the proper stage of ovarian development. For species of fish which undergo normal gonad development but fail to spawn in captivity, identification of this stage is critical.

To date, selection of recipients has been largely subjective. External anatomical characteristics have been described and used, e.g., depth and fullness of belly, colour and state of swelling of the cloaca, softness and resilience of the belly, roughness of pectoral fins or presence of head tubercules. More complex descriptions include the microscopic appearance of oocytes (Sundararaj & Goswami, 1969), histological structures of eggs (Chen, Chow & Sim, 1969), or other histological data. Other physiological parameters associated with sexual maturity, such as elevated plasma proteins and calcium concentration, have been used but are of little practical use. Shehadeh, Kuo & Milisen (1973*a*) described a method for the assessment of ovarian maturity *in vivo*, which was accurate and reliable and could replace all subjective methods for Mugilidae.

Kuo, Nash & Shehadeh (1974*a*) described standard procedures developed and applied regularly to induce spawning of *M. cephalus* under controlled conditions. Methods for determination of the stages of egg development and required dosage of partially purified salmon gonadotropin (SG-100) for spawning were illustrated and emphasized so that the procedures could be readily used by other culturists. The potency of the gonadotropin was described by Donaldson, Yamazaki, Dye & Philleo (1972) as 1 mg equivalent to 2150 IU human chorionic gonadotropin (HCG).

Liao, Lu, Huang & Lin (1971), from experience gained spawning *M. cephalus* freshly taken from coastal waters, recommended a first injection within 1 h after stocking in the tanks and again within 24 h. The total dosage for each fish was 2.75–5 mullet pituitary glands mixed with 20–50 Rabbit Units of Synahorin. Vitamin E was also injected into the fish. The fish were later stripped and the eggs fertilized by the dry method after careful observation for the correct spawning time. The best response was judged by distension of the belly after about 10 h, a loose and soft belly, and finally the release of 'water eggs' particularly when the spawner was handled.

Anatomical characteristics are not always the most reliable indication of maturity, particularly for fish held captive all year round. Many females with soft and enlarged bellies can be in the early stage of oocyte development even during the natural breeding season. Enlarged abdomens are often the result of engorged intestines and accumulation of visceral fat.

Ovarian maturity, that is the stage of development of intra-ovarian oocytes, is most accurately obtained by the method of Shehadeh *et al.* (1973*a*). Intra-ovarian oocytes are removed *in vivo* from an unanaesthetized female through a polyethylene cannula. The cannula is inserted into the oviduct for a distance of 6–7 cm from the cloaca, and oocytes sucked orally into the tube by the operator as the cannula is withdrawn. Oocyte samples from the mid-portion of the ovary are the most representative and sampling error is minimized by avoiding the extremities.

The oocytes are removed from the cannula and washed and preserved in a solution of 1% formalin in 0.6% sodium chloride solution. They are then placed on a small 'Plexiglas' plate and measured with an ocular micrometer. Fine grooves cut in the plate align the oocytes and facilitate measurement. Egg diameters are measured along the horizontal axis and the measurements grouped into 50 μm class intervals. The sexual maturity of the fish is expressed in terms of mean egg diameter, calculated from the egg diameter frequency distribution.

The oocytes of *M. cephalus* develop in synchrony. Ovarian development therefore is determined accurately and quickly without sacrificing female fish. The method also provides a means to observe and record oocyte development in individual fish and thus precludes variation between females in the broodstock. Furthermore it replaces the need for any histological processing and examination of oocytes.

Eggs

Fecundity

Thomson (1963) reviewed a number of papers which reported the fecundity of *Mugil cephalus* in terms of total egg numbers. He quoted estimates of 1.2–2.8 million eggs per fish. Sebastian & Nair (1974) estimated the fecundity of *Mugil macrolepis* at 1.2–4.0 million eggs per fish depending on size. Hickling (1970) listed the fecundity of several species of grey mullet as follows:

Species	No. of eggs (thousands/kg)	Reference
Mugil cunnensis	15–57	After Sarojini (1958)
Aldrachetta forsteri	126–650	Thomson (1957)
Mugil parsia	200–600	Sarojini (1957)
Crenimugil labrosus	372–745	Hickling (1970)
Liza ramada	581–1243	Hickling (1970)
Mugil cephalus	1200–2800	Thomson (1963)
Mugil cephalus	3600–7200	Nikolskii (1954)

Kuo, Shehadeh & Milisen (1973*a*) determined the fecundity of *M. cephalus* to be about 648 eggs/g body weight of three-year old fish. Nash *et al.* (1974) quoted 849 eggs/g for older individuals. Liao, Cheng, Tseng, Lim, Hsieh & Chen (1972) reported 0·7–1.9 million/fish.

In Hawaii and Taiwan, where induced spawning of *M. cephalus* is regularly performed, IV-group females weighing between 1 and 2 kg each are preferred for breeding. Consequently over 1 million eggs per female fish are released for fertilization following the induced hormone treatment.

Morphology and quality

The morphology of eggs before fertilization was described for several of the Mugilidae by the early naturalists, most of whom made reference to the characteristic large oil globule. The availability of eggs at all stages of development during the induced breeding procedure has resulted in full descriptions for *Mugil cephalus* by Sanzo (1930), Tang (1964), Yashouv (1969), Liao *et al.* (1971), Kuo *et al.* (1973*a*) and Tung (1973); by Sanzo for *M. chelo* (1936) and *M. labeo* (1937); by Anderson (1957) for *M. curema*; by Perceva-Ostroumova (1951) and Dekhnik (1954) for *M. saliens.*

Eggs of the Mugilidae are spherical and transparent. The surface of the egg shell is smooth and unsculptured. The yolk appears unsegmented and there is predominantly one large oil globule making the eggs extremely buoyant. The eggs are not adhesive.

Eggs from *M. cephalus* and *M. capito* were observed by Yashouv & Berner-Samsonov (1970) to have more than one oil globule and these subsequently developed and hatched. During development they observed the droplets to merge. On hatching the larvae had one oil globule (rarely two) located in the yolk sac.

Sanzo (1936) described eggs of *M. chelo* with one large and several smaller globules, and Perceva-Ostroumova (1951) noted the same for *M. saliens*. Kuo *et al.* (1973*a*) stated that the frequency of multiple oil droplets in eggs of *M. cephalus* increased with the manual pressure of artificial stripping. Spontaneous release of the eggs by the females produced eggs with a single oil globule and Nash *et al.* (1974) considered that to be normal and desirable. Although the small oil globules were observed to coalesce during development the survival of eggs which initially contained multiple oil droplets was always low.

Yashouv & Berner-Samsonov (1970), in an extensive contribution to the knowledge of eggs and early larval stages of Mugilidae, reviewed data of egg and oil globule diameters in samples from *Mugil saliens*, *M. cephalus*, *M. capito*, *M. auratus* and *M. chelo* at a variety of locations. They included some data from the synopsis on *M. cephalus* prepared by Thomson (1963). The comprehensive data revealed a wide range of diameters reported for the same species in different locations.

Kuo *et al.* (1973*a*) reported the mean egg diameter of fertilized eggs of *M. cephalus* as 930 μm, with a range of 880–980 μm. The single large oil globule had a uniform diameter of 330 μm. Tung (1973) quoted a mean egg diameter of 0.89 mm for the same species, and oil globule diameter of 0.39 mm. Nash *et al.* (1974) specified a mean egg diameter of 0.93 mm.

A question posed by many workers culturing either Mugilidae or other species by induced breeding is whether the quality of individual oocytes is inferior to those produced in nature. Induced breeding does accelerate final development and therefore the egg and subsequent embryo may be deficient in certain biochemical constituents necessary for total development.

Bromhall (1954), Sarojini (1958), Erman (1961) and Hickling (1970) examined the gonads of several species of mullet including *M. cephalus* and observed that their state of development was often consistent with a seasonal production of more than one batch of eggs, possibly within a period of a month. Kuo, Shehadeh & Nash (1973*b*) demonstrated that *M. cephalus* could be induced to spawn more than once a year. It is possible that artifically induced spawning, with its positive climax and total release of oocytes from the ovary, may be an unnatural forced reaction impeding complete development of the embryonic stages.

Little work has been accomplished on the biochemical composition of the oocytes before and after spawning. Kuo (unpublished data) examined changes in the biochemistry of eggs of *M. cephalus* during hydration and changes in mean egg diameter, water content and osmolarity. He showed that both soluble (glucose) and insoluble (glycogen) carbohydrates gradually decreased through hydration with a distinct drop prior to spawning. Total lipids increased from 16 to 25%. The major polar lipid was lethicin with decreasing amounts of phosphatidyl/ethanolamine and lysolecithin, plus other unidentified trace components. The main nonpolar lipids were cholesteryl-esters and triglycerides. The proportion of free fatty acids was small. Palmitoleic acid ($C_{16:1}$), palmitic acid ($C_{16:0}$) and oleic acid ($C_{18:1}$) were the major fatty acids, with decreasing amounts of myristic acid ($C_{14:0}$), linoleic acid ($C_{18:2}$) and some higher fatty acids containing 20, 22 and 24 carbon atoms. He recorded a conspicuous increase in the amount of soluble nitrogen-containing ninhydrin compounds. There was a general increase in amino acids particularly the neutral ones such as alamine, serine, leucine and isoleucine. There was some change in the level of total protein but less pronounced than that of the soluble nitrogen containing compounds.

Kuo (unpublished) also showed the pre-ovulatory eggs were isotonic with sea water and the ovulatory eggs became hypotonic due to the rapid intake of external water during hydration. In the 4 h postovulatory period a remarkable increase in the osmolarity adjusted the osmotic pressure from hypotonic to isotonic to sea water at the time of spawning. A change of the main individual electrolytes sodium, potassium, calcium and magnesium ions was found variable due to the increase in water content. No comparable data exist for the analysis of eggs from *M. cephalus* developing naturally without induced treatment.

Although the need to identify the biochemical composition of normal eggs of Mugilidae still exists, artificial propagation of certain species following induced breeding has been reasonably successful in Taiwan, Hawaii, Israel and India. It is improbable that the quality of eggs produced by external hormone stimulation is inferior. The failure of the larvae to survive in large numbers is probably due to subsequent mishandling because of poor techniques.

Behaviour

The vertical migratory behaviour of the eggs of Mugilidae has been variously described. Sanzo (1936) observed that all fertile eggs of *Mugil chelo* sank soon after fertilization, while Yashouv (1969) reported sinking towards the end of incubation for eggs of *M. cephalus*. Kuo *et al.* (1973*a*) recorded that the majority of eggs of *M. cephalus* which sank within the first twelve hours were undeveloped or unfertilized (absence of perivitelline space). They noted that when eggs were prevented from settling by strong aeration in small vessels survival was increased, and the incubation period reduced by two hours at the same temperature. They concluded that water temperature and turbulence had a decided effect on incubation time, and this was supported by Tung (1973).

Tang (1964) noted that eggs of *M. cephalus* sank in standing water and so incubated them in suspension by circulating water. Yashouv & Berner-Samsonov (1970) described the development of *M. capito* eggs floating on the surface. Eggs of *M. cephalus* lost their buoyancy after 20 h and sank, but hatched successfully. Sebastian & Nair (1974) observed that eggs of *M. macrolepis* floated in shallow hatching trays, and Liao (1974) reported that the fertile eggs of *M. cephalus* were buoyant and floated near the surface. He maintained them in suspension during incubation by aeration as he noted that some did sink later. The rearing practices of Nash *et al.* (1974) attempted to maintain the fertilized eggs of *M. cephalus* in suspension by circulating water, particularly when incubating at a density of 250 eggs/l.

Although incubation is apparently continued for eggs of *M. cephalus* which sink below the surface, the artificial environment of a container for hatching obviously introduces problems not normally encountered by eggs liberated in the sea. Density of eggs during incubation is a key factor influencing hatching rates and overcrowding produces problems of agglutination, bacterial contamination, oxygen depletion and increased metabolite production. Suspension of the eggs by water movement counters these problems, and few workers attempt to incubate fish eggs without the minimum of circulation or water movement through aeration, particularly when operating at high density.

Environmental conditions for incubation

Egg development and hatching time are both temperature dependent. Tang (1964) reported that hatching of *M. cephalus* took place in 59–64 h at temperatures ranging from 20.0 to 24.5 °C and salinity from 24.39 to 35.29‰. Fertilization was low (32%) and the rate of hatching was below 10%. Yashouv & Berner-Samsonov (1970) noted that under laboratory conditions the eggs of *M. cephalus* and *M. capito* developed and hatched within 36–44 h at 22–32 °C. Kuo *et al.* (1973*a*) stated that hatching of *M. cephalus* eggs was evident 36–38 h after fertilization at 24 °C, and 48–50 h at 22 °C. Total length of the newly hatched larvae was 2.65 ± 0.23 mm. Salinity was 32‰.

Liao (1974) stated that hatching of *M. cephalus* eggs took place in 34–38 h at 23–24.5 °C, and at 49–54 h in 22.5–23.7 °C, with salinities of 30.1–33.8‰. Tung (1973) described the relationships between mean water temperature (θ) and duration of incubation period (T) as $T_e^{0.135\theta} = 1262$ in still water, and as $T_e^{0.037\theta} = 106$ in running water. Finally, Sebastian & Nair (1974) recorded the incubation time of *M. macrolepis* as 23 h at 26–29 °C and 29–31‰ salinity.

Nash *et al.* (1974), Nash & Sylvester (unpublished data) and Nash & Kuo (1975) report the survival of eggs of *M. cephalus* within broad ranges of temperature, salinity and dissolved oxygen. Minimal mortalities of eggs occurred at 22 °C for normal sea water (32‰), and an effective temperature range for incubation was 11–24 °C.

Most workers prefer a working temperature range of 18–24 °C for *M. cephalus*. Above 25 °C incubation is inhibited although some eggs will hatch at 30 °C; the mortality beyond 25 °C is usually above 90% and often total. Optimal salinities for incubation are 30–32‰ under ambient temperature conditions (19.5–20.5 °C), and significant decreases in egg survival occur with eggs incubated in mean oxygen concentrations below 5.0 ppm.

Good temperature control during incubation is essential. Present work on the propagation of *M. cephalus* is conducted within the most desirable temperature range and mostly at the optimum level. Although individual daily temperature fluctuations may be responsible for some egg mortality, it is believed that temperature *per se* is not influencing the high mortalities experienced in the early stages of development. However the working range of 18–24 °C is suitable for rapid bacterial growth and therefore indirectly a causative factor of environmental instability during incubation.

Kuo (unpublished data) recorded the change in osmolarity of the eggs prior to spawning and that they were isotonic to sea water at the time of spawning. This fact together with the migratory behaviour of the adults at breeding time moving out into oceanic water indicate that salinity level is most important for egg development, and incubation should be conducted in full sea water.

Incubation facilities

Liao (1974) reported that fertilized eggs of *Mugil cephalus* prior to 1968 were incubated and hatched in two types of hatching equipment. The first was a flowing water type. It consisted of a fine mesh net hanging in the water and provided with a continuous slow exchange of water. It was similar in design to that used for hatching eggs of Chinese carp. The second was a static system of simple containers each with aerated sea water. After hatching the larvae were transferred to other rearing tanks. The systems were later modified to avoid loss of larvae during transfer. All development stages were then completed in either plastic or large concrete tanks indoors with good environmental control.

Kuo *et al.* (1973*a*) incubated eggs in well aerated static sea water (32‰) in 140 l fibreglass tanks. Incubation temperatures were 22–24 °C. They worked on a series of improvements and with Nash *et al.* (1974) developed a modified circular kreisel for incubating eggs of *M. cephalus*. The advantages of the kreisel were that the eggs could be pretreated with antibiotics to reduce bacterial growth and the density of eggs was high for the size of the apparatus. Also separation of the emergent larvae from empty shells and inviable eggs could be made during transfer of the larvae to the rearing containers. This reduced potential fouling in the rearing tanks.

The kreisel has proved to be an effective rearing apparatus and can be constructed in various dimensions. Aeration and circulation of the water are maintained by an airlift pump located in the centre column. The water is drawn by the airlift pump into the column through a large filter stone, aerated and passed back down the outside of the pump and redirected into the container. A connection on one of the discharge arms of the column directs a fraction of the water to waste, and that volume is replaced by an incoming supply located above the column.

Nash *et al.* (1974) used ambient sea water (32‰) in the kreisel. Before use it was treated by filtering, irradiating and treating with antibiotics. Both penicillin (10 IU/ml) and streptomycin (0.01 mg/ml) were added daily and effectively reduced marine bacterial growths. The eggs were suspended in sea water until after hatching and the larvae safely transferred by siphon into larger prepared rearing containers.

Yashouv (1969) used specially designed incubators to prevent the washing out of eggs with the changing sea water. Sebastian & Nair (1974) used shallow trays similar to those used for the incubation of salmonid eggs.

Oppenheimer (1955) demonstrated bacterial growth on fish egg shells by photomicrographs and found that the percentage hatch of a number of marine species increased by controlling marine bacteria with a mixture of penicillin and streptomycin. Shelbourne (1964) credited the successful culture of marine flatfish to the use of antibiotics. Using controlled amounts of the two he demonstrated that the survival of young flatfish was significantly increased from 40% to 60% at metamorphosis.

The antibiotic treatment in incubation tanks reduces bacterial activity around the shell and prevents agglutination of the eggs. Physical damage to the egg shell, or the covering of the shell with bacterial slime or mucus, breaks down the osmotic balance between the eggs and the surrounding sea water. The osmotic regulation between the two appears to be a key factor in successful propagation of marine and brackish water species, and needs a great deal of further attention and research.

Although the effectiveness of antibiotics can be replaced by ultraviolet sterilization and bacterial filters in the water system of a hatchery, the usefulness of antibiotic treatments cannot be underestimated (Struhsaker *et al.* 1973). Nash *et al.* (1974) utilized antibiotic washes and prolonged treatments during egg incubation. Buoyant viable eggs at the gastrula stage were removed from the spawning tank, washed under irradiated and filtered sea water, and then dipped for one minute in a sea water bath containing potassium penicillin G (80 IU/ml) and streptomycin sulphate (0.05 mg/ml). They were then distributed into the kreisels at a density of about 250 eggs/l and incubation was continued for the remainder of the period. Low levels of antibiotic were added daily and reduced all bacterial contamination.

Oppenheimer and Shelbourne theorized on the ways that bacterial activity affected and damaged eggs (Costlow, 1969). Penicillin is effective against Gram-positive organisms, and streptomycin against Gram-negative organisms. The combination therefore sterilizes sea water if there are no resistant organisms present. Antibiotics have also proved effective in the culture of shellfish (Walne, 1958). However they are known to unbalance the mechanisms for calcium transfer in some species and could not be used for the culture of *Haliotis tuberculata* (Nash, unpublished data). Although the work of Nash *et al.* (1974) did not demonstrate fully the need for antibiotic treatment in the culture of *Mugil cephalus*, the need for bacterial control is vital for successful incubation of many marine and brackish water species.

Little mention is made in the literature on construction materials used for incubation tanks. Shelbourne (1964) recommended the use of inert materials and the incubators used in the first flatfish hatchery were made of black polyethylene. The colour was important to provide a nonreflective surface against which the larvae could clearly see living food particles. Other workers mentioned the use of fine mesh net containers of terylene or nylon, but omitted to specify the materials of the larger container in which the net bag was suspended. Both cement and fibreglass incubators have been used after prolonged leaching of metallic ions or resins but plasticizers have proved damaging. A great deal of information is lacking on the suitability of certain constructional materials. Until good data are available most workers practise prolonged leaching of containers either in a heated atmosphere or submerged under water.

Larvae and larval culture

Hubbs (1943) defined the terminology for the young stages of fishes and separated prelarval, postlarval and juvenile stages on observed criteria. He considered that the postlarval stage began immediately on absorption of the yolk sac, and lasted as long as the structure and form were unlike that of the juvenile. The juvenile he considered to be the young stage similar to the adults in all essentials.

For most teleosts the formation of the scales signifies the end of the postlarval stage. The Mugilidae however develop scales in the early postlarval stages (at 8–10 mm in length) and are soon well developed (at 12–14 mm in length). Hubbs criteria do not therefore apply.

Roule (1917) divided the postlarval stage of Mugilidae into two successive periods. The first had rudimentary scales as the diagnostic criterion followed by the second stage with true scales. Anderson (1958) regarded the formation of the third spine of the anal fin as signifying completion of the postlarval stage and classed individuals as juveniles if the third anal ray had fused into a spine. Tung (1973) described five stages of larval development and morphology.

Young mullet which first appear in small schools along the coasts and in the estuaries are fully scaled. They measure 18–28 mm in length. Extrapolated growth data for the species indicate that young of this size are between 30 and 45 days old. The transition stage from postlarvae to juveniles used by Anderson (1958) does not occur until the young are 35–45 mm in length or 45–60 days old. Thomson (1963) regarded the transition complete at about 50 mm when the third anal spine formed from the anterior ray and the adipose eyelid started to form.

For the purpose of this text the artificial propagation of the mullet must include the culture of individuals to a stage of development when they can withstand relocation from hatchery to nursery pond. Heavy losses will occur if the young fish are mishandled or transferred too soon. This treatise considers young mullet up to 50 days old as the responsibility of the hatchery. They are technically larvae until that time. Young fish will be designated as juveniles when they are transferred from the hatchery to small ponds and they should be at least 50 days old. This arbitrary classification closely fits the morphological definition used by Anderson (1958). The use of the term 'fry' for young fish is mostly avoided in the text. It is commonly used to describe all the resources of postlarvae and juveniles collected along the coastlines and transferred to nursery ponds.

Morphology

Some of the first descriptions on the morphology of the larvae of Mugilidae were made by Sanzo. He observed and illustrated the early stages of *Mugil chelo* and *Mugil cephalus* (1936) and *Mugil labeo* (1937). He considered their prolarvae to be poorly developed. They measured only 2.2–2.5 mm in length. The mouths were closed and there were no traces of a branchial skeleton. Characteristic of the larvae were the voluminous yolk sac and large oil globule often accompanied by smaller oil droplets.

The first complete morphologic descriptions of larvae of Mugilidae were made by those workers involved in the induced spawning of the adults by hormone injection. Mostly they described the development of *M. cephalus*. Among them were Tang (1964), Yashouv (1969), Yashouv & Berner-Samsonov (1970), Liao *et al.* (1971), Kuo *et al.* (1973*a*) and Tung (1973). Sebastian & Nair (1974) described the development of *Mugil macrolepis*.

Newly hatched larvae of *M. cephalus* vary in length between 2.2 and 3.5 mm. The oil globule (and any additional small droplets) is situated in the posterior part of the yolk. Tung (1973) recorded 24 myotomes. He described the anterior half of the body bent on the yolk sac. The larvae had five or six pairs of cupulae on the body side, from eye to tail, and several pairs on the front of the head. Feeding began three to five days after hatching. Yashouv & Berner-Samsonov (1970) gave full descriptions of the keys to the eggs and larvae of five mullet species: *M. cephalus*, *M. capito*, *M. saliens*, *M. chelo* and *M. auratus*.

Thomson (1963) reviewed work on *M. cephalus* and quoted in detail the embryonic development described by Sanzo (1936) and Anderson (1958). A full and accurate description of development and behaviour of *M. cephalus* was made by Liao (1974), and is reprinted in Table 8.2.

Behaviour

Liao (1974) described the larvae of *Mugil cephalus* as having weak swimming activity with the posture of belly up and head down, sometimes moving with a jerky motion up and down. Kuo *et al.* (1973*a*) reported that newly hatched larvae were inactive and usually remained upside down suspended in the water column in an inclined position with the ventral side oriented towards the water surface. Occasionally each larva would go into a jerky motion and right its position. It would then dart upward rapidly then sink passively back to its resting position. They noted that sustained larval activity increased after the second day.

The presence of the oil globule influences the activity of the larvae during early development. Kuo *et al.* (1973*a*) dealt at length with the vertical distribution of the larvae and changes in specific gravity. They noted that larvae gained sustained swimming powers between the tenth and twelfth days after hatching and documented vertical distribution and measured specific gravity before that period. They observed that during the first two days the larvae were passively suspended in the water column and tended to become evenly distributed in depth as their specific gravity increased from 1.0263 g/ml at 12 h to 1.0339 g/ml at 36 h after hatching.

They described a vertical migration of the larvae between the second and third days. At the end of 60 h over 87% of the larvae came to rest on the bottom of the 80 cm experimental column, with the remainder suspended in the lower 50 cm of the column. After 72 h the pattern was reversed with over 85% of the larvae aggregated at the surface. During this migratory period the larval specific gravity decreased from 1.0310 g/ml at 60 h to 1.0264 g/ml at 96 h by which time all the larvae were back at the surface. The larvae remained at the surface until the sixth or seventh day.

Between the sixth and seventh day after hatching a second migration occurred. It was accompanied by a similar increase in the specific gravity of the larvae on the seventh day. All the larvae were on the bottom of the tank

Table 8.2. *Development of mullet larvae and their behaviour (after Liao, 1974)*

Days after hatching	Total length (mm)	Development and behaviour
1	2.56–3.52	Newly hatched larvae had a large yolk and oil globule. The front part of notochord being curved along the yolk sac and the curve degree related to the duration of hatching. At low temperature, this duration was long and the curve more distinct. Weak swimming activity with the posture of the belly up and head down, sometimes with jerky motion slightly up and down. Pigmentation dependended on individual. Eye not coloured. Mouth not developed. Digestive tube not well-developed.
2	2.64–3.28	Formation of organs was in progress. More pigmentation was found in eye and body. The total length was shorter than before. Mouth was under development. Bud of pectoral fin appeared. Nostrils were obvious.
3–4	3.11–3.53	Opening of mouth. Good development of upper and lower jaws. Irregular peristalsis of stomach and intestines, able to take food. Yolk was diminished, being $\frac{1}{4}$ of original size. Oil globule was also reduced. It was the first critical period, always accompanied by serious numbers of deaths. Gill clefts appeared. Being attracted by and tending to concentrate at 600–1400 lux area. Distributed at upper level during night-time.
5–7	3.06–3.40	Digestive tube was well developed. Movement up and down individually both during day and night. Feeding was easy to observe but only limited in the day-time. Formation of abnormalities in the bladder and swelling of hyoid could be inhibited by freshening of rearing water. Formation of stomach, intestine, gall bladder, pancreas, gas bladder and continued reduction of oil globule.
8	3.35–3.80	Complete disappearance of oil globule. Formation of gill filament. This is the flexing point of the growth curve, with the growth starting to accelerate.
10–13	3.45–5.10	Finfold moved backward. Gill filament well developed. Body surface became dark in colour. Strong phototaxis. Formation of hypural bone. It was the second critical period with very low survival rate.
14–15	3.85–5.70	Commencement of swimming in schools. Formation of urostyle. 7–9 ray bases were found in each of anal fin and second dorsal fin. Gill lamellae formed on gill filament.
16–19	5.40–6.60	Seventeen soft rays in caudal fin. Black spots scattered on the whole body. Shiny silver white complexion appeared from gill cover along ventral part of anus. Schooling in upper levels, sometimes into middle level.
20–21	6.00–7.65	Showing phototaxis during day-time while floating during night-time. Appearance of brown colour sometimes and silver green at other times, higher variety found in healthy larvae. Four soft rays on the first dorsal fin.
22–24	8.25–10.9	Formation of complete 20 soft rays in caudal fin, 11 rays in anal fin, 6 rays in pelvic fin, 15 rays in pectoral fin. Fin membrane of dorsal fin and pelvic fin almost entirely degenerated. Appearance of scales with 1–3 ridged circles. The size of largest scales reached 400×250 μm. Silver-white complexion. In day-time, larvae swam in upper levels in schools and against aeration and stream. At night they floated and scattered on surface of water and liked to gather under the light.
25–28	8.80–15.0	All scales and fin rays were well formed. Silver-green complexion. Appearance of teeth. Two nostrils separated.
29–32	16.6–20.7	Very sensitive. Gathering in small schools. In day-time, swam in middle and lower levels, at night-time, continued to be floating but easily startled.
34–35	22.2–26.2	In day-time, larvae swam in large schools along the circle of rearing tank in the middle and lower levels. In night-time, floating individually. Grass-green in colour, sometimes silver-white on the dorsal part. Some diseased fish were found with 'pop-eye' symptom.
37–40	23.1–29.3	Some changes in feeding behaviour, feeding at late afternoon. Sensitive to light, and no longer gathering under light.
	27.5–32.8	Strongly resistant to environment. Suitable for stocking.

on the eighth day, with a complete reversal on the ninth day. Conditions at the time were full sea water (32‰) and ambient temperature (24 °C).

Kuo *et al.* (1973*a*) concluded that the first sinking was probably related to the rapid absorption of the yolk sac and the resulting change in specific gravity. Planktonic larvae are normally able to distribute themselves through the surface waters. This ability may be more pronounced in Mugilidae because of the continuous references to the spawning of the adults over deep water (see pp. 270–5) and the rarity of eggs and larvae taken in plankton tows.

The second descent could not be fully explained. Although the larvae were not capable of sustained swimming at the time, they were capable of swimming to the surface of the water column. Kuo *et al.* (1973*a*) could not account for the sudden and transient increase in specific gravity of the larvae on the seventh day. Morphological observations of histologic sections of the larvae revealed that the pneumatic duct of the air bladder was occluded between the sixth and seventh days, but the relevance of this was not understood. The migration was known not to be a phototropic response.

During the mass propagation of the larvae Kuo *et al.* (1973*a*) experienced heavy mortality at the critical stages associated with the two migratory periods. They attributed the mortality to mechanical and physical damage during prolonged contact with the bottom of the container as the larvae had no control over their ability to escape. The mortality was reduced slightly by using deeper containers (1.5 m) for rearing.

Nash *et al.* (1974) used a rearing kreisel and increased survival significantly during the first migration. The kreisel provided mechanical circulation but the larvae were protected from the damaging effects of strong aeration by a simple central device. Summarizing the migratory behaviour they believed that the larvae were responding first to changes in utilization of the yolk and sank to the bottom. After ascending the larvae had improved musculature control and lost the indiscriminate floating activity much earlier than previously stated.

The second vertical migration was associated again by Nash *et al.* (1974) with the changes in specific gravity of the larvae and suggested that two factors were concerned. The first was associated with the physiological and morphological changes, the second with nutrition. They observed that there were distinct differences between feeding and nonfeeding larvae as early as the seventh day after hatching. Established larvae were longer and more active, and were intensely pigmented. Unestablished or nonfeeding larvae remained in the surface water layers and used the increased surface tension along the sides of the container to support themselves. Development of the pigment was slow and little or no growth was evident. Such larvae died between day 7 and day 10 depending on the water temperature. Coincidental with the migratory movement of the larvae to the bottom, many undeveloped individuals sank because they were unable to sustain themselves further and were moribund. Hence the earlier association of heavy mortality with the migratory period. Developed larvae also had to undergo the changes associated with the migration which must be critical to further development. As with other larval forms undergoing metamorphosis the mullet larvae became inactive for a period and sank.

All workers now believe that those larvae which survive the second migration are capable of full development, providing that the conditions and food are suitable and if they are not mishandled. Established larvae after the second migration soon develop scales and then school together.

Nutrition

The dearth of larval Mugilidae in plankton tows is responsible for the lack of information on their natural foods during this critical stage of development. This scarcity of knowledge during the first formative fifty days of life is to some extent compensated for by the wealth of data on the food and feeding habits of juveniles and adults of Mugilidae from regions throughout the world, e.g., Hiatt (1944), Jacob & Krishnamurthi (1948), Thomson (1954), Sarojini (1957), Hickling (1970) and Zismann, Berdugo & Kimor (1974). However, this is not in the scope of this chapter.

The data on acceptable foods for the larvae of Mugilidae have been produced by workers developing culture techniques, usually following induced spawning practices. The variety of larval foods which have been tried and evaluated by fish culturists working with many marine and brackish water species has been reviewed by May (1970). Aspects of food and nutrition of marine fish larvae important to the best culture practices have been developed by Houde (1972).

Following the format of May (1970), Table 8.3 lists the food organisms and prepared foods which have been used by culturists in their search for an adequate food for the larvae of the Mugilidae. The sources are predominantly the publications of workers in Taiwan, Israel, Hawaii and India and in most cases were tried on the larvae of *Mugil cephalus*.

Emergent larvae of the grey mullet are believed to be entirely carnivorous during early development, becoming omnivorous and capable of digesting plant material some time before metamorphosis. The most successful larval rearing results of Liao *et al.* (1971), Kuo *et al.* (1973*a*), Tung (1973), Nash *et al.* (1974) and Sebastian & Nair (1974) indicated the use of a mixed dietary regime during the first ten days of life.

Many workers include wild or natural zooplankton in their regimes. Shehadeh & Kuo (unpublished data) had some success with the nauplii of *Artemia salina* fed alone on the seventh day after hatching, but the results were later improved with a phytoplankton supplement (Kuo *et al.*, 1973*a*).

Table 8.3. *Experimental foods for larvae of grey mullet species, tried at one time or another*

1. Wild plankton	
2. Protista	
Dunaliella sp.	*Prorocentrum micans*
Chlorella sp.	*Oxyrrhis* species
Gymnodinium splendens	*Coscinodiscus* species
Isochrysis galbana	*Chaetoceros species*
Monochrysis lutherii	*Biddulphia mobiliensis*
Skeletonema costatum	*Ditylum brightwelli*
Thalassiosira species	
Nitzschia species	and unidentified phytoplankton,
Platymonas subcordiformis	diatoms and green water
Cryptomonas maculata	
3. Metazoa – planktonic forms	
Brachionus plicatilis	*Arbacia* species
Artemia salina	
Crassostrea gigas	and unidentified Copepods
4. Prepared formulations, including at one time or another	
Cod liver oil	*Artemia salina*
Powdered oil-cake	Copepods
Bean cream	Oyster flesh
Fish ovaries	Fish meal
Egg albumin	Urea
Boiled egg yolk	Rice bran and flour
Liver juice and enzymes	Milk powder
Amino acids	Powdered oats
Marine vitamins	
Brewer's yeast	and commercially available
Freeze-dried phytoplankton	tropical fish feeds
Freeze-dried zooplankton	

Sebastian & Nair (1974) noted that *Mugil macrolepis* fed exclusively on copepods but grew best on copepods prepared in association with an algal bloom, particularly *Chlorella* species.

Many culturists rearing marine and brackish water species of finfish and shellfish believe that the presence of phytoplankters in the rearing containers is beneficial to the technique, but not always of direct value as food. Although some phytoplankters are found in the gut of prolarvae it is not thought generally that they are nutritionally sufficient. The indirect benefits of phytoplankters are probably the stabilization of the rearing environment through the removal of metabolites, or the supplementation of necessary marine vitamins or amino acids in solution, or for maintaining the nutritional level of the zooplankters before they themselves are consumed.

Several artifical diets compounded from both natural and synthesized materials have been fed to the larvae of mullet with limited success. Most have the disadvantage on artificial diets in association with natural organisms.

Against the poor larval survival data, which may be the result of other factors, the true nutritional value of many artificial diets cannot yet be estimated satisfactorily.

Nash & Kuo (1975) hypothesized that the larvae of the grey mullet were able to utilize at least one or several of the live or inert dietary organisms listed in Table 8.3. It was their belief that it was the preparation, feeding procedures and practices which, among other things, were at fault and not the types of organisms tested. They stated several instances for the human influence to be an unintentional cause of larval mortality during larval feeding. For example, separation of the food organism from its own rearing medium was not always adequately performed, and the separation of the nauplii of *Artemia salina* from the cysts was often inefficient.

The selection of the food organism by the larva is made, in all probability, on the simple criteria of movement and size in the first instance. The ability of larvae to feed is regulated by feeding behaviour and vision. Blaxter (1969*a*) described the visual thresholds and spectral sensitivity of a number of marine species. As with many marine and brackish water species the eyes of the larvae of Mugilidae are large at hatching and become rapidly pigmented within three days. The eyes are only capable of coarse movement perception and, as with other fish species, are poorly equipped visually with a single type of visual cell. They have little ability to adapt to dark or light situations during the first hours after hatching. Most culturists follow the early techniques of Shelbourne (1964) and try to improve conditions for feeding by providing nonreflective dark surfaces for the rearing containers. The food organisms are clearly silhouetted as they move around.

Houde (1972) believed that food organisms within the size range 50–100 μm were preferred by fish larvae with relatively large mouths. The acceptable sizes of food increased rapidly as the larvae grew. The recent successes of a number of culturists using the rotifer *Brachionus plicatilis* (Theilacker & McMaster, 1971), in association with a number of phytoplankters, infers that the size of about 100 μm is adequate for the first two days of feeding the larvae of the Mugilidae.

The ratio of food or organismic density to the larval density is another key factor to successful larval culture irrespective of species. The optimum food density must be maintained continuously through to the end of the larval period.

A superabundance of food can be as harmful as too little. A larva must not be intimidated by large or quick-moving organisms or totally dominated by a number of them in one location; for example organisms collect in a corner where the light intensity is often increased by the high light reflection of an interface. Conversely the density of larvae can be too high. In addition to the problems of inhibitors produced by the larvae when overcrowded (Blaxter, 1969*b*), the physical contact, the increased competition for space and food and production of metabolites all lower the survival rate. Kuo (unpublished data) showed a higher survival of larvae of *Mugil cephalus* at 12/l than either 8 or 16/l. Shelbourne (1964) operated at higher densities for *Pleuronectes platessa*; a similar survival of about 70% was obtained from two populations of established feeding larvae at 28 and 56/l. During development to metamorphosis the survival decreased to 30% and the density to 12 and 24 larvae/l respectively. May, Popper & McVey (unpublished data) successfully reared the larvae of *Siganus canaliculatus* at a density of 5/l. The data in Liao (1974) indicated effective rearing at a density of less than 10 larvae/l. Each individual larval form must be provided with the opportunity to observe and attack live food particles. Many failures occur, particularly during the first crucial days of the learning process, and a larva without success cannot survive for long.

Riley (1966) and Rosenthal & Hempel (1970) concluded that a higher food concentration was necessary at the time larvae initiated feeding than subsequently, probably because the younger stages were less capable of capturing food. Houde (1972) recommended a food level using wild zooplankton (copepod nauplii and copepodites) of 3.0/ml for the first two days of feeding larvae, but a level of only 1.5/ml on subsequent days. If rotifers were used he suggested a higher concentration. If only nauplii of *Artemia salina* were fed then he recommended a lower concentration.

Nash & Kuo (1975) specified an organismic density of 20–30/ml during the first feeding stages of *Mugil cephalus* which included all organisms. Liao *et al.* (1971) described the use of fertilized oyster eggs and cultured diatoms on the third day of development. Liao (1974) stated that the density of oyster eggs or trochophore larvae was maintained as high as 400–500 organisms/ml. Yeast and albumin were then added as supplements on day 4 and rotifers and copepods on day 6.

Kuo (unpublished data) showed that feeding larvae of *M. cephalus* first on days 4 or 5 was better in terms of survival and management than that on either days 3 or 6. Furthermore, copepods smaller than 150 μm were not utilized until day 5. He also performed four series of experiments feeding emergent larvae with *Isochrysis*, *Brachionus*, *Chlorella* or natural zooplankton. Observations of stomach contents during development between days 5–11 were made. *Isochrysis* was taken first of all food organisms tested and ingested in quantity on day 5 after hatching. The food preference of mullet larvae for natural zooplankton or *Isochrysis* as an initial food was further examined. Again the food organisms were given singly and in combination and the gut contents examined daily. No food preference was indicated except *Isochrysis* was once more consumed readily on day 5.

The logistics of aquaculture for the hatchery operation for marine and brackish water fish are considerable. If certain species of fish require individual organisms over a long period of their development, it is possible that the larval food production system alone will be greater than that of the fish larvae themselves. Shelbourne (in Costlow, 1969) quoted a daily requirement of 200 million nauplii for a flatfish hatchery producing 0.5 million juveniles. That is why the dehydrated cysts of *Artemia salina* are such an advantage at present as they can be conveniently stored and prepared with a minimum of delay and acclimation (Nash, 1973). It also explains the strong interest in artificial foods.

The present methods of larval culture which are proving most successful are those using a mixed dietary regime, often by the creation and stabilization of an ecological system. The 'green water' technique has been used successfully by Fujimura & Okamoto (1972) for the culture of *Macrobrachium rosenbergii*, and by several Japanese in the culture of penaeid prawns.

The use of the mixed dietry regime has several advantages. It requires culture of several species in one container before the relase of the fish larvae. Following the preparation of the system with cured or conditioned sea water the environment becomes stablized. There are few rapid and diverse environmental changes which can upset the delicate biochemical balance of the larvae. The mixed regime is also economic in decreasing the numbers of individual organisms which have to be produced on a daily basis. The wide choice of organisms probably provides a better qualitative as well as quantitative diet for the larvae and also permits individual larvae to develop at their own rate by feeding on organisms relative to their size.

Pelagic copepods according to House (1972) were probably ideal food for marine fish larvae. Liao (1974), Nash *et al.* (1974), Kuo *et al.* (1973*a*) and Tung (1973) all used natural zooplankton in their respective feeding regimes. It was the copepods which proved to be the most significant and beneficial organisms for the diet of the larvae of *Mugil cephalus*. The Japanese culturists regard copepods as the best larval fish food (Harada, 1970) and consequently there have been many attempts to rear them intensively. In many cases large natural populations can be found in the filamentous green algae which grow in the enclosures retaining the broodstock or maturing juveniles.

Environmental conditions for rearing

Salinity is probably the most unregulated and uncontrolled major parameter which influences the incubation and larval rearing of marine species. The majority of workers conclude that natural high saline waters (32–35‰) are optimum. The eggs of most marine and brackish water species are liberated into oceanic waters and, with the emergent larvae, are adapted to develop at high salt concentrations. Holliday (1969) concluded that survival of embryos and larvae of many species could be increased at low salinities (10–16‰) because those levels were iso-osmotic with body fluids. Houde (1972) found that many species had high survival rates over a wide range of salinities. He did not consider salinity as critical as some other rearing tank conditions which affected growth and survival.

In Israel, Taiwan and Hawaii, where the majority of work has been accomplished on the rearing of *Mugil cephalus*, offshore salinities and experimental conditions are all very similar, namely 32–35‰ with up to 39.5‰ for the eastern Mediterranean. Sebastian & Nair (1974) operated at slightly less (29–31‰) for the culture of *M. macrolepis*. However, the effect of salinity on larval survival and development is possibly more significant than that for incubation of the eggs.

Liao *et al.* (1971) reduced the salinity from 32 to 26‰ during larval development in three stages commencing on the sixth day after hatching. They concluded that there was an advantage rearing the larvae of *M. cephalus* in diluted or sweetened sea water. Nash and Sylvester (unpublished data) provided tolerance levels of larvae of *M. cephalus* to varying salinities. They showed that the larvae could only withstand prolonged exposures to salinities of 25–34‰ at 20 °C during the first week of development, with an optimum at 26–28‰ for 96 h exposure.

The operational procedures for the culture of *M. cephalus* as outlined by Nash *et al.* (1974) in Hawaii did not include techniques for reducing the salinity of the water during the first fourteen days. They used full saline conditions until completion of the second migration and then made dilutions. Better results were reported by Liao (1974) in Taiwan with the dilution technique.

Nash & Kuo (1975) theorized on the value of reducing salinity during early development and hypothesized that there was a link between the need to change salinity and the second vertical migration of the larvae (see pp. 287–90). They believed that the unexplained rise in specific gravity at the start of the second migration was unnatural and a result of osmotic imbalance. The larvae had not sufficient fresh water internally to maintain the balance and consequently sank. The larvae must therefore be cultured in sea water which is changed to suit the osmotic regulation. Such fine control was an external compensating reaction against the change in specific gravity. They also contended that the fresh water consumption of the larvae needed increasing and both cultured and artificial food preparations required greater consideration for fresh water content.

Larval development is temperature dependent. Shelbourne (1964) demonstrated the need for optimum temperature control for the culture of marine flatfish. He also noted differences between the survival of larvae in natural conditions and those in the intensive hatchery environment where bacterial activity was potentially more dangerous. The reasoning of Shelbourne is particularly relevant to the culture of fish and shellfish in the tropical and subtropical latitudes. There the ambient temperatures are highly conducive to bacterial growth, and the optimum rates for yolk utilization by the larvae are probably narrowly defined and close to a critical level.

Strict temperature control for the incubation of the eggs of the Mugilidae is important (see pp. 282–3). Emergent and developing larvae up to metamorphosis tolerate an ever widening temperature range and their growth rate responds accordingly.

Liao (1974) reported the successful culture of the larvae of *Mugil cephalus* over a number of preceding years within the ambient temperature range of 19–24 °C. Nash and Sylvester (unpublished data) recorded minimum mortality of the larvae of *M. cephalus* between 18.9 and 25.3 °C, although some larvae survived temperatures as low as 15.9 °C and as high as 29.1 °C. Sebastian & Nair (1974) operated within a higher range of 26–29 °C for the culture of *Mugil macrolepis*.

It is believed by Nash & Kuo (1975) that, of the two development stages of egg incubation and larval growth, the former was more critical and required careful temperature regulation. Nash & Shelbourne (1967), working in the thermal discharges of coastal electrical generating plants, exposed the eggs and larvae of marine flatfish at various stages of development to the elevated temperatures. Emergent and developing larvae were able to withstand substantial thermal shocks and grew rapidly at the higher temperatures. The early start to growth produced significant savings in time taken to reach market size. However, there was no benefit in attempting to use elevated temperatures to increase the rate of egg development as survival was in fact decreased.

A great deal of data on survival and temperature is available for juveniles of the Mugilidae, but is beyond the scope of this chapter which deals with the production of larvae up to 50 days.

Little information exits on the levels of dissolved oxygen suitable for the rearing of the larvae of the Mugilidae. Although the levels of dissolved oxygen are measured regularly as part of many culture operations, the data are often excluded from reports. Nash and Sylvester (unpublished data) determined that the survival of the larvae of *Mugil cephalus* was significantly changed for mean oxygen concentrations below 5.4 ppm.

The size of the culture system does not necessarily influence the success of a mass propagation effort. Many workers operating in carefully monitored small units, and with high density feeding but low larval density, have produced exceptional survival figures. However that is not aquaculture. In terms of mass propagation the larger units are more appropriate for rearing the numbers of larvae which are necessary for a culture practice to become economic.

The successful culture system of Fujinaga (1963) with *Penaeus japonicus* encouraged both Liao *et al.* (1971) and Nash *et al.* (1974) to use larger containers for the culture of *M. cephalus*. Survival of the larvae in them was definitely increased but the reasons were obscure. At present the usefulness of the larger container may be in the lowering of larval density and increasing spatial freedom. It would also decrease the influence of any specific inhibitors released by the larvae, and prevent a size hierarchy. However, they stocked the containers with eggs or larvae at 50–250/l initially, with a final density of between 5 and 50/l on day 21.

The disadvantage of the larger container is the great demand on available larval food, particularly as food density is a key factor for survival. The logistics to supply the larval food on a single species daily supply basis would be tremendous for a farm of any size. Shelbourne (in Costlow, 1969) operated an intensive *Artemia salina* nauplii system capable of producing over 200 million nauplii per day. However in the context of a mixed feeding regime (see pp. 293–4), populations of differing organisms can make up the organismic content to the required density in the simplest way.

The advantage of the large container as theorized by Nash & Kuo (1975) is the stabilizing of the rearing environment. Although good tank hygiene is necessary and water has to be replenished and salinity reduced, the effects of the exchanges or the additions of food are buffered by the size of the system. The larvae are therefore protected from sudden shocks or exposure (albeit for a short period) to adverse conditions.

Little data are available on the suitability of materials for containers for rearing the larvae. A study of the effects of materials on several small marine organisms was made by Bernhard & Zattera (1970). Plastic, polyethylene, fibreglass, concrete, vinyl and wood have all been used at some time for Mugilidae with little comment on their value or suitability. Most workers leach potentially toxic chemicals and plasticizers from moulded or fibreglass units.

The colour of the rearing units can be important. Nonreflective black polyethylene tanks to prevent areas of high light intensity have been widely adopted. An even light intensity over the tanks prevents localized gathering of the larval food and the larvae. Exposure of larvae to direct sunlight has to be avoided and all rearing operations are preferably conducted indoors. Liao (1974) noted that the larvae were sensitive to light. Four-day old larvae exhibited phototaxis and six-day old larvae migrated up and down according to the time of day, but fed only in the day. Larvae avoided strong illumination but were attracted by dim light intensity of about 600–1400 lux.

Post larval development

Between the present critical stage of early development (day 12–14) and the juvenile stage there appear to be few problems. The larvae which have survived beyond day 14 are usually hardy, feed well and grow rapidly. By the end of the third week the scales begin to appear and the larvae school together.

Growth is rapid. The larvae feed voraciously on nauplii of *Artemia salina*, phytoplankters, diatoms, copepods and artificial dry preparations.

Water flow through the rearing tanks can be increased or replaced by strong aeration. The most damaging effect is handling. Many larvae undergo a handling trauma and become inactive and die after violent quivering. Handling is not advised until the larvae are juveniles and ready for transfer.

Culture technique

From the results obtained by Kuo *et al.* (1973*a*), Tung (1973), Nash *et al.* (1974), together with a decade of information from Taiwan summarized by Liao (1974), the following culture technique combines the best practices for the artificial propagation of the grey mullet, *Mugil cephalus*. It can probably be applied to other Mugilidae with minor modifications.

Although the survival rate of established larvae has been reported as high as 19.5% in Taiwan and 25.5% at 14 days in small tanks in Hawaii, the technique holds no guarantees. However a considerable amount of good information has been assembled, far more in fact than was at the disposal of the White Fish Authority of the United Kingdom before it successfully developed the first marine fish hatchery for flatfish in 1963 (Shelbourne & Nash, 1966). There are certainly sufficient data for construction of a pilot scale hatchery to improve the methods and demonstrate the techniques, and also to learn of the new problems produced by the increased scale of operation.

Broodstock

Four-year old broodstock should be used, maintained healthily in captivity in sea water of 32–35‰ salinity and at 20–22 °C. The resources of broodstock can be either migratory fish about to spawn or a resident pond population. The latter is preferable but the stock should be replenished each year with additional individuals. The population should be sexed and the two maintained separately.

Egg samples should be taken regularly from the females and measured to determine development. The eggs must be above the critical stage (preferably 650 μm in diameter) before spawning is induced.

If purified salmon gonadotropin is used a total dose of between 12 and 21 μg/g body weight is necessary per female and is applied in two injections. The time interval between injections is 24–48 h depending on observed egg development after the priming dose. Alternatively, fish can be induced to spawn by injecting a total dose of 2.5–6 pituitary glands of mullet, 10–60 RU of Synahorin, and 0–300 mg vitamin E. The time interval between two injections is 24 h. The females usually spawn 12 h after the second injection.

Spawning in the natural season

Spawning is prefaced by hydration of the oocytes characterised by sudden enlargement of the abdomen and deposition of calcium. Two hours before spawning the males can be released with the females in a ratio of 3:1. The females can be spawned either as individuals in isolated aquaria or within a population in a large tank. The spawning fish should be maintained in seawater (32–35‰) and at 20–22 °C. In small aquaria the seawater should be flowing rapidly through the system and the water strongly aerated. A few seconds before spawning the seawater is shut off but the aeration continued. In the large tanks an exchange of sea water is maintained but the outlet is protected by a suitable fine mesh screen.

Preferably all spawning should be performed indoors with environmental controls for temperature and subdued light, and with a minimum of disturbance. Natural spawning is preferred but fertilization can be completed if necessary by either the standard dry or wet methods.

About one million eggs per female are released. Fertilization can be estimated one hour after spawning by microscopic examination of a sample of eggs. The eggs are extremely buoyant and can be removed from the surface of the spawning tank with a soft fine mesh net and transferred to the incubators. If individual females are spawned in small aquaria then the eggs should be dispersed among other tanks and aerated.

Spawning out of season

Spawning out of the natural season can be induced if environmental control facilities are available. The fish should be conditioned for a period of about 120 days before the desired spawning time. A combination of a retarded photoperiod (6 L/18 D) and constant temperature of 21 °C is effective for the development of oocytes. Both males and females should be exposed to the conditions and can be kept together.

Samples of eggs should be taken from the females regularly to determine the state of oocyte development. If the fish are in the refractory stage at the start of conditioning, the first signs of development (stage II) should be visible after about 60 days. Oocytes in the tertiary yolk globule stage (stage III) can be anticipated after about 120 days, and will be ready (above 650 μm) for hormone injection any time thereafter. The methods of inducement are the same as described in the previous section, although an additional injection may be necessary.

Egg incubation

Egg incubation is performed as a separate hatchery process in specially designed containers to maintain strict control over the environmental parameters. This avoids any overloading of the rearing kreisels, as they have to be prepared with larval food.

One hour after spawning and fertilization the aeration in the fish spawning tanks is stopped. The fertile buoyant eggs rise to the surface of the tank and can be removed in small numbers with a soft fine-mesh hand net. The eggs in the net are washed gently in running irradiated and filtered sea water and dipped for one minute in a sea water bath containing the antibiotics potassium penicillin G (80 IU/ml) and streptomycin sulphate (0.05 mg/ml). The eggs are then distributed throughout several incubators at a density of no more than 400 eggs/l. The incubators are circular in design and fabricated with an inverted conical base to prevent eggs settling on the bottom during incubation. The aerator is located at the very base of the incubator in the cone to provide maximum circulation.

The incubators are filled prior to egg transfer with sea water of 32–35‰ salinity and maintained at 20–22 °C. The water in the incubators is also sterilized and filtered during filling, and thereafter treated with antibiotics at a dose rate of 10 IU/ml of penicillin and 0.01 mg/ml of streptomycin per day.

Under such conditions incubation will take between 50 and 60 h to complete. Incubation should always be performed indoors under subdued indirect lighting (1400 lux) and maintained at constant temperature.

Larval rearing

After 60 hours incubation is complete. The emergent larvae are then ready to be transferred slowly by siphon to the rearing kreisels. Before siphoning, the circulation in the incubators is stopped when the viable larvae soon occupy the upper levels of water. The empty egg cases and unhatched eggs sink, and care should be taken to avoid transferring them with the larvae.

The rearing kreisels are prepared several days in advance. Conditioned or aged sea water is used to fill each one. Alternatively the sea water is sterilized and filtered during the filling operation. The sea water must be 32–35‰ salinity and temperature maintained at 20–22 °C before receiving the larvae. Before the larvae are transferred into the kreisels, each kreisel is inoculated with cultures of either *Dunaliella* or *Chlorella*, or other phytoplanktonic species capable of supporting the rotifer, *Brachionus plicatilis*. *Chlorella* species are preferred.

The organismic content of the kreisels should be monitored daily before and after loading with larvae. The objective is to achieve a density of 10^4–10^5 particles/ml of which 10/ml should be rotifers of suitable size. The preferred size of established rotifers is 100–250 μm. Given these organismic densities in the first three days, the food ecosystem will maintain itself and its numbers to support the increasing demands of the developing larvae over the first twenty-one day period. Thereafter, beginning on day 14, the diet can be supplemented with the day-old nauplii of *Artemia salina*, continuing until day 40. Older nauplii and artifical food can then be used. The density of the larvae in the rearing kreisels at transfer should be no more than 6/l, with an anticipated larval density at day 21 of about 0.33/l.

The rearing kreisels should be as large as possible, but of dimensions in keeping with the hatchery building, environmental controls, available resources of water and air, and particularly the resources of larval food. Concrete, fibreglass, polyethylene and butyl lined containers are all acceptable materials for the kreisels, following prolonged leaching in water or a heated atmosphere. The sides should be dark in colour and it is useful to paint the internal base of the tanks white. Depth is not important and 1 m is an average dimension so that tanks can be used for other purposes after the rearing season. If the tanks are fabricated specially for the purpose of rearing, the design should include protected inlets and outlets in addition to features which facilitate capture of the juveniles after 50 days.

Each tank should be fitted with an overhead diffused light giving an intensity of no more than 1400 lux at the surface. Regular illumination with separate controls should be available.

Temperature controls should be available to maintain the air and water temperature at 20–22 °C although there are indications of improved growth at 24 °C from day 1. The initial salinity of the sea water in the tanks should be 32–35‰. The water provided for the tanks should preferably be from a reservoir where it is passed continuously through filters and sterilized by ultraviolet lamps.

On day 4 after hatching, the water in the rearing tanks should be diluted with fresh water at a rate which decreases the salinity by 0.5‰ per day to produce a salinity of 30‰ by day 7. Thereafter it should be decreased regularly each day to 20‰ by day 30 and then maintained for the remaining 20 days of operation.

Daily records should be kept of temperature, salinity, dissolved oxygen and pH levels. Daily determinations of nitrite, nitrate, ammonia, sulphide and phosphate are also useful. Until the exchange of water through the rearing kreisels becomes substantial, these metabolites accumulate in the first 21 days. Although the tolerance limits to the individual metabolites are not yet known, as yet unpublished data indicate that the following levels can be tolerated and are presently considered acceptable but not desirable. Nitrite up to 3.00 μg, nitrate up to 90.00 μg, ammonia up to 30.00 μg, phosphate up to 9.00 μg and hydrogen sulphide up to 5.00 μg at day 1. Preferably the levels of metabolites should be as low as possible and not fluctuate radically. Dissolved oxygen levels in the kreisels should preferably be above 7.0 ppm, and pH between 7.9 and 8.3.

The rearing kreisels should be cleaned daily with a small siphon and dead larvae and uneaten food removed. With large tanks cleaning is not always easy as parts of the tank may be inaccessible. Circular tanks have the advantage of self cleaning either by moving the debris to the centre where it can be extracted by siphon, or by exhausting through the central standpipe. The latter is most effective when the rate of water exchange is high and the flow rate can only be increased when the larvae are established and strong (over 30 days old).

Between days 30 and 50 the growth of the larvae is rapid and they readily accept moistened artificial food, older nauplii of *Artemia salina* and larger copepods.

Mullet younger than 40 days old should not be moved or handled. Almost all react traumatically to handling and die. In consequence the hatching and rearing activity should be considered as a 50 day operation, although the young fry can be handled safely after 40 days and demonstrate schooling behaviour.

Speculation for the future

The Mugilidae probably have the brightest future of all the marine and brackish water finfish in the developing technology of aquaculture. The majority are desirable fish with good flesh texture and taste, particularly when captured or taken from water of high salinity. They are distributed widely and have a capacity for tolerating extreme conditions of temperature, salinity and dissolved oxygen. They are naturally hardy animals which thrive on good husbandry but are also capable of withstanding poor farming practices.

The many species of Mugilidae give them genetic potential unparalleled by few other aquatic species. It will be the manipulation of the genes and the selection of strains and breeding lines which will make them the aquatic equivalent of the first fatstock land animal.

Sufficient knowledge on the artificial propagation of *Mugil cephalus* has been produced to justify an investment in a pilot scale hatchery operation. The purpose of the hatchery would be to improve the existing techniques by operating at the increased scale, demonstrate the system and produce a large number of juveniles annually to use for farming experiments in management and production.

The logistics of such a hatchery need specifying so that the involvement is understood. Indoor tank facilities covering 3000 m^2 of floor space are estimated for a pilot-scale hatchery designed for the instantaneous production of one million 50 day old juveniles. Estimates are based on technical data from the previous section with anticipated overall survival of 5% of the original number of eggs after 50 days. For example, fifteen females and forty males will be the minimum requirement for the broodstock. At least twice that number should be held in reserve in indoor tanks or small ponds, each with a capacity of 1 m^3 per fish. The fifteen females will provide a minimum of 18 million eggs of which 5% are expected to survive development to the juvenile stage. Spawning can be achieved in several small aquaria located in a breeding room. It should be phased over a 16 day period with three females induced every fourth day. This increases efficiency through the hatchery and reduces the demand on space and facilities and on the larval food resources.

The eggs from each female will be transferred after twelve hours into the circular and conical incubators, each 1.5 m deep and 1.25 m^3 capacity. After hatching all the tanks will be sterilized and cleaned in preparation for the next batch of eggs.

Three hundred circular larval rearing tanks will be required, 1 m deep and 10 m^3 each in capacity. Previously the tanks will have been prepared and conditioned in groups with each group ready for larval reception every fourth day.

The food production unit will be considerable. Although continuous culture systems are being developed for the production of larval foods, batch cultures at present are more reliable. For the mixed phytoplankton culture system required to inoculate the larval rearing tanks for initial cell density of 10^4–10^5 cells/ml at stocking, a total of nine tanks of 10 m^3 capacity will be required for each of the groups of fish larvae being processed. The culture would continue to enable this density to be maintained if additions were required. A total of four tanks of the same capacity will be required to supply batch cultures of *Brachionus plicatilis* at an initial stocking rate of 10/ml and then to supplement daily with the same number if required. It is estimated that sufficient supply of small copepods (2/ml) could also be obtained from four 5 m^3 capacity outdoor ponds, suitably inoculated and with a good growth of micro-algae.

In addition to the culture units it will be necessary to have an *Artemia salina* nauplius incubator capable of a daily output of 200 million nauplii per day.

The overall water requirement for the system excluding the larval food unit is nearly 3200 m^3. A storage reservoir of 50 m^3 should be available together with a service reservoir of 20 m^3 in which the water is continuously circulated and treated. Together with service laboratories and accommodation, the entire facility could be housed in a single story building occupying 4000 m^2 of floor space. An estimate team of twelve staff and technical aides could operate the hatchery.

As demonstrated by the White Fish Authority at their marine flatfish hatchery in Britain, the pilot operation is necessary to define problems not apparent in the enlarged laboratory practices, and to learn the economies of scale. After twelve years of development with hatchery production, a basis for the commercial production of flatfish is now very much refined. Larval rearing density has been increased considerably through management experience and consequently has made a significant saving in capital construction requirement.

For the mullet hatchery, tripling the final rearing density to 1 larva/l by increasing survival from 5 to 15%, the same facilities could be used to produce 3 million juveniles over a period of less than $2\frac{1}{2}$ months; alternatively the facilities and capital cost could be reduced by two-thirds for the original production of one million larvae. With environmental manipulation facilities to provide spawning fish on a year round basis, the same facilities could produce 15 million juveniles per year. Tang (1974) estimated that the aquaculture fishery of 2000 metric tons of grey mullets in Taiwan is based on the collection of 10 million juveniles annually.

In addition to the advantages of genetics for increased growth and production in ponds, genetic engineering and cross-breeding within the Mugilidae may result in the production of larvae with hybrid vigor sufficient to increase survival and for rearing at increased density. The benefits of genetic engineering were described by Purdom (1972). He said that gynogenesis promised to be a rapid way of producing inbred lines, and the production of artificial triploids already showed greater growth rates than the diploids. Triploids also had the advantage of being sterile, which has always been a significant factor in fatstock production.

Liao *et al.* (1972) reported the problem of obtaining sexually mature males throughout the breeding season. He described the ease with which they could be caught and used early in the season, but by the end they were spent leaving several female fish unfertilized. Therefore large populations of males are necessary to support a hatchery production system so that this does not occur. Environmental control should also be used for males retained indoors. The shortage of males demonstrates the need for more work on the cryogenic preservation of sperm, first developed for fish by Blaxter (1953), and followed by Hwang, Cheng, Lee & Liao (1972) and Chao, Cheng & Liao (1974) for the males of *Mugil cephalus*. Unpublished work by Watters in Hawaii demonstrated the differences in activity of sperm in a number of media. The motility and viability of the sperm suspended in ambient sea water (32‰) lasted for about one hour and much longer than that in other media. However, there were differences between sperm of different males.

These factors demonstrate once more the need to define the quality of eggs and sperm from mature fish, and to establish that the quality is maintained year by year. Fish tend to get treated as part of a population, used and returned to the group. The first process of marking or isolating good breeding fish can commence now without waiting for the biochemistry of egg quality to be established. Much fundamental work is possible now which will lay the foundation for improvement by genetics and stock selection.

The Mugilidae have the greatest potential for becoming the most important supplier of aquatic animal protein for mankind. However, this potential can only be realised by the successful artificial propagation of juveniles from hatcheries. The wealth of information that is available on the culture of the mullet is more than enough to justify the establishing of a coastal hatchery for the pilot scale production of juveniles. Many enterprises have succeeded with much less basic knowledge and with less at stake.

References

Abraham, M., Blanc, N. & Yashouv, A. (1966). Oogenesis in five species of grey mullets (Teleostei, Mugilidae) from natural and land locked habitats. *Israel J. of Zool.* **15** (3–4), 155–72.

Abraham, M., Blanc, N. & Yashouv, A. (1968). Persistent yolk nuclei in the oocytes of *Mugil cephalus* when confined to fresh water environment. *Ann. Embriol. Morphol.* **1**, 169–78.

Adams, W., Montalban, H. R. & Martin, C. (1931). Cultivation of bangos in the Philippines. *Philippine J. Sci.* **47**, 1–38.

Ahsan, S. N. (1966). The control of cyclical changes in the testicular activity of the lake chub, *Couesius plumbeus* (Agassiz). *Can. J. Zool.* **44**, 149–59.

Anderson, W. W. (1957). Early development, spawning, growth and occurrence of the silver mullet (*Mugil curema*) along the South Atlantic coast of the United States. *Fish. Bull. US Fish Wildl. Serv.* **57** (119), 397–414.

Anderson, W. W. (1958). Larval development, growth and spawning of striped mullet (*Mugil cephalus*) along the South Atlantic coast of the United States. *Fish. Bull. US Fish Wildl. Serv.* **58** (144), 501–19.

Angelos, H. G. (1971). A preliminary report on the observation and possibilities of induced spawning in mullet and milkfish. *Occas. Pap. Indo-Pac. Fish. Counc.* **71** (8), 1–12.

Anon. (1940). Mullet experiment successful. *Fish. Gazet.* **57**(8), 21.

Apyekin, V. S. & Tronina, T. M. (1972). Experiments on stimulation of maturing and spawning in mullet (tentative results). *Gidrobiol. Zh.* **8** (1), 82–9.

Arnold, E. L. & Thompson, J. R. (1958). Offshore spawning of the striped mullet, *Mugil cephalus*, in the Gulf of Mexico. *Copeia* (2), 130–132.

Avanesov, E. M. (1972). Present spawning conditions of mullets (genus Mugil) in the Caspian Sea. *J. Ichthyol.* (*Vopr. Ikhtiol.*, English Edition), **12** (3), 419–25.

Babaian, K. E. (1958). Caspian mullet. *Zool. Zh.* **36**, 1505–13. (In Russian, English summary.)

Babaian, K. E. (1960). Vyrashichivanie kefali v karpovykh prudakh (Rearing mullet in carp ponds). *Rybovod Rybolov.* **6**, 23. (In Russian.)

Babaian, K. E. & Zaitsev, Y. P. (1964). New data on the biology of the grey mullet and prospects for increasing grey mullet rearing in the USSR. *Zool. Zh.* **43** (9), 1342–54. (In Russian, English summary.)

Baggerman, B. (1957). An experimental study on the timing of breeding and migration in the three-spined stickleback (*Gasterosteus aculeatus* L.). *Arch. néerl. Zool.* **12**, 105–317.

Basu, S. P. (1946). Possibilities of mullet culture in India. *Indian Farming* **7** (11), 517–22.

Belloc, G. (1938). Biologie et pêche de l'étang de Biguglia. *Rapp. P.-V. Comm. Int. Explor. Sci. Mer Mediterr.* **11**, 405–31.

Bernhard, M. & Zattera, A. (1970). The importance of avoiding chemical contamination for a successful cultivation of marine organisms. *Helgoländer Wiss. Meeresunters.* **20**, 655–75.

Blanco, G. J. & Acosta, P. A. (1958). The propagation of the grey mullet in northern Luzon brackish water fishponds. *Philipp. J. Fish.* **6** (1), 1–5.

Blaxter, J. H. S. (1953). Sperm storage and cross-fertilization of spring and autumn spawning herring. *Nature, Lond.* **172**, 1172–90.

Blaxter, J. H. S. (1969*a*). Visual thresholds and spectral sensitivity of flatfish larvae. *J. Exp. Biol.* **51**, 221–30.

Blaxter, J. H. S. (1969*b*). Development: eggs and larvae. In *Fish Physiology*, vol. 3, ed. W. S. Hoar and D. J. Randall, pp. 178–252, Academic Press, New York.

Breder, C. M. (1940). The spawning of *Mugil cephalus* on the Florida West Coast. *Copeia* (2), 139–39.
Broadhead, G. C. (1953). Investigations of the black mullet, *Mugil cephalus* L., in northwest Florida. *Tech. Ser. Florida State Board Conserv.* (7), 1–33.
Bromhall, J. D. (1954). A note on the reproduction of the grey mullet, *Mugil cephalus. Hong Kong Univ. Fish. J.* **1**, 19–34.
Bullough, W. S. (1940). The effect of the reduction of light in spring on the breeding season of the minnow (*Phoxinus laevis* L.). *Proc. Zool. Soc. Lond.* (*A*) **100**, 149–57.
Campbell, A. Y. G. (1921). Madras Fisheries Administration Report for the Year 1919–1920. *Madras Fish. Bull.* (13), 1–34.
Cervigón, F. & Padrón, M. (1974). Hypersaline lagoons as an ideal environment for mullet culture. In *Proceedings of the IBP/PM International Symposium on the Grey Mullets and their Culture, Haifa* 2–8 June 1974.
Chao, N. H., Chen, H. P. & Liao, I. C. (1974). Study on cryogenic preservation of grey mullet sperm in Taiwan. In *Proceedings of the IBP/PM International Symposium on the Grey Mullets and their Culture, Haifa* 2–8 June 1974.
Chen, F. Y., Chow, M. & Sim, B. K. (1969). Induced spawning of the three major Chinese carps in Malacca, Malaysia. *Malays. Agric. J.* **47**, 211–38.
Chepurnov, V. S. & Dmitriyev, Y. I. (1962). Results of an investigation on the rearing of mullet in the estuaries of the Odessa Region and measures recommended to increase productivity. *Uch. Zap. Kishinev Un-ta* **62** (1).
Chow, I. (1958). Growth characteristics of four species of pond fish in Hong Kong. *Hong Kong Univ. Fish. J.* **2**, 29–36.
Costlow, J. D. (1969). *Marine Biology*, vol. 5. Gordon and Breach, New York.
D'Ancona, U. (1955). Les lagunes Méditerranéenes et leur exploitation (Mediterranean lagoons and their utilization). *Proc. Tech. Pap. Fish. Counc. Mediterr.* **3**, 145–49.
De Angelis, R. (1969). Mediterranean brackish water lagoons and their exploitation. *Stud. Rev. Gen. Fish. Counc. Mediterr.* **12**, 1–41.
Dekhnik, T. V. (1953). Reproduction of *Mugil cephalus* in the Black Sea. *Dokl. Akad. Nauk. SSSR* **93**, 201–4. (In Russian, English summary.)
Dekhnik, T. V. (1954). Reproduction of the anchovy and mullet in the Black Sea. *Tr. Vses. Nauchno-Issled. Inst. Morsk. Rybn. Khoz. Okeanogr.* **28**, 34–48. (In Russian.)
Demir, N. (1971). On the occurrence of grey mullet postlarvae off Plymouth. *J. Mar. Assoc. UK* **51**, 235–46.
Devanesan, D. W. & Chacko, P. I. (1943). On the possibility of culture of certain marine mullets in fresh water tanks. *Proc. Nat. Inst. Sci. India* **9**, 249–50.
Donaldson, E. M., Yamazaki, F., Dye, H. M. & Philleo, W. W. (1972). Preparation of gonadotropin from salmon (*Onchorhynchus tshawytscha*) pituitary glands. *Gen. Comp. Endocr.* **18**, 460–81.
El-Zarka, S. El-D. (1968). Rehabilitation of the fisheries of an inland saline lake in the United Arab Republic. *Stud. Rev. Gen. Fish. Coun. Mediterr.* (35), 21–44.
Erman, F. (1958). About the artificial feeding of grey mullets in Israel. *Balik Balikcilik* **6** (3), 9–12.
Erman, F. (1961). On the biology of the thick-lipped grey mullet (*M. chelo* Cuv.). *Rapp. P.-V. Commn Int. Explor. Sci. Mer Mediterr.* **16**, 276–85.
Faouzi, H. (1936). Successful stocking of Lake Qarun with mullets (*Mugil cephalus* and *Mugil capito* Cuv. and Val.) from the Mediterranean. *Int. Rev. Gesamten Hydrobiol. Hydrogr.* **35** (5/6), 434–9.
Fitch, J. E. (1972). A case for striped mullet, *Mugil cephalus*, spawning at sea. *California Fish & Game* **58** (3), 246–8.

Fujimura, T. & Okamoto, H. (1972). Notes on progress made in developing a mass culturing technique for *Macrobrachium rosenbergii* in Hawaii. In *Coastal Aquaculture in the Indo-Pacific Region* (ed. T. V. R. Pillay), Fishing News (Books) Ltd., London.

Fujinaga, M. (1963). Culture of Kuruma shrimp (*Penaeus japonicus*). *Curr. Aff. Bull. Indo-Pac. Fish. Counc.* **36**, 10–11.

Ganapati, S. V. & Alikunhi, K. H. (1949). Experiments on the acclimatization of salt water fish seed to fresh water. *Proc. Indian Sci. Congr.* **36** (3), 161 (Abstract).

Gandolfi, G. & Orsini, P. (1970). Some observations on the reproductive biology of *Mugli saliens* from the Venice Lagoon. *Boll. Pesca Piscic. Idrobiol.* **25** (1), 85–93. (In Italian).

Gravely, F. H. (1929). Madras Fisheries Administration Report for the year 1927–1928. *Madras Fish. Bull.* **23** (1), 1–86.

Hamis, A. H. H. (1972). On young mullet from the Black Sea and the southeast Mediterranean sea. *Tr. Vses. Nauchno-Issled. Inst. Morsk. Rybn. Khoz. Okeanogr.* **90**, 142–7.

Harada, T. (1970). The present status of marine fish cultivation research in Japan. *Helgolander Wiss. Meeresunters.* **20**, 594–601.

Harrington, R. W. Jr. (1959). Photoperiodism in fishes in relation to the annual sexual cycle. In *Photoperiodism and Related Phenomena in Plants and Animals*, vol. 55, pp. 651–67, Am. Assoc. Adv. Sci. Pub., Washington DC.

Heldt, H. (1948). Contribution à l'étude de la biologie des muges des lacs Tunisiens. *Bull. Stn. Oceanogr. Salammbô* (41), 1–50.

Henderson, N. E. (1963). Influence of light and temperature on the reproductive cycle of the eastern brook trout, *Salvelinus fontinalis* (Mitchill). *J. Fish. Res. Board Can.* **20**, 859–97.

Hiatt, R. W. (1944). Food chains and food cycle in Hawaiian fish ponds. I. The food and feeding habits of mullets (*Mugil cephalus*), milk fish (*Chanos chanos*) and the ten-pounder (*Elops machnata*). *Trans. Am. Fish. Soc.* **74**, 250–61.

Hickling, C. F. (1970). A contribution to the natural history of the English grey mullets (Pisces, Mugilidae). *J. Mar. Biol. Assoc. UK* **50** (3), 609–33.

Hines, R. & Yashouv, A. (1971). Some environmental factors influencing the activity of spermatozoa of *Mugil capito* Cuvier, a grey mullet. *J. Fish Biol.* **3**, 123–7.

Hoar, W. S. (1959). Endocrine factors on the ecological adaptation of fishes. In *Comparative Endocrinology*, ed. A. Gorbman, pp. 1–23. Wiley, New York.

Holliday, F. G. T. (1969). The effects of salinity on the eggs and larvae of teleosts. In *Fish Physiology*, vol. 1, ed. W. S. Hoar & D. J. Randall, 293–311. Academic Press, New York.

Hora, S. L. & Nair, K. K. (1944). Suggestions for the development of salt water bheris or Bhasabadha fisheries in the Sunderbans. *Fish. Dev. Pam., Dep. Fish. Bengal* (1), 1–12.

Hornell, J. (1922). The Madras Fisheries Administration Report for the Year 1920–1921. *Madras Fish. Bull.* **15** (1), 44 pp.

Hotta, H. (1955). On the mature mugilid fish from Kabashima, Nagasaki Prefecture, Japan, with additional notes on the intestinal convolution of Mugilidae. *Jap. J. Ichthyol.* **4**, 162–9. (In Japanese, English summary.)

Houde, E. D. (1972). Some recent advances and unsolved problems in the culture of marine fish larvae. In *Proceedings of the Third Annual Workshop of the World Mariculture Society*, St Petersburg, Florida.

Hubbs, C. L. (1943). Terminology of early stages of fishes. *Copeia* (4), 260.

Artificial propagation

Hwang, S. W., Chen, H. P., Lee, D. L. & Liao, I. C. (1972). Preliminary results in the cryogenic preservation of grey mullet (*Mugil cephalus*) sperm. *J. Fish. Soc. Taiwan* **1** (2), 1–7.

Jacob, P. K. & Krishnamurthi, B. (1948). Breeding and feeding habits of mullets *Mugil* in Ennore Creek. *J. Bombay Nat. Hist. Soc.* **47**, 663–8.

Job, T. J. & Chacko, P. I. (1947). Rearing of salt water fish in fresh waters of Madras. *Indian Ecol.* **2** (1), 1–9.

Johnson, M. C. (1954). Preliminary experiments on fish culture in brackish water ponds. *Prog. Fish-Culturist* **16** (3), 131–3.

Kennedy, M. & Fitzmaurice, P. (1969). Age and growth of thick-lipped grey mullet, *Crenimugil labrosus* (Risso) in Irish waters. *J. Mar. Biol. Assoc. UK* **48**, 683–99.

Kesteven, G. L. (1953). Further results of tagging sea mullet, *Mugil cephalus*, on the eastern Australian coast. *Aust. J. Mar. Freshwater Res.* **4** (2), 251–306.

Kuo, C. M. & Nash, C. E. (1975). Recent progress on the control of ovarian development and induced spawning of the grey mullet (*Mugil cephalus* L.). *Aquaculture* **5**, 19–29.

Kuo, C. M., Shehadeh, Z. H. & Milisen, K. K. (1973*a*). A preliminary report on the development, growth and survival of laboratory reared larvae of the grey mullet, *Mugil cephalus* L. *J. Fish Biol.* **5**, 459–70.

Kuo, C. M., Shehadeh, Z. H. & Nash, C. E. (1973*b*). Induced spawning of captive grey mullet (*Mugil cephalus* L.) females by injection of human chorionic gonadotropin. *Aquaculture* **1**, 429–32.

Kuo, C. M., Nash, C. E. & Shehadeh, Z. H. (1974*a*). A procedural guide to induce spawning in grey mullet (*Mugil cephalus* L.). *Aquaculture* **3**, 1–14.

Kuo, C. M., Nash, C. E. & Shehadeh, Z. H. (1974*b*). The effects of temperature and photoperiod on ovarian development in captive grey mullet (*Mugil cephalus* L.). *Aquaculture* **3**, 25–43.

Kurian, C. V. (1974). Mullets and mullet fisheries in India. In *Proceedings of the IBP/PM International Symposium on the Grey Mullets and their Culture, Haifa*, 2–8 June 1974.

Le Dantec, J. (1955). Quelques observations sur la biologie des muges des réservoirs de Certes à Audenge. *Rev. Trav. Off. (Sci. Tech.) Pêches Marit.* **19** (1), 93–7.

Liao, I. C. (1974). The experiments on the induced breeding of the grey mullet in Taiwan from 1963 to 1973. In *Proceedingsof the IBP/PM International Symposium on the Grey Mullets and their Culture, Haifa*, 2–8 June 1974.

Liao, I. C., Cheng, C. S., Tseng, L. C., Lim, M. Y., Hsieh, L. S. & Chen, H. P. (1972). Preliminary report on the mass propagation of grey mullets, *Mugil cephalus*. *Fish. Ser. Chin.-Am. J. Comm. Rural Reconstr.* (12), 1–14.

Liao, I. C., Lu, Y. J., Huang, T. L. & Lin, M. C. (1971). Experiments on induced breeding of the grey mullet, *Mugil cephalus*. *Fish. ser. Chin.-Am. J. Comm. Rural Reconstr.* (11), 1–29.

Lin, S. Y. (1940). Fish culture in ponds in the new territories of Hong Kong. *J. Hong Kong Fish. Res. Stn.* **1** (2), 161–93.

Linder, D. R., Strawn, K. & Luebke, R. W. (1974). The culture of striped mullet (*Mugil cephalus* L.) in ponds receiving heated effluent from a power plant. In *Proceedings of the IBP/PM International Symposium on the Grey Mullets and their Culture, Haifa*, 2–8 June 1974.

Lunz, G. R. (1951). A salt water fish pond. *Contributions from the Bears Bluff Laboratories, no. 12*, Bears Bluff Laboratories, Wadmalaw Island, South Carolina.

May, R. C. (1970). Feeding larval marine fishes in the laboratory: a review. *Rep. Calif. Coop. Oceanic Fish. Invest.* **14**, 76–83.

Medlen, A. B. (1951). Preliminary observation on the effect of temperature and light upon the reproduction in *Gambusia affinis*. *Copeia* 148–53.

Merriman, D. & Schedl, H. P. (1941). The effects of light and temperature on gametogenesis in the four-spined stickleback, *Apeltes quadracus* (Mitchell). *J. Exp. Zool.* **88**, 413–49.

Mohanty, S. K. (1973). Temperature tolerance of grey mullet. *FAO Aquaculture Bull.* **5** (2), 9.

Mookerjee, H. K., Ganguly, D. N. & Sircar, A. (1946). On the composition of food of the Indian mullet *Mugil parsia* Ham. with suggestions to culture them in fresh water ponds of Bengal. *Proc. Indian Sci. Congr.* **33** (3), 132.

Nash, C. E. (1973). Automated mass-production of *Artemia salina* nauplii for hatcheries. *Aquaculture* **2**, 289–98.

Nash, C. E. & Kuo, C. M. (1975). Hypotheses for problems impeding the mass propagation of grey mullet and other finfish. *Aquaculture* **5** (2), 119–34.

Nash, C. E., Kuo, C. M. & McConnell, S. C. (1974). Operational procedures for rearing larvae of the grey mullet (*Mugil cephalus* L.). *Aquaculture* **3**, 15–24.

Nash, C. E. & Shelbourne, J. E. (1967). Power station effluent as an environment for flatfish culture. *ICES Fish. Imp. Comm.* CME: 10.

Nikolskii, G. (1954). *Special Ichthyology*. Sovietskaya nauka. Moscow (2nd edn. translated from Russian by J. I. Lengy & Z. Krauthamer, Jerusalem, IPST Cat. no. 233, 1961).

Oppenheimer, C. H. (1955). The effect of marine bacteria in the development and hatching of pelagic fish eggs, and the control of such bacteria by antibiotics. *Copeia* (1), 43–9.

Oren, O. H. (1971). International biological programme: coordinated studies on grey mullets. *Mar. Biol.* **10** (1), 30–3.

Paget, G. (1923). The determination of the rate of growth of Bouri, *Mugil cephalus*, in Lake Mariyut. *Rep. Fish. Serv. Egypt* 1922, 43–9.

Panikkar, N. K. (1951). Problems of marine and estuarine fisheries. In *Indian Fisheries*, 79–87, Ministry of Agriculture.

Perceva-Ostroumova, T. A. (1951). Reproduction and development of grey mullets introduced in the Caspian Sea. *Tr. Vses. Nauchno-Issled. Inst. Morsk. Rybn. Khoz. Okeanogr.* **18**, 123–34.

Perlmutter, A., Bograd, L. & Pruginin, J. (1957). Use of the estuarine and sea fish of the family Mugilidae (grey mullets) for pond culture in Israel. *Proc. Tech. Pap. Gen. Fish. Coun. Mediterr.* **4**, 289–304.

Pillay, T. V. R. (1947). Possibilities of mullet culture in India. *Indian Farming* **8** (11), 544–9.

Pruginin, Y. & Kitai, H. (1957). Cultivating mullets as a secondary fish in the carp ponds. *Bamidgeh* **9** (4), 70–5.

Purdom, C. E. (1972). Genetics and fish farming. *Laboratory leaflet* (New Ser.) No. 25, MAFF, Fisheries Laboratory, Lowestoft, Suffolk, UK.

Riley, J. D. (1966). Marine fish culture in Britain. VII. Plaice (*Pleuronectes platessa*) post-larval feeding on *Artemia salina* L. nauplii and the effects of varying feeding levels. *J. Cons. Int. Explor. Mer* **30**, 204–21.

Rosenthal, H. & Hempel, G. (1970). Experimental studies in feeding and food requirements of herring larvae (*Clupea harengus* L.). In *Marine Food Chains*, ed. J. H. Steele, University of California Press, Berkeley.

Roughley, T. C. (1916). Fishes of Australia and their technology. *Tech. Educ. Ser. Publ. Instr. Dep. N.S. Wales* (21), 1–296.

Roule, L. (1917). Sur le developpement larvaire et post-larvaire des poissons du genre *Mugil. C.R. Hebd. Seances Acad. Sci., Paris* **164** (5), 194–6.

Sanzo, L. (1930). Uova e larve di *Mugil cephalus* Cuv. ottenute per fecondazione artificiale. *Mem. R. Com. Talassogr. Ital.* **179**, 3.

Sanzo, L. (1931). Uova e larve di mugilidi. *Atti Soc. Ital. Prog. Sci.* **20**, 324–5.

Sanzo, L. (1936). Contributi alla conoscenza della sviluppo embrionario e post-embrionario nei Mugilidi. I. *Mugil cephalus* Cuv., II. *M. chelo* Cuv. *Mem. R. Com. Talassogr. Ital.* **230**, 1–11.

Sanzo, L. (1937). Uova e larve di *Mugil labeo* Cuv. *Boll. Pesca Piscio. Hidrobiol.* **13** (5), 506–10.

Sarojini, K. K. (1957). Biology and fisheries of the grey mullets of Bengal. I. Biology of *Mugil parsia* Hamilton with notes on its fishery in Bengal. *Indian J. Fish.* **4** (1), 160–207.

Sarojini, K. K. (1958). Biology and fisheries of the grey mullets of Bengal. II. Biology of *Mugil cunnesius* Valenciennes. *Indian J. Fish.* **5** (1), 56–76.

Sebastian, M. J. & Nair, V. A. (1973). The induced spawning of the grey mullets. *Seafood Export J.* **5** (1), 1–4.

Sebastian, M. J. & Nair, V. A. (1974). The induced spawning of the grey mullet, *Mugil macrolepis* (Aguas) Smith, and the large scale rearing of its larvae. In *Proceedings of the IBP/PM International Symposium on the Grey Mullets and their Culture, Haifa*, 2–8 June 1974.

Sharpe, B. (1945). Five thousand pounds of fish a year per acre. *Ctry. Gentleman*, **115** (12), 26–8.

Shehadeh, Z. H. & Ellis, J. N. (1970). Induced spawning of the striped mullet, *Mugil cephalus* L. *J. Fish Biol.* **2**, 355–60.

Shehadeh, Z. H., Kuo, C. M. & Milisen, K. K. (1973*a*). Validation of an *in vivo* method for monitoring ovarian development in the grey mullet (*Mugil cephalus* L.). *J. Fish Biol.* **5**, 489–96.

Shehadeh, Z. H., Kuo, C. M. & Milisen, K. K. (1973*b*). Induced spawning of grey mullet *Mugil cephalus* L. with fractionated salmon pituitary extract. *J. Fish Biol.* **5**, 471–8.

Shehadeh, Z. H., Kuo, C. M. & Nash, C. E. (1973*c*). Establishing brood stocks of grey mullet (*Mugil cephalus* L.) in small ponds. *Aquaculture* **2**, 379–84.

Shehadeh, Z. H., Madden, W. D. & Dohl, T. P. (1972). The effect of exogenous hormone treatment on spermiation and vitellogenesis in the grey mullet, *Mugil cephalus* L. *J. Fish Biol.* **5**, 479–87.

Shelbourne, J. E. (1964). The artificial propagation of marine fish. *Adv. Mar. Biol.* **2**, 1–83.

Shelbourne, J. E. (1969). In *Marine Biology*, vol. 2, ed. J. D. Costlow, Gordon and Breach, New York.

Shelbourne, J. E. & Nash, C. E. (1966). Sea fish culture in Britain. *Proc. Nutr. Soc.* **25** (2), 133.

Shireman, J. V. (1974). Preliminary experiments on striped mullet culture in fresh water catfish ponds. *Prog. Fish-Cult.* **26** (2), 113.

Smith, J. L. B. (1935). The fishes of the family Mugilidae in South Africa. *Ann. South Afr. Mus.* **30** (19), 587–644.

Stenger, A. H. (1959). A study of the structure and development of certain reproductive tissues of *Mugil cephalus* L. *Zool.*, NY **44** (2), 53–68.

Struhsaker, J. W., Hashimoto, D. Y., Girard, S. M., Prior, F. T. & Conney, T. D. (1973). Effects of antibiotics on survival of carangid fish larvae (*Caranx mate*), reared in the laboratory. *Aquaculture* **2**, 53–88.

Sundararaj, B. I. & Goswami, S. W. (1969). Techniques of hypophysation of catfishes. In *FAO/UNDP Regional Seminar on Induced Breeding of Cultivated Fishes*. FRi/IBCF/15, 1–23 FAO.

Tamura, T. (1966). *Marine Aquaculture*. Translation from the Japanese of the 2nd edn. (This revised and enlarged edition translated by M. I. Watanabe, National Science Foundation, Washington DC, PB 194 051T, Part 1.)

Tang, Y. A. (1964). Induced spawning of striped mullet by hormone injection. *Jap. J. Ichthyol.* **12** (1/2), 23–8.

Tang. Y. A. (1974). Collection, handling and distribution of grey mullet fingerlings in Taiwan. In *Proceedings of the IBP/PM International Symposium on the Grey Mullets and their Culture, Haifa*, 2–8 June 1974.

Theilacker, G. H. & McMaster, M. F. (1971). Mass culture of the rotifer *Brachionus plicatilis* and its evaluation as a food for larval anchovies. *Mar. Biol.* **10**, 183–8.

Thomson, J. M. (1954). The organs of feeding and the food of some Australian mullet. *Aust. J. Mar. Fresh-water Res.* **5** (3), 469–85.

Thomson, J. M. (1955). The movements and migrations of mullet (*Mugil cephalus* L.). *Aust. J. Mar. Fresh-water Res.* **6** (3), 328–47.

Thomson, J. M. (1957). Biological studies of economic significance of yellow-eye mullet, *Alderichetta forsteri* (Cuv. and Val.) (Mugilidae). *Aust. J. Mar. Fresh-water Res.* **8** (1), 1–13.

Thomson, J. M. (1963). Synopsis of biological data on the grey mullet *Mugil cephalus* L. 1758. *Fish. Synop. Div. Fish. Oceanogr.* CSIRO, Australia (1).

Thomson, J. M. (1966). The grey mullets. *Oceanogr. Mar. Biol. Ann. Rev.* **4**, 301–35.

Thong, L. H. (1969). Contributions à l'étude de la biologie des Mugilidés (poissons téléostéens) des côtes du massif Armoricain. *Trav. Fac. Sci. Univ. Rennes* (*Sér. Océanogr. Biol.*), (2), 55–182.

Tung, I. H. (1973). On the egg development and larval stages of the grey mullet, *Mugil cephalus* Linnaeus. *Rep. Inst. Fish. Biol. Minist. Econ. Aff. Nat. Taiwan Univ.* **3** (1), 187–215.

Vodyanitskii, V. A. & Kazanova, I. I. (1954). The identification of pelagic eggs and larvae of Black Sea fishes. *Tr. Vses. Nauchno-Issled. Inst. Morsk. Rybn. Khoz. Okeanogr.* **28**, 240–325.

Wallace, J. H. (1974). Aspects of the biology of *Mugil cephalus* in a hypersaline estuarine lake on the east coast of South Africa. In *Proceedings of the IBP/PM International Symposium on the Grey Mullets and their Culture, Haifa*, 2–8, June 1974.

Walne, P. R. (1958). The importance of bacteria in laboratory experiments on rearing the larvae of *Ostrea edulis*. I. Feeding experiments. *J. Mar. Biol. Assoc. UK* **43**, 767–84.

Wimpenny, R. S. & Faouzi, H. (1935). The breeding of a grey mullet, *Mugil capito* Cuv., in Lake Qarun, Egypt. *Nature, Lond.* **135** (3425), 1041.

Yang, W. T. & Kim, U. B. (1962). A preliminary report on the artificial culture of grey mullet in Korea. *Proc. Indo-Pac. Fish. Counc.* **9** (2/3), 62–70.

Yashouv, A. (1966). Breeding and growth of grey mullet (*Mugil cephalus* L.). *Bamidgeh* **18** (1), 3–13.

Yashouv, A. (1969). Preliminary report on induced spawning of *Mugil cephalus* L. reared in captivity in fresh water ponds. *Bamidgeh* **21** (1), 19–24.

Yashouv, A. & Ben-Shachar, A. (1967). Breeding and growth of Mugilidae. II. Feeding experiments under laboratory conditions with *Mugil cephalus* L. and *M. capito* Cuv. *Bamidgeh* **19** (2/3), 50–66.

Yashouv, A. & Berner-Samsonov, E. (1970). Contribution to the knowledge of eggs and early larval stages of mullets (Mugilidae) along the Israeli coast. *Bamidgeh* **22** (3), 72–89.

Yoshioka, H. (1962). On the effects of environmental factors upon the reproduction of fishes. 1. The effects of day-length on the reproduction of the Japanese killifish *Oryzias latipas. Bull. Fac. Fish. Hokkaido Univ.* (13), 123–36.

Zismann, L., Berdugo, V. & Kimor, B. (1974). The food and feeding of early stages of grey mullets in the Haifa Bay region. In *Proceedings of the IBP/PM International Symposium on the Grey Mullets and their Culture, Haifa*, 2–8 June 1974.

Zviagina, O. A. (1961). The distribution of eggs of mackerel *Pneumatophorus japonicus* (Houttuyn) and mullet (*Mugil so-iuy* Basilewsky) in Peter the Great Bay. *Tr. Inst. Okeanol.* **43**, 328–36.

9. Methods of capture of grey mullets

M. BEN-YAMI & E. GROFIT

Introduction

Grey mullets (family Mugilidae) are distributed around the world, mainly in tropical and subtropical waters. Some of them, notably, the striped mullet *Mugil cephalus* (L.) are truly cosmopolitan and of considerable commercial importance. Their behaviour differs from that of most other fishes. Because of this, specialized fishing methods some of them similar all over the world, were developed. These differ mostly in details and/or in the degree of sophistication. In the present chapter we wish to review, in general terms, the gear and methods used in the mullet fishery.

The habitat and behaviour of grey mullets

Notes on the habitat, behaviour, feeding, migration and tolerance of mugilids are given in various chapters of the present volume, while for *Mugil cephalus* specifically much information can be found in the synopsis on this species by Thompson (1966) as well as in papers by Ben-Yami (1974), Berg Bogdanov, Kojin & Rass (1949), Kristensen (1964), Babayan (1965) and Bograd (1961). Here we shall mention only factors which affected the development and the design of the equipment and methods employed in grey mullet fishery.

The general opinion prevailing among fishermen and biologists on the behaviour and habitat of grey mullets can be summed up as follows: grey mullets are euryhaline, eurythermic, tending to warm waters, resistant to oxygen deficiency, extremely agile, schooling, leap from the water (especially when frightened or encountering near-surface obstacles), shun light and are sensitive to noise. They enter estuaries, lagoons and rivers, and perform catadromous spawning migration. This generalized statement is true for many mullet species, but especially for the largest, most common and commercially most important, the striped mullet *Mugil cephalus* (L.).

Some mullets, on the other hand, may deviate from this generalized pattern, e.g., be totally marine like the *Oedalechilus labeo* in the eastern Mediterranean (Zismann, personal communication); not venturing too far into fresh or brackish water like the Indo-Pacific *Myxus elongatus* or the cosmopolitan *Liza aurata* (Bograd, 1961); or, as the South African *Strializa canaliculatus*, not standing out as a jumping fish. These common behavioural traits had, even in ancient times, led to the invention of the very special mullet fishing methods.

When they land after jumping from the water the frightened mullets may be caught on rafts, mats, horizontally floating verandah nets or, even inside a rowing boat. They are even shot at when leaping. Sometimes, for increased efficiency, two or more methods are combined.

Capture of adults

Angling, spearing and cast-netting

Triple hooks are sometimes used to catch mullets. A hooked line is thrown in the midst of a school and given a sharp jerk to hook a fish in the body. This method seems to be known everywhere. Cooper (1937) mentions, that in England, to facilitate angling, bait is applied to chum the mullets. Grey mullets, as many other shallow water fish, are caught with cast-nets and speared all over the world.

Seines

Schooling habits enable the use of diverse seines (beach seine, purse seine, etc.). Migratory schools may be intercepted on their run by entangling gear, such as gill nets or trammels, and by several types of fixed or mobile traps. Mullets are commercially fished in their normal habitats, such as inshore shallow waters, estuaries, creeks, bays and coastal lagoons.

Beach seining is one of the oldest methods for catching mullets, as well as other inshore fish. In its simplest form, a beach seine is a long, narrow strip of netting, floated at the corkline and weighted at the footrope. The net is set from the shore in a semicircle to enclose a school of fish. It is then hauled from both ends to the shore, gradually reducing the enclosed area, until the catch is concentrated in the bunt of the seine and hauled onto the beach, or into the boat (Fig. 9.1). This method was already in use in ancient Egypt, as shown in early wall drawings (Nun, 1964).

Nowadays, at least five species of grey mullet (*Mugil cephalus*, *Liza ramada*, *L. saliens*, *L. aurata* and *L. provensalis*) form about one-third of the fish caught along the Mediterranean coast of Egypt, in bagless beach seines, which are 200–400 m long, and 3.5 m deep at the wings and 7 m deep at the centre. The mesh size decreases from 17 mm in the wings to 8.5 mm, in the bunt. Wooden spreaders 1.5 m long, are sometimes used at the wing tips. These nets are used with 100 m or more long sweeps (Rafail, 1968, 1971).

Mullet fishing with beach seines is known also in many other places. The seines may differ in construction, having either a long bag with a trawl-type

Fig. 9.1. Beach seining in the Red Sea (Ethiopia).

overhang as in Italy, and Iskanderun Bay (Aasen & Akyuz, 1956), they may be bagless as in Ethiopia and the southern Red Sea (Ben-Yami, 1964), or they may have a small bag without overhang like the Spanish 'boliche' (Cabo, 1959). Bagless and bagged seines are used in the Far East by USSR fishermen, to fish the Pacific mullet (*Mugil so-iuy*) both in the sea and estuaries (Berg *et al.*, 1949).

In the Black Sea, fire and other light sources are used to drive light-shunning mullet towards the shore where they are surrounded with a beach seine (Nikonorov, 1963).

Mullet leaping over the bunt of the beach seine, during the final stage of operation, may be re-captured using a 'backing net', set just behind the bunt. Such a net may be a separate net, or a section of the same beach seine wing already drawn in (Kesteven, 1942).

Beach seines in association with verandah nets are specialized grey mullet gear. They were reported from the USSR (Berg *et al.*, 1949) and Israel (under the name 'tahweek' Perlmutter, 1956), but have probably been used in many other countries, notably, in the Middle East. In the Lagoon of Bardawil (North Sinai), e.g., bagless drag seines up to 2 km long, 6–7 m deep at the bunt are used with a 1 km long sweep rope at each side. They are set from diesel-propelled work boats or towed dories, and operated by teams of up to 30 men, depending on the size of the net. The verandahs consist of 2–2.5 m wide trammels, kept afloat horizontally by means of transverse reed canes,

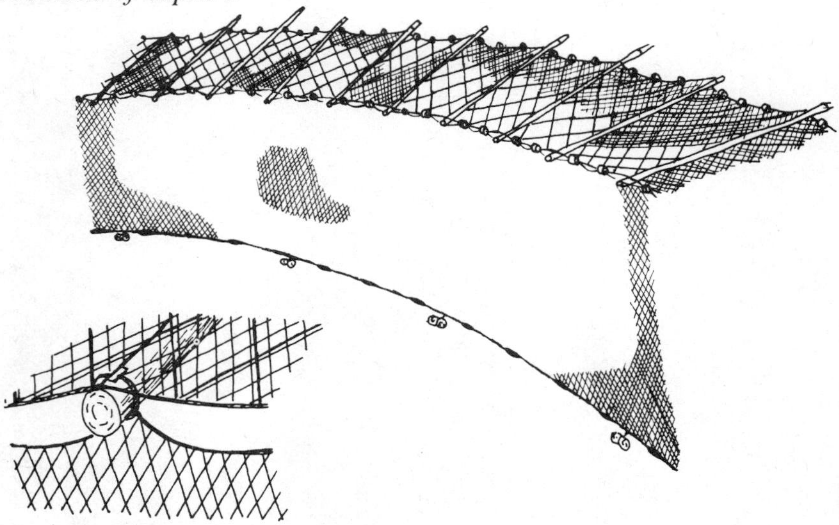

Fig. 9.2. Verandah net (von Brandt, 1964*a*). The horizontal verandah may be attached to a gill-net, trammel, beach seine, purse seine or stationary barrier net (weir, stop net, etc.).

(Fig. 9.2). They are attached to the central part of the beach seine before the final stage of the operation, when the crowded fish start leaping out of the water.

Israeli fishermen who have introduced a mechanized beach seine in this lagoon, replaced the traditional reed struts in the verandah with sections of PVC electric conduit plugged at both ends.

A Russian beach seine used mainly in the Sea of Azov and in the Straits of Kerch has a 700 m long corkline and a 690 m long leadline. It is fitted at its centre with a 50–75 cm wide verandah supported by bamboo canes, (Babayan, 1965).

Roundhaul nets

Roundhaul nets are, generally, oblong sheets of netting, floated at the top, ballasted at the bottom and set off-shore, mostly around mobile schools of fish. The best known roundhaul nets are purse seines (bunt at one end), ringnets (bunt at the centre) and lamparas (leadline shorter than corkline, mesh size decreasing towards the centrally-situated bunt, no purse rings), (Scofield, 1951), and they appear in numerous variants (Ben-Yami, 1974). They have been developed from the beach seines and represent filtering, not entangling gear.

A simple form of such gear is the 'Lawa ni volavola' of Fiji, (Hornell, 1940). It is employed in a community fishing method, where a team is composed of several pairs of women, each providing a rectangular section of net. These are tied together to form a single sheet with poles at both ends. While wading in water, the whole net, footrope included, is held above water. The women holding the net move in a slight curve, while a group of beaters wade in close attendance at both extremeties. When fish are sighted, the footrope is lowered and the beaters rush and scare the fish from the ends towards the centre, by beating the water with sticks. Meanwhile, the net is brought to form a circle around the school. Fish which strike the net are caught by hand and put into palm-leaf baskets.

A roundhaul fishing method that incorporates some elements of culture was reported from Dahomey. It is called 'akadja' and consists in fencing large areas (as much as $2\frac{1}{2}$ acres) by driving long twigs, tree branches, etc. into the lagoon bed. The aledja is left for 4 to 6 months to become a habitat for tilapia, catfish and grey mullet, etc., who find shelter there and feed on algal growth that develops. The fish are harvested by encircling the akadja with nets, removing the brush, and capturing the fish (Pillay, 1968).

Most roundhaul nets used for the capture of mullets are, of course, more sophisticated, having pursing arrangement along their leadline.

A Japanese 500–550 m purse seine, known as 'Chudaka ami', is a rather specialized gear for mullet fishing. Twelve boats participate in setting the net around a sighted school. Then, while two of them are pursing, the other ten distribute around the perimeter and drop anchors. The numerous fishermen hold the floatline high enough to prevent the escape of the mullet, (Anon., 1960). Obviously, this type of operation requires numerous personnel and craft.

An English method of throwing straw into the water over the encircled area surrounded by a seine, to keep the mullet from leaping from the net, was described by Cooper (1937).

In Israel, in the Mediterranean, the bulk of the mullets are caught with roundhaul nets, either Italian-style 'cianciolo' purse-seines, (FAO, 1973) or Turkish-style 'gir-gir' ringnets. The 'gir-gir' ring-nets are set from two rowing boats towed to the fishing spot by a third, motorized, boat. When fish are located, the motor-boat moves aside, the ring net being silently set by the rowing boats. After they complete the setting, the motor-boat joins the operation to play the role of a power-skiff and prevent both rowing-boats from drifting into the netting wall, while they first purse then haul the seine.

In Lake Kinneret, mullets are fished with purse-seines used for bleak and St Peter fish, (FAO, 1975). To prevent the agile mullets from escaping beneath the leadline, their sinking speed is improved by adding ballast to the leadline and, recently, by re-designing the nets using lower hanging ratios and a tapered and gavelled end (i.e. mounted on breastlines, FAO, 1975).

Similar techniques are employed in many other countries, e.g., in the USSR, where *Liza aurata* are caught in two-boat lampara seines (locally called 'alaman') and in purse-seines, (Berg *et al.*, 1949). Since all these are regular seines, without verandah or other arrangements, some mullets manage to escape, by jumping over the floatline. One method to prevent grey mullet from escaping, is to fit the floatline with light rings and to purse the top of the seine at an early stage of the operation. An additional sophistication is to mount at the normal floatline an additional strip of netting, several metres in width, its other edge being mounted to a second floatline fitted with purse rings. This facilitates fast pursing of the top of the net, therefore preventing the mullets from escaping upwards, (Babayan, 1965).

Entangling gear including gillnets and trammels are universally used for the catch of mullets. They may be used passively or set on sighting a school, across its path or around it. Only some of them are specialized mullet gear.

Gill-nets

Small-meshed 25–50 m long, 1 m deep gill-nets are used in Lake Quarun (Egypt) for the capture of *Liza saliens*. They are set on moonless nights and hauled in the early morning. During a single 1961/2 season (from August to May) these nets yielded 335 tons of *L. saliens* (El-Zarka & El Sedfy, 1970).

Davies (1936) reported on a $93 \times \frac{3}{4}$ m gill-net, known as mullet-net of Cadgwith, used on the east coast of England. It was set between rocks over high water. Mullets were frightened into the net by splashing.

A more advanced method of setting gill-nets is reported from Florida (USA). There, gill-nets are set sometimes in a spiral fashion, from skiffs, towed at 15 knots by motor launches, to surround a sighted school. Additional gill-nets may be set in a zig-zag pattern inside the surrounded area. The fish are frightened into the nets by noise. In some cases, light aircraft are used for spotting. Catches of up to 5 ton per set were reported (Siebenholer, 1955). Use of fire spread on water to scare mullets into gill-nets was reported from the same area, (Sutton, 1950). A somewhat similar method, called 'make sashi ami', is applied by Japanese fishermen in two-boat fishing for mullet and young yellowtail, (Fisheries Agency of Japan, 1960).

Surrounding gill-nets have also been used for mullet fishing in the Black and Azov Seas (VNIRO, 1949), by Arab fishermen in the eastern Mediterranean, and by Margarita fishermen in Venezuela, (Mendex-Arocha, 1963).

Roughley (1951), reported on a floating mullet gill-net that is used in Australia.

Trammels

For fishing grey mullet, trammels are usually set so as to surround a school. In the Caspian Sea, USSR fishermen use very large trammels, e.g., 500 m × 12 m, made of synthetic twine, with 80 mm stretch mesh lint, 600 mm stretch mesh armouring (Andreev, 1966). A trammel with only one armouring, called 'Chole' or mullet net, has been reported from England. It is 60 m long and set around a school with the armouring at the outside. The fishes are driven into the net by noise, (Davies, 1936). The fishermen of the eastern Mediterranean and Lake Kinneret (Israel), use trammels in a manner rather similar to the Florida gill-nets fishermen (see above).

Babayan (1965) described a fishery in the Caspian Sea (USSR) where mullet is caught in surrounding trammels. The fishing unit consists of ten men manning one motorized vessel and two work boats in tow. The vessel is equipped with a searchlight. During the night, the searchlight beam causes the fish to leap out of the water and, thus, to be spotted by the fishermen. The work boats now surround the school with a 500–600 m long and 5.5 m deep trammel. Subsequently the larger vessel enters the encircled area and using the searchlight and noise stampedes the fish into the net. Catches may reach 3 to $3\frac{1}{2}$ metric tons per haul. This method is also used in day fishing when fish are discovered by sight and driven into the net by noise.

In Israel, a grey mullet fishery has been developed in the Lake Kinneret (Lake Tiberias) after stocking this lake with some 20 million grey mullet (*Liza ramada* and *Mugil cephalus*) fry since 1960. Since these fish cannot spawn in fresh water, they try to migrate down the Jordan River, which however, is barred near its exit from Kinneret with a sluice-gate controlling the water level in the lake. In autumn, when the *L. ramada* spawning migrations begin, fish assemble at that sluice-gate in large schools and are caught there in floating trammels. This lucrative fishery is limited to a small area of the Jordan River between the lake and the sluice-gate, so that the fishermen have to take turns in setting their nets in order not to interfere with each other's fishing.

Framed gill-nets

A framed gill-net consists of a gill-net with a crosswise pattern of horizontal and vertical twines inserted into the netting so that netting bags are created in the rectangular 'windows' in which fish get both tangled and gilled, (Andreev, 1966).

Konarbaev & Kubashev (1961) found framed gill-nets to be more efficient in grey mullet fishing in the Caspian Sea than the usual trammels. This, probably, because the framed nets could be set around a school almost twice as fast as the trammels. Disentangling fish from the framed nets was three times faster than from the trammels, weight per weight. Also the former nets, length per length, are cheaper and easier to assemble than the latter. The framed mullet net used in the Caspian consisted of 50 m (rope length) sections, with a working depth of 6 m. Hung at hanging ratios of $E' = 0.50$ and $E' = 0.87$, each section was made of 100×7 m polyamide (Kapron) netting panel. The recommended size of a 'window' in the net is 25×25 cm.

Methods of capture

Verandah nets

Verandah nets were described above in association with the Bardawil Lagoon beach seines. A basic type of verandah net is the 'lava ni varanda' of Fiji, where it was probably introduced by Japanese fishermen. It consists of two parts, a vertical section, probably a gill-net, with floats and sinkers and a horizontal verandah, probably a trammel, attached to the floatline of the former, (Hornell, 1940).

In the eastern Mediterranean, verandahs consist of a trammel with armouring at one side only, (von Brandt, 1964*a*). The use of verandah nets of various degrees of sophistication is reported from the Gulf of Suez, Mauritius, Japan (Hornell, 1950) and Spain (Cabo, 1959).

To conclude, verandah nets are used in association with vertical trammels, gill-nets, stop nets, fence nets (barrier seines), boat seines, beach seines and stationary traps and barriers and appear in most, though not all, mullet fisheries, (Fig. 9.2).

Fishing in the air

Under the heading of 'fishing in the air', we include methods designed to induce fish to leap into the air, this leading to their capture. The most basic form of 'air fishing' is undoubtedly, shooting at leaping mullets, a practice reported from Turkey (von Brandt, 1964*a*), and from the Malabar Coast, India by Hornell (1950).

Another simple method was reported from Lake Vrana, Yugoslavia: mullet entering the lake through a channel from the sea encountered a transversely anchored boat, then while trying to clear this obstacle they fell into it (von Brandt, 1964*a*).

An early Chinese aerial trap consists of a stake-held barrier-net protruding from the water. To both sides of this wall and along its length, net-bags are attached into which the mullets fall when trying to clear the barrier, (von Brandt, 1964*a*). A similar principle is applied in the Italian 'saltarello', a fixed installation, set in a circular fashion, with an opening facing the direction of the run of the fish. It consists of a vertical net barrier reaching from the bottom to the surface of a verandah, of which the outer circumference is held above water-level by poles driven into the ground, (Fig. 9.3). Thus, the verandah in this case, forms a concave collar around the barrier net, rather than a flat surface-floating platform. Similar nets are also operated in Egyptian lagoons, and are reported also from southern India and Indo-China, (von Brandt, 1964*a*; Hornell, 1950).

USSR fishermen catch *Mugil cephalus* mainly in the Black Sea and the Sea of Azov and *Liza aurata* in the Caspian, on mats or basket-work raft traps, spread in a fleet surrounding shoals of grey mullet (Berg *et al.*, 1949, and see

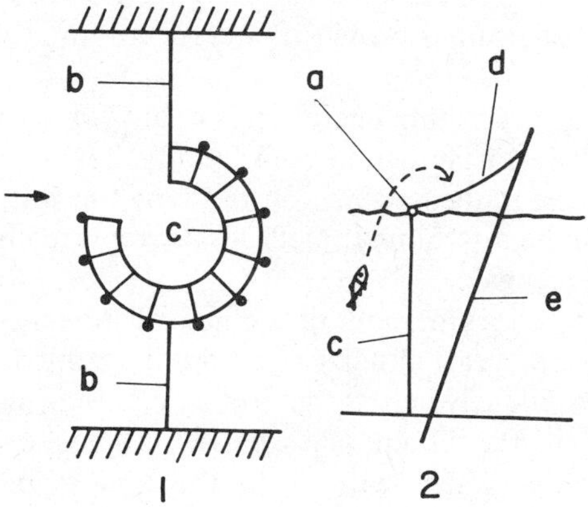

Fig. 9.3. The 'saltarello' – an Italian 'jump-in' net after von Brandt (1964*a*). 1, top view; 2, side view; a, float line; b, wing (barrier net); c, vertical netting wall; d, verandah netting; e, pole.

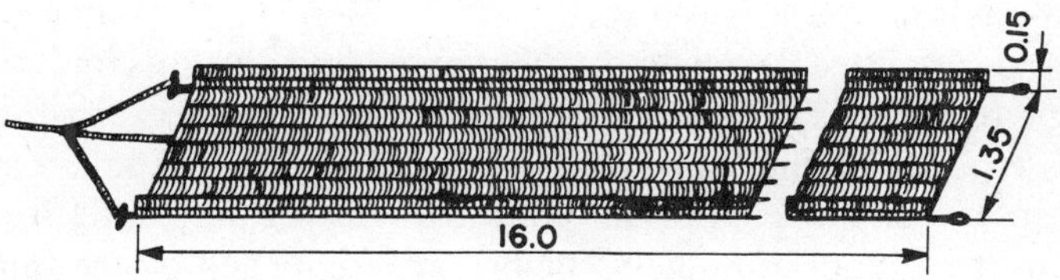

Fig. 9.4. Russian raft-traps, after Babayan (1965).

Fig. 9.4). Such a fleet consists of separate rafts, 12 m long and 1.2 m wide fitted with 15 cm high vertical fringes, tied to each other by lugs of rope or rope bridles. The fish, driven towards the raft-traps fleet by noise made by fishermen in several boats, jump in panic and land on the rafts. Apparently, however, this behaviour depends upon the circumstances, for in the Caspian Sea in depths exceeding 8 m, the mullet rather dive down than jump on the rafts, (Babayan, 1965). There is also some disagreement between different authors as to whether this fishing is best in moonless (Berg *et al.*, 1949; Babayan, 1965) or in moonlight conditions.

Variants of this method have also been reported from China, India, Burma and Thailand. To create the impression of an obstacle, bright, light-reflecting objects may be tied alongside the raft or the raft itself amy have an extension submerged at a downward angle (Hornell, 1950).

Another method is to use solid rafts. The Maltese 'raft fishery' is based on the fact that on moonlit nights, rafts, anchored and floating on the surface, cast a shadow, which is regarded by mullets as an obstacle. They leap and land on the rafts. The surface of the raft is covered by brushwood or netting in which they get entangled and may then be collected. Sometimes, a section of vertical netting is provided underneath, to prevent the fish from swimming under the raft.

Another mode of raft-fishing operation is to anchor across the entrance of a creek (Anon, 1956). Fishermen in boats paddle down the creek, making a noise and driving the mullets, which on reaching the raft and the net hung from it, jump onto the raft. Similar methods are reported from the Mondego Rover region in Portugal.

Hornell (1950) describes methods of fishing with moving rafts from SE Asia and China. A canoe or a raft, fitted with a brightly painted, partly submerged board, is paddled obliquely across the probable path of mullets. At the side of this craft, opposite the school, a low barrier is erected. Fish scared by the drag of the board in the water, leap to clear the canoe or raft and are retained by the barrier. The raft's surface is covered by branches and bushes, so that the fish cannot return to the sea. A very similar method is the 'changodam' reported from Cochin and Travancore (India). Two very narrow dugout canoes are connected laterally with poles, so that the distance between their bows is about 4 m and between their sterns 2.75 m. To the outer gunwales of each canoe are fitted nets held by sticks extending upwards and outwards at an angle of about 25°. The net is hung so as to bag between the sticks. A 'tickler' type device, made of a rope weighted by pieces of chain is attached between the bows. The canoes are ballasted, leaving very little freebroad. Paddling over a school of mullets, the fish are frightened by the turbulence caused by the 'tickler', jump and land on either canoe or the slanting nets. Variants of this method have been reported from Madagascar and East Africa, (von Brandt, 1959, 1964*b*).

Dip-nets

Dip-nets are used in inshore waters, in harbours, estuaries and lagoons in most parts of the world, being particularly common in South Asia, Japan and the Philippines. Dip-nets may still be seen also in many European coastal areas, e.g., in the Venetian lagoons. The dip-net usually consists of a square piece of netting hung from a frame or crossed poles, so that the netting forms a bag and is suspended from a single long boom, (Fig. 9.5). In brooks and narrow rivers, however, the dip-net may be suspended from masts at both sides of the river, so as to fish its whole width. The net is lowered to, or near to, the bottom and when fish are sighted over the net, it is lifted out of the water by pulleys.

The French 'mugilière' has been designed especially for mullets. Used in harbours, it consists of a very large piece of netting, hung with much slack between four ropes, attached at two corners to the shore and at two others to anchored boats. The sides attached to the shore are held above the water

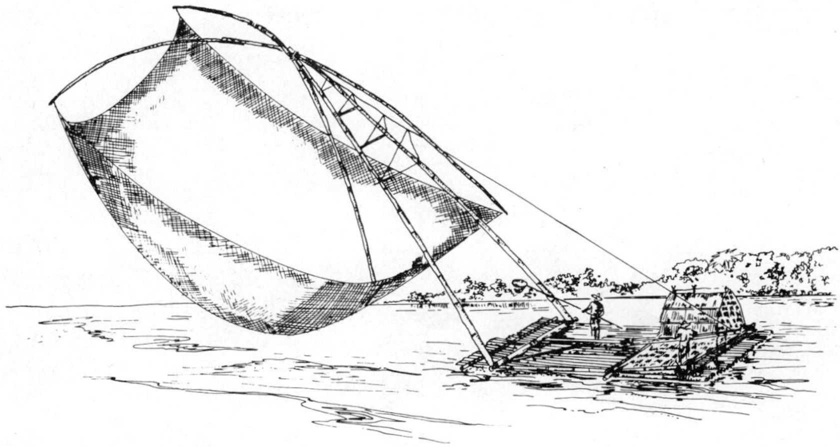

Fig. 9.5. A dip-net from the Phillippines used for catching mullet in rivers.

and the sides attached to the two boats are submerged. The net bags downwards, nearly touching bottom and is lifted when fish appear over the net, (Gourret, 1934).

Similar in pattern and operation is the Turkish 'tcheukertme', except that three boats are involved, each hauling the net by two bridled haul ropes from those three sides of the net bag that do not face the shore, while the fourth side is suspended from the shore or jetty by similar ropes (Devedjan, 1926).

Fish barriers and traps

Mullets are caught in many countries using stationary barriers fencing off a creek, estuary, small bay or lagoon. Most are so designed that fish may enter but not leave. Traps are large, stationary installations, set to guide fish into a closed, or partially closed crib, sometimes called the 'death-chamber' (Fig. 9.6 and 9.7).

The Turkish 'dahlian' is a large trap-net consisting of a leader (a long fence net set between beach and the trap) and a stake-mounted chamber-trap with a large gate left open at the seaward side of the leader. During day-time, watchmen are placed on wooden towers, stituated at opposite ends of the opening. They close the gate by lifting a submerged net section when a school of fish enters the enclosed area. Designed for any migratory fish, the 'dahlian' is reported also to catch mullets. Under other names, the 'dahlian' also is reported to be used by Bulgarian and Russian Black sea fishermen, (Fig. 9.8).

It appears that similar gear has also been developed independently in Madagaskar (von Brandt, 1964*b*). In the USSR, the development of intensive fishery has brought to a near extinction a variant of a mullet trap-net called

Fig. 9.6. A stationary fish barrier that leads the fish into a fyke net was illustrated by Roule (1914). Note the height of the leader above the water surface.

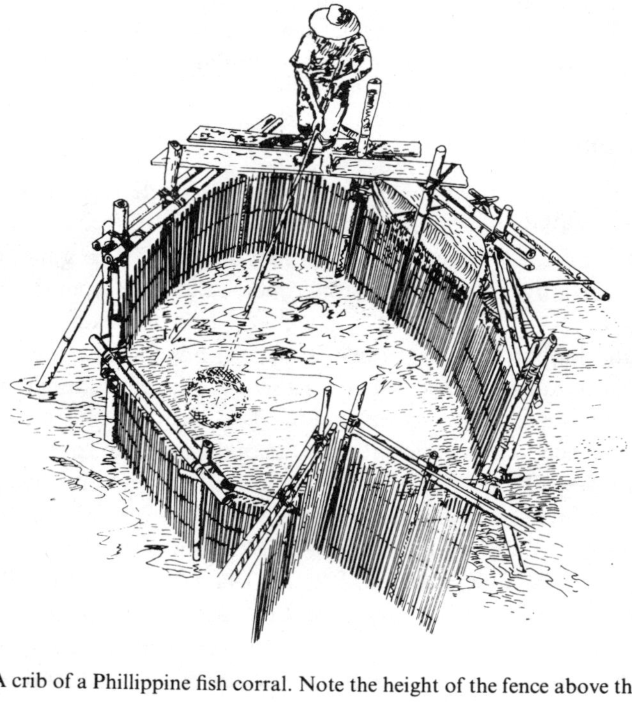

Fig. 9.7. A crib of a Phillippine fish corral. Note the height of the fence above the water surface.

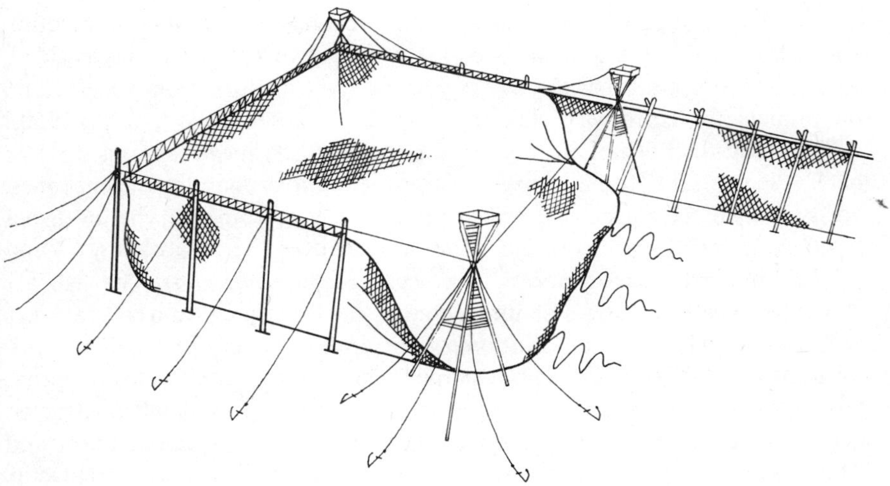

Fig. 9.8. The crib of a 'dahlian' type of trap-net. Note the watch-towers, after von Brandt (1964*a*).

'karaviya', native to the Crimea and the Eastern Black Sea, (Babayan, 1965).

Tide-fishing barrier nets, also called stop nets, are used for the capture of mullets and other fish in the intertidal area. Such a method, e.g., has been introduced in Ethiopia by Yemenite fishermen and practised off Assab and Massawa. A stake net is set parallel to the shore at the fringe of the inter-tidal strip. The stakes are high enough to prevent the grey mullet from jumping over. The net is lowered into the water at high-tide. The fish, cut off from the sea by the net, are collected manually at low-tide (Ben-Yami, 1964). Tide barriers of this kind, varying slightly in detail, are known in Fiji as 'lava ni vola' (Hornell, 1950), in Malaya as 'ampang' (Kesteven, 1949), in the Philippines as 'takop' (Rasalan, 1953), in the USA as a stop net (Sutton, 1950) and are also found in the Red Sea area of the Arabian peninsula and many other countries. A tide gill-net was graphically described by Roule (1914) from Provence (France).

Fishing mullets in enclosed and semi-enclosed water bodies

The tendency of many grey mullet spp., among them such commercially important ones as *Mugil cephalus*, *Liza ramada*, *L. aurata* and *L. saliens*, to enter estuarine brackish and hypersaline lagoon waters, has been exploited over the ages. Actually the methods applied extend over the whole range between the two extremes of fishing and fish farming.

A very simple operation is the Japanese 'yorino'. It takes advantage of the habit of mullets to form (during the winter months) large, nearly inactive schools in small, warm-water bays. The bay is closed by a vertical net wall, leaving openings by which the mullets may reach the place of concentration and as long as schooling is in progress, any distrubance in the area is avoided. At the appropriate time, the bay is completely sealed off from the sea and the mullets are fished with gill-nets set in the enclosed area, (Anon., 1960).

The traditional fishing installations in the Mediterranean lagoons, that go back at least to the Roman period, but probably, to even more ancient times, are based on the migratory behaviour patterns of some euryhaline fishes (mullets; gilthead seabream; *Sparus aurata*; seabass *Dicentrarchus* spp.) and of the classicially catadromous, river eel (*Anguilla anguilla*). The fish are captured, usually in narrow channels connecting the lagoons and coastal lakes with the sea where various stationary entrapping systems are constructed. Most of these systems involve V-shaped 'non-return valves', closing gates, and cages of diverse shapes where the fish are finally captured, all constructed of grilled screens. The construction material, previously canes, reeds and wood has, in many places, been replaced by iron, aluminium alloy and plastic. The operation of these systems is in many cases correlated with tides, owing to the tendency of the fish to swim against the current. Where the water level in the lagoon can be controlled by means of sluice gates, the operation depends less on the tides.

In general, large fish are captured in the cage-like traps on their attempt to leave the lagoon in a seawards direction, while there are V-shaped entrances left at the channels' sides to let the fry in. All these installations protrude high enough above the water level to prevent mullets from jumping over. The captured fish are collected from the cages using scoop-nets. Detailed descriptions of the fishing installations in Italian and some other Mediterranean lagoons were published by De Angelis (1959, 1960).

In Spain, a variety of gear is used to harvest grey mullet in coastal lagoons. It includes beach seines, small trap-nets, surrounding nets fitted with verandahs and diverse types of installations such as the 'encanizada'. This is a reed barrage, separating a lagoon from the open sea. Young fish can enter the lagoon by passing between the reeds, but adult fish become too large to get out. They are captured in a variety of entrapping devices connected with the main barrage, (Cabo, 1959). Similar methods are also used in the bay of Iskanderun, Turkey, where the barrier consists of bamboo sticks, bound together with straw to form a small 'meshed' mat, held upright by wooden poles (Aasen & Akyuz, 1956), as well as in the lagoon of Lessina (Italy) and probably many other parts of the world.

In Italy, the capture of grey mullets is associated with lagoon and coastal lakes fishery and fish culture (vallicultura) and with intermediate stages between both. In all these cases, however, the methods of capture are based on the principle described above.

A description of a barrier-net mullet fishery in a saline lagoon as practised by Egyptian fishermen in Sinai prior to the 1967 Middle East War is given by Ben-Yami (1974): the Lagoon of Bardawil, in the North Sinai Peninsula, is a natural, 600 km^2, depression, cut off from the sea by a bar and kept flooded through two artificial and one natural channel. The latter channel alone is insufficient to keep the lagoon permanently flooded, although during winter gales the sea breaks through from time to time. These breaks, however, soon silt up, the lagoon becoming a drying, salt-encrusted pan (Jarvis, 1941). The cutting of the channels must be considered an inseparable part of the fishery of Bradawil, for there is no other apparent reason for maintaining the Lagoon in a flooded state. For a long period the bulk of the Bardawil catches was composed of grey mullets. Only some 20 years ago, the share of gilthead bream (*Sparus aurata*) started approaching that of the mullets, probably, as a result of the digging of permanent channels and of stocking the lagoon with imported fry (A. Abu-Zuqry, personal communication).

A method, practised in the twenties and thirties, was described by Jarvis (1941). A huge net barrier is constructed across the channel. The mullet attempting to leave the lagoon, nose against the meshes and finding their way barred, jump to clear it and are captured on a verandah created by strut-supported netting. The fish are then collected by men who are constantly passing up and down in boats. Females were stripped for roe, subsequently processed into a salted product, a delicacy named 'b'tarikh' (Arabic for roe).

In the early fifties, permanent channels were dug and stationary fence nets with verandahs became the main fishing gear in the lagoons. These were rather huge installations set in semi-circles around the lagoon-side ends of the channels (Fig. 9.9). The length of such a barrier exceeded 2 km. Most of it was made up of two rows of wooden or concrete piles, driven into the lagoon bed, while some sections had only single row of piles. The distance between the piles in the row facing the lagoon was amazingly small (between 1 and 2 m), while the width of the space between both rows was approximately 5 m. The second row of piles (facing the channel) consisted of piles spaced approximately every 5 m (for every fifth pile in the first row there is one pile in the second row). Here and there, wider spaces were left between the piles to enable passage of boats. The principle of the action of such nets is similar to that of the Italian saltarello (von Brandt, 1964*a*), from which it probably originates, and to the Indian 'kattuvalai' and consists in the mullets encountering a netting wall and attempting to jump over it, only to land in a horizontal or a bagging verandah spread between the two rows of piles.

Cotton, and more recently, low quality synthetic netting, 40 mm mesh, was used for the fencing walls at the Bardawil channels. The lower edge of the netting wall was mounted on a heavy 1″ iron chain, at a hanging ratio of 0.83, approximately. This heavy ballast, was evidently, stipulated by the strong currents and wave action at the channels, which could lift any, but heavily ballasted netting off the bottom, and thus produce an escape route for fish.

The netting wall frequently had to be cleaned of seaweed and algae and

Fig. 9.9. Two rows of piles, driven into the lagoon bed, enclose in a 2 km long semi-circle, the entrance to a man-made channel connecting the Bardawil Lagoon (North Sinai Peninsula) with the Mediterranean (courtesy of Dr D. Ben-Shay).

other sedentary life forms (marine fouling) which by choking the net's meshes, were endangering its correct action. Choked netting presents a very formidable resistance to the water flow, which under the above mentioned conditions of the channels might lead to any kind of disaster (acccelerated silting along the net barrier, falling of piles, breaking the netting, etc.). To maintain the net in working condition, during that part of the year when the net remained submerged, i.e., between spring and early winter, a special crew of three to four men was employed. They passed along the net wall cleaning the netting with brooms or brushes. Net sections made of natural fibres were replaced every fortnight and dried on shore.

Nevertheless, the densely set rows of piles were causing gradual silting up of the whole area along the barrier net and between it and the channel. Also the very channels, due to long-shore transport, were migrating along the bar and silting up. Therefore, from time to time, major dredging works have been necessary to maintain the whole system in operational condition.

Obviously this type of fishery involving digging and maintenance of channels to exploit the fish population in the whole lagoon can only be run by a co-operative organization (e.g., as in the USSR or in the Lessina lagoons in Italy), private land-owners (e.g., the valli in the Venetian Lagoon) or a government licenced contractor with monopoly rights (e.g., as in the Bardawil Lagoon).

Capture of fry

Mullet fry, particularly of species farmed in lagoons and ponds are being anxiously sought by fish farmers and by specialized 'fry-fishermen'. They are captured by specially developed gear and methods. Most fry-fishermen exploit the tendency of the mullet fry to enter lagoon channels, creeks, brooks and rivers, where, during the appropriate seasons, they wait for the run.

In Egypt, fry are collected in canals connecting brackish lakes with the Mediterranean Sea, at pumping stations, etc. At one station, a rectangular, framed, 50 × 50 cm, scoop-net is used. It is equipped with a long handle and a hand rope attached to the frame opposite the handle to assist in pushing the scoop-net through water and to dry it. The netting is mosquito-netting, the bag is elongated, conical and the codend is transversely marked with coloured threads, each mark indicating the level of about accumulatively 1,000 fry collected in the codend. This marking of the bag must be seasonally calibrated according to the changing fry size (El-Zarka & Kamel, 1967).

In Israel, fry are collected with small mosquito-netting beach seines, and with hand-operated scoop-nets made, mostly, of stiff window-screen netting. The best collecting spots are at the sea-shore near estuaries, in the estuaries, in rivers, especially at drainage outlets of fish ponds, in the cooling basins of sea-shore electric power plants, etc. Fry accumulate in small indentations and in quiet spots in rivers, where they rest before pushing upstream. Such indentations can be made artificially (Perlmutter, Bograd & Pruginin, 1957). For fry survey and research purposes, a comparatively large, two-man operated fry beach seine was designed by M. Ben-Yami (Fig. 9.10 and 9.11).

In the Far East (China, Hong Kong and Taiwan) fry of *Mugil cephalus* are captured at low tide by means of large dip-nets and 2–4 m^2 umbrella nets, with a mesh of 6 mm or less. Dip-nets are usually operated by pairs of fishermen, who wade against the tide and drive fry into the net with their feet. Umbrella nets are shore-operated, using a 8 m long bamboo pole (Bardach, Ryther & MacLarney, 1972).

A modern Chinese beach seine reported from Taiwan is illustrated in Tang (1975, Fig. 19). This two man operated beach seine is equipped with sledge-type danleno-runners at the wing tips and with wooden vertical spreaders distributed every 0.5–1.0 m along the whole length of the net, securing the proper position of the net, notwithstanding the shallowness of the water. A short-bagged and wingless light drag-net is used on the coast of Taiwan to fish mullet fry where the water is deeper than 1 m. For this purpose a small boat or a raft is employed to carry the net and its sweeps off the beach and to set the gear in a beach-seine fashion, first one rope then the net then the other rope. Finally, the end of the second rope is brought ashore and the net hauled. Such nets can also be towed for a while parallel to the shore, prior to hauling (Tang, 1975).

Fig. 9.10. A fry-survey beach seine in action.

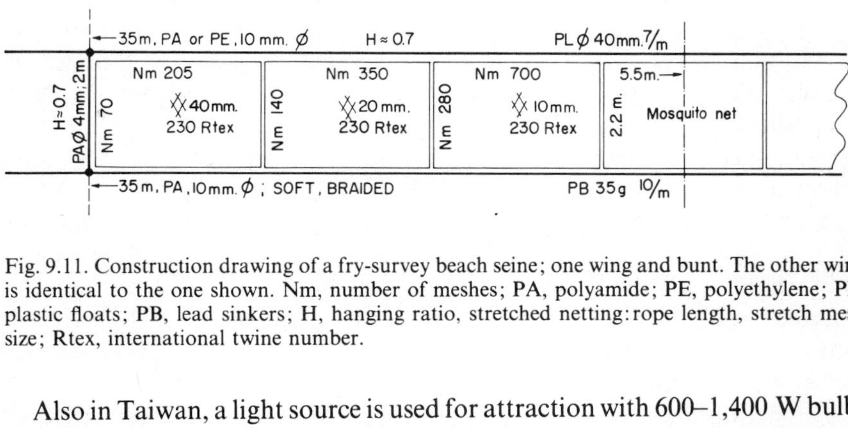

Fig. 9.11. Construction drawing of a fry-survey beach seine; one wing and bunt. The other wing is identical to the one shown. Nm, number of meshes; PA, polyamide; PE, polyethylene; PL, plastic floats; PB, lead sinkers; H, hanging ratio, stretched netting: rope length, stretch mesh size; Rtex, international twine number.

Also in Taiwan, a light source is used for attraction with 600–1,400 W bulbs usually in association with scoop-nets, (I. C. Liao, personal communication).

A native method of capturing mullet fingerlings, by means of a 1.5 × 0.9 m net woven of thin sea grass, has been reported from China, (VNIRO, 1949).

In Southern India, fry are collected from swamps by means of hand-operated 1 m deep bottom-seines with mesh sizes decreasing from 2.5 cm at the mouth to 6 mm at the bunt. The net is kept open by two men, while the fry are driven into it using a 30–60 m long scare-line made up of rope and palm leaves. When the scare-line is brought up to the net, the wings are brought together and the fry are thus entrapped (Bardach *et al.*, 1972).

In the USSR a three-stage method was applied when fry fishing in the Caspian Sea for stocking *Liza aurata* and *L. saliens*, in the Aral Sea. The fry were captured in coastal inlets, lagoons and tidal lakes using 70 m long, 3 mm stretch mesh knotless beach seines. After setting such a big seine and concentrating the fry in a limited area, they were fished out in small batches by means of shorter, 10–12 m, and 25–30 m long seines made of the same material. Finally, the fry were taken from the smaller seines with scoop-nets and put in portable containers, (Mityushkin, 1956).

Apart from these rather active methods, when the mullets are cultured in sea-shore ponds, embanked low-lying lands and lagoons that can be separated from the sea, the dry may be trapped by various, mostly V-shaped grills. In Bengal (India), e.g., tidal water is admitted into enclosed flooded areas through sluice gates. The fry are brought into these fish farms by the incoming tides. In the sluice gates there are fitted split V-shaped bamboo gratings. The two arms of the V do not quite meet at the apex, so that a very narrow passage is left for the fry to gain entry during high tides. During low tide, the arms of the V are pressed together by the pressure of the out-flowing water, thus preventing the entrapped fry from escaping. The V-shaped funnels are fitted with split-bamboo trap boxes where the fish are collected, sorted and released into the farm, (Sarojini, 1951).

Basically similar techniques are known from the Italian valli and from many other Mediterranean lagoons, (De Angelis, 1959, 1960), although nowadays, more advanced materials, e.g., non-corosive metals and plastics are used for grills.

Capture of larvae and post-larvae

Mullet larvae and post-larvae can be occasionally caught with regular plankton and larva nets, and with scoop-nets in association with light attraction. A net suitable for sampling larvae of grey mullets seems to be a neuston net that can be towed off the propeller wake as, e.g., the side-tracking neuston net (Ben-Yami *et al.*, 1970).

Acknowledgements

The authors' thanks are extended to Mrs L. Zismann and Mrs Avigail Gilboa for reading the manuscript and to Mrs Irit Idan who prepared the drawings.

References

Aasen, O. & Akyuz, E. (1956). Some data concerning fisheries in Iskanderun Bay. *Rep. Fish. Res. Centre, Istanbul* **1** (4), 1–8.

Andreev, N. N. (1966). *Handbook of Fishing Gear and its Rigging*. Israel Programme for Scientific Translations, Jerusalem. (Translated from Russian: *Spravidinik po orudiyam lova, setenes tnym materialem i promyslovomu snaryazheniyu*. Pishche-promizdat, Moscow.)

Methods of capture

Anon. (1956). Unusual method for catching mullets. *Commerc. Fish. Rev.* **18** (17), 65.

Anon. (1960). *Illustration of Japanese Fishing Boats and Fishing Gear.* Fisheries Agency of Japan, Tokyo.

Babayan, K. E. (1965). *Kefali, biologiya sposoby lova i vyrashchivaniya.* (Grey mullets, their biology, methods of capture and culture.) Pishchevaya Promyshlennost, Moscow. (In Russian.)

Bardach, J. E., Ryther, J. H. & MacLarney, W. O. (1972). *Aquaculture: the Farming and Husbandry of Freshwater and Marine Organisms.* Wiley-Interscience, New York.

Ben-Yami, M. (1964). *Report on the Fisheries of Ethiopia.* Ministry of Foreign Affairs, Dept. of International Cooperation, Jerusalem.

Ben-Yami, M. (1974). A note on the grey mullet fishery in the Bardawil Lagoon (North Sinai). In *Proceedings of the IBP/PM International Symposium on the Grey Mullets and their Culture, Haifa* 2–8 June 1974.

Ben-Yami, M., Herzberg, A., Pizanti, S. & Lourie, A. (1970). A side-tracking newston net. *Mar. Biol.* **6** (4), 312–16.

Berg, L. S., Bogdanov, A. S., Kojin, N. J. & Rass, T. S. (1949). Promysl. ryby SSSR. (Commercial Fishes of the USSR.) Food Industries Press, Moscow.

Bograd, L. (1961). *Mugil* species in the rivers of Israel. *Fish Bull., Haifa* **3** (7), 8–10. (In Hebrew, English abstract.)

Cabo, F. L. (1959). Fishing gear used in Spanish coastal lagoons. *Stud. Rev. Gen. Fish. Counc. Mediterr.* (9), 1–7.

Cooper, E. (1937). *Modern Sea Fishing (from Bass to Tunny).* A. & C. Black Ltd, London.

Davies, F. M. (1936). An account of the fishing gear of England and Wales. *Fish. Invest. Series II*, **15** (2), 1–55.

De Angelis, R. (1959). Fishing installations in saline lagoons. *Stud. Rev. Gen. Fish. Counc. Medit.* **7**, 1–16.

De Angelis, R. (1960). Mediterranean Brackish water lagoons and their exploitation. *Stud. Rev. Gen. Fish. Counc. Mediterr.* **12**, 1–41.

Devedjan, K. (1926). *Pêche et pêcheries de la Turquie.* Istanbul.

El-Zarka, S. El-D., El Maghraby, A. M. & Abdel-Hamid, Kh. (1970). Studies on the distribution, growth and abundance of migrating fry and juveniles of mullets in brackish coastal lake (Edku) in the UAR. *Stud. Rev. Gen. Fish. Counc. Mediterr.* **46**, 1–19.

El-Zarka, S. El-D. & El-Sedfy, H. M. (1970). The biology and fishery of *Mugil saliens* (Risso) in Lake Quarun, UAR. *Bull. Inst. Oceanogr. Fish., Cairo* (1), 1–26.

El-Zarka, El-D. & Kamel, F. (1967). Mullet fry transplantation and its contribution to the fisheries of inland brackish lakes in the UAR. *Pro. Tech. Pap. Gen. Fish. Counc. Mediterr.* (8), 209–26.

Fao (1973). *Catalogue of Fishing Gear Design.* Fishing News (Books) Ltd, London.

Fao (1975). *Catalogue of Small Scale Fishing Gear* (ed. C. Nedelec), Fishing News (Books) Ltd, London.

Gourret, P. (1934). *Les Pêcheries et les Poissons de la Méditerrané (Provence).* J. B. Baillic, Paris.

Hornell, J. (1940). *Report on the Fisheries of Fiji, Suva.* F. W. Smith, Govt. Printer.

Hornell, J. (1950). *Fishing in Many Waters*, Cambridge University Press.

Jarvis, C. S. (1941). *Yesterday and Today in Sinai.* William Blackwood & Sons Ltd, London.

Kesteven, G. L. (1942). Studies in the biology of the Australian mullet. I. Account of the fishery and preliminary statement of the biology of *Mugil cobula*, Günther. *Bull. Counc. Sci. Ind. Res. Melb.* (157).

Kesteven, G. L. (ed.) (1949). *Malayan Fisheries*. Singapore Publishing House Ltd.

Konarbaev, V. & Kubashev, K. V. (1961). Lov kefali v Kaspiiskom more ramovoy set'yu. (Mullet catching in the Caspian Sea with a box net.) *Rybn. Khoz.* (5), 65–8.

Kristensen, I. (1964). Hypersaline bays as an environment for young fish. *Proc. Gulf Carribb. Fish. Inst.* **16**, 139–42.

Mendez-Arocha, A. (1963). *La Pesca en Margarita*. Estacion de Investigaciones Marinas de Margarita.

Mityushkin, V. A. (1956). Dvukhletnii opyt rabou po perevozke kefali iz Kaspiiskogo morya v Aral'skoe. (Two years of experience in transporting grey mullets from the Caspian Sea to the Aral Sea.) *Ryb. Khoz.* **32** (10), 60–2.

Nikonorov, I. V. (1963). *Lov ryby na svet*. (*Fishing with Light*.) Pishchevaya Peomyshlennost, Moscow.

Nun, M. (1964). *Hadayig Ha'yvri Hakadum*. (*Ancient Jewish Fishery*.) Hakibutz Hameuchad Publishing House Ltd, Tel-Aviv.

Perlmutter, A. (1956). *Sea Fisheries*. USOM Agricultural Report 13. USOM, Tel-Aviv.

Perlmutter, A., Bograd, L. & Pruginin, J. (1957). Use of the estuaries and sea-fish of the family Mugilidae (grey mullets) for pond culture in Israel. *Proc. Tech. Pap. Gen. Fish. Counc. Mediterr.* **4**, 289–304.

Pillay, T. V. R. (1968). Estuarine fisheries of West Africa. In *Estuaries* (ed. G. H. Lauf). American Association for the Advancement of Science.

Rafail, S. Z. (1968). Investigations of mullet fisheries by beach seine on the UAR Mediterranean coast. *Stud. Rev. Gen. Fish. Counc. Mediterr.* (35), 5–19.

Rafail, S. Z. (1971). Investigation on the Scieanidae and Moronidae catches and on the total catch by beach seine on the UAR Mediterranean coast. *Stud. Rev. Gen. Fish. Counc. Mediterr.* (48), 1–26.

Rasalan, S. B. (1953). Fishing gear commonly used in the Philippine fishery. In *Philippine Fisheries*, ed. D. V. V. Villadolid, pp. 53–62. Bureau of Fisheries, Manila.

Roughley, T. (1951). *Fish and Fisheries in Australia*. Angus & Robertson, Sydney.

Roule, L. (1914). *Les poissons dans la nature et dans leurs relations avec la pisciculture et la pêche*. (*Fishes in Nature and their Relation to Fish-Breeding and Fishery*.) J. B. Baillière, Paris.

Sarojini, K. K. (1951). The fishery and biology of the Indian grey mullets (a review). *J. Zool. Soc. India* **3** (1), 159–79.

Schofield, W. L. (1951). Purse seines and other round haul nets in California. *Fish. Bull. Calif.* (81), 1–83.

Siebenholer, L. B. (1955). Commercial fishing gear and fishing methods in Florida. *Mar. Lab. Univ. Miami Tech. Series* (13), 1–15.

Smith, J. L. B. (1948). A generic revision of the mugilid fishes in South Africa. *Ann. Mag. Nat. Hist.* **14** (11), 833–43.

Sutton, O. W. (1950). Gear studies on the Florida mullet fishery. *Proc. Gulf Caribb. Fish. Inst.* **2**, 47–50.

Tang, Y. A. (1975). Collection, handling and distribution of grey mullet fingerlings in Taiwan. *Aquaculture* **5** (1), 81–4.

Thompson, J. M. (1966). The grey mullets. In *Oceanography and Marine Biology*, ed. H. Barnes, vol. 4, pp. 301–55. Allen & Unwin, London.

VNIRO (1949). *Promyslovye Ryby SSSR.* (*Commercial Fishes of the USSR.*) Pishchepromizdar, Moscow.
von Brandt, A. (1959). Classification of Fishing Gear. In *Modern Fishing Gear of the World*, ed. H. Kristjonson, pp. 274–96. Fishing News (Books) Ltd., London.
von Brandt, A. (1964*a*). *Fish Catching Methods of the World.* Fishing News (Books) Ltd., London.
von Brandt, A. (1964*b*). Madagaskar: Fischereiliche Reisenotizen. *Prot. Fish. Techn.* **9** (41), 148–96.

10. Handling, transportation and stocking of fry

M. BEN-YAMI

Introduction

One of the restraints to the development of the culture of grey mullets and in some areas, a true bottle-neck, is the fry supply. The increasing difficulties the culturists are faced with are due both to the increasing fishing pressure and to the pollution of estuaries and inshore areas. Even when hatchery-grown mullet fry may become commercially available in the foreseeable future, good care and appropriate treatment of the captured fry will remain an important and decisive factor in stocking operations.

The fry, however, whether captured from nature or resulting from induced spawning, must be transported to their final destination, sometimes for considerable distances. The mortality rate during the transportation may critically affect the feasibility of the whole operation. The fry should arrive at the pond, lagoon or lake in good condition, so that their survival rate is as high as possible. The shocks, experienced by the fry during capture and transport affect them even when stocked in a favourable environment thereafter, and the mortality may continue for some time afterwards.

The grey mullet fry is very sensitive to careless handling, and drying in air and wind. Abrasion by nets causes loss of scales and injuries. An injured fish does not survive, although its death is not necessarily immediate, because its direct cause is not in the injury itself but rather the consequential infection. Shocked fry may also become more vulnerable to predation than unshocked ones (Yocom & Edsall, 1974).

Another apparent cause of death are haemorrhages in and around the brain area. These may be a result of the agitated fry bumping into the hard walls of the container (Zismann, personal communication). Therefore the handling methods employed with more hardy species such as tilapia and carp, are inadequate for grey mullets and must be improved, their survival rates should improve with increasingly careful handling (Yashouv, 1972).

Besides the shocks and injuries due to initial mistreatment, mortality may also occur as a result of lack or excess of oxygen, sudden changes in temperature and salinity, as well as other environmental shocks, diseases, parasites and starvation. In spite of all these problems and the extent to which grey mullets are stocked in ponds, lagoons and lakes in so many countries of the world, little research has actually been done in the field of fry handling, transporation and stocking. The existing techniques have been established mostly by trial and error which is, undoubtedly, a sound basis for practice, but trial and error have their limits, beyond which only systematic scientific research and experimentation are able to help.

In the present chapter I shall try to deal with practical problems involved with improving the survival rate of the mullet fry, review and discuss pertinent literature and suggest a few basic, practical rules. The problems that will be discussed are: the definition and estimation of survival rates, the main causes of fry mortality, the main environmental conditions affecting the survival of the fry (temperature, salinity, oxygen content, diseases and parasites), food and feeding, acclimatization and nursery procedures and the use of anaesthetics.

I shall use the term fry as a definition of fish from their postlarval stage until, in the case of grey mullets, they reach the length of about 60 mm, at which stage they are considered as juveniles or finglerlings. This is in agreement with the definition used by Zismann (pp. 25–7). this is also the size range within which grey mullets are captured in the rivers, estuaries and the sea for stocking purposes.

Fry survival in stocking operations

Little information has been published, so far, on the survival rates of grey mullet fry in stocking operations. Most authors are rather ambiguous in their use of the terms of *survival* and *mortality* with respect to the stocking operations, during which fry are captured, transported, acclimatized and in some cases, nursed for prolonged periods and, finally, stocked into the new environment.

To sort out the available information and for further use in this chapter, the following definitions will be applied:

(1) Total survival (S) = the ratio between the number of fish yielded (in ponds) (C), or the estimated number of recruits (R) in an extensive culture or fishery (e.g., in lakes and lagoons where not all fish can be harvested), and the number of the fry captured (F). Hence, expressed as a percentage:

$$S(\%) = \frac{100R}{F} \quad \text{or} \quad S(\%) = \frac{100C}{F};$$

and the total mortality (M):

$$M(\%) = 100 - S(\%).$$

(2) Stocking survival (S_s) = the ratio between C or R and the number of fry introduced into the pond or lake, etc. (F_s). Hence:

$$S_s(\%) = \frac{100C}{F_s} \quad \text{or} \quad \frac{100R}{F_s};$$

and the stocking mortality (M_s): $M_s(\%) = 100 - S_s$.

(3) Handling survival (S_h) = the ratio between F_s and F:

$$S_h(\%) = \frac{100F_s}{F};$$

while the handling mortality (M_h): $M_h(\%) = 100 - S_h$.

Hence

$$S = \frac{S_h \cdot Ss}{100}.$$

Where nursery procedure is applied as an intermediate stage, the stocking survival ought to be considered separately for the nursery and the final, rearing stages. Hence, intermediate stocking survival S'_s and the final stocking survival S''_s.

Thus in this case,

$$S = \frac{S_h \cdot S'_s \cdot S''_s}{10^4}.$$

The information available on the survival rates achieved in various countries can be divided into that stemming from estimation and calculations based on large-scale stocking operations, and that resulting from laboratory and other controlled studies. Estimates of the stocking survival in lake fishery stocking operations were included in reports from Egyptian brackish lakes of the Nile Delta (El-Zarka & Kamel, 1967) and from Lake Kinneret, Israel (Bar-Ilan, 1975). Neither report indicates the mortality of the fry between capture and stocking, i.e., the handling survival. As in both cases no acclimatization or nursery techniques were applied, and the stocking took place very soon, if not immediately after the capture, the mortality due to handling probably occurred after stocking. Thus, we can consider their data as representing, practically, the total survival.

In Lake Mariut, the survival deteriorated from approximately 20% in the 1920s to 0.123% in the '50s. In Lake Qarun, the survival is apparently still less (El-Zarka & Kamel, 1967). Although, these authors attributed the different mortality rates to the 'survival conditions' in the lakes, it seems that much of it could be ascribed to environmental shock at stocking and inadequate handling practice. The Egyptian data involves over 500 million fry stocked over several decades. In Lake Kinneret, up to and including 1973, approximately 19 million fry were stocked and nearly 1800 tons of mullet (*Mugil cephalus* and *Liza ramada*) landed (Bar-Ilan, 1975). The average weight of a grey mullet marketed from Lake Kinneret having been estimated at about 1 kg (M. Gino, personal communication), and considering that a certain number of the fish introduced avoided capture and are still living in the lake (a condition occurring also in the Egyptian lakes), it can be concluded that the total survival exceeded 10%.

Data on handling survival have been published in the USSR. Between 1954 and 1956 millions of *Liza aurata* and *L. saliens* fry captured in the Caspian Sea were stocked in the Aral Sea. In the spring when 70% of the fry were 30–40 mm and the rest 60–100 mm long fingerlings, the mortality due to handling and transportation was 14%, i.e., 86% survival. In the autumn when all the fish consisted of 60–100 mm fingerlings, the mortality was still less and over 90% of the captured fish were still alive on stocking in the Aral Sea. This low mortality rate was achieved in spite of the fact that the fry were transported by rail for five to six days. Unfortunately, no estimation was given of the stocking survival (Babayan, 1965; Mints, 1956; Mityushkin, 1956).

In pond culture, since all surviving fish are harvested, the survival estimation should be more reliable, as should the estimation on stocking, as the total survival appears to be much higher. This is, evidently, because of the various acclimatization and nursery techniques applied by the pond farmers. According to Liao (see Chap. 11), the handling survival of the *Mugil cephalus* fry in Taiwan is approximately 30%, and the stocking survival less than 50%. Hence, the total survival ought to be 15%. Since, according to Tang (1975), some 10 million fry are needed in Taiwan to produce approximately 2000 tons of marketable mullet, the average weight of a fish at harvesting (w) must be approximately 1.33 kg. This is because

$$S = \frac{100Y}{F_w} \qquad (Y = \text{total weight of catch})$$

and consequently,

$$w = \frac{100Y}{FS}, \quad \text{hence} \quad w = \frac{100 \times 2 \times 10^6}{10 \times 10^6 \times 15} = 1.333.$$

If w is less than 1.333 kg, S is higher than 15%, and vice versa.

The total survival of *Mugil cephalus* and *Liza ramada* fry collected by Israeli fish farmers is, generally, much higher approaching 50% (Mires, 1970) and over. This, however, is still very variable, probably due to different handling practices and different pond conditions. Where the fry are transferred directly from the place of capture to the ponds, without intermediate treatment, the survival may be much lower. At Narakal, India, e.g., only 5% of fry usually survive. They are caught in canals in a swamp area and transferred in baskets or in the nets in which they were captured into nearby ponds (Thomson, 1955). There seems to be a similar situation at Serov's ponds (Egypt), where the stocking survival of *M. cephalus* and *L. ramada* fry is 10–20% (Eisawy, Ishak

& Hamza, 1974). The reasons given for this low survival rate are predation and, in spite of the high salinity gradient and the possibility of other environmental stress, the lack of acclimatization procedures.

Obviously, under controlled conditions in research institutions, the highest survival rates can be achieved. In a survival experiment carried out in 29 ponds at the Dor (Israel) Fish Culture Research Station, the average mortality rate did not exceed 11% (89% survival), while the maximum mortality recorded for only one of the 29 ponds, was between 31 and 38% (62 and 69% survival, Yashouv, 1972). The high survival rates achieved in some cases seem to indicate that careful handling and the application of suitable acclimatization and nursing procedures may reduce mortality considerably, especially where it is very high initially.

Salinity and temperature

These are, apparently, the two main factors affecting the fry survival. This generally accepted view, ought to be qualified, as salinity and temperature are easy to control and hence their influence is the easiest to study. Perhaps this could be the reason that more information exists from experiments and experience involving these two factors than on any other aspects of fry handling, transporting and stocking.

Gradual adaptation (acclimatization) is an obvious procedure to reduce mortality due to osmotic and thermal shocks, wherever there are salinity and temperature differences between the water where the dry have been captured and those in the pond, lagoon or lake of destination. Sarojini (1951) summarized the pertinent information available then, concluding that gradual lowering of salinity minimized mortality. The casualty rate, e.g., in the fry of *Mugil cephalus* and *M. seheli* was reduced below 1% by acclimatization. *Mugil troscheli* fry collected from the sea, were succesfully acclimatized by the following procedure: on the first day after capture they were conditioned in sea water, on the second in 50% sea water/50% fresh water, and on the third day in 100% fresh water. Liao (see Chap. 11) notes that *M. cephalus* fry are more sensitive to salinity than to temperature changes, with larger fingerlings (35–42 mm long) being more resistant to low salinities than the smaller fry (23–31 mm).

Mires (1970) reported on field trials under controlled conditions with *Liza ramada* fry. He found them resistant to sudden changes of salinity, results that were not substantiated in a latter study (Mires, Shak & Shilo, 1974). This disparity could have been caused, however, by factors other than the salinity and temperature, the only parameters monitored. One possible reason for the difference in the results could be, e.g., the condition of the fry after the stress endured at capture. The value of acclimatization procedures for the survival of mullet fry was emphasized in the same study, when *L. ramada* fry, acclimatized to fresh water and kept there for a fortnight were then subjected to extreme salinity changes (from fresh water to sea water and vice versa), no serious losses occurred. This seems to indicate that fry that once having undergone adaptations are much more resistant to environmental stresses than fry newly arrived from open sea.

Mugil cephalus fry experimented with during the same study was critically affected by sudden salinity changes. An interesting phenomenon of major practical value was recorded: while the mortality in sudden transfer of the fry from 100% sea water to approximately 28% sea water (*c.* 11‰) was negligible, a sudden tranfer from 11% sea water (*c.* 4‰) to fresh water resulted in total death of all fry (Mires *et al.*, 1974).

Chervinsky (1974), in somewhat similar study investigated the reaction of *Liza aurata* fry (20–40 mm long) to both abrupt and gradual changes from sea water to fresh water. Although, the species was different, the results were strikingly similar. No mortality occurred when the salinity was decreased abruptly from 100% sea water to 10% sea water, while all fish died when transferred directly from sea water into fresh water. When, however, the change from 10% sea water to no sea water (100% fresh water) was executed gradually, all fish survived. This feature of the fry adaptation process should be borne in mind. In nature, the salinity preferences of mullet fry differ from species to species. In the Egyptian Lake Borollus, for example, only two of the five species of mullet fry, *Mugil cephalus* and *Liza ramada* enter the fresh water proper, while the other three species, *L. saliens*, *L. aurata* and *Chelon labrosus*, also euryhaline species, prefer sea and brackish water, especially at the canal between the sea and the lake, where the waters mix (El-Maghraby, Hashem & El-Sedfy, 1974).

This natural avoidance of fresh water areas does not preclude, however, the capability of all three of these species to survive and grow in fresh water. Single, commercial-size specimens of these three species have been found in the catch of the fresh water Lake Kinneret grey mullets, where they must have been incidentally introduced along with the *M. cephalus* and *L. ramada* fry – the regular stocking material of this lake (L. Zismann & Y. Chervinski, personal communication).

Fry of *Liza aurata*, *L. saliens* and *Chelon labrosus* have all been successfully acclimatized in fresh water, at high handling survival rates in current laboratory tests at the Haifa (Israel) Fisheries Technology Unit, the last species completing one year in a fresh-water tank. Similar results have been obtained by Y. Chervinski (personal communication) in the ponds of the Dor (Israel) Fish Farming and Aquaculture Station. He obtained a 100% handling survival after abrupt transfer of *C. labrosus* from 100% sea water to 100% fresh water. He experimented with 18–32 mm fry captured on the beach and kept 3–10 days in sea water. The experiment was concluded after seventeen repetitions using ten fish each time with the first batch surviving for 12 days and the last one for 24 h.

All these facts indicate that, especially where the fry of the faster growing *M. cephalus* and *L. ramada* are in short supply, the fry of the other three species can be used as stocking material for fresh-water lakes and for extensive culture. They grow too slowly to be used for intensive pond culture. As far as *C. labrosus* is concerned, the above results are in apparent contradiction to the findings of Lasserre & Gallis (1975), who established a pattern of irreversible decrease in plasma sodium and chloride and in plasma osmolality in these fish after three months in fresh water. This pattern, according to these authors indicates inevitable death of the fish.

These differences in results may stem from the fact that the fresh water in Israel, while still potable at $S = 400$ ppm, is several times more saline than that used in France for Lasserre's and Gallis' experiments. Hence, the Israeli results should, for the time being be considered applicable only for salinities of about 400 ppm and over. Another reason for such discrepancy in the results of the French and Israeli studies may be the great difference in temperature ranges: 5–15 °C and 15–30 °C respectively. Further studies and experience will, undoubtedly, provide a solution.

The resistance of the different species of mullet fry to salinity changes may, however, be critically affected by a simultaneous change of temperature. Also the fry osmotic adaptability at threshold values of the salinity are affected by the temperature, and vice versa. In the above mentioned study, e.g., the mortality of *Liza ramada* fry increased when they were exposed to the combined effect of salinity and temperature changes, as compared with salinity changes alone. The *Mugil cephalus* fry demonstrated a lower mortality rate at 17 °C than at 10 °C, although in the first case the salinity change was sharper than in the latter instance (Mires *et al.*, 1974). These results seem to confirm the conclusions of Sylvester (1974), who, in Hawaii (USA), investigated the thermal response of juvenile, 80–120 mm SL, *Mugil cephalus* to acclimatization time and fluctuating low temperatures. He found that longer acclimatization times are necessary for lower than for higher temperatures.

Most species of mullets are evidently eurythermic and so are their fry. They can withstand considerable temperature variations, provided these are gradual. At the lower end of the range, this was demonstrated during the transportation of *Liza aurata* and *L. saliens* fry from the Caspian to the Aral Sea. The water temperature was gradually lowered from 17 °C to 10 °C during the 5 to 6 days when the fish were on the train, without adverse effect (Mints, 1956).

Still, mullet fry of different species occurring in the same area and even during the same season, may react differently to temperature changes. Such a possibility is implied by a study of ecology, distribution and abundance of two mullets *Mugil cephalus* and *M. curema* in southern Texas, USA (Moore,

1974). Of these two, *M. curema* seems to prefer sub-tropical conditions, being more sensitive to cold and more resistant to heat, while *M. cephalus* prefers more temperate conditions and shows opposite reaction to the temperature extremes. Another aspect of the mullet thermal adaptability, namely, the effect of the acclimatization temperature was studied by Sylvester (1975). Two grey mullets, *M. cephalus* and *Chelon engeli* kept for some time at different acclimatization temperatures, within the 20–29 °C range, were exposed to temperature increase until the critical maximum was achieved, at which the first lost their equilibrium. The results were that the higher the acclimatization temperature, the higher was the critical thermal maximum. The experiments carried out by Auld (1972) may prove to be of some relevance to the mullet fry acclimatization question. He studied the effect of calcium on the osmotic tolerance of juvenile (25–150 mm) *M. cephalus*. He found that the fry survived better with calcium present and that increased content of calcium had a beneficial effect on the survival rate. Undoubtedly the application of this knowledge in practice requires further studies. Most euryhaline and eurythermic grey mullets are marine species, which is one of the main reasons for their being adapted for inland water aquaculture and stocking. They survive in temperatures ranging from less than 10 °C to over 32 °C and in salinities from zero to over 100‰. Nonetheless, there is no better way to kill the fry and fingerlings than exposing them to thermal and osmotic shocks. With gradual adaptation, however, which may be faster with some and slower with other species, their survival rate can be kept high, thus improving the feasibility of stocking operations.

Oxygen

Suffocation is undoubtedly one of the main reasons for massive mortality during the transportation stage. Since, for obvious reasons, relatively large fry biomass must be crowded in containers of manageable size, their handling survival depends upon a constant and adequate supply of oxygen, either in its pure state or in the form of air. Fry are transported in plastic bags inflated with oxygen under pressure, or in a variety of vessels where air or oxygen is supplied by thin tubes from either pressure bottles or air pumps driven, e.g., by the car battery.

During the previously mentioned operation of stocking the Aral Sea with *L. aurata* and *L. saliens* from the Caspian Sea a study was carried out aimed at the determination of the optimal density of the fry during transportation in railroad cars (Mints, 1956). For this purpose, oxygen consumption by the mullet fingerlings was measured by weight groups and at different water temperatures. The results, in mg O_2/h/kg fish weight, were as follows:

Weight of fish (g)	15 °C	20–21 °C	24–25 °C	30 °C
0.1–0.95	0.08	0.09	0.12	0.22
1.1–4.45	0.06	0.06	0.09	0.13
24–45	0.03	0.04	0.06	—

The increase of the oxygen consumption with temperature was very sharp. In particular, in the small fry, of less than 1 g, it almost tripled when the temperature increased from 15 °C to 30 °C, more than doubled between 20 °C and 30 °C, and increased by 50% within the narrow temperature range of 20 °C to 24 °C. The larger were the fingerlings, the lower was the consumption of oxygen per kg of the fish weight. In motion, the oxygen consumption raised by 33–37%, and the author noticed that during the journey, the fish were moving constantly.

Mints (1956) also collected data on the lethal levels of oxygen content for different weight groups. For example, fry less than 5 g in weight died when the oxygen content was less than 3.76 mg/l of water at 43% saturation level. For the larger fry, the threshold was 2.17 mg/l at 25% saturation. The lowest oxygen content, critical to all fry was found to be 1 mg/l of water. On the other hand many Israeli fish culturists have blamed super-aeration or super-oxygenation for certain cases of *M. cephalus* fry mortality.

Diseases and parasities

One factor contributing to the mortality of the grey mullet fry upon their capture and transportation to a new environment are diseases and parasites. Obviously, the fry subjected to physiological stresses lose much of their natural resistance to some diseases and parasites. Mass mortality may occur as a result of an outbreak of infection and/or infestation due to non-hygienic transportation vessels, acclimatization tanks and nursery ponds. While death caused by suffocation and by thermal or osmotic shocks may be immediate or nearly so, death from diseases and parasites may occur after the fry have already been introduced into the final destination basin. If this happens to be a large lake or lagoon, the mortality may remain undetected for a long time. In more intensive culture, e.g., fish ponds, where there is a good follow-up after the ponds have been stocked, the epidemic may be discovered in time and remedial action taken.

A general review, bibliography and counter-measures with respect to fish parasites and diseases were given by Huet (1971) and are referred to later (see Chap. 13). Besides reviewing parasites and diseases, including environmental diseases, Huet included instructions for tank and pond disinfection and how to prepare and send samples for laboratory analyses of diseases and mortality. He also suggests some preventive measures and treatments, but there are hardly any specific references to grey mullets in this textbook. The same can be said about Sinderman's (1970) review, which deals particularly with fish kept in captivity, but nonetheless includes valid general information. As Paperna and Overstreet (Chap. 13) deal in great detail with mullet diseases and parasites, only a few words will be said here on grey mullet fry and fingerling diseases and parasites.

Lahav & Sarig (1972) used formalin to prevent and control infestation of *Mugil cephalus* fingerlings (5–30 g) by unicellar parasites, *Chilodonella* sp., *Costia* sp. and *Trichodina* sp. Their experiments resulted in eradication of the protozoan parasites during 6 h treatment in 30–40 ppm technical formalin. The authors recommend such treatment for ponds, nursery ponds and tanks, but warn that in non-aerated ponds, oxygen deficiency may develop during the night, due to the killing of phytoplankton. While salinity has no effect on formalin toxicity, the latter increases with temperature. Smaller fish are more sensitive to prolonged exposure to the formalin treatment, although the mullets were generally less sensitive than tilapia and carp (Lahav & Sarig, 1972). This treatment was made, however, with comparatively large fingerlings of *M. cephalus* only. Our own preliminary experiments with *Chelon labrosus* fry seem to indicate that that this species is more sensitive to formalin than *M. cephalus*, although its smaller size might have been another contributing factor to the mortality that occurred under the same treatment recommended by Lahav & Sarig (1972).

Yashouv & Ben Shachar (1967) reported that mortality of *M. cephalus* fingerlings (0.16–3.5 g) due to Saprolegnia infection was overcome by adding 1 part of sea water to 9 parts of fresh water in which the fish were kept, in one case, and to 4 parts of fresh water in another.

Anaesthetics

Anyone who has participated in fishing for mullet fry will have noted the great amount of energy these little creatures expend during the process of capture. Surrounded and crowded by the net, often in shallow, warm water, thick with mud stirred up by the net's leadline, they swim in circles, jump and break the water surface and push against the netting. Generally, besides the mechanical damage they can suffer from their own rough behaviour, they exhaust so much of their energy that one wonders how they are able to survive. In this exhausted condition they still have to overcome the combined shock of handling and transportation, which involves besides the possible severe physical ordeal, the need for fast osmotic, thermal and metabolic adaptation.

Obviously, the question has been posed as to whether, by the use of drugs that will anaesthetize or tranquilize the fry, the metabolic processes in the fish be slowed down, so that the energy waste is reduced and much of the exhaustion avoided. In addition, the oxygen requirements of the fry might be reduced, enabling increased density during transportation. If all this could be achieved without side effects on the fry, the survival rate should improve considerably.

The drugging and poisoning of fish is an ancient fishing method which rather unfortunately, is still applied in some areas, mostly by poachers who sometimes use quite dangerous substances. Also ichthyologists use drugs and poisons in their survey operations, especially when they are interested in fishing out all the creatures existing in a limited water space. Total fish poisoning is applied also as a common measure in the management of small lakes and artificial reservoirs. In almost all these instances lethal doses are employed. The problem has thus been to use the drugs in sub-lethal doses so as to keep fish alive but anaesthetized or tranquilized during the critical stages of handling. Surprisingly enough, very few scientists have investigated this possibility for grey mullet fry. Much research has been done, especially in the USA, on other fish.

An early report on the use of Thiouracil as an aid in holding and transporting fish was published by Osborn (1951). He succeeded in lowering the oxygen consumption of the fish by some 20%. This enabled him to increase their density in the transporting tanks by 15 to 20%. Dick engaged on the British White Fish Authority programme of research in aquaculture, studied various aspects of the use of commercially produced anaesthetic MS-222 Sandoz (tricaine methanesulphonate) for use in fish transportation. Among the fish studied were the grey mullet (*Chelon labrosus*) a species chosen as a 'test case' because from previous experience it was known that the results obtained with this fish were applicable to most other species of marine fish (Dick, 1975). Although the studies were not carried out with fry, they may also be valid for fry or at least give a very good starting point for the introduction of anaesthetic practice in mullet fry handling. Dick used solutions of 1:75000 to 1:100000 and found them sufficient to slow down the fish so that they could be captured and handled with ease. Actually, no significant difference was found between the effects of these two concentrations, while, obviously 1:100000 was less expensive. Tranquilized fish were observed to consume less oxygen, and no outward signs of shock were observed. However, when fish were kept in the tranquillizing solution for 48 h, the pH dropped from 8 to between 5 and 4, which created a possible stress factor. Buffering the solution was suggested as a remedy. On the basis of his own and experiments of other workers carried out with different fishes and described by Dick (1975), it can be summed up in a preliminary way, that solutions 1:100000 and 1:75000 of the MS-222 are not lethal even during prolonged immersion, but they do not result in complete anaesthesis. Solutions of 1:50000 and stronger are lethal on prolonged immersion, though they may be applied for total anaesthesis when it is necesssary to render fish immobile, e.g., for weight and length measurements. Such solutions are applicable only under the condition that the fish are not subjected to such concentrations (1:30000 – 1:50000) for more than, say, 30 min (Dick, 1975).

The reader should bear in mind that these results are not necessarily directly applicable to the fry of the various species of mullet under various climatic conditions and water salinities, and some local experimentation seems highly necessary before they are applied. Such a regional and very thorough study was performed by Durve in India during the last decade (Durve & Dharma Raja, 1966; Durve, 1970, 1975). The fish used in these experiments were fry and fingerlings of *Liza tade*, a mullet species that is commercially cultivated in brackish-water ponds in the Tamil Nadu province, with *Mugil cephalus* and *L. dussumieri*. The experiments were carried out in fresh water (temperature, 27–28 °C; pH, 8.1), and the fry he used were acclimatized to fresh water and kept in that medium for 8–10 days before experimentation. Durve's findings are summed up in Table 10.1. Of the thirteen drugs tested, only seven were found suitable and appear in the Table, but three of them require special solvents to render them water soluble. Thus, four drugs can be used conveniently: MS-222, chloral hydrate, tertiary amyl alcohol and teritiary butyl alcohol. As far as the MS-222 is concerned, the dose recommended by Durve (1975) in the Table is twice that recommended by Dick (1975), while Durve's lethal dose is 4½ times that of Dick's. The diversity of these results emphasises the necessity for local tests.

Drugs are expensive and the feasibility of their application in the mullet fry handling and transporting must be evaluated separately for each particular case. Anaesthetics may be used in any or in all of the following stages of fry handling:

(1) Sorting, weighing, measurement of samples, etc. – full anaesthesia for a short time.

(2) Transportation – tranquillizing for, sometimes, extended periods.

(3) Acclimatization in water from the stocked basin – tranquillizing for periods up to 24 to 48 h.

(4) Capture – tranquillizing for a short time when the fish are still in the seine. Quick action drugs are needed such as tertiary amyl alcohol, chlorobutane or quinaldine (will not be effective in current and waves).

Food and nursing

Grey mullet fry routes, from the sea shore or river mouth where they are captured to their final destination in the pond or lake, are diverse in terms of both distance and duration. In some instances, e.g., in Lake Kinneret or some of the Nile Delta lakes stocking operations, the fry are released in the lake within a few hours after their capture (Bar-Ilan, 1975; El-Maghrabi

Table 10.1 *Anaesthetics and their concentrations for the transporation of mullet seed (Durve, 1975)*

No.	Anaesthetic	Solubility	Concentration (g or ml/100 ml)	Tolerance tested (h)	Lethal dose (g or ml/100 ml)	Remarks
1	MS-222	Water, very	0.003	24	0.015	Expensive
		soluble	0.004	5		
			0.005	5		
2	Chloral hydrate	Water, very	0.02	24	—[a]	—
		soluble	0.025	5		
3	Tertiary amyl alcohol	Water, very	0.05	5	0.50	—
		soluble	0.08	24		
			0.10	5		
4	Chlorobutanol	Hot water or	0.002	24	0.02	Requires vehicle to go
		organic solvent	0.003	5		into aqueous solution
5	Quinaldine	Organic solvent	0.0009	24	0.005	Requires vehicle to gp
			0.001	5		into aqueous solution
6	Paraldehyde	Organic solvent	0.01			Requires vehicle to go
			0.05	24	0.35	into aqueous solution
7	Tertiary butyl alcohol	Water	0.30	24		
			0.35	5	—[a]	—

[a] N.B. The lethal dose for these anaesthetics, being very high, could not be determined.

et al., 1974), while at the other extreme, as e.g., in Taiwan or Italy, the fry are nursed, i.e., kept in special nursery ponds for prolonged periods, where they are fed and acclimatized to their new water medium (Korringa, 1975; Tang, 1975). The effect of various methods of nursing and feeding of the fry on their survival, both *en route* and in their new environment (handling and stocking survival, respectively), has yet to be investigated and the feasibility of such procedures evaluated. It is preferable that the investigations be made independently in the different countries under their specific climatic and other conditions.

As long, however, as well proved procedures are not available, it seems to be advisiable for mullet culturists whenever possible, to do their own practical experimentation by comparing results obtained with nursed and non-nursed fry. For these culturists, the following summary of the present state of the art may be of interest. Small fry (below 50–60 mm) of all investigated grey mullet species were reported to be carnivorous microphages, although their diet preferences may vary from species to species (Albertine-Berhaut, 1973; Bapat & Bal, 1951; Egusa, 1950; Nash & Kuo, 1975; Odum, 1970). This had been demonstrated under laboratory conditions by Yashouv & Ben Shachar (1967). Kept in the same environment, *Mugil cephalus* preferred planktonic crustaceans and rotifers over chironomid larvae, while the preference of *Liza ramada* was just the opposite. Stomach content of fry of several mullet species, 4–23 mm long, captured in open sea, was investigated by Zismann, Berdugo & Kimor (1974) and found to consist mostly of copepods. The comparison of the stomach content with plankton samples indicated that the fry select the copepods, preferring them over planktonic organisms of comparable size found in the samples.

At some stage of their life, mullet fry start feeding also on vegetable food (diatoms and other micro-algae) and within a short time relinquish partly or totally, zooplankton hunting. This was found for *Liza aurata*, *L. ramada* and *L. saliens* by Albertine-Berhaut (1973). She established a distinct change in the diet in fingerlings of about 50 mm SL. This general pattern of shift from zooplankton to algal diet was observed also in earlier investigations reviewed by Sarojini (1951). Chervinski (1975) examined the stomach contents of *L. aurata* fingerlings and young (50–250 mm long) raised in sea water ponds and found them feeding on diatoms and detritus when they were between 50 and 100 mm long, and again including some crustaceans in their diet when the fingerlings were between 100 and 200 mm long.

Current studies indicate that although algae are frequently found in the stomach of mullet fingerlings of the size range when they are due to pass from the zooplankton to the phytoplankton diet, they are not necessarily digested. Some algae are found at the end of the intestinal tract, intact (U. Pollingher, personal communication). Crustaceans and polychaetes, however, were digested (L. Zismann, personal communication). One may hypothesize that the fingerlings' diet transition is gradual and that they commence eating algae before their alimentary system is adapted to digest vegetable food. This ability is developed later, perhaps even only after being stimulated by the very presence of the algae in the alimentary tract. This hypothesis ought to be verified experimentally under strictly controlled conditions. According to Stickney and Shumway (1974) cellulose activity in fishes is probably produced by microflora of the alimentary tract. It is possible that the very presence and/or development of such cellulose processing bacteria is stimulated by occurrence of algae in the intestine. Nursery practices in mullet culture were recommended and tested by several authors. Yashouv & Karamusta (1967) observed that *Mugil cephalus* and *Liza ramada* fingerlings trapped in a beach side rain-water pool which abounded with natural food (mostly small crustaceans), grew during the winter months much faster than fry introduced during the autumn into fish ponds which during the same period were poor in natural food. This observation led them to a laboratory experiment with artificial feeding of fry of *M. cephalus* and *L. aurata*. Fed on fish meal and flour the fish grew well, but after some time mortality started among the *L. aurata*. Since no parasite infection or disease could be found in the dead fingerlings, the possibility of avitaminosis was suggested. Subsequently, the fish were given an addition of fresh food consisting of algae and live chironomid larvae, and the mortality dwindled till, after a fortnight, it ceased. In other laboratory experiments, *Liza ramada* fry, approximately 40 mm long and 0.5–0.7 g weight were fed on several different diets, including sorghum and pellets composed of wheat, fish meal and soya flour, containing 21% protein. The pellets alone produced better growth in the terms of weight than sorghum alone, but the addition of plankton to both, noticeably improved the growth in the terms of both weight and length (Yashouv & Ben-Shachar, 1967). Experiments with fry of *M. cephalus* were less conclusive, although the author's impression was that also in this case the addition of plankton and other natural food may stimulate better growth than artificial feed alone (Yashouv & Ben-Shachar, 1967). Liao (1974) fed *Mugil cephalus* fry from induced spawning with rice bran and flour when they were 30 to 45 days old and 15–33 mm long. This feed complemented and partly replaced zooplankton (copepods, rotifers, *Artemia* larvae, etc.).

While the above mentioned studies dealt with complementing artificial feeds fed to mullet fry in artificial environment by the addition of natural food, the following information is concerned with addition of artificial feeds to the natural food found by the fry in nursery ponds. Feeding of mullet fry in Italian valliculture with artificial feed was recommended by D'Ancona (1954), and became according to Korringa (1975) the latest development in the valliculture. In the nursery ponds in the Venice area fry are offered artificial feeds, prior to their release into the valli, where they continue to grow on natural food only. In 1970, during a visit to a valli in Chioggia in the southern part of the Gulk of Venice, I had the opportunity to observe such nursery practice: a row of narrow shallow ponds, some 20 m long, were stocked with mullet fry purchased from fry fishermen. The owner of the valli was at that time experimenting with an automatic feeder activated by an alarm-clock mechanism. The feeder was supplying a constant dribble of flour into the water where crowding fry were heartily feeding. There was a very good control over the salinity of the water in the ponds by means of two separate inflows, one from the lagoon sea water, the other from a river-water canal and a salinity level of approximately 15‰ was maintained.

In Taiwan, nursing the mullet fry is the normal practice (Tang, 1975). This is done in two stages by the fry fishermen who place the freshly caught fry in acclimatizing ponds, generally made of bricks and situated at or near the beach. These ponds are filled with water in the salinity range of 10–15‰. The stocking density is 300–500 fry per m^2 of pond area. After feeding them with wheat flour twice a day, for a period of 7–10 days, they are sold to commercial fish nurseries. There the second stage of nursing starts. These nurseries are located at or near fish farm centres. Fingerlings are cultivated in fresh water for a period of 40 days until they attain an average weight of 2–3 g, when they are finally distributed to fish farmers.

In Israel the keeping of mullet fry in nursery ponds, before they are introduced into polyculture ponds along with carp, was recommended by Pruginin & Kitai (1957), and later became a rather common practice. The mullet fry are nursed in 1–1½ ha ponds, stocked at a density of *c.* 30000 fish/ha. To prevent the growth of filamentous algae in the ponds, they are also stocked with 200 to 300 medium size carp, each. After 60–100 days the mullets reach 1–3 g and are transferred either to the final rearing ponds, or to larger, secondary nursing ponds (Pruginin, Shilo & Mires, 1975).

The influence of temperature on the nursing process was studied by Yashouv (1967), who experimented with *Mugil cephalus* fry captured on the Dor (Israel) sea-shore during winter when the water temperature ranged from 13 °C to 18 °C. He nursed the fish for 2 months in 500 l asbestos tanks with running brackish water (3.5–9.5‰) and intensive feeding on a rich diet. Where the water was heated and the temperature maintained at a minimum of 20 °C, the fish grew at least twice as fast as those kept in water of ambient temperature, although the density of the fry in the heated tank was 6–10 times that in the other tanks. He concluded that *M. cephalus* fry captured in winter can be nursed at a density of 500–1000 fish/m^2 in large tanks or small ponds. Exchange of water should be provided for the removal of the metabolic products. The basins should be covered with plastic sheets and, if possible, heated to maintain a relatively high temperature; the fish should be fed with suitable food.

Some authors have discussed the feeding of the mullet fry and fingerlings even during the short acclimatization periods, when they are kept in tanks or small ponds. Sarojini (1951) quoted a report from Madras (India) describing a practice of feeding *Mugil seheli* fry during an 18 h acclimatization period. Perlmutter, Bograd & Pruginin (1956) recommended placing freshly caught fry in holding ponds having a rich culture of algae growing on the bottom and in the water. Acclimatization practices become more and more popular also with the Israeli fish farmers, one of the reasons being the increasing scarcity of the fry. Many of them reached the practical conclusion that they can improve the survival rate by keeping the freshly caught fry for 24–48 h in tanks where the 'water of capture' from the sea or estuary is gradually diluted with the 'water of destination' from the fish pond. Mires (1973) suggests the use of fish hatchery facilities for the acclimatization of mullet fry, a practice feasible only where such facilities are available near the water basins which are to be stocked with the fry.

The effect of feeding the fry and fingerlings during the acclimatization process ought to be investigated further. Liao (see Chap. 11) assumes that the smaller, zooplankton feeding fry, would probably survive better if fed after capture with copepods, *Artemia* nauplii, etc.

In our current but not yet concluded observations of the fry mortality of five species of grey mullet (*Mugil cephalus*, *Liza ramada*, *L. aurata*, *L. saliens* and *Chelon labrosus*) my associates and I are under strong impression that the mortality is lower where the fry kept in plastic tanks, find some natural food either developing in the tanks or added from fish ponds with 'conditioned' water abounding with algae and zooplankton. Unidentified 'green water' was satisfactorily used at the Dor (Israel) Fish Culture Research Station in rearing *L. ramada* larvae from induced spawning (Yashouv, personal communication), and in growing prawn and milkfish, thus supporting the opinion of some culturists that the development toward clinical techniques for finfish culture may be wrong (Nash & Kuo, 1975).

Sebastian & Nair (1975) reported on the beneficial effect of 'seasoned greenwater' while rearing *Mugil macrolepis* larvae from induced spawning at the Azhikode experimental station (Kerala, India). The greenwater included unicellular green algae *Chlorella* sp., diatoms, dinoflagellates, euglenoids and copepods, while blue-green algae, diatoms and vorticellids grew on the inner side of the concrete tanks used for the culture. Liao (1974) comes to a similar conclusion in his report on much similar studies carried out in Taiwan, during which survival rates of *M. cephalus* larvae was higher in water inoculated with algae (green algae and diatoms). The algae, although not an important food component at this stage of the fish life, may be fulfilling a function of maintaining the water in the required condition (Liao, 1974; Yashouv & Ben-Shachar, 1967).

An interesting possibility of using urea both as a complimentary component of artificial food for mullet fry, and as a partial substitute for protein was investigated by French scientists (Albertine-Berhaut & Vallet, 1972). Their studies involved *L. aurata* fry 32 and 50 mm long. Fry nursed on high-protein food (albumen and complementary amino-acids) almost maintained their growth rate when 15% of that feed was replaced by urea. When they were kept on a low-protein, deficient diet, addition of urea considerably reduced their loss of weight. In the larger, 50 mm fingerlings, the utilization of the urea was still better.

All the above information seems to indicate two possible patterns of the fry acclimatization and nursing: the first is by adding natural food (plankton, algae, 'greenwater', etc.) where the fish are kept in tanks with 'clean' sea, river, tap or filtered water; the other, as practised in the Italian valli and in Taiwan, by adding artificial food where the fish are kept in nursery ponds where plankton and algae can develop naturally.

Conclusions and recommendations

A juvenile mullet captured in the sea or in a brackish estuary is subject to a series of environmental stresses, coming in a rather rapid sequence, before it can finally adjust to its new habitat. Its reaction to these stresses will vary with species and environmental conditions. Well planned experimentation may result in establishing procedures which may reduce mortality both during the transportation and handling process, and after the stocking. For example, for many years *Liza ramada* fry caught upstream in the Na'aman River (Israel) was considered 'adapted' to fresh water and, thus, released directly after capture in the Lake Kinneret. However, in preliminary experiments, started in 1975 in the wake of the fry shortage, considerable reduction of stocking mortality (30 to 100%) was obtained by 24 h acclimatization of the Na'aman fry in concrete ponds, where the fry were initially introduced in water from the place of capture and the water from the lake was added gradually (S. Pisanty, in preparation).

Current studies and research in the field of the culture of grey mullets, as well as the accumulation of the mullet farming experience will, undoubtedly, continue to contribute also towards the formulation of rational procedures of the fry handling and transportation, including acclimatization and nursery practices. It seems, none the less, highly recommendable to carry on with specific experimental work in at least the main countries where grey mullets are farmed. The knowledge gained will be applicable not only where the fry are captured in their natural environment, but also where they are the product of induced spawning and have to be nursed to a stocking size. As the final rearing stage may take place far from the hatchery and nursery, the problem of transportation remains actual in both cases.

The main aspects requiring further investigation seem to be:

(1) The effect of feeding on fry survival at various stages, while different feeds are experimented with under different conditions.

(2) The biological influence and economical feasibility of nursery ponds and tanks, thermal insulation and heating during cold weather periods.

(3) Preventive practices against infection, and treatment of diseases and parasites infestation and, particularly, the best dosage for the different developmental stages, species and environmental conditions.

(4) Practical application of anaesthetising and tranquilizing drugs and their economical feasibility. Where, for example, fry are scarce and fetch a very high price if bought, or require prohibitive investment of expensive manpower if fished, the use of drugs even during the capture operation ought to be evaluated.

(5) Application of non-conventional practices, for example, addition of calcium to the water and urea to the food.

At this stage, it is possible to extract from the existing results, recommendations and experience a sizable number of hints for the culturists who deal with fry. The following is the author's attempt to sum up those hints that, if they will not help with all species and under all conditions, will not hurt either.

Handling the fish at capture

When fry are plentiful, several short hauls are preferable to one long haul. The fish captured can be stored in approximately 1 m^3 floating cages, made of mosquito netting or of saran window screen. When enough fish are collected transfer them, preferably in water, to the aerated tanks, or the plastic bags in which they will be carried to their destination. The plastic bags should be inflated with oxygen under pressure as soon as the fish are put in. Tie the bags well, so that the pressure is maintained. The water in these transportation vessels should be taken from the place where the catch has been made.

Where floating net containers (cages) cannot be used, e.g., on shallows and in waves, keep the fish, before they are put into the transportation vessels, in a portable container as large as feasible, filled with clean water taken from the place where the fry were caught. Sometimes fishing is done at an estuary, and one haul may be carried out in the river (fresh water), another in its mouth (brackish water) and still another in the nearby sea (all sea water or almost sea water). Under such conditions, if possible, keep the fish in separate vessels, each batch in the water taken from the respective spot of capture. The same is true for the containers or plastic bags in which the fry are transported to their destination.

At the end of a haul, do not lift the whole bunt of the seine out of the water with the catch in it to carry it to the holding vessel. Keep the bunt in the water so that the fry may swim in it, and bail them out in small batches using a soft scooping net (dipper) made of mosquito netting, or similar material. Try to keep the fry out of the water for as short a time as possible. Again, tranquilizing the fish in the seine bunt before bailing may prove advantageous.

Controlling the temperature

Always carry a thermometer and use it to make sure that the newly captured fry are not subjected to temperature shock. The vessel, into which the fry are put for transportation, should contain water of the same temperature as the water in which they were caught. After that, a gradual change in the temperature, even to a considerable degree, will be less harmful than an initial thermal shock of any extent. In summer, carry insulated containers (ice boxes) with ice cubes or flakes to cool the water, if necessary, under field work conditions.

During prolonged transportation, survival is better at lower ambient temperatures. The lower the temperature of the water, the more fry can be kept in the transportation containers. Temperatures between 15 and 18 °C can be conveniently tolerated by most, if not all, species of grey mullets. Where fish are carried in air-tight tied plastic bags, ice can be inserted between the bag and the stiff container in which the bag is kept. When transferring the fry from transportation vessels to acclimatization tanks and from those to nursery ponds, avoid thermal shocks, this is achieved best by slow mixing of water from both basins involved.

Aeration and oxygen supply

Where plastic bags are used for transportation, they can be inflated, after having been filled with approximately half of their capacity with water (and fry), with oxygen from a portable container (see Fig. 10.1). The bag should be inflated to a degree that it resists a moderate finger pressure, then tied securely to prevent drop in the pressure. Truck or jeep carried, or towed containers, can best be aerated by the car-battery operated aquarium-type air pumps of the larger types.

Salinity

Mugil cephalus, *Liza ramada*, *L. aurata* and, probably also other cultured mullet species, can withstand major salinity fluctuations within the 50–5‰ range without damage. The critical stage seems to be the transition from approximately 5–4‰ to fresh water. Therefore, keep fry caught in sea or brackish water at that minimum salinity for at least 24 h before transferring them, preferably, gradually to fresh water (see also Chervinski, 1974; Mires *et al.*, 1974).

Salinometers may be too expensive for some conditions and budgets. Only a few models are suitable for field conditions. The Haifa Fisheries Technology Unit uses specially designed hydrometers for on-the-spot salinity measurement. A series of such hydrometers can conveniently cover the required salinity

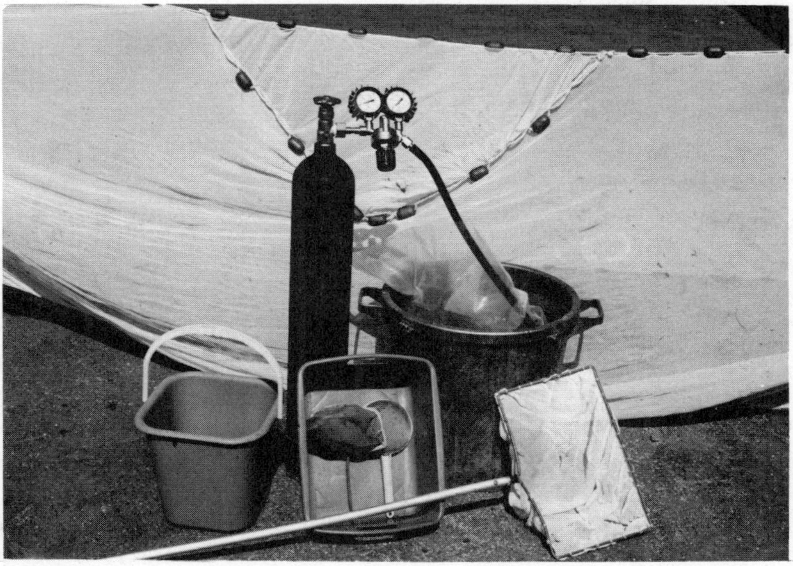

Fig. 10.1. A set of simple instruments used for fry capture and handling. These include: mosquito-netting beach seine, oxygen bottle, plastic bag in a container, bucket, plastic box with a hand scoop-net and a kitchen strainer, and mosquito-netting hand-net.

range. Using the hydrometer readings, temperature calibration table and special nomograms, absolute salinity can be determined at an accuracy sufficient for our purpose. Relative salinity can be compared for the same temperatures by direct reading of the hydrometer's scale (S. Pisanty, in preparation).

Density of fry in the transportation and acclimatization vessels and in nursery ponds

For high fry density (number of fish per volume of water) the following conditions must be met: a good oxygen supply, a relatively low temperature and the fry should be small and, preferably, tranquillized. The larger the fry the less must be the density. Although less than directly proportional, oxygen consumption is a function of biomass and a 55 mm long fingerling, e.g., may be three times heavier than 35–40 mm fry. The correct density, therefore, must be determined using common sense, experience and a rule of thumb, developed locally. For the beginner, stocking a 100 × 60 cm plastic bag with 10–15 thousand of 20–25 mm fry (total length) is quite safe under oxygen pressure, and at temperature of less than 20 °C. According to Mints (1956) the density at 24 °C ought to be half that at 17 °C.

Handling, transportation and stocking of fry

Estimation of fry numbers is of a considerable importance. For this purpose, a good method is that of marking a dipper net with graduations (see Chapter 9). Undoubtedly, and under all conditions, the lower the density the better is the survival of the fry.

Acclimatization

The need for large amounts of water from the place of capture for good acclimatization may present the main difficulty to be overcome. Pour this water in the acclimatization tanks or ponds before the introduction of the fry, so that the process starts at a fry density less than during the transportation. Transfer the fry from the transportation vessels by gently pouring the fry into the latter, water with fish together, rather than by bailing the fish out. Dilute the water from the place of capture gradually (see the recommendations on salinity) with water from the pond or lake to be stocked with the fry. Also, bring the water, gradually, to the temperature of that in the pond or lake to be stocked. Be very careful at temperatures less than 12 °C. Algal growth on the walls of the acclimatization tanks and addition of natural food ('green-water', copepods, chironomid larvae, rotifers, cladocera, etc.) should improve the survival. Rearing of chironomid larvae as food for fish fry has been described by Konstantinov (1954), Wirszubski & Elchunes (1952) and Yashouv (1970), and rearing of rotifers by Theilacker & McMaster (1971) and Raymond, Baxter & Brandt (1973). Production of 'greenwater' and of copepod cultures was described by Sebastian & Nair (1975).

Acclimatization and nursery techniques may considerably improve the total survival of the fry; however, they involve additional efforts and investment in manpower and equipment. In some areas, such efforts have so far been deemed unnecessary owing to the apparently unlimited supply of the fry. Where the supply of fry is becoming scarce, these techniques may make the continuation of stocking feasible, even at a low rate of the mullet fry capture.

Acknowledgements

Much of the experience and information given in this Chapter is the result of fruitful and pleasant co-operation and team work with Dr J. Chervinski (Fish and Aquaculture Research Station, Dor), Messrs. E. Grofit and S. Pizanty (Fisheries Technology Unit, Haifa), Mrs L. Zismann (Israel Oceanographic and Limnological Research Ltd) and the crew of the Tiberian Stocking Unit of the Fisheries Department. Thanks are also due to Dr O. H. Oren, without whose initiative and encouragement this chapter would not be written and, again, to Mrs L. Zismann for critical reading of the manuscript.

References

Albertine-Berhaut, J. (1973). Biology of juvenile stages of teleostean Mugilidae *Mugil auratus* Risso 1810, *M. capito* Cuvier 1829, and *M. saliens* Risso 1810. I. Feeding habits. *Aquaculture* **2** (3), 251–66.

Albertine-Berhaut, J. & Vallet, F. (1972). Alimentary utilization of urea in mullets. *Tethys* **3** (3), 677–80.

Auls, A. M. (1972). The effect of calcium on the osmotic tolerance of juvenile striped mullet (*Mugil cephalus* L.). In *Proceedings of the 26th Annual Conference of Southeastern Association of Game and Fish Commissioners*, 22–25 Oct. 1972, Knoxville, Tennessee (ed. A. L. Mitchell), pp. 591–97.

Babayan, K. E. (1965). Kefali, Biologiya, Sposoby lova i vyrashchiva-shchivaniya. (Grey mullets, their Biology, Methods of Capture and Culture.) Pishchevaya Promyshlennost, Moscow. (In Russian.)

Bapat, S. V. & Bal, D. V. (1951). The food of some young fishes from Bombay. *Proc. Indian Acad. Sci.* (*B*), **35** (2), 78–92.

Bar-Ilan, M. (1975). Stocking of *Mugil capito* and *M. cephalus* and their commercial catch in Lake Kinneret. *Aquaculture* **5** (1), 85–90.

Chervinsky, J. (1974). *Liza aurata* (Risso) (Pisces, Mugilidae) and its adaptability to various saline conditions. In *Proceedings of the IBP/PM International Symposium on the Grey Mullets and their Culture, Haifa*, 2–8 June 1974.

Chervinsky, J. (1975). Experimental raising of golden grey mullet (*Liza aurata* (Risso)) in saltwater ponds. *Aquaculture* **5** (1), 91–8.

D'Ancona, U. (1954). Fishing and fish culture in brackish water lagoons. *FAO Fish. Bull.* **7** (4), 147–72.

Dick, G. L. (1975). Some observations on the use of MS-222 Sandoz with grey mullet (*Mugil chelo* Cuvier). *J. Fish. Biol.* **7** (2), 263–8.

Durve, V. S. (1970). Effect of anaesthetics on the behaviour of mullet fingerlings and the scope of using these in different fishery procedures – II. *Indian J. Fish* **13** (1), 158–82.

Durve, V. S. (1975). Anaesthetics in the transport of mullet seed. *Aquaculture* **5** (1), 53–63.

Durve, V. S. & Dharma Raja, S. K. (1966). Effects of anaesthetics on the behaviour of mullet fingerlings and the scope of using these in different fishery procedures – I. *J. Mar. Biol. Assoc. India* **8** (1), 28–55.

Egusa, S. (1950). Some notes on the feeding habit of the young of *Mugil cephalus* L. *Bull. Jap. Soc. Sci. Fish.* **15** (11), 715–20.

Eisawy, A. M., Ishak, M. M. & Hamza, A. (1974). Experimental rearing of two mullet species *Mugil cephalus* and *M. capito* in Egyptian fish farms. *Bull. Inst. Oceanogr. ARE* **4**, 57–96.

El-Maghraby, A. M., Hashem, M. T. & El-Sedfy, H. M. (1974). Species composition and distribution of the mullet fry in Lake Borollus. *Bull. Inst. Oceanogr. Fish., ARE* **4**, 1–13.

El-Zarka, S. El-D. & Kamel, F. (1967). Mullet fry transplantation and its contribution to the fisheries of inland brackish lakes in the UAR. *Proc. Tech. Pap. Gen. Fish. Coun. Mediterr.* (8), 209–26.

Huet, M. (1971). *Textbook of fish culture. Breeding and cultivation of fish.* Fishing News (Books) Ltd, London.

Konstantinov, A. S. (1954). Opyt poluproizvodstvennogo razvedeniya motylya. (Experimental semi-industrial breeding of midge.) *Ryb. Khoz.* **30** (11), 41–3.

Korringa, P. (1975). Mullet from Italian lagoons. *Fish. Farming Int.* **2** (1), 37–8.

Lahav, M. & Sarig, S. (1972). Control of unicellular parasites using formalin. *Bamidgeh* **24** (1), 3–11.

Lasserre, P. & Gallis, J.-L. (1975). Osmoregulation and differential penetration of two grey mullets, *Chelon labrosus* (Risso) and *Liza ramada* (Risso) in estuarine fish ponds. *Aquaculture* **5** (4), 323–44.

Liao, I.-C. (1974). Experiments on induced breeding of the grey mullet in Taiwan from 1963 to 1973. *Aquaculture* **2** (2), 1–24.

Mints, A. G. (1956). O normakh posadki molodi kefali pri ee transportirovke. (Density norms of grey-mullet fry during transporation.) *Ryb. Khoz.* **32** (10), 62–4. (In Russian.)

Mires, D. (1970). Preliminary observations on the effects of salinity and temperature of water changes on *Mugil capito* fry. *Bamidgeh* **22** (1), 19–24.

Mires, D. (1973). A hatchery for breeding and forced spawning at kibbutz Ein Hamifratz. *Bamidgeh* **25** (3), 72–84.

Mires, D., Shak, Y. & Shilo, S. (1974). Further observations on the effect of salinity and temperature changes on *Mugil capito* and *M. cephalus* fry. *Bamidgeh* **26** (4), 104–9.

Mityushkin, V. A. (1956). Dvukhletnii opyt rabot po perevozke kefali iz Kaspiiskogo morya v Aral'skoe. (Two years of experience in transporting grey mullets from the Caspian Sea to the Aral Sea.) *Ryb. Khoz.* **32** (10), 60–2. (In Russian.)

Moore, R. H. (1974). General ecology, distribution and relative abundance of *Mugil cephalus* and *Mugil curema* on the South Texas coast. *Contrib. Mar. Sci.* **28**, 241–55.

Nash, C. E. & Kuo, C.-M. (1975). Hypotheses for problems impeding the mass propagation of grey mullet and other finfish. *Aquaculture* **5** (2), 119–34.

Odum, W. E. (1970). Utilization of the grazing and plant detritus food chains by the striped mullet *Mugil cephalus*. In *Marine Food Chains*, ed. J. H. Steele, pp. 222–39. Oliver & Boyd, Edinburgh.

Osborn, P. S. (1951). Some experiments on the use of Thiouracil as an aid in holding and transporting fish. *Prog. Fish Cult.* **13** (2), 75–8.

Perlmutter, A., Bograd, L. & Pruginin, J. (1956). Use of the estuarine and sea fish of the family Mugilidae (grey mullets) for pond culture in Israel. *Proc. Tech. Pap. Gen. Fish. Counc. Mediterr.* (4), 289–304.

Pruginin, Y. & Kitai, H. (1957). Cultivating mullets as a secondary fish in the carp ponds. *Bamidgeh* **9** (4), 70–5.

Pruginin, Y., Shilo, S. & Mires, D. (1975). Grey mullet: a component in polyculture in Israel. *Aquaculture* **5** (3), 291–8.

Raymond, L. P., Baxter, G. A. & Brandt, W. (1973). Preliminary experimentation with larval feeds. In *The Grey Mullet* (*Mugil cephalus* L.): *Induced Breeding and Larval Research, 1972–1973* vol. II. Oceanic Institute, Waimanalo, Hawaii.

Sarojini, K. K. (1951). The fishery and biology of the Indian grey mullets – a review. *J. Zool. Soc. India* **3** (1), 159–79.

Sebastian, M. J. & Nair, V. A. (1975). The induced spawning of the grey mullet, *Mugil macrolepis* (Aguas) Smith and the large-scale rearing of its larvae. *Aquaculture* **5** (1), 41–52.

Sinderman, C. J. (1970). Disease and parasite problems in marine aquaculture. In *Marine Aquaculture*, ed. W. J. McNeal. Oregon State University Press, Corvallis.

Stickney, R. R. & Shumway, S. E. (1974). Occurrence of cellulose activity in the stomachs of fishes. *J. Fish Biol.* **6** (6), 779–90.

Sylvester, J. R. (1974). Thermal response of juvenile Hawaiian mullet *Mugil cephalus* (L.) to acclimation time and fluctuating low temperatures. *J. Fish. Biol.* **6** (6), 791–6.

Sylvester, J. R. (1975). Critical thermal maxima of three species of Hawaiian estuarine fish: a comparative study. *J. Fish Biol.* **7** (2), 257–62.

Tang, Y. A. (1975). Collection, handling and distribution of grey mullet fingerlings in Taiwan. *Aquaculture* **5** (1), 81–4.

Theilacker, G. H. & McMaster, M. F. (1971). Mass culture of the rotifer *Brachionus plicatilus* and its evaluation as a food for larval anchovies. *J. Mar. Biol.* **10**, 183–8.

Thomson, J. M. (1955). Brackish water fish farming. *Fish. Newsl. Aust.* **14** (3), 13, 15 and **14** (4), 17, 19.

Yashouv, A. (1967). Al iklum v'gidul intensivi shel degigei hakipon gedal harosh. (On acclimatization and intensive rearing of grey mullet (*M. cephalus*) fry.) In *Report No. 3/4 on the studies in Mullets Mugilidae culture in fish ponds*. Min. Agric., Dept. Fish., Fish Farming Research Station, Dor (In Hebrew).

Yashouv, A. (1970). Propagation of chironomid larva as food for fish fry. *Bamidgeh* **22** (4), 101–5.

Yashouv, A. (1972). Efficiency of mullet growth in fishponds. *Bamidgeh* **24** (1), 2–25.

Yashouv, A. & Ben-Shachar, A. (1967). Breeding and growth of Mugilidae. II. Feeding experiments under laboratory conditions with *Mugil cephalus* L. and *Mugil capito* Cuvier. *Bamidgeh* **19** (2–3), 50–66.

Yashouv, A. & Karamusta, K. (1967). B'khipus akharei takhlifim l'mazon tivyi b'gidul degigei kiponim b'meikhalim. (In search for substitution feeds to natural food in raising mullet fingerlings in tanks.) In *Report No. 3/4 on the studies in mullets Mugilidae culture in fish ponds*. Min. Agric., Dept. Fish., Fish Farming Research Station, Dor (In Hebrew), pp. 41–8.

Yocom, T. G. & Edsall, T. A. (1974). Effect of acclimation temperature and heat shock on vulnerability of fry of lake whitefish (*Coregonus clupeaformis*) to predation. *J. Fish. Res. Board Can.* **31**, 1503–6.

Wirszubski, A. & Elchuness, M. (1952). A preliminary report on the breeding of Chironomidae in concrete ponds. *Bamidgeh* **4** (1–2), 15–16.

Zismann, L., Berdugo, V. & Kimor, B. (1974). The food and feeding habits of early stages of grey mullets in the Haifa Bay region. In *Proceedings of the IBP/PM International Symposium on the Grey Mullets and their Culture, Haifa*, 2–8 June 1974.

11. Cultivation methods

I.-C. LIAO

Introduction

Mullet culture has a history of several hundred years both in the Orient and in the West. However, the cultivation methods of mullet are still of the primary extensive pattern and little research on this topic has been done. There are several advantages of mullet so that they are considered a suitable group for culture. Firstly, mullet are euryhaline and can be stocked in sea water, brackish water and even fresh water areas which are still not utilized on a broad scale in many places in the world. Secondly, they are eurythermal and distributed throughout the warm waters of the world. Thirdly, they feed on low priced food and grow rapidly.

Among the common problems of human beings, food supply is one of the serious crises. It is much to be regretted that mullet farming is not contributing as it should be to world food supplies. However, as a Chinese proverb says, 'It is not too late to repair the fold after the sheep have strayed'. On June 2–8, 1974, the 'IBP/PM International Symposium on the Grey Mullets and Their Culture' was held in Haifa, Israel with the support of the Food and Agriculture Organization of the United Nations (FAO, the United Nations Educational, Scientific and Cultural Organization (UNESCO), the International Biological Programme (IBP) and other organizations. It was certainly a wise and significant act.

Since the research on mullet culture is far from complete, further studies are needed in many respects. Therefore, it is not possible to summarize the very perfect or ideal cultivation methods of mullet for the time being. But for the reason of the considerable importance of mullet in aquaculture, the data concerned have been collected as far as possible and summarized here in this chapter. It is hoped that many colleagues with the same interest will find this chapter a useful reference when they plan to look for further research subjects.

General biology

The family Mugilidae contains three to five genera and about seventy seven species. Among them, the grey mullet, also called striped or jumping mullet, *Mugil cephalus* L. is one of the most important species, and it is the main species for cultivation. The following paragraphs are mainly concerning this species.

The grey mullet are eurythermal and euryhaline, therefore they are widely distributed in the world. They are present both in the areas of low temperature as the northern part of Japan, Hokkaido, and areas of high temperature as

South Africa, Cuba and India. Oshima (1957) indicated that the critical thermal minimum for 27 mm juvenile mullet is 3 °C and that mullet of 180 mm or more in length are able to endure 0 °C. Being euryhaline, the fingerlings collected from the estuarine areas can be stocked in fresh-water ponds by gradually diluting the sea water they originally inhabited down to fresh water. The larger the mullet are, the more resistant to the fluctuation of salinity they will be. Wallace (1974) found that mullet were cultured successfully even in conditions with a salinity as high as 60–70‰.

Mullet look graceful with a greyish-green colour on the dorsal surface and silver-white on the ventral surface. They are highly esteemed for the mature ovaries. The dried roe is a gourmet food of very high price. The cultured mullet are fat and tasty, especially in winter. The ovaries of mullet reared in fresh-water ponds have been found maturing but never reached the ripe or spawning stage (Eckstein, 1974). Liao, Lee, Lim & Lo (1971) found that with hormone treatment, mullet which had been transferred from fresh water into sea water for three months before the spawning season, were successfully induced to spawn and the fertilized eggs hatched well.

Three kinds of migration are found in the life history of mullet. One is the osmoregulatory migration, a phenomenon of juveniles anadromously migrating toward estuaries. Another is the offshore, or seaward migration, a phenomenon of adult mullet migrating toward the open sea after being in estuaries for several years. The other is the spawning migration. For example, mullet migrate southward in schools from the East China Sea every year during the period of December to February. After entering the Taiwan Strait, they head towards the middle region of the west coast of Taiwan and then proceed southward along the coast. There are several well-known spawning grounds in the world, e.g., the east coast of the United States from North Carolina to Florida (Anderson, 1958), the waters 40–50 miles south east of the estuary of the Mississippi River and 900–1500 m deep over the continental slope (Arnold & Thompson, 1958) and the Chi-Hsing Chiao near the south west point of Taiwan. All these grounds are located in the regions usually affected by warm currents. The surface water temperature in the spawning season is 20–23 °C.

In general, the fecundity of mullet ranges from 800 thousand to 2.7 million eggs. The mullet egg is round, averaging 0.93–0.95 mm in diameter with a yellowish-brown oil globule measuring about 0.38 mm in diameter. The eggs are conditionally pelagic, in nature, they may drift with the current into deeper water and float at mid-depths until hatching. In the case of artificial propagation, the larvae just after hatching measure 2.08–3.50 mm in body length and feed on tiny zooplankton such as fertilized oysters eggs. The ten-day-old larvae measure about 4.0 mm in body length and are fond of rotifera. The twenty-day-old larvae measure about 7.0 mm in body length and are fond of copepods and *Artemia.* The one-month-old larvae measure about 20.0 mm in body length and are fairly resistant to environmental changes, hence, are suitable for stocking in ponds.

Just like yellow-tail, tuna, common sea bass and perch, mullet in Japan are called 'hussegiyo' whose names are changed as the fish grow. 'Haku' or 'Kirara' are the mullet of 23–30 mm, 'Oboko' are those of 31–180 mm, 'Ina' are those of 181–300 mm, 'Bora' are those above 300 mm, 'Todo' are those of extra large size and 'Karasumibora' are those ripe in spawning season. It is convenient for description to give different names for mullet at various growth stages, because their habitats, feeding habits and the fishing methods are quite different.

Fingerlings

Ninety-nine per cent of fingerlings for stocking are collected from the sea. This amount will not be inadequate as long as the culture techniques, particularly for the fingerlings collected from the sea and not yet stocked, can be improved and the survival rate of that stage can be raised. Usually the survival rate of fingerlings before stocking is below 30% and after stocking is below 50%, so the survival rate of fingerlings is very low.

Because the survival rate of fingerlings has not been increased and environmental pollution probably will cause the decline of fingerlings recruitment, the study of artificial propagation of mullet has been carried out by scientists in many places of the world, and the greatest success in rearing fingerlings has been achieved in Taiwan. There the techniques of artificial propagation are improved every year and are fairly advanced now. The survival rate of artificially propagated fingerlings has a record of 19.35% in 1972/73 (Liao, 1974). The solution of rearing food organisms for mullet larvae and of the proper control of culture environments will soon help to break the record and achieve near perfect techniques.

The survival rate of artificially propagated fingerlings after stocking in the fish ponds is 95% or above, while that of the so called 'Haku', wild mullet fingerlings of 23–30 mm, is 30% or below. There are many different fishing methods of collecting wild mullet fingerlings. Simple scoop nets (Fig. 11.1), skimming nets (Fig. 11.2) and hand-operated seine nets (Fig. 11.3) are all common gear for this purpose. In addition to these, attracting with lights at night on a motor raft (Fig. 11.4) is also effective for collecting mullet fingerlings.

The so called haku with body length less than 30 mm (Fig. 11.5) generally swim weakly in schools, therefore they are easily collected from estuaries. However, they are so fragile that the survival rate in culture is often low. In other respects, the oboko with body length above 50 mm are more suitable

Fig. 11.1. The primary traditional method for collecting mullet fingerlings with a simple scoop-net.

Fig. 11.2. Collecting mullet fingerlings with a skimming-net.

Fig. 11.3. Collecting mullet fingerlings with a hand-operated seine net.

Fig. 11.4. Night-time method for attracting mullet fingerlings with electric light and catching with a simple scoop-net.

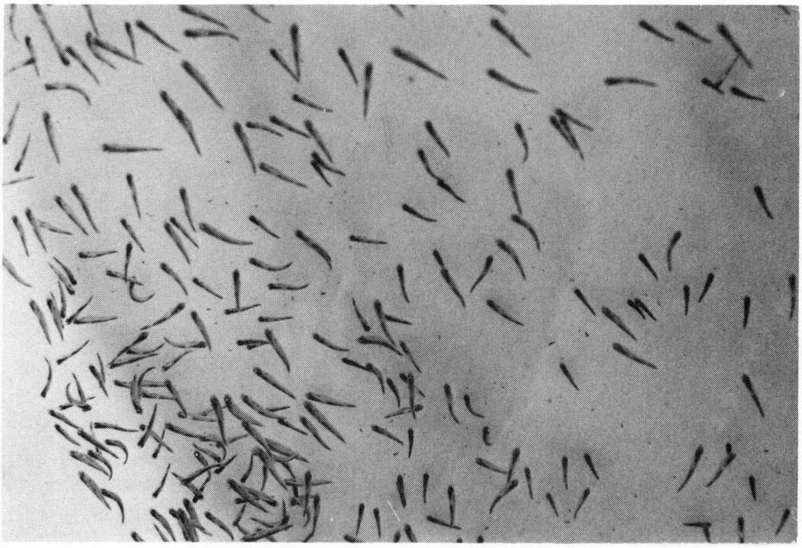

Fig. 11.5. Artificially propagated fingerlings 2.3–3.0 cm in body length, corresponding to the size of 'haku' wild mullet fingerlings.

for stocking but there is a difficulty in collecting them owing to their active swimming. Haku, laterally flattened with a silver-white complexion, can be obtained from the coastal area, for two or three months during the peak of the mullet spawning season. Oboko look like the adult 'fusiform' and can't be collected easily because they swim in a scattered fashion over the muddy bottom of estuaries. Ordinarily they can be got within the sluice gate by taking advantage of their characteristics of swimming against the fresh-water current (Fig. 11.6).

Young mullet, especially those less than 30 mm, are fond of zooplankton. Growing from Haku to Oboko, mullet change their feeding habits and the shapes of their digestive tract as well (Fig. 11.7). As Egusa (1950) reported, the main food for Haku is small benthic crustaceans such as Amphipoda and harpacticoid Copepoda, while Oboko are herbivorous and detritophagic.

Table 11.1 shows the salinity tolerance of Haku and Oboko. It is noted that Oboko can withstand low salinity while Haku cannot. This should be considered in stocking haku. Nakamura (1967) did a further experiment on this problem. Haku were first stocked in brackish water with 6‰ chloride, and the rearing water was then gradiently diluted to fresh water over 2–4 days, 7 days and 9–11 days. The resulting survival rates were 50%, 80% and 100% respectively. The conclusion is the longer the period for gradual freshening, the higher the survival rates.

In some places of the subfrigid zone such as Northern Japan, the

Fig. 11.6. Collecting mullet fingerlings which swim against the current from a sluice gate opened at low tide, owing to their fondness for low salinity water (courtesy Mr Masayoshi Uno, Aichi Prefecture Fisheries Experimental Station).

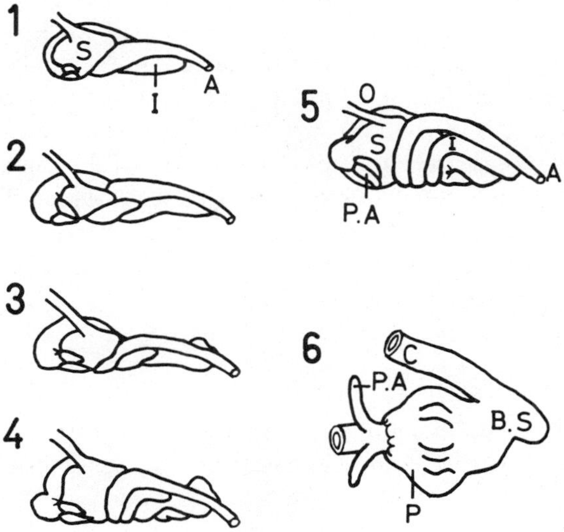

Fig. 11.7. Variation in curvature of the digestive tract of different sized mullet (after Egusha, 1950). 1–5, digestive tract of 'Haku'–'Oboko'; 6, stomach of adult mullet. O, oesophagus; BS, blind sac; P, pyloric portion; I, intestine; C, cardiac portion; S, stomach; PA, pyloric appendages; A, anus.

Table 11.1. *The salinity tolerance of two different sized mullet fingerlings (after Nakamura, 1967)*

	'Haku' (BW = 0.10–0.32 g; TL = 23–31 mm)			'Oboko' (BW = 0.71.4 g; TL = 35–42 mm)	
Salinity (‰)...	0.08	2.50	4.87	0.08	3.64
No. of fingerlings...	30	32	34	20	20
2 hrs	30	5	1	0	0
1 day		17	13	11	1
2		2	1	1	1
3		1	0	1	0
4		7	3	0	0
Mortality (%)	100	100	52.9	65.0	10.0

overwintered fry, in addition to haku and oboko, are also good for stocking. The over-wintered fry are the yearling haku or oboko of unusually small size, weighing 25–45 g only. When stocked in spring, they usually grow up to bora of edible size in the end of the same year with a high survival rate; for this reason they are ideal for cultivation. To get the overwintered fry, one has to increase the stocking density of haku or oboko and wait until the next spring.

In warmer places, the one-year-old mullet already weigh 200–300 g or more and have reached their market size. There seldom or never do the over-wintered fry appear. But it is worth trying to stock haku or oboko densely when they are available, with the hope that they may turn out to be the over-wintered fry through growth inhibition and can be stocked for further cultivation, if the natural supply of fingerlings is not available later. This is one way to assure a source of fingerlings. Mullet fingerlings kept in polyethylene bags inflated with oxygen can be transported alive easily. For long distance transportation, fewer fingerlings are maintained in each bag than for short distance transportation. In summer, ice cubes are added to lower the water temperature in the bag.

Cultivation methods

Three different types of cultivation methods, namely: extensive culture, polyculture and intensive culture or monoculture, are described in detail as follows.

Extensive culture

Most of the mullet culture in the world belongs to the extensive culture type. The fish ponds are commonly located on the shore, near estuaries for the

Fig. 11.8. General view of mullet pond of Saji Fishery Farm at Fukue, Aichi Prefecture, Japan (courtesy Mr Masayoshi Uno, Aichi Prefecture Fisheries Experimental Station).

convenience of letting mullet fingerlings get into the sluice gate by themselves (Figs. 11.6 and 11.8). Feed is seldom given because there is natural food in the ponds. Growth depends on the stocking density and the pond condition. Currently the yearling mullet attain 15 cm in body length, and 50–60 g in weight. The Italian 'valli' and 'lagoon' fisheries are the extensive culture patterns of mullet, and the total annual yield ranges from 90 to 200 kg/ha of fish (Bardach, Ryther & McLarney, 1972). Besides, there is another well-known method, the 'kawa culture' in Aichi Prefecture, Japan. 'Kawa culture' in Japanese means the culture method of stocking fishes in waterways.

The water temperature of kawa culture ranges 29–31 °C in summer and 4–6 °C in winter. Mullet is the main species stocked in the waterways and some other minor species are carp, crucian carp, common sea bass, perch and eel. The stocking density of mullet is quite high, about 15800–19750/ha. Oshima (1957) reported that the average production was 480 kg/ha (Table 11.2). According to Nakamura's investigation (1967), there was a total production of 167.25 tons from a total area of 427 ha, averaging 392 kg/ha. The report of the Aichi Prefecture Fisheries Experimental Station, Japan, showed that a total production of 245830 kg was obtained from thirty two management units with a total area of 371 ha, averaging 663 kg/ha.

The best result of kawa culture was that two-year-old cultured mullet measured 27.6–30.3 cm and weighed 490–578 g, averaging 29.3 cm in body length and 522 g in body weight (Nakamura, 1967). They were a little slower

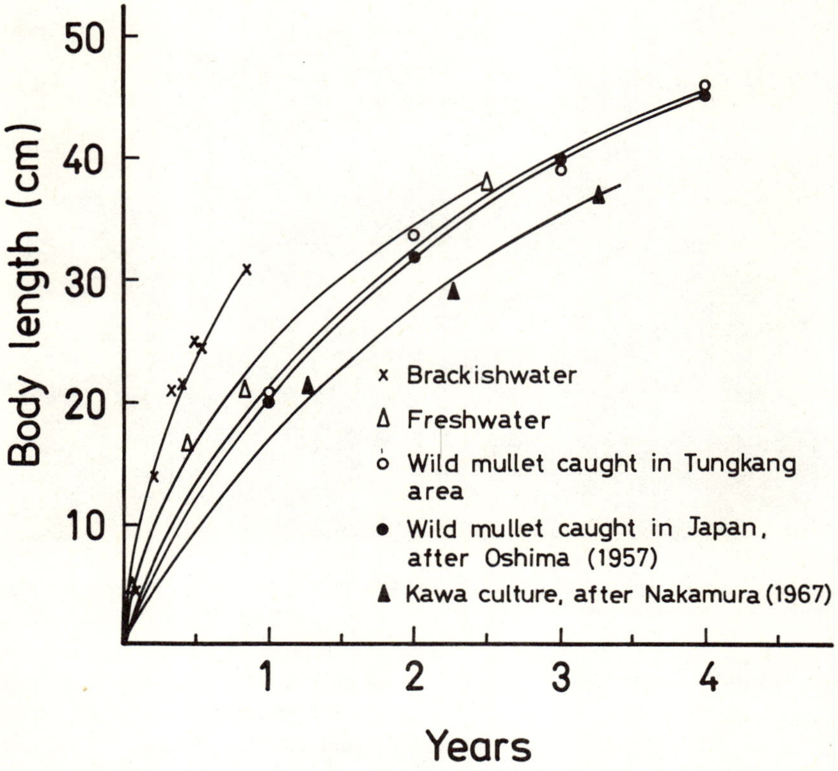

Fig. 11.9. Comparison of growth between wild mullet and cultured mullet.

Table 11.2. *The production of 'Kawa culture' (after Oshima, 1957)*

	Stocking density (number/ha)	Survival rate (%)	Average weight (g)	Production (kg/ha)
A	16500	6	528	520
B	33250	5	262	461
C	24500	10	225	588
D	8240	15	450	555

A, B, C and D represent four different scales of management.

in growth than the wild mullet (Fig. 11.9). Moreover, the survival rate in kawa culture was as low as 5–10%, calculated at the harvest of two-year-old mullet. It might be that in kawa culture there was a great loss of mullet fingerlings, including being eaten by perch and escaping through the bamboo fence. Generally speaking, the production of mullet was still higher than that of carp and crucian carp under the same condition of no feed except the

Fig. 11.10. An intensive polyculture pond in Israel. Note that the pond is equipped with aeration.

natural food in the waterways. A conclusion can be drawn that mullet is really a good culture species for the purpose of making most use of the natural environment and getting quite a high production in extensive culture.

Polyculture

Polyculture is also called mixed fish culture (Yashouv, 1968). It sounds like a modern cultivation method, yet this is not so for the mullet farming, since mullet have been mix-cultured with other fishes for a long time.

Pruginin & Kitai (1957) and Yashouv (1968) suggested that it was fairly good to introduce mullet or tilapia as a secondary fish in the medium sized carp ponds of four to six hectare (Figs. 11.10 and 11.11). In Israel, although the carp–mullet combination did not give the expected increase in total fish production, it was still proved practicable because of the high value of the mullet. Yashouv (1968) reported that mullet, when weighing 30–70 g in their second year, were stocked in carp ponds at the density of 500–800/ha and were marketed when weighing 400–700 g after 120–150 days. Also in Israel a method of bi-yearly rearing, which is to stock the second year mullet with average body weight of 200 g and to raise them to a larger size (1 kg or more) at the end of the second year, was being developed with great potentialities (Pruginin, Shilo & Mires, 1974). This practice, as well as stocking the overwintered fry and culturing them to the bora size in Japan, is based upon

Fig. 11.11. Mullet pond in Israel showing the feeder.

Table 11.3. *The survival rate of polycultured mullet under conditions with nearly no feed (after Nakamura, 1949)*

	No. of fish ponds					
	3	4	3	4	3	4
Total mullet stocked	3200	9960	4880	7200	1600	8520
Stocking density (no./m^2)	1.2	2.5	1.8	1.8	0.6	2.2
Survival rate (%)	77.3	55.3	74.9	37.5	46.0	35.8
Average weight (g)	64.5	40.5	38.0	92.5	212.2	124.0
Production (kg/ha)	563	560	520	620	575	940

Pond no. 3, 2640 m^2; pond no. 4, 3960 m^2.

the same principle and brings about large sized mullet with a high survival rate. Large sized mullet are high in price, hence this kind of management is economic.

Mixed-cultures of mullet, carp and eel, with only 57–227 g/m^2 of rice bran as feed was investigated by Nakamura (1949). Table 11.3 shows the results with an average survival rate of 50% and an average body weight of 94 g. Comparing the results of pond nos. 3 and 4 in each set of experiments, it is obvious that the survival rate decreases and the average weight reduces as the stocking density is raised. A similar conclusion is drawn from experiments providing sufficient feed and no feed (Fig. 11.12).

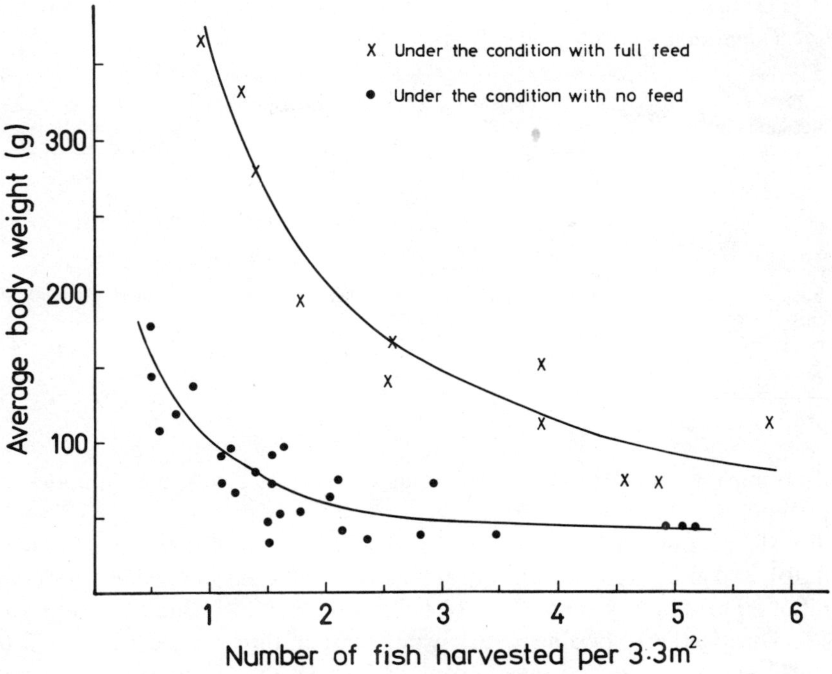

Fig. 11.12. Relation between the number of fish harvested/3.3 m² and average body weight (after Nakamura, 1967).

It is difficult to determine the optimal stocking density since the species for mixed-culture are different and the water quality and the bottom quality vary in each pond. Yashouv (1968) reported that no more than 800 mullet/ha were stocked in the case of carp-mullet mixed-culture. Recently, Pruginin *et al.* (1974) indicated that the stocking density of mullet could be increased up to 1200/ha in the polyculture system of tilapia, carp and silver carp because the equipment and the techniques had been improved.

In Taiwan, mullet are customarily stocked in milkfish ponds and in mixed-cultures with milkfish and grass prawn. In 1971, biologists at Tungkang Marine Laboratory, Taiwan, monocultured 1000 artificially propagated 45-day-old mullet fingerlings, averaging 2.5 cm length, in a milkfish pond of 2000 m². Only a small quantity of flour and rice bran was given as feed. Seventy-two days later, the survival rate was 92.9%. The fingerlings were then transferred to a pond of one hectare and mix-cultured with milkfish and grass prawn. The water in the ponds was brackish water with a constant salinity of 8–10‰. Feed was mainly for the milkfish, the main species in the polyculture pond. The growth data of mullet in this experiment are listed in Table 11.4. For a secondary fish in polyculture, mullet grew fairly well. The

Table 11.4. *Growth of mullet mixed-cultured with milkfish and grass prawn in a milkfis. pond at Tungkang, Taiwan*

Date of measurement	25 Feb. 1971	27 Mar. 1971	8 May 1971	26 June 1971	25 July 1971	19 Aug. 1971	9 Sept. 1971	24 Dec. 1971
Days of culture	0	30	72	121	150	175	196	302
Days after hatching	45	75	117	166	195	220	241	347
Sampling size	10	11	21	7	8	3	11	2
Average body length (cm)	2.5	4.82	13.91	20.98	21.36	24.96	24.46	30.80
Average body weight (g)	0.36	2.51	49.49	153.96	180.32	284.67	245.62	470.95

only thing regrettable was that the final survival rate was not estimated at that time.

For the purpose of further testing the feasibility of mixed-culturing mullet, milkfish and grass prawn, 100, 200, and 200 of each were stocked in a concrete tank of 20 m × 10 m × 1.5 m (Fig. 11.13) at Tungkang Marine Laboratory in 1974. During this experiment, the water temperature ranged from 18.3 to 32.9 °C and the salinity ranged from 10.5 to 19.8‰. Feed was artificially prepared.

As shown in Table 11.5, mullet grew better than milkfish in this mixed-culture experiment. The mean daily weight gain of three consecutive periods was 0.70 g, 1.33 g and 1.27 g for mullet, and 0.60 g, 0.65 g and 0.47 g for milkfish. In the period from January to February 1975, the mean daily weight gain for mullet was 1.15 g but that for milkfish is a negative value. Apparently, the growth of milkfish had slowed down since the middle of November and declined since January (Fig. 11.14). Milkfish is a tropical fish whereas mullet is fairly resistant to low temperature, it is no wonder that in winter milkfish did not grow as fast as mullet. It implies that to rear mullet in the cold season is rather beneficial. This experiment is still progressing and it is hoped that more information will be available for the following years.

Intensive culture

There are a few reports about the intensive culture of mullet. From the viewpoint of feeding habits, it seems as if mullet are not ideal for intensive culture. In fact, intensive culture by no means implicates single species culture. Intensive culture of mullet, with some other species as secondary fishes which will not cause interspecific competition, is thus recommended. Given 0.3–2.0 kg/m² silkworm pupae as feed, an intensive culture of ina with

Fig. 11.13. Harvesting the cultured mullet in an experimental concrete tank (20 m × 10 m × 1.5 m), at Tungkang Marine Laboratory, Taiwan.

Table 11.5. *Growth of mullet mixed-cultured with milkfish and grass prawn in a concrete tank (20 m × 10 m × 1.5 m)*

Date of measurement...	1 Aug. 1974	20 Sept. 1974	10 Nov. 1974	30 Dec. 1974	18 Feb. 1975
		Grey mullet (*Mugil cephalus*)			
Sampling size	10	10	10	10	10
Average weight (g)	5.30	40.10	107.71	171.06	228.51
Mean daily weight gain (g)	—	0.70	1.33	1.27	1.15
		Milkfish (*Chanos chanos*)			
Sampling size	20	20	20	20	20
Average weight (g)	7.63	37.60	70.77	94.17	89.67
Mean daily weight gain (g)	—	0.60	0.65	0.47	−0.09

a density of $2/m^2$ resulted in Japan, in a production of 1290 kg/ha of mullet (Table 11.6).

Another case was reported by Linder, Strawn & Luebke (1974). They fed mullet with commercially prepared feed and took advantage of heated effluent from a power plant to maintain the optimal water temperature. In 1972, the mean daily weight gain was 0.71 g and 0.66 g in two ponds and the mean daily standard length gain was 0.65 mm and 0.59 mm respectively. In 1973, in four ponds the mean daily weight gain was 0.64 g, 0.36 g, 0.67 g and 0.59 g, with

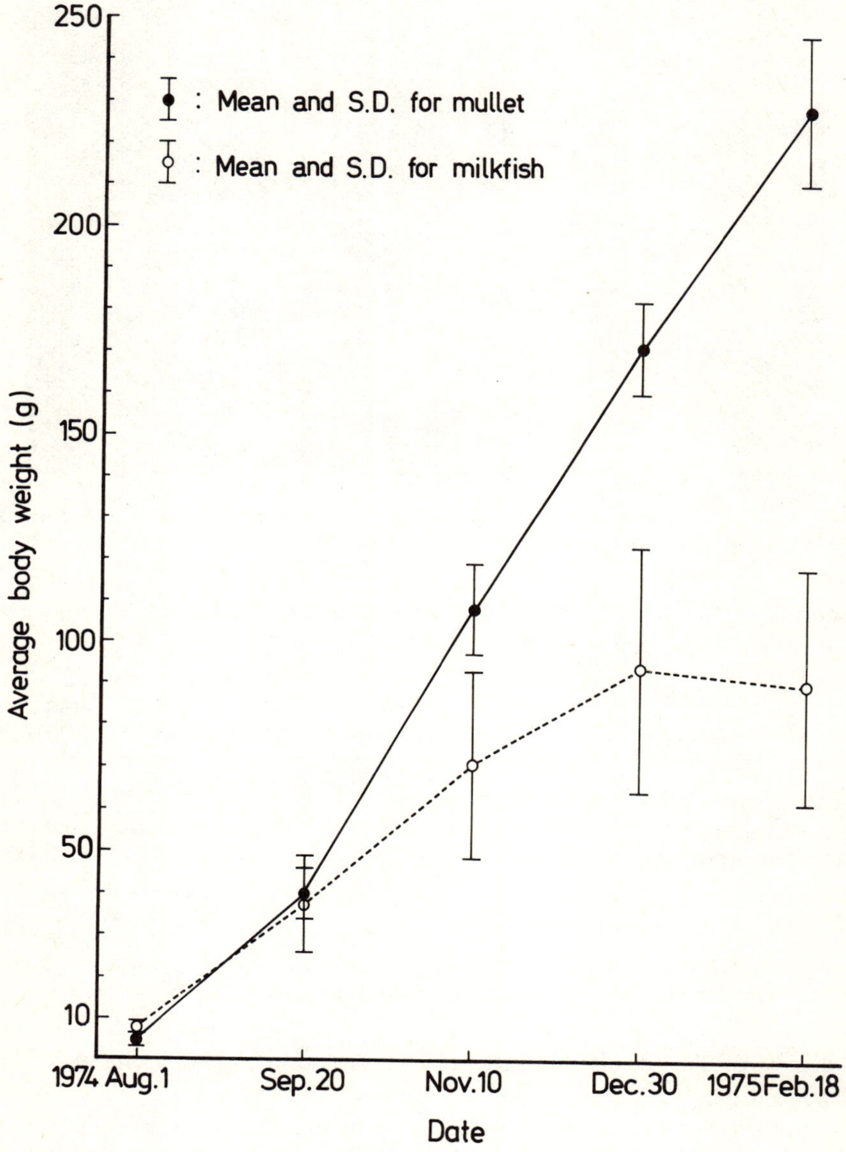

Fig. 11.14. Growth of mullet and milkfish mixed-cultured with grass prawn in a concrete tank (20 m × 10 m × 1.5 m).

Table 11.6. *Growth of mullet intensively cultured in Aichi, Japan (after Nakamura, 1967)*

Stocked source	Survival rate (%)	Body weight (g)	Times of weight increase	Production (kg/ha)
1-year-old fry	48	75–360	220	1290
2-year-old fry	35–85	290–490	3.7–7.6	

the mean daily standard length gain 1.03 mm, 0.81 mm, 0.71 mm and 0.17 mm in the order mentioned. The resulting survival rates were 50–85%. Production was 293–804 kg/ha with food conversion at 2.24–3.31. The cost of commercial feed for the yield of 1 kg of mullet was 0.61–0.87 US dollars.

Discussion and prospects

Fingerlings are the most essential need in culturing fish. At present, 99% of mullet fingerlings come from the sea in about three months during the peak of the spawning season. Notwithstanding that the fingerling supply fluctuates with many factors, it may not be insufficient provided that more care be devoted to fingerlings during and after collection.

The method of collecting mullet fingerlings varies from place to place. Haku, with a body length less than 30 mm, are poor swimmers and can be caught easily with simple scoop nets, skimming nets or hand-operated seine nets, or can be attracted by light at night as mentioned earlier. Of late people have started to use motor boats along the coast to trawl the fingerlings thus saving much labour and increasing the harvest. This method, inflicting much injury on the fragile fingerlings, is not recommended. Another method is to obtain fingerlings at a sluice gate where the fingerlings swim against the water of low salinity.

Some fingerlings obtained at certain time are called 'dead fry' because most of them die before stocking from unknown causes. For rationally utilizing the natural resources, it is absolutely necessary to find out the reason for their low survival rate which is seldom higher than 10%. In general, the fluctuation of salinity has more effect on mullet fingerlings than that of water temperature. Therefore, the sudden drop of salinity should be avoided and only gradual freshening of stocking water is recommended.

As for feeding, since the newly established techniques of rearing food organisms on a large scale have not been learned by the mullet farmers so far, they feed fingerlings only with flour and rice bran. From experiments on artificial propagation, biologists at Tungkang Marine Laboratory ascertained that mullet fingerlings in this stage are fond of zooplankton. Therefore, it is

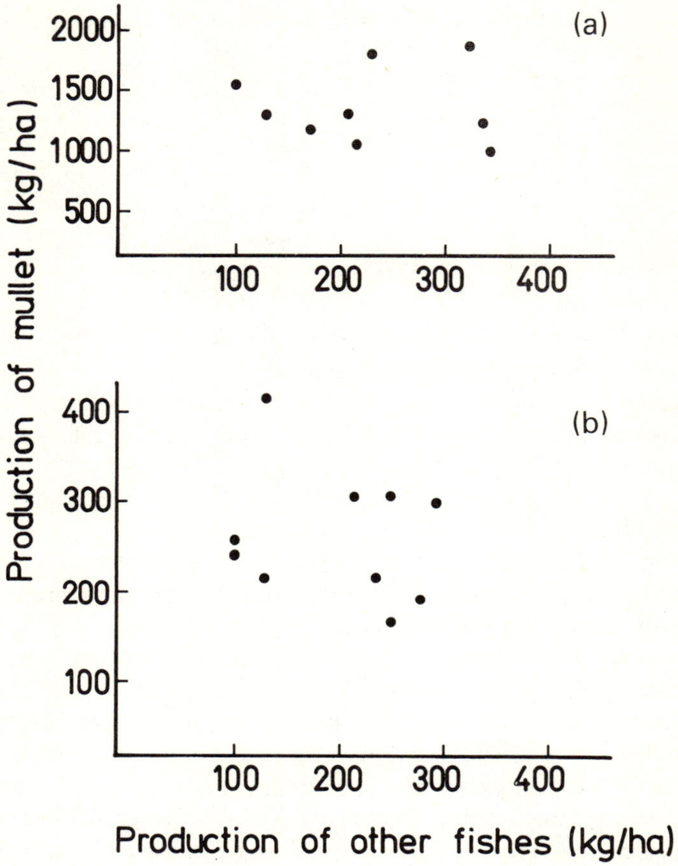

Fig. 11.15. Mullet production in two polyculture ponds (after Nakamura, 1967), a, polycultured with carp and eel; b, polycultured with carp and crucian carp.

possible to raise the survival rate if the farmers have copepods and *Artemia* nauplii to feed to the fingerlings. Efforts in artificial propagation have been made to attain the objective of assuring a supply of fingerlings, in particular fingerlings free of parasites. The greatest success in rearing mullet fingerlings, including the control of water quality and the thorough solution of feed for larval mullet, has been achieved at Tungkang Marine Laboratory. Mass production of mullet fingerlings is now possible in the laboratory and some private commercial companies, following the same techniques, also have started to produce the fingerlings on a large scale. There is no doubt that mullet fingerlings will be produced in millions within a few years, just as the fingerlings of grass carp, silver carp and marine prawns.

Mullet, feeding entirely on natural feed, are most suitable for polyculture.

Table 11.7. *Frequency distribution of the average daily carp production in commercial ponds of the fish farm Maagan-Michael in monoculture (A) and mix-cultured with* M. cephalus *(B) (after Yashouv, 1968)*

Pond	5.5–8.5 (kg/ha)	8.6–11.5 (kg/ha)	11.6–14.5 (kg/ha)	14.6–17.5 (kg/ha)	17.6–20.5 (kg/ha)	Average daily carp production (kg/ha)
A	—	7	16	5	3	13.8
B	7	27	12	—	—	10.7

In Japan, when they were mixed-cultured with carp and eel or with carp and crucian carp, no relation could be found between the total production and the production of mullet (Fig. 11.15). The reason was that feeding habits for mullet were quite different from those for the others. Provided that the maximum stocking density for each species was allowed, the stocking density for mullet in polyculture could be decided without considering the stocking ratio of mullet to other fishes, because the production of mullet is independent.

In Israel, *Mugil cephalus* and *Tilapia aurea* were found to be the most suitable for the requirements of polyculture, however, the presence of mullet tended to check carp growth in the case when the stocking density of mullet was 800 to 1200/ha (Table 11.7). Yashouv's conclusion is at variance with Nakamura's. A possible reason is that the stocking density of carp in Israel was almost equal to the maximum stocking density allowed, much higher than the level in Japan, and thereby the advantage of polyculture is not revealed.

To date, polyculture attracts much attention and has become very advanced. For example, tilapia and perch are mix-cultured because they play the roles of prey and predator in the culturing system. The optimal stocking ratio of one species to another should be figured out for an ideal polyculture. This is an interesting and important problem in both theoretical and practical respects. Polyculture has many advantages, such as high efficiency in utilizing natural feed and culturing space, and more variety in harvesting. What is needed in polyculture at present is to work out the optimal stocking ratios of various species combinations in order to achieve stable ecosystems. In addition, salinity is also an important parameter in mullet polyculture with other species. Nakamura (1967) found that the chlorinity level would affect the total production in the polyculture of carp, crucian carp and mullet (Fig. 11.16).

Where chloride concentrations were below 5‰, which is the maximum limit of chloride tolerance for carp and crucian carp, the total production decreased as the chlorinity increased, whereas in the instances of levels over 6 or 7‰, which is beyond the maximum limit of chloride tolerance for carp and crucian

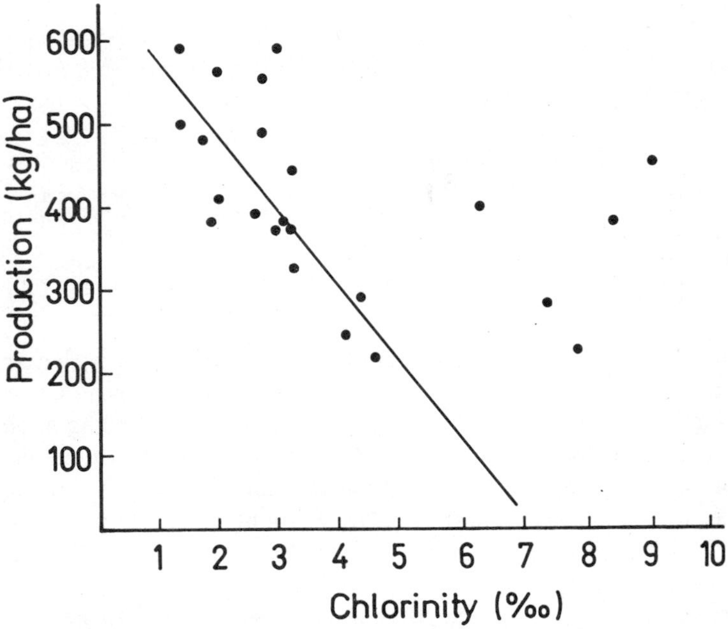

Fig. 11.16. Relation of chlorinity in the rearing conditions to the total production of mullet, carp and crucian carp (after Nakamura, 1967).

Fig. 11.17. The experimental concrete tanks (8.2 m diam.) at Tungkang Marine Laboratory, Taiwan.

Table 11.8. *Comparison of the growth of mullet cultured in sea, brackish and fresh water. Water temperature, 17.1–31 °C*

Date of measurement	5 Oct. 1974	15 Nov. 1974	25 Dec. 1974	3 Feb. 1975
Sea water				
Sampling size	50	10	10	10
Average weight (g)	113.77	207.04	257.51	278.59
Mean daily weight gain (g)	—	2.28	1.26	0.53
Brackish water				
Sampling size	50	10	10	10
Average weight (g)	113.42	208.69	261.52	286.96
Mean daily weight gain (g)	—	2.32	1.32	0.64
Fresh water				
Sampling size	50	10	10	10
Average weight (g)	114.15	187.35	207.40	226.98
Mean daily weight gain (g)	—	1.79	0.50	0.49

Salinity of sea water, 16.4–32.7‰; salinity of brackish water, 11.8–20.6‰.

carp, only the mullet survived. Thus, the chloride concentration in such a polyculture pond should be maintained under 3‰.

Mullet can also be cultured in fresh water as the extensive culture in Lake Suwa reported by Nakamura (1967).

As for mullet culture in sea water of high salinity, there is the well-known farming in the vicinity of the Dead Sea in Isarel (Bardach *et al.*, 1972). In 1964, Israeli biologists began on a series of experiments which showed these areas might be brought into food production by stocking various combinations of *Mugil cephalus* and *Tilapia nilotica.* The best results thus far obtained were from an 0.8 ha pond stocked with 50 g mullet at 214/ha and tilapia of the same size at 139/ha. The total yield of the pond at the end of a 109 days growing season was 1155 kg/ha, of which 512 kg/ha, or 44.3% was mullet. This example indicates that mullet can be cultured in sea water with a high salinity of 36–145‰.

For further investigation of favourable salinities for mullet, an experiment on the effect of salinity on growth of mullet was done at Tungkang Marine Laboratory in 1974. Three round concrete tanks, each with a diameter of 8.25 m (Fig. 11.17), were filled with fresh, brackish and sea water respectively. Fifty young mullet were stocked in each tank and fed with artificially prepared feed everyday. The water was aerated and changed as needed. The results of 120 days culture are shown in Table 11.8.

Comparing the growth of mullet in the three tanks (Fig. 11.18), it is apparent that mullet in the brackish water grew almost as fast as in sea water, but growth seemed to be retarded in fresh water. To test if there is a significant difference in the average weights between treatments a t-test was applied. It

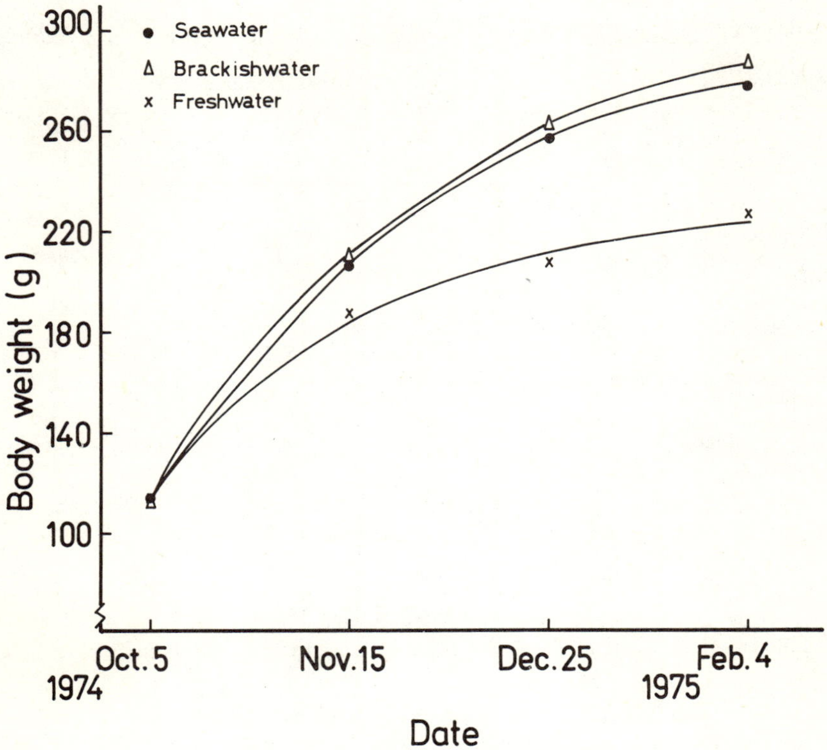

Fig. 11.18. Growth curve of mullet cultured in sea, brackish and fresh water.

was found that the difference in growth of mullet cultured in brackish or sea water and those in fresh water had been significant ($P < 0.05$) before Nov. 15, and that of mullet cultured in brackish water and in sea water was not significant ($P > 0.05$). During the cultivation period, nine mullet cultured in the fresh-water tank died but none were found dead in the other two tanks. It is also worthy of note that mullet in the fresh-water tank were not docile or tame and were injured easily when measurements were taken. It seems that for mullet monoculture, fresh water was by no means as good as brackish water or as sea water. As to the best range of salinity for mullet farming, it still needs further study.

Under the conditions for cultured in brackish water, mullet, weighing 113.42 g on average, would grow up to 286.96 g within 120 days, with the highest mean daily weight gain of 2.32 g during the first 40 days.

Results from different experiments prove that the cultured mullet (Figs. 11.19–11.24) grow much faster than the wild ones, as shown in Fig. 11.9.

Fig. 11.19. Harvesting cultured mullet in a typical polyculture pond of medium size in Taiwan.

Fig. 11.20. Harvesting cultured mullet in a big fishpond in Taiwan.

Fig. 11.21. The mullet being caught.

Fig. 11.22. Mullet harvesting machine at a fish farm, Maagan-Michael, Israel.

Fig. 11.23. One-year-old cultured mullet measuring 434 g in body weight and 28.6 mm standard length.

Fig. 11.24. Fourteen-month-old cultured mullet.

It is certain that in the near future when the methods of rearing wild fingerlings can be improved as well as establishing techniques for artificial propagation of fingerlings, millions of fingerlings will thus be sustained. It is suggested that then after being cultured for a while, the fingerlings be stocked into the boundless ocean where they will depend on natural feed and grow up without cultivation. It will be possible to harvest the adult mullet in course of their migration, if only their life history, migration and its mechanisms can be elucidated. As soon as such an economic idea comes true, it will be a pioneer approach to utilize the extra broad open system for aquaculture.

Considering the extravagantly high price of the dried mullet roe, it is profitable to culture mullet until their ovaries are at the secondary or tertiary yolk stage, which is most suitable for the dried roe. It takes at least two years, usually more than three years, for mullet to grow and develop up to these stages. The ovaries of pond-reared mullet can become quite mature with the GSI of 6 or even higher (Fig. 11.25). Generally there are more females than males in mullet culture, a phenomenon quite different from that found in the wild population. The mechanism involved is not yet known.

In general, mullet grow much better in a eutrophic pond with low transparency than in an oligotrophic pond with high transparency. However, serious oxygen depletion which may have a lethal effect on mullet frequently occurs in the eutrophic pond. Therefore a special effort to control the water quality should be made.

As to the dissolved oxygen requirement, Chiba (1974) reported that oxygen concentration should be maintained above 50% saturation. No feed specifically prepared for mullet has ever been reported. Chiba & Lee (1974) tried to produce some experimental purified compounds as feed for young mullet; Furuichi & Yone (1974) initiated the study on the utilization of carbohydrate of young mullet. But they did not reach a final conclusions. Linder *et al.* (1974) tried some commercially prepared feed for mullet. Biologists at Tungkang Marine Laboratory got better results by following a recipe of 20% fish meal, 30% rice bran, 40% yeast, 10% wheat flour and 0.5 g/kg vitamin E. Customarily, mullet farmers let mullet feed on what is left over in the pond by other mix-cultured fishes or occasionally they feed mullet only with a little bit of rice bran. Under such a condition, the pond-reared mullet still grow faster than the wild ones. It is quite possible that growth will be more increased and time for maturation can be shortened, if the artificial feed fully satisfying the nutritional requirements of mullet can be developed.

In addition to stocking mullet fingerlings into the ocean, transplantation should also be applied to mullet if ecological surveys showed this is advisable. Among a variety of mullet species, interspecific crosses should be tested with the attempt to get the better hybrids for aquaculture.

As a matter of fact, only a few studies of mullet culture have been done, no doubt, advanced researches are needed in every respect. For instance, it

Fig. 11.25. Three-year-old cultured mullet measuring 1019.7 g in body weight, 38.65 cm standard length, of 61.2 g ovary weight and 6.0 GSI.

has been found that cultured mullet grow faster, maturing one or one and a half months earlier than the local wild ones, probably because the water temperature falls at an earlier date in ponds than in the ocean when winter comes. This phenomenon is good not only for the dried roe dealers to sell at higher prices and earlier but also for the artificially propagated fingerlings to survive easily in warmer weather as compared with later, and thus implies that cultivation of adult mullet is of profit. Yet, no details on the physiological or the ecological aspect, the control of water quality, the development of proper feed, or the analysis of capital and costs, have been obtained. Besides this, research into diseases and parasites is also an important topic with respect to intensive culture. A study on this aspect indicated that 0.2 ppm Bromex is good for getting rid of the mullet parasite *Ergasilus sieboldi* (Lahav & Sarig, 1967).

In conclusion, the problems to be solved in the field of mullet culture are specified for readers' reference:

(1) The physiology and the ecology of mullet fingerlings should be made clear so as to raise the survival rate and make the best use of the millions of wild fingerlings available during a short season. Low-temperature and overcrowding treatments should be applied to inhibit mullet growth for the sake of supplying fingerlings when they are out of season.

(2) Some economic types of prepared feed, useful for different stages of mullet growth should be developed for the benefit of intensive culture.

(3) The optimal salinity range for various growth stages and other factors concerning water quality should be investigated.

(4) A selective breeding programme for mullet is required.

(5) The optimal stocking density should be worked out for various combinations in polyculture or for the purpose of reaching the expected size within a set period.

(6) The optimal conditions for rearing adult mullet should be established with a view to providing spawners for the artificial propagation.

(7) Study on the diseases and parasites of mullet should be intensified in order to prepare against such problems.

(8) A more effective system for exchanging research information should be instituted as soon as possible, and more bilateral or multilateral research programmes or workshops should be held for solving mutual problems.

It is my sincere hope to invite more scientists to study the economically important fish – mullet – expecting that the researches will lead to the early establishment of perfected cultivation techniques and the mass production of mullet.

References

Anderson, W. W. (1958). Larval development, growth and spawning of striped mullet (*Mugil cephalus*) along the South Atlantic coast of the United States. *US Fish. Wildl. Serv. Fish. Bull.* **58**, 501–19.

Arnold, E. L., Jr. & Thompson, J. R. (1958). Offshore spawning of the striped mullet, *Mugil cephalus*, in the Gulf of Mexico. Copeia **2**, 130–2.

Bardach, J. E., Ryther, J. H. & McLarney, W. O. (1972). *Aquaculture: The farming and husbandry of freshwater and marine organisms.* Wiley-Interscience, New York.

Chiba, K. (1974). A study on the influence of oxygen concentration on the growth of young grey mullet. In *Proceeding of the Spring Meeting of Japanese Society of Scientific Fisheries, 1979*, (in Japanese).

Chiba, K. & Lee, D. L. (1974). Study of the purified experimental diet for young grey mullet. In *Proceeding of the Spring Meeting of Japanese Society of Scientific Fisheries*, 1974 (in Japanese).

Eckstein, B. (1974). Possible reasons for the infertility of grey mullets confined to fresh water. In *Proceedings of the IBP/PM International Symposium on the Grey Mullets and their Culture, Haifa*, 2–8 June 1974.

Egusa, S. (1950). Some notes on the feeding habit of the young of *Mugil cephalus* L. *Bull. Jap. Soc. Sci. Fish.* **5** (11), 715–20 (in Japanese, English abstract).

Furuichi, M. & Yone, Y. (1974). Study on the utilization of carbohydrate of young grey mullet. In *Proceeding of the Autumn Meeting of Japanese Society of Scientific Fisheries*, 1974 (in Japanese).

Lahav, M. & Sarig, S. (1967). *Ergasilus sieboldi*, Nordman infestation of grey mullet in Israel fish ponds. *Bamidgeh* **19** (4), 69–80.

Liao, I.-C. (1974). Experiments on induced breeding of the grey mullet in Taiwan from 1963 to 1973. In *Proceedings of the IBP/PM International Symposium on the Grey Mullets and their Culture, Haifa*, 2–8 June 1974.

Liao, I.-C., Lee, D. L., Lim, M. Y. & Lo, M. C. (1971). Preliminary report on induced breeding of pond-reared mullet, *Mugil cephalus* L. Reports of Fish Culture Research Supported by Rockefeller Foundation, *JCRR Fish. Ser.* **11**, 30–5 (in Chinese, English abstract).

Linder, D. R., Strawn, K. & Luebke, R. W. (1974). The culture of striped mullet (*Mugil cephalus* L.) in ponds receiving heated effluent from a power plant. In *Proceedings of the IBP/PM International Symposium on the Grey Mullets and their Culture, Haifa*, 2–8 June 1974.

Nakamura, N. (1949). Some notes on the pond-culture of mullets. *Bull. Jap. Soc. Sci. Fish.* **14** (4), 211–5 (in Japanese, English abstract).

Nakamura, N. (1967). Grey Mullet. In *Particulars on Fish Culture, Suisangaku series 23*, (ed. N. Y. Kawamoto) Kousesha Kousekaku, Tokyo, pp. 212–28 (in Japanese).

Oshima, Y. (1957). Cultivation of Mullet. In *Suisankouza Series 3: Inland Aquaculture*, pp. 431–48, Dai Nippon Suisan Kai. Tokyo, (in Japanese).

Pruginin, Y. & Kitai, H. (1957). Cultivating mullets as a secondary fish in the carp ponds. *Bamidgeh* **9** (4), 70–5.

Pruginin, Y., Shilo, S. & Mires, D. (1974). Grey mullet – a component in polyculture, in the fishponds of Israel. In *Proceedings of the IBP/PM International Symposium on the Grey Mullets and their Culture, Haifa*, 2–8 June 1974.

Wallace, J. H. (1974). Aspects of the biology of *Mugil cephalus* in a hypersaline estuarine lake on the East coast of South Africa. In *Proceedings of the IBP/PM International Symposium on the Grey Mullets and their Culture, Haifa*, 2–8 June 1974.

Yashouv, A. (1968). Mixed fish culture – an ecological approach to increase pond productivity. In *Proceedings of the World Symposium on Warm-water Pond Fish Culture*, FAO Fisheries Reports no. 44, vol. 4, 258–73.

12. The Mugilidae in polyculture in fresh and brackish water fishponds

S. SARIG

Introduction

The Mugilidae are spread over wide areas of the world. Several important properties make this family, and especially the species *Mugil cephalus* and *Liza ramada* (*M. capito*), suitable for polyculture in fishponds and in stocking plankton-rich lakes and reservoirs. These properties are mainly: the feeding habits, the easy adaptation to fresh waters, abundance of fry in large schools in the estuaries and easy long-distance transportation.

The natural food of juvenile *M. cephalus* consists mainly of diatoms and occasionally copepods mixed with sand grains, which often fill up to 60% of its stomach. Adult fish prefer copepods before spawning migration and filamentous algae during the spawning season (Tung, 1971; Ling, 1967). From observation it is evident that detritus is an important component in the food of Mugilidae, therefore, they integrate well with carp and tilapia in breeding ponds in a free niche in this biotope. In the fertilized ponds where supplementary food is added, mass developments of algae frequently appear. These algae are not fully utilized by carp or tilapia, but serve as excellent food for Mugilidae, thus increasing the total yield and return from the ponds without additional input.

Not much information has been published to date on the role of *Mugil* in commercial aquaculture. The main sources of Mugilidae for the markets are the seas, lagoons, reservoirs or brackish water ponds where the *Mugil* fingerlings live naturally, having entered with the water current during high tides or floods. In some countries Mugilids are stocked by man in monoculture and raised without additional feeding or fertilization. In most such cases the yields are very low and sometimes do not justify the investment in pond construction, waterways and labour.

The value of *Mugil* on the market is a decisive factor in the development of more intensive methods of cultivation. Uchida (1973) claims that this is the main reason why this industry has not been developed until now in USA despite the good conditions for *Mugil* cultivation, especially in Florida and Texas.

Experiments which were carried out in Egypt (Zarka & Fahmy, 1968), where *Mugil* is in great demand, gave yields of 192–350 kg/ha of *M. cephalus*

Table 12.1. *Fish cultivation in a pond and in reservoirs in Taiwan and the part of* M. cephalus *in the total yield (calculated per hectare)*

	Fish pond				Reservoirs			
	Stocking		Yield		Stocking		Yield	
Fish species	Number	%	kg	%	Number	%	Number	%
Common carp	3900	29	910	12.5	80	15	28	6.6
Chinese carps	3610	26	3705	50.8	400	73	366	86.3
Sea perch (predator)	70	1	170	2.3	12	1	13	3.1
M. cephalus	6030	44	2502	34.4	58	11	17	4.0
Total	13610	100	7287	100.0	550	100	424	100.0

in fertilized or manured ponds as against 131 kg/ha without fertilization. This means an increase in yields of 36–77% as a result of fertilization. In addition to the yield of *Mugil*, 162–219 kg/ha more of other fish species, which entered freely into the ponds, were harvested. Chen (1972) mentioned also that fertilization of ponds in Taiwan and Hong Kong where *Mugil* were raised increased the yields of the ponds by 96%. General information on growth of *Mugil* in ponds with average yields of 300–350 kg/ha is available also, from France (Hickling, 1962), India, Vietnam, Indonesia, Hawaii and Pakistan (Ling, 1972). According to Dang (1972) the catch of *M. cephalus* and *Liza* sp. from ponds and lagoons in Vietnam reached 5000 tons per year. Tanaka (1972) includes *M. cephalus* in the eighteen most important fish species raised in ponds in Japan.

Very good results were obtained in Taiwan and Hong Kong, in polyculture with Chinese carp, in ponds in which fish were fed intensively with rice bran, peanut meal or soyabean cakes. Pig manure and night soil, which is also added to these ponds, are partly eaten by *Mugil* (Ling, 1967).

In Hong Kong, *Mugil* are stocked in ponds in densities of 10000 fish per hectare together with 1500 Chinese carp. They are fed with rice bran and peanut cakes in addition to repeated applications of animal manure at 2–5 day intervals. The larger fish are continuously fished out during the season. At the end of a growth season of 300 days the gain of these ponds is 3500 kg/ha with a food conversion rate (FCR) of 1:1.6.

Tang (1970) gives interesting data on *M. cephalus* in polyculture in both irrigation reservoirs of 3.5 to 22 ha, and in a 6 ha fishpond in Taiwan. The fishpond received 10000 kg manure and 1700 kg rice bran calculated as dry matter per annum per hectare. The manure contained 12% chicken droppings, 11% pig manure and 77% night soil. The data in Table 12.1 present the total

Table 12.2. *Pondfish species marketed in Israel in 1968–74*

Year	Carp		*Tilapia*		*Mugil*		Silver carp		Total
	Ton	%	Ton	%	Ton	%	Ton	%	Ton
1968	8572	85.0	931	9.3	597	5.7	—	—	10100
1969	8362	81.9	1178	11.5	680	6.6	—	—	10220
1970	10140	85.1	1051	8.8	675	5.7	48	0.4	11914
1971	10705	86.4	1079	8.6	513	4.1	236	1.9	12533
1972	11105	84.7	1211	9.2	547	4.2	243	1.9	13106
1973	11444	83.3	1140	8.1	490	3.5	713	5.1	13787
1974	9322	75.7	1557	12.6	409	3.3	1026	8.4	12314

yield and the percentage of *Mugil* in the yield. Especially noteworthy is the part of *M. cephalus* in the fertilized and artificially fed pond. Here the rich production of algae, thanks to fertilization, gives excellent growth conditions for the fish. In Taiwan very good results have been achieved with artificial propagation of *M. cephalus*. One group of fingerlings, obtained by artificial spawning in a hatchery, was introduced in a brackish water pond (Liao, Lu, Huang & Lin, 1972). Unfortunately not enough data on the density of stocking, fertilization, feeding, etc. were given. But the data on the individual growth rates of these fingerlings were very encouraging. Fingerlings which were introduced 54 days after hatching reached 82 g after 97 days and 217 g after 146 days.

Raising of *Mugil* (*M. cephalus* and *Liza ramada*) in polyculture in Israel

Mugil are raised in Israel only in polyculture. The number of *Mugil* in the total stocking per hectare depends mainly on the number of fry caught during the season in the estuaries. The abundance of fry in the Mediterranean, along the Israeli coast, has decreased dramatically over the last few years. Table 12.2 gives data for each fish species in the total of marketed fish from ponds in Israel (Sarig, 1970, 1973, 1975). The proportion of *Mugil* varied from 3 to 6%. One has to remember that the price that the producer obtains for *Mugil* on the market is three times higher than that for silver carp, twice as high as for carp and 50% higher than for tilapia.

Fish culture in Israel is highly intensive. The ponds are fertilized twice a month with phosphates and nitrogen during the season from March to November (250 days of good growth). Once a week poultry droppings are added and the fish are fed daily. In the beginning of the season they are fed with sorghum then later with pellets containing 25% protein and 3000 Kcal.

Table 12.3. *The various sources of* Mugil *marketed in Israel in 1965–74*

Year	Fish ponds		Reservoirs		Lake Kinneret		Mediterranean		Total
	Ton	%	Ton	%	Ton	%	Ton	%	Ton
1965	253	62	—	—	56	13	109	25	418
1966	429	58	7	1	178	24	121	17	735
1967	346	51	6	1	221	32	106	16	677
1968	597	58	21	2	242	24	130	16	990
1969	680	64	38	4	183	17	147	15	1048
1970	675	66	32	3	182	18	134	13	1023
1971	513	53	21	2	307	31	127	14	968
1972	547	61	17	2	197	23	128	14	889
1973	490	56	17	2	201	23	157	19	865
1974	409	50	16	2	195	24	195	24	815

The calculation of feeds added to the pond are based only on feeding the carp and partly the tilapia, leaving silver carp and *Mugil* to depend on natural food like plankton, on detritus and whatever is left from the additional food given to the carp.

Fish culture in Israel is one of the branches of agriculture of the 'mixed agriculture system'. The size of a fish farm in such a system is from 25 to 200 ha. Ponds vary in size from 3 to 10 ha (Fig. 12.1). A growth season in Israel, during which carp and tilapia reach marketable size, is approximately 120 days. The ponds are usually drained and fish harvested twice during the season, from March to November. The fish which have not reached marketable size during the first season, especially the *Mugil*, are restocked for a second season. The temperature of the water during the season varies from 18 to 32 °C. The pH is 8.0–9.0. The diurnal oxygen regime is very variable due to the heavy water blooms and fluctuates in summer between 5–10% saturation at sunrise to 150% in the afternoons. Consequently, fish kill due to oxygen depletion can occur frequently if aeration is not applied. The different sources of *Mugil* for the local market are given in Table 12.3. Almost 60% of the *Mugil* marketed comes from fishponds. The marketable size of *Mugil* is 400–1000 g. The price of the fish on the market increases with the weight of the fish. The demand is very high and all the sources put together do not cover the market potential.

Catching of fry

The main factor which governs the production of *Mugil* in fishponds and in Lake Kinneret in Israel is success in catching fry along the sea shore in the short season of their appearance. Bograd (1961) described six species of Mugilidae abundant along the Mediterranean coast of Israel: *M. cephalus*,

Fig. 12.1. A typical view of a concentration of fish ponds in one of the regions of Israel.

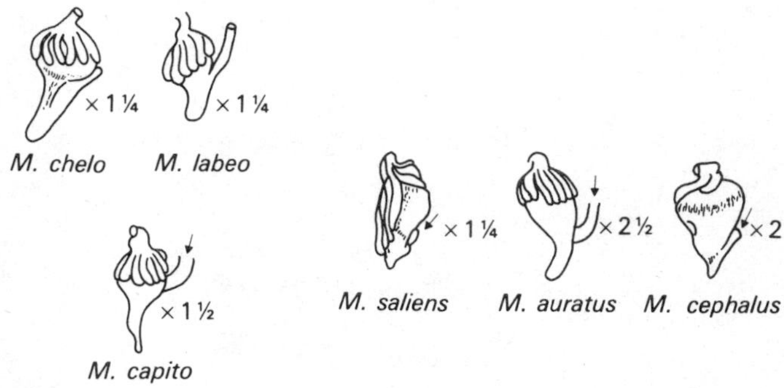

Fig. 12.2. A key for the identification of Mugilidae by their pyloric caeca (after Bogard, 1955).

M. auratus, *M. saliens*, *M. chelo*, *M. labrosus* and *M. capito* (*Liza ramada*). Only *M. cephalus* and *L. ramada* adapt easily to fresh water and have a satisfactory individual growth rate. Therefore, only these two species are cultivated in ponds and in the lake. Their distribution along the streams has also verified that *M. cephalus* and *L. ramada* are, naturally, the most tolerant to fresh water. *M. saliens* prefer brackish water and *M. labrosus* and *M. auratus* are the least tolerant to fresh water. Bograd (1955) has published a key which enables quite easy identification of the various *Mugil* species by their pyloric caeca while the fingerlings are still of the size of 18–25 mm (Fig. 12.2).

Great progress in *Mugil* culture in fish ponds began after Perlmutter, Bograd & Pruginin (1957) defined the correct seasonal appearance of each of the species along the sea shore, enabling the fish farmers to collect the preferred two species during the season when they appear in exclusive schools in sizes of 18–22 mm, i.e., *M. cephalus* from September to December and *L. ramada* from January to March. This saved trouble, wasted time and the disappointment resulting from catching and stocking the ponds with undesirable species.

Because of the pollution of the Mediterranean the catch of *Mugil* fry is restricted now to four rivers only: Naaman, Daliah, Yarkon and Rubin (Fig. 12.3). The catching of the fry is done by means of special hand nets or short seine nets depending on the local conditions (Fig. 12.4: see also Figs. in Ben-Yami's Chapter 10). The fry are immediately transferred in buckets of river water to transportation tanks especially constructed for this purpose and equipped with oxygen containers and aeration pipes. Fry are carried in these tanks for distances up to 300 km and more, the tanks contain river water mixed with fresh water. Arriving at the farm the water in the tank is slowly mixed with pond water in order to adjust salinity and temperature. After half an hour the fry, together with the water, are emptied into the nursery pond.

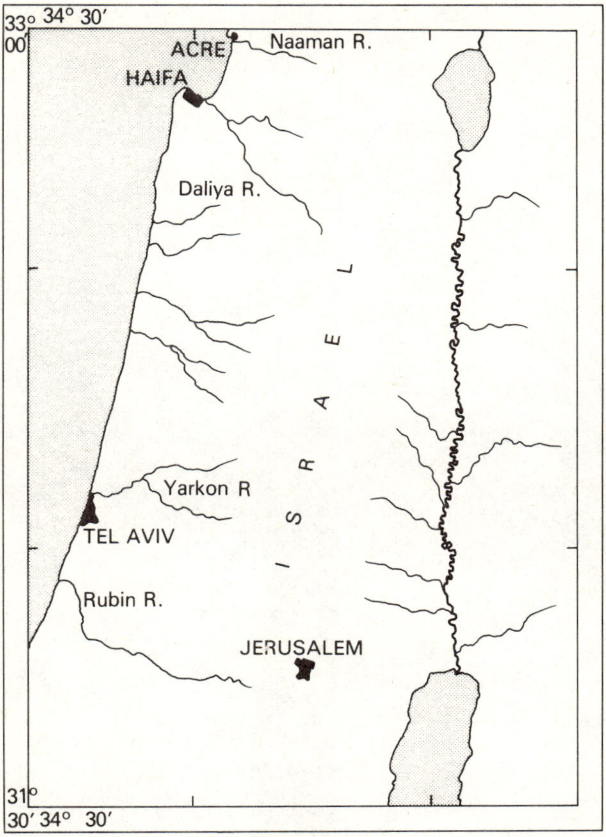

Fig. 12.3. Map of Israel showing main catching centres of *Mugil* fry.

The salinity of the water in the estuaries varies from 6000 to 10000 mg/Cl^-/l. With the first mixing in the tank the salinity decreases to 3000–5000 mg/Cl^-/l. Pond water contains, according to the region, from 100 to 1000 mg/Cl^-/l. Through these three steps of transfer, the *Mugil* fry are adjusted to their new habitat in a very short time.

Several experiments have been carried out in Israel to find the influence of temperature and salinity changes on the survival of *Mugil* fry. Mires (1970) found that immediate transfer of *Liza ramada* from 25 °C to 11–12 °C killed all the fry in 5 min, and an immediate transfer from 25 °C to 30 °C killed 39% of the fry. When the water was gradually cooled and the temperature reduced during a period of 45 min from 25° to 12 °C the mortality was only 5%. Regarding salinity, the immediate transfer from 60 to 15800 mg/Cl^-/l or vice versa, did not harm the fry. *M. cephalus* behaved differently. Mires, Shak & Shilo (1974) found that an immediate transfer from about 20600 mg/Cl^-/l

Fig. 12.4. Collecting *Mugil* fry in the mouth of a river with a special hand-net.

Table 12.4. *Results of first nursing of* Mugil *fry brought from estuaries to fish ponds (initial size: 18–25 mm; 0.25 g)*

Pond no.	Size (ha)	Days of growth	Number of fry in 1000		Survival (%)	Size at harvest (g)	Increment (kg/ha)			Daily increment (g/ha)			% of *Mugil* in total yield
			Stocked	Harvested			*Mugil*	Carp	Total	*Mugil*	Carp	Total	
1	3.5	211	265	101	38	38	1050	2312	3362	4976	10957	15933	31
2	3.6	207	110	63	57	35	604	1695	2294	2918	8188	11106	26
3	2.7	234	78	39	49	67	956	1570	2526	4085	6709	10794	38
4	3.0	164	50	35	70	40	383	1120	1503	2335	6829	9164	25
5	2.7	138	125	57	46	7	139	765	904	1007	5543	6550	15
6	3.0	145	80	56	70	7–33	414	788	2855	2579	2579	5434	53
7	2.7	257	98	51	52	25	436	696	1132	1697	2708	4405	39
8	3.0	190	74	51	69	15	250	476	726	1316	2505	3821	34

Table 12.5. *Daily individual increment of* Mugil *fry during the effective period of growth (excluding the cold winter days of November 15–March 15)*

Pond no.	Density/ha at harvest	Effective days of growth	Individual increment (g/ha)	Daily increment (g/ha)
1	14444	174	0.39	5633
2	11667	134	0.30	3500
3	28857	151	0.25	7214
4	17520	147	0.24	4205
5	17000	70	0.21	3570
6	18889	197	0.13	2456
7	21111	108	0.06	1267

(sea water) to 690 mg/Cl^-/l caused a 97% mortality, and a complete kill if the decreases in salinity and temperature were simultaneous.

Nursing of Mugil *fry*

The most common method used in Israel is to nurse fry of 0.2 g, the size at which they are caught in the estuaries (first stage), to 15–50 g when suitable for restocking (second stage of nursing), in polyculture. This method has two disadvantages: the fry, which become weakened during the catch and transport are under stress and suffer high mortalities and it is impossible to evaluate exactly the number of fry in the pond.

Another method, which was adopted recently, is to acclimatize the fry brought from the estuaries, for 5–15 days in concrete or fibreglass tanks. During this time the fry are disinfected with formalin, cleaned of protozoan parasites (Lahav & Sarig, 1972), and fed a little with fish or soya meal, or ground fish feed pellets. Only the stronger fry survive for stocking the ponds, their number is known more exactly, and the loss at harvest is normally from 5 to 10%.

The young *Mugil* fry are nursed in ponds together with carp of 200–300 g size. The main purpose of the introduction of carp into these ponds is because while it feeds on the pond bottom it mixes the mud and thus prevents the development of filamentous algae which interfere with the development of plankton and decrease the growth of the fry. At the same time the carp increase the yield of the pond by utilizing different natural food from the *Mugil* fry. The remains of small particles from the supplementary food given to the carp as well as its excreta serve as additional food for the *Mugil*. Table 12.4 gives data on the results of this technique in different regions of the country. From these we can see that the survival of fry, until they reach the size of 7–60 g, is 38–70%. The difference in size at harvest is influenced mainly by the natural fertility of the pond, more than by the length of the growth period or density of stocking.

The effective individual daily increment of these fry must be calculated by deducting from the actual period of holding the fry in the pond, the cold winter days when there is no growth at all. This calculated actual individual daily increment can be found in Table 12.5. The total yield of the ponds varied from 726 to 3362 kg/ha. The proportion of *Mugil* fry in the total yield was 15–53%. Even under local conditions of fish culture with yields of 2.5–4.0 tons/ha, these results from nursing ponds are encouraging, especially as in such a way the *Mugil* fry are ready for restocking after not more than 6 months.

Growth of Mugil *in polyculture*

After the *Mugil* fingerlings reach a size of 10–30 g they are stocked in low densities, as supplementary fish in polyculture, together with carp, tilapia and sometimes also with silver carp. At this stage the management of the pond varies, depending on the local factors which influence the planning of the farm. The following are the main steps of cultivation.

Nursing Mugil *fingerlings*

This phase includes the growth of nursed fingerlings (7–50 g) to their third phase of growth, and possibly to marketable size, in fattening ponds (Table 12.6). Table 12.6 shows that after 70–160 days, when the 'age' of the fish (from the day when they were brought to the farm) was between 9 to 18 months, the *Mugil* reached the size of 130–400 g. This is still not the preferred, most profitable, marketable size, as the best prices obtained on the local market are for fish over 500 g. The fish are therefore, in most cases, restocked for another period of growth. The yields at this phase of growth also gave encouraging results. In general, the yields varied between 1186 to 3539 kg/ha with an FCR of 2.0–3.7.

The proportion of *Mugil* in the total yield (except pond no. 6) was 10–26%. Thanks to the increment of the *Mugil* which were raised without additional feed, the FCR calculated for all the fish in the pond went down by 0.2–1.0 units. This reduced by 10–25% the part played by food in the cost of production of the fish. If we add to this the higher price which the producer gets for *Mugil*, the value of *Mugil* in the total planning of polyculture becomes much higher than the additional 10–26% in the total yield.

Raising of Mugil *to marketable size*

At this third phase, nursed fingerlings of 100–300 g size are restocked in the polyculture ponds and raised to the desired marketable size. Table 12.7 shows

Table 12.6. *Nursing of* Mugil *fingerlings from 7–50 g to 100–400 g size*

Pond no.	1	2	3	4	5	6
Size of pond (ha)	3.0	2.7	4.4	4.4	3.4	3.6
Days of growth	70	75	98	136	59	97
Number of *Mugil* at harvest/ha	3730	560	4000	920	1170	1690
Initial size (g)	35	50	15	33	7	15
Size at harvest (g)	373	360	100	400	130	165
Yield of *Mugil* (kg/ha)	369	163	305	333	144	230
Daily increment (g/ha)	5270	2173	3112	2448	2440	2370
Individual increment (day/g)	1.41	3.88	0.78	2.66	2.09	1.44
Yield of other species:						
carp (kg/ha)	1060	463	2265	2487	924	1755
Tilapia (kg/ha)	—	—	—	719	118	—
Total yield (kg/ha)	1429	626	2570	3539	1186	1985
Proportion of *Mugil* in total yield (%)	26	35	12	9	12	12
Food conversion rate without *Mugil*	3.7	6.1	3.0	2.4	2.4	2.1
Food conversion rate with *Mugil*	2.7	4.5	2.6	2.2	2.1	1.7

the results of this phase in various ponds from different regions of the country. From this we see that regarding techniques of cultivation, especially as to stocking rates and length of the growth period, the management is very flexible and adjusted to the local zootechnical conditions. The proportion of the *Mugil* in the total yield, except pond no. 3, was from 13 to 35%. In pond no. 3, despite the high individual daily growth, the proportion of *Mugil* in the total was low because of the low rate of stocking.

The individual daily increment in all the other ponds was 0.9–3.2 g. With densities of 400–1200 *Mugil*/ha no significant influence of the rate of stocking on the growth was observed. Combining all the phases of growth from nursing of fry to marketing we find that from the size of 0.2 g they reach marketable size in approximately 450 days, by feeding only on the natural food and detritus found in the ponds and without any additional feed.

Another method, which is mainly applicable in ponds which are not drained during the year, recommends raising of the *Mugil* in two phases only. In the first phase the fry are nursed to the size of 40–100 g. Later they are stocked in ponds in polyculture and kept there continuously until they reach marketable size (Table 12.8). In this case the proportion of *Mugil* in the total yield was even higher than in the method described above. From Tables 12.7 and 12.8 we can see that with increased individual weight of the *Mugil* their daily individual increment also increases. Where there is no need to empty and refill the ponds during the summer, this method has the advantage that

Table 12.7. *Raising of* Mugil *to marketable size in polyculture*

Pond no.	1	2	3	4	5	6	7	8	9	10
Size of pond (ha)	3.4	3.4	3.5	4.2	3.5	2.8	4.5	3.5	4.4	3.4
Days of growth	249	341	174	308	150	196	166	163	164	109
Number of *Mugil* at harvest/ha	2350	1230	160	850	800	570	400	480	580	1190
Initial size (g)	90	80	300	100	100	250	250	250	250	330
Size at harvest (g)	440	400	1500	800	580	730	720	700	750	600
Yield of *Mugil* (kg/ha)	900	363	167	572	372	268	178	189	286	390
Daily increment (g/ha)	3615	1065	1044	2750	2480	1367	1072	1160	1554	3578
Individual increment (days/g)	1.54	0.87	6.52	3.24	3.10	2.40	2.68	2.42	2.68	3.01
Yield of other species: carp (kg/ha)	2687	1272	1930	1335	1257	950	785	860	710	524
tilapia (kg/ha)	980	417	712	356	181	260	425	280	290	179
silver carp (kg/ha)	891	842	15	—	—	—	—	—	—	20
Total yield (kg/ha)	3058	2894	2824	2263	1810	1478	1388	1329	1286	1113
Part of *Mugil* in total yield (%)	29	13	6	25	21	18	13	14	22	35
Food conversion rate without *Mugil*	2.4	3.5	2.4	3.4	2.2	1.1	0.6	0.6	0.8	3.6
Food conversion rate with *Mugil*	2.0	3.1	1.9	2.6	1.7	0.9	0.5	0.5	0.6	2.3

Table 12.8. *Raising of* Mugil *fingerlings to marketable size during one season from nursing to marketing*

Pond no.	1	2	3	4	5	6	7
Days of growth	287	262	244	325	304	270	277
Number of *Mugil* at harvest/ha	540	580	560	780	670	600	670
Initial size (g)	45	45	45	120	120	120	120
Size at harvest (g)	780	570	620	1000	1100	1000	1100
Yield of *Mugil* (kg/ha)	400	300	320	700	700	520	650
Daily increment (g/ha)	1360	1160	1330	2100	2160	1950	2380
Individual increment (day/g)	2.5	2.0	2.4	2.7	3.2	3.3	3.5
Yield of other species: carp (kg/ha)	1920	2630	2130	1590	1400	1540	1550
Total yield (kg/ha)	2320	2930	2450	2290	2100	2060	2200
Part of *Mugil* in total yield (%)	17	10	13	31	33	25	30

it saves the losses and stresses caused by handling during transfer of fingerlings.

The influence of the initial weight of the stocked *Mugil* is obvious from Table 8. If fingerlings had reached 45 g in their nursing stage, then they reached a marketable size of 570–780 g at this stage. But, if they achieved the size of 100–120 g at the nursing stage then they were marketed at 1000–1100 g, a size which gives much higher profits to the producer. In ponds showing higher individual weights the yields of *Mugil* per hectare were also higher by 200–380 kg/ha compared with the ponds which were stocked with smaller fingerlings.

The influence of the greater individual weight of the *Mugil* on the growth of the carp in the ponds is evident and reached 47%, but, the total yield of both species together decreased only by 4%. Bearing in mind that the *Mugil* were raised without cost of food and obtain a high price on the market, the final result is certainly in favour of this polyculture.

It is of interest to compare the individual growth of *Mugil* in ponds with data on the growth of *M. cephalus* in nature. Thompson (1963) gives information on weight of 4-year-old *M. cephalus* in various parts of the world, which are from 229 g in the Black Sea up to 1217 g in the Mediterranean. Rafail (1968) measured the total length of *Mugil* found along the Mediterranean coast of Egypt. If we correct these data using the length/weight relations of Rangaswamy (1973), we find that *M. cephalus* on the Mediterranean coast of Egypt reached 300 g at the age of 2 years and 900 g at the age of 3 years. *L. ramada* reached 150 and 300 g at the same ages. Rangaswamy (1973) had found in Lakes Pulicat, Adyar and Ennor in India, 2-year-old *M. cephalus* of 300 g, 3-year-old fish of 600 g, and 4-year-old fish of 600 g.

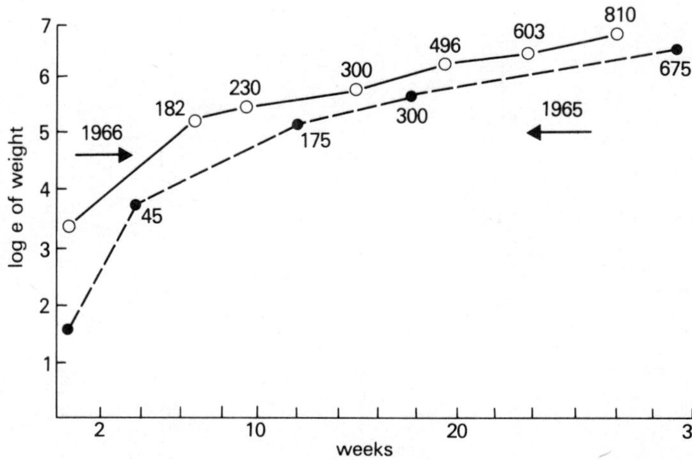

Fig. 12.5. Growth curve of *Mugil cephalus* in experimental ponds in 1965–66. Figures are the average weight at sample weighings (after Yashouv, 1972).

Yashouv (1972) gives two typical growth curves of *M. cephalus* in polyculture in 1965 and 1966 (Fig. 12.5). In these conditions, the fish reached 675 and 810 g in 30–32 weeks. If we add to this period 28 weeks more, for the time needed for nursing the fingerlings in their first year of life to reach the stocking size of 15–35 g, the initial weight of the fingerlings in these trials, we will find that just a little more than one year was needed in this case in order to reach the weight of 3-year-old *M. cephalus* in nature. Also the various results from the previous tables show that *Mugil*, in polyculture, needed only half the time taken to reach the same weight in nature.

The influence of the addition of *Mugil* to the total yield in polycultures in Israel is clear from the data of Yashouv (1972) (Fig. 12.6). These figures are results of various experiments carried out in 1966 in 0.1 ha ponds of the Fish and Aquaculture Research Station DOR (Table 12.9).

Comparison of growth of M. cephalus *and* L. ramada *in polyculture*

It is a well known fact, that in nature *M. cephalus* grows much faster than *L. ramada*. According to Rafail (1968), based on the formula of Rangaswamy, *L. ramada* at the age of 2 years reached 150 g and in 3 years 300 g, while *M. cephalus* reached at the same time achieved 300 and 900 g respectively. Table 12.10 gives comparative data on the growth of these two species grown in the same pond, together with carp and tilapia. The experiments were carried out in the Dor Fish and Aquaculture Research Station. All the experiments continued for 197 days of growth. The initial weight of *M. cephalus* was 30 g and that of *L. ramada* 80 g.

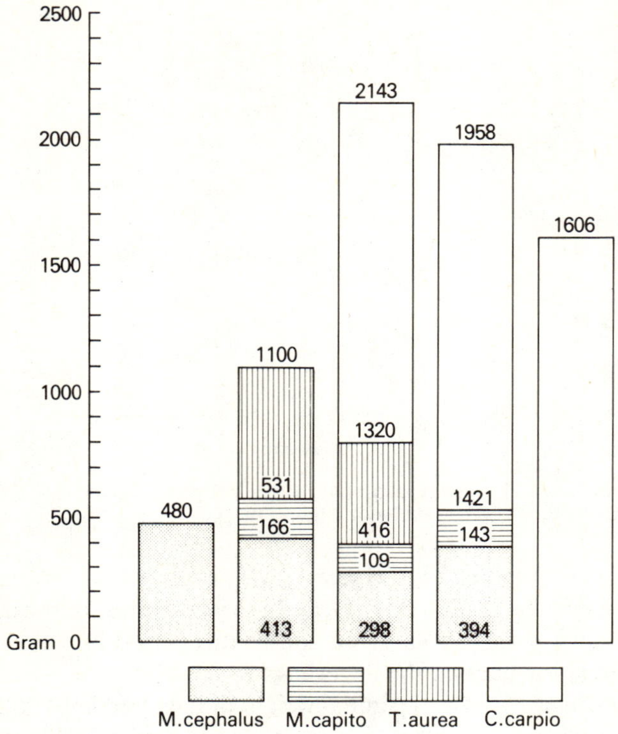

Fig. 12.6. Daily growth increment (g/0.1 ha) of the different fish species in various stocking combinations (after Yashouv, 1972).

The figures in Table 10 are averaged from ten ponds. These results show that the growth of *L. ramada* was only 59% of that of *M. cephalus*. This in spite of the fact that the initial weight of *L. ramada* was almost three times more than *M. cephalus* and the density of stocking was the same.

The growth of Mugil in Lake Kinneret

Lake Kinneret (Sea of Galilee or L. Tiberias) is a very plankton-rich lake. It was therefore decided to try to introduce Mugilidae into the lake. The size of the lake is approximately 16000 ha. The stocking of the lake was carried out in the same way as for fishponds, with newly caught *Mugil* fry of 0.2 g as described above. The number of fry stocked each year was counted, but nothing is known of the rate of survival of this fry in the lake. The first stocking started in 1958. In the first five years only some 190000 fry were stocked. It is estimated that 90% of the stocked *Mugil* were *Liza ramada*. The first catch of 1600 kg was caught in 1962, individuals being of 1–2 kg. These results

Table 12.9. *Daily increment of* Mugil, *carp and tilapia in various culture combinations in 1966*

Yield of *Mugil* in total (%)	Total daily increment (g/ha)	Daily increment of *Mugil* (g/ha)	Daily increment per species (g/ha)				Type of culture
			L. ramada	*M. cephalus*	tilapia	carp	
—	16060	—	—	—	—	16060	Monoculture
—	4860	4860	—	4860	—	—	Monoculture
52.16	11100	5790	1660	4130	5310	—	Polyculture
28.20	19020	5370	1430	3840	—	13650	Polyculture
18.99	21430	4070	1090	2980	4160	13200	Polyculture

Table 12.10. *Growth of* M. cephalus *in polyculture with* L. ramada, T. aurea *and* C. carpio (*all data are calculated to 0.1 ha*) (*after Yashouv, 1972*)

	L. ramada	*M. cephalus*
Number at stocking	1110 (580–1720)	1118 (640–1850)
Individual increment (day/g)	1.33 (1.07–1.70)	3.25 (2.65–3.68)
Daily increment (g/ha)	1572 (820–2930)	3586 (2160–4890)
Size at harvest (g)	342 (296–390)	669 (553–754)

Table 12.11. *Stocking and catch of* Mugil *in Lake Kinneret in 1958–74*

Year	Stocking (× 10000)	Catch	
		Tons	Number
1958	1	—	—
1959	5	—	—
1960	4	—	—
1961	9	—	—
1962	36	1.6	1070
1963	95	5.9	4000
1964	2	22.1	14730
1965	103	55.5	37000
1966	147	177.7	118450
1967	562	220.9	147260
1968	151	242.4	161600
1969	182	183.1	122070
1970	362	182.0	121330
1971	63	306.0	204000
1972	1	197.0	131330
1973	3	201.0	134000
1974	—	195.0	130000
Total	1726	1990.2	1326840

Percentage of catch from the total stocking number, 7.7%; percentage of *Mugil* in the total catch of the lake, 9.9%.

encouraged continuous and expanded stocking in the coming years. Table 12.11 shows the stocking and catch of *Mugil* during 1958–74 in Lake Kinneret. The number of fish caught is an estimate based on an average individual weight of 1.5 kg. Samples which were taken each year from catches have shown that the average weight of *L. ramada* was 1.0–1.2 kg and of *M. cephalus* 2–4 kg. Consequently, the total catch during these years, in numbers, was about 8% of the fish introduced into the lake.

During the fifteen years since the first catch of *Mugil* from the lake in 1962, the proportion of *Mugil* in the total catch of fish from the lake became almost 10%. The importance of this 10% in the income of the fishermen is better evaluated if we consider that the price that the fishermen get on the markets for these fish is double that of the other fish species from the lake, and ten times more than the bleak (which is canned) and which makes up almost 60% of the total catch of the lake.

References

Bograd, L. (1955). Identification of the grey mullets on the Israel Mediterranean coast. *Comm. Fish. Bull.* **5**, 22–4 (in Hebrew.)

Bograd, L. (1961). Occurrence of *Mugil* in the rivers of Israel. *Bull. Res. Counc. Israel, B*, **9** (4), 169–90.

Chen, T. P. (1972). Fertilization and feeding in coastal fish farms in Taiwan. In *Coastal Aquaculture in the Indo-Pacific Region*, pp. 2–25. Fishing News (Books) Ltd., London.

Dang, L. V. (1972). Coastal Aquaculture in Vietnam. In *Coastal Aquaculture in the Indo-Pacific Region*, pp. 105–108. Fishing News (Books) Ltd, London.

Hickling, C. F. (1962). *Fish culture.* Faber and Faber, London.

Lahav, M. & Sarig, S. (1972). Control of unicellular parasites using formalin. *Bamidgeh* **24** (1), 3–11.

Liao, I. C., Lu, Y. J., Huang, T. L. & Lin, M. C. (1972). Experiments on induced breeding of the grey mullet, *Mugil cephalus* L. In *Coastal Aquaculture in the Indo-Pacific Region*, pp. 213–243. Fishing News (Books) Ltd, London.

Ling, S. W. (1967). Feeds and feeding of warm-water fishes in ponds in Asia and the Far East. *FAO Fish. Rep.* **44** (3), 291–309.

Ling, S. W. (1972). A review of the status and problems of coastal aquaculture in the Indo-Pacific region. In *Coastal Aquaculture in the Indo-pacific Region*, pp. 2–24. Fishing News (Books) Ltd, London.

Mires, D. (1970). Preliminary observations on the effects of salinity and temperature of water changes on *Mugil capito* fry. *Bamidgeh* **22** (1), 19–24.

Mires, D., Shak, Y. & Shilo, S. (1974). Further observations on the effect of salinity and temperature changes on *Mugil capito* and *Mugil cephalus* fry. *Bamidgeh* **26** (4), 104–9.

Perlmutter, A. L., Bograd, L. & Pruginin, J. (1957). Use of estuarine and sea fish of the family Mugilidae (grey mullets) for pond culture in Israel. *Proc. Gen. Fish. Counc. Mediterr.* **4**, 289–304.

Rafail, S. Z. (1968). Investigations of mullet fisheries by beach seine on the UAR Mediterranean coast. *Stud. Rev. Gen. Fish. Counc. Mediterr.* **35**, 1–19.

Rangaswamy, C. R. (1973). Studies on the age and growth and food habits of the grey mullet, *Mugil cephalus* L. of the Lake Pulicat. *J. Inl. Fish. Soc. India* **5**, 9–22.

Sarig, S. (1970). Fisheries and fish culture in Israel in 1969. *Bamidgeh* **22** (3), 51–68.

Sarig, S. (1973). Fisheries and fish culture in Israel in 1972. *Bamidgeh* **25** (2), 35–55.

Sarig, S. (1975). Fisheries and fish culture in Israel in 1974. *Bamidgeh* **27** (2), 25–39.

Shak, Y. & Shilo, S. (1974). Further observations on the effect of salinity and temperature changes on *Mugil capito* and *Mugil cephalus* fry. *Bamidgeh* **26** (4), 104–9.

Tanaka, J. (1972). Cultured fin-fish of temperate waters in Japan. In *Coastal Aquaculture in the Indo-Pacific Region*, pp. 150–151. Fishing News (Books) Ltd, London.

Tang, Y. A. (1970). Evaluation of balance between fishes and available fish foods in multispecies fish culture ponds in Taiwan. *Trans. Am. Fish. Soc.* **99** (4), 708–18.

Thompson, J. M. (1963). *Synopsis of biological data on the grey mullet, Mugil cephalus L. 1758.* CSIRO, Fisheries, Synopsis no. 1.

Tung, I. H. (1971). On the food habit of grey mullet, *Mugil cephalus* L. in Taiwan. *Rep. Inst. Fish. Biol., Taipei, Taiwan* **2** (4), 67.

Uchida, R. N. (1973). Review of recent progress in coastal aquaculture in the United States. In *Coastal Aquaculture in the Indo-Pacific Region*, pp. 84–105. Fishing News (Books) Ltd, London.

Yashouv, A. (1972). Efficiency of mullet growth in fishponds. *Bamidgeh* **24** (1), 12–25.

Zarka, S. & Fahmy, F. K. (1968). Experiments in the culture of the grey mullet, *Mugil cephalus* in brackish-water ponds in the UAR. *FAO Fish. Rep.* **44** (5), 255–66.

13. Parasites and diseases of mullets (Mugilidae)*

I. PAPERNA & R. M. OVERSTREET

Introduction

The following treatment of parasites, diseases and conditions affecting mullet hopefully serves several functions. It acquaints someone involved in rearing mullets with problems he can face and topics he should investigate. We cannot go into extensive illustrative detail on every species or group, but do provide a listing of most parasites reported or known from mullet and some pertinent general information on them. Because of these enumerations, the paper should also act as a review for anyone interested in mullet parasites or the use of such parasites as indicators about a mullet's diet and migratory behaviour. Unfortunately, limited space prohibits us from presenting all the references used. The paper also deals with the public health aspects of eating or selling mullet, whether the product is to be raw, cooked, salted or smoked.

As interest in culturing mullet in fresh water, brackish water, and sea water increases, the importance of diseases and parasites in confined fish becomes apparent. History has already shown that diseases and parasites play a significantly detrimental role in aquaculture. In fact, disease-outbreaks have formed one of the major barriers against expanding the industry.

Comprehensive data on mullet culture and associated parasites depend mainly on the wide-scale polyculture of mullet with carp and tilapia in fresh water and brackish water ponds in Israel (Paperna, 1964*a*; Lahav & Sarig, 1967; Sarig, 1971; Paperna & Lahav, 1971; Paperna, 1975). Information concerning parasites and diseases of mullets is minimal from other countries where these fishes are cultured. Moreover, knowledge defining the parasitofauna of mullets from their natural environment is still limited to a few regions, in spite of the world-wide distribution of mullets. The parasites have been studied in some detail from the northern Black Sea and the Sea of Azov (Reshetnikova, 1955), from Israel (Paperna, 1964*a*, 1975; Paperna & Lahav, 1971, 1975), and from the south-eastern United States (Rawson, 1973*b*;

* Compilation of data from the eastern Mediterranean and studies in the Red Sea were supported by Grant No. 94 of the United States–Israel Binational Science Foundation. Other aspects were conducted in cooperation with the US Department of Commerce, NOAA, National Marine Fisheries Service, under PL 88-309, Project No. 2-174-R and NOAA, Office of Sea Grant, under Grant No. 04-3-158-53. The US Government is authorized to produce and distribute reprints for governmental purposes notwithstanding any copyright notation that may appear hereon.

Skinner, 1974; and this paper). Our knowledge about parasites of mullet from these and other regions is derived mainly from taxonomic accounts of specialized groups; too many papers on mullet parasites have been published to cite them all.

Even though spawning can be induced on an experimental basis, mullet-farming still depends on fry or fingerlings stocked from natural waters. This fact, combined with the behaviour of mullets, promotes disease. Mullet in Israel illustrate this point; they crowd into coastal streams and drainage channels that serve as water supplies and drains for fishponds. Their presence provides a constant dissemination of disease-causing agents between cultured and natural populations of mullet; passage can occur in the water or through intermediate hosts. Infection may also be introduced with individuals being stocked. Once in the system, many agents can easily perpetuate themselves. Thus, we need comprehensive studies on parasites from natural stocks.

Members of the Mugilidae occur throughout the world, and *Mugil cephalus* ranges throughout both the Indo-Pacific and Atlantic regions. Therefore, we are not surprised to find among mugilid parasites a considerable number of identical or similar species from widely-separated areas. The similarities imply that results from a parasitological investigation in one region can be applied to others.

Recent revisions have altered the names of several mullets cited in this paper. To avoid confusion, however, we list the names as they appear in most current literature.

Bacteria

Epizootics with infectious bacteria can result in mass mortalities of fishes. Typically such a disease persists for long periods, spreads from area to area and affects selected species of hosts. Few die-offs, however, have been studied comprehensively. An account of two of these, involving *Mugil cephalus*, follows. The first epizootic occurred in November 1968 and restricted itself to *M. cephalus* and a menhaden in Galveston Bay, Texas, USA (Lewis, Grumbles, McConnell & Flowers, 1970). Considerable mucoid material covered the gill filaments, and purulent material presided in abdominal cavities of infected fish. Blood from hearts of individuals of the two dying species revealed an abundance of bipolar, thick, pleomorphic rods with rounded ends characterized as a *Pasteurella*-like bacterium by the authors. With diluted pure cultures of the organism, the authors caused death to all aquatically-exposed mullet and inoculated white mice. In addition the organism was compared with and found different from *Pasteurella piscicida* Janssen & Surgalla, which caused mortalities in the white perch, *Roccus americanus*, in Chesapeake Bay, USA during the summer of 1963 (Snieszko, Bullock, Hollis & Boone, 1964).

In the second case, *Mugil cephalus* was among several species of fishes affected in acute to chronic outbreaks from Florida to Alabama, USA during August and September of 1972. Plumb and coworkers (Plumb, Schachte, Gaines, Peltier & Carroll, 1974) isolated a species of *Streptococcus* from mullet and other moribund fishes and suggested that it had caused the deaths. Cook & Lofton (1975) supported this assumption by obtaining the same bacterium from a moribund menhaden and experimentally producing mortalities in all five tested fishes including *M. cephalus*. Dying fish displayed erratic swimming behaviour. They had haemorrhagic lesions externally, and internally their peritoneal cavities and intestines were filled with a bloody fluid.

Other bacteria infect and cause lesions in mullet. *Achromobacter aquamarinus* and two strains of *A. superficialis* were isolated from *Mugil cunnesius* with fin-rot, whereas *Escherichia intermedia*, *Achromobacter* sp., and four strains of *Aeromonas salmonicida* were isolated from fish without symptoms of disease (Almeida, Da Silva & Freitas, 1968). We also have observed mullet with severe lesions which could be either directly or secondarily involved with bacterial organisms.

Donald H. Lewis of Texas A & M University takes a special interest in bacterial infections in mullet. While sampling seasonally for bacteria in mullet from near Galveston, Texas, Lewis (personal communication) found a high incidence of mullet acting as carriers of *Vibrio anguillarum* during early spring. When under the stress of being transported to his laboratory, the fish developed signs typical of a vibrio infection. They had petechial haemorrhages in and at the base of the fins, around the vent, and in the oral cavity. Pure cultures were isolated from the liver of afflicted fish. The presence of *V. anguillarum* was apparently seasonal because Lewis did not encounter it from mullet in midsummer and early autumn.

Periodically, Lewis also observed a condition in mullet characterized by loss of scales and large lesions on the abdominal wall. Most often, *Pseudomonas* sp. was isolated from the lesions, the liver, and frequently from the blood. The bacterium closely resembled one he encountered from epizootics in the Atlantic croaker, *Micropogon undulatus*.

Other diseases in Texas could also be associated with or caused by bacteria. Lewis encountered several large-scale mullet-kills during the spring of 1974. Moribund fishes exhibited signs typical of whirling disease in salmonids. As in the case of whirling disease, a myxosporidan was present, a finding also reported by Iversen, Chitty & VanMeter (1971) from dead mullet in Florida. In this instance, however, a fastidious anaerobic Gram-variable rod bacterium was recovered from the meninges of afflicted fish. A graduate student working under Lewis is studying this condition.

When maintained in ponds or other crowded conditions, mullet will be more susceptible to bacterial diseases than when in their natural habitat. Consequently, the importance of maintaining good environmental conditions cannot be overstressed. Bullock & co-workers (Bullock, Conroy & Snieszko, 1971) discussed several bacteria and variables in infectious bacterial diseases, some of which can probably affect mullets. Bullock (1971) and Lewis (1973) provided methods on how to culture and identify suspected pathogens.

Secondary bacterial infections often follow infestations with ectoparasites. Although not documented from mullets, bacteria infected Atlantic salmon, which had a copepod and was called 'white spot disease' (White, 1940), and yellowtail in Japanese cage culture, which had monogeneans and caligids (Kubota & Takakuwa, 1963).

In addition to harming mullet, bacteria in or on mullet can also cause disease in man when he touches or eats the fish. This possibility is well reviewed by Janssen (1970), who also points out the need for further research in this important area of public health. A few of the bacteria recovered from fishes are *Aeromonas hydrophilia*, *Mycobacterium marinum*, *M. fortuitum*, *Vibrio parahaemolyticus*, *Erysipelothrix rhusiopathiae* and *Leptospira icterohaemorhagiae*. All of these cause disease in man. Mullet could also act as vectors for cholera, salmonellosis, shigellosis, and presumably several other diseases in addition to those caused by the above-mentioned bacteria.

One of the commonly encountered and better-studied marine organisms, *Vibrio parahaemolyticus*, causes bacterial food-poisoning in temperature climates throughout the world. Symptoms include abdominal pain, nausea, diarrhoea and mild fever. Both the organisms and enterotoxins responsible for the disease can be destroyed by heating at 60 °C for five minutes. Most bacterial diseases that could be acquired from mullet also can be prevented by cooking the fish.

Fungi

Sarig (1971) reported the water-mould, *Sparolegnia* sp., infecting mullets stocked in fresh-water ponds. The infection was implicated in mortalties of some of the infected fish. This non-specific fungus attacks most fresh-water fishes, particularly those with an injured integument, such as happens following handling. Infection also occurs mainly in ponds rich with organic matter, such as heavily-fertilized ponds. It can be eliminated with a low concentration of salt or by a variety of other treatments.

Other fungal diseases, presumably secondary infections, occur in *Mugil cephalus* and probably other mullets in low saline waters, but have not been adequately studied.

Protozoa

Protozoans infect internally or infest externally all mullets and occasionally cause diseases which may be followed by death. Forming a large, complex

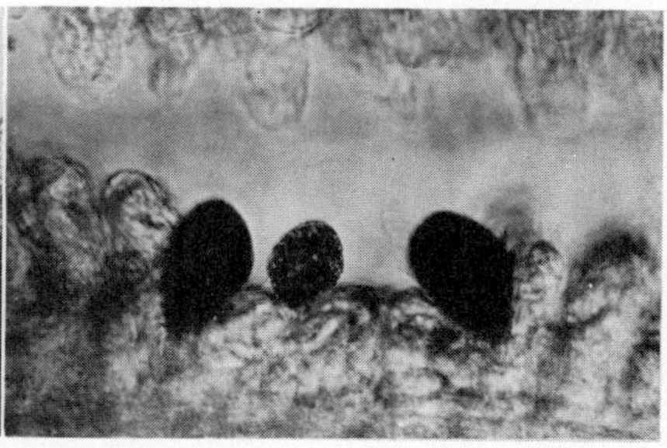

Fig. 13.1. A moderate infestation of *Amyloodinium ocellatum* on the gills of *Micropogon undulatus* from the Mississippi, ×25, unfixed and unstained.

Fig. 13.2. *Amyloodinium ocellatum*, enlarge sector of Fig. 13.1, ×450.

group, protozoans have in common only that they are single-celled. Even that character does not hold true for members of the Myxosporida, which develop from a multicellular origin. In fact, new taxonomic schemes separate most of the Protozoa into separate phyla. The first group considered will be the flagellates.

Flagellates

Flagellates and amoebae comprise the Sarcomastigophora, but only flagellates have been observed in or on mullets. The parasitic dinoflagellate *Amyloodinium ocellatum* (Brown) (Figs. 13.1, 13.2), or a closely related species, occasionally infests *Mugil cephalus* in Mississippi and can easily kill most fishes stocked in ponds. It has also been reported from *M. cephalus* in the Aquarium of the Zoological Society of London (Brown, 1934). Adrian Lawler (unpublished data) has been able to kill experimentally that host and most other tested estuarine fishes from Mississippi. *Oodinium cyprintum* [sic] was reported from *M. cephalus* in Georgia (Rawson, 1973*b*), but since *O. cyprinodontum* Lawler appears highly specific for cyprinidontid fishes, the record should be confirmed.

Amyloodinium ocellatum and related species become harmful to confined fish because of their reproductive capabilities. After reaching a certain size, being stressed, or following other conditions, the relatively large, up to 150 μm in diameter, usually chalky-coloured trophont stage, attached by filamentous rhizoids to gills or integument, will drop from its host, withdraw its rhizoids and stomopode-tube and encyst within a cellulose covering. Several divisions take place, ultimately forming anywhere between 128 and 2048 flagellated swarmers (depending on the species) ready to infest a host. If it were not for this *Gymnodinium*-like swarmer, the identity of the trophont on the gill

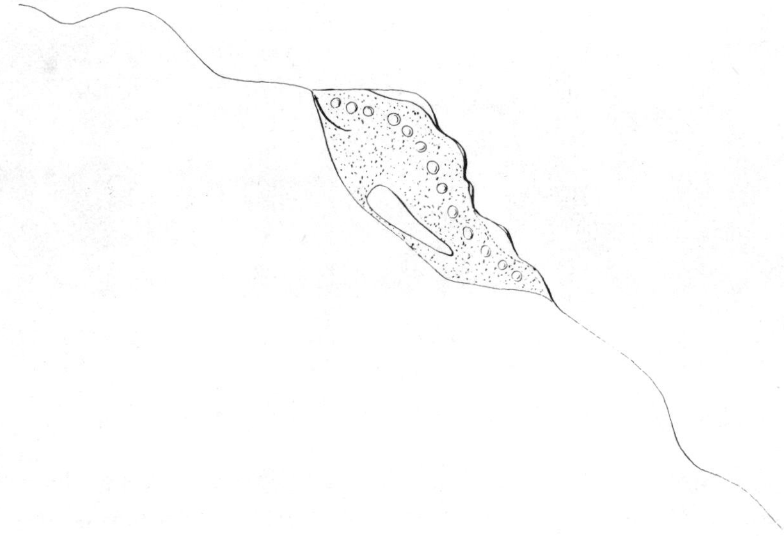

Fig. 13.3. *Bodomonas* sp. from *Valamugil seheli*, Red Sea.

filament would never have been realized. A large dinocaryonic nucleus can be used to distinguish the trophont as a dinoflagellate.

Hosts in their natural environment seldom possess heavy infestations, but the swarmers, or dinospores, easily infest confined fish in large numbers because of the easy access to the host. After time and proper conditions, gills of confined fish become covered, and the fish ultimately dies. In this aquarium-trade, the disease, caused by several species, can apparently be treated with a number of remedies. Preliminary studies at the Gulf Coast Research Laboratory (Lawler, in preparation) show fish heavily infested with *A. ocellatum* are seldom saved by most tested chemicals. Fresh-water baths did effectively rid attached parasites.

While on the host, some dinoflagellate species damage tissue during feeding, whereas others possess chlorophyll, manufacturing their own food, and cause minimal host-reponse (Lom & Lawler, 1973).

Other flagellate groups infecting mullet include inconspicuous species. A species of the biflagellated *Bodomonas* Davis (also known as *Cryptobia* sp.) infests the gills of *Valimugil seheli* and *Mugil cephalus* from the Red Sea (Fig. 13.3), but could only be demonstrated on the fish after hosts had been maintained for two to three months in sea-water tanks. Infestations never reached epizootic proportions. Fantham (1919) reported 'Trichomonas' from the large intestine of *Mugil capito* in South Africa.

Possibly more serious for a pond situation would be *Hexamita* sp. Members of this genus have three pairs of flagella anteriorly and one pair posteriorly, thus differing considerably from most other flagellates. *Hexamita* sp. occurred in the intestinal epithelial tissue of a few examined juvenile *Mugil subviridus* from the Gulf of Elat. Pathological aspects of infections have not been studied. Other species, such as *H. salmonis* (Moore), can be pathogenic to young hatchery-reared trout and salmon. When environmental factors and physiological conditions of the oyster-host permit, *H. nelsoni* Schlict & Mackin harms its host. The species from mullet could presumably become pathogenic given proper cues or possibly different mugilid hosts.

Neave, in 1906 (according to Wenyen, 1926, quoted by C. Dale Becker, personal communication) reported an unidentified trypanosome from *Mugil* sp. in the Sudan. In Mississippi, *Trypanosoma mugicola* infects the blood of *Mugil cephalus* (see Overstreet, 1974; Becker & Overstreet, in preparation). In Alabama, infections are apparently more common (Williams, Gains & Phelps, 1974). Overstreet (1974) and Sawyer, Lawler & Overstreet (1975) suggested that the leech *Myzobdella lugubris* Leidy, or some other species such as *Calliobdella vivida* (Verrill), transmits the infection. This trypanosome seems to have no effect on the host, and few do adversely affect their hosts. The related *Trypanoplasma salmositica* (Katz) also apparently causes no harm in naturally-infected Pacific salmon, but emerges as a serious pathogen in hatchery-reared rainbow trout.

Ciliates

Ciliates, members of Ciliophora, often infest stressed fish in large numbers. One cosmopolitan group consists of species of *Trichodina* Ehrenberg. Lom (1962) studied species from the Black Sea, including those from the gills of mullet. *Trichodina domerguei* f. *partidisci* Lom infested *Mugil saliens*, *T. lepsii* Lom infested *M. auratus*, and *T. puytoraci* Lom infested both and, in addition, *M. cephalus*. In the Adriatic Sea, *T. jadranica* (Raabe) glides about on and attaches to *M. cephalus* and in South Africa *T. mugilis* Fantham occurs on *M. capito*.

Unidentified trichodinids occur in numerous areas. They infest *M. capito* and *M. saliens* in coastal streams of the Mediterranean coast of Israel (Lahav, unpublished data) and in the Bardawil lagoon and *Crenimugil crenilabis* in the Gulf of Elat in the Red Sea (Fig. 13.4). Skinner (1974) considered a single unidentified trichodinid on *M. cephalus* from Florida most similar to *T. halli* Radnos and Nigrelli. What appear to be two species live on *Mugil cephalus* and *M. curema* from at least Louisiana to Florida. In addition to the gills, they also occur commonly on the integument. One or both parasites infested *M. cephalus* being cultured experimentally in ponds at the Rockefeller Refuge, Grand Chenier, Louisiana, (Overstreet, unpublished data) and in cages in

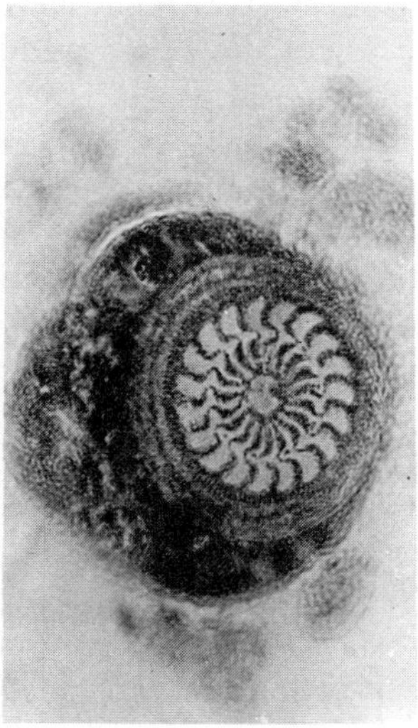

Fig. 13.4. *Trichodina* sp. from *Mugil capito*, Bardawil lagoon, North Sinai. Silver impregnated (Klein procedure), ×450.

mixed-culture with Florida pompano at Dauphin Island, Alabama (Williams *et al.*, 1974; Overstreet, unpublished data). Often another peritrich, *Scyphidia* sp., also occurred on the gills and integument. The prevalence of all species occasionally reached high levels (Figs. 13.5–13.8).

Unless influenced by environmental conditions, stress of the host or other parasites, the numbers of individuals usually remain low. Given the proper conditions, the commensals-turned-pathogens cover most of the gill tissue. These ciliates have a characteristic ring of interlocking denticles which help support the disc, which, with a sharp cutting border, can pinch off pieces of host epithelial tissue.

Mortalities usually result when young fish are heavily infested. But even though susceptible to trichodinid infestation, mullet kept with pompano in Alabama had fewer parasites than their heavily-infested cage-mates. Deaths have been reported for hatchery-reared flatfish (Purdom & Howard, 1971; Pearse, 1972). Meyer (1970) mentioned that fry of fresh-water channel catfish died from *Trichodina* sp. within an hour after hatching and that heavily-infested golden shiners usually failed to feed. He also reported disease occurring in crowded fish held in ponds over the winter. In some die-offs, monogeneans and copepods have accompanied the ciliates and appeared to promote multiplication of the protozoan.

Additional peritrichs and presumably other ciliates occur on mullet. Seen

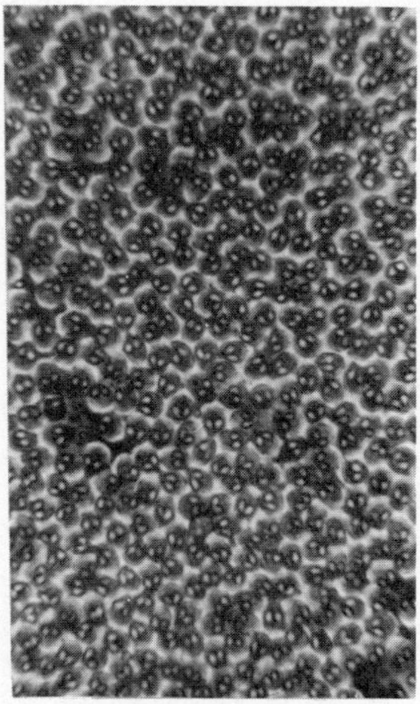

Fig. 13.5. *Myxobolus cf. parvus*, spores from a cyst in the wall of the intestine of *Crenimugil crenilabis*, Red Sea, unfixed, unstained, phase illumination, ×450.

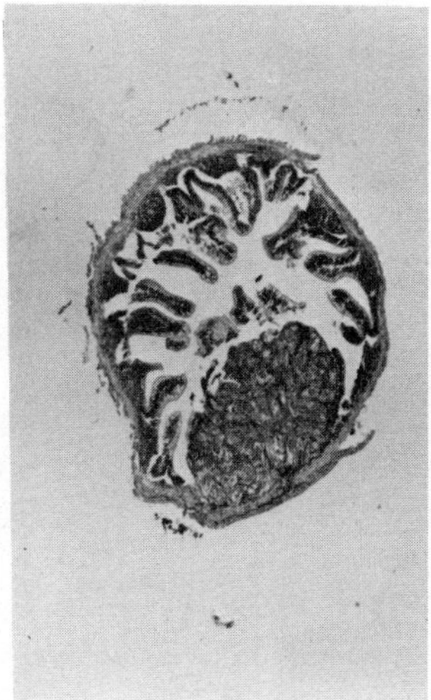

Fig. 13.6. *Myxobolus cf. parvus*, cyst in the intestinal wall of *Crenimugil crenilabis* extending into the lumen of the intestine, 7 μm section of paraffin embedded tissue, stained in Harris' haematoxylin-eosin, ×25.

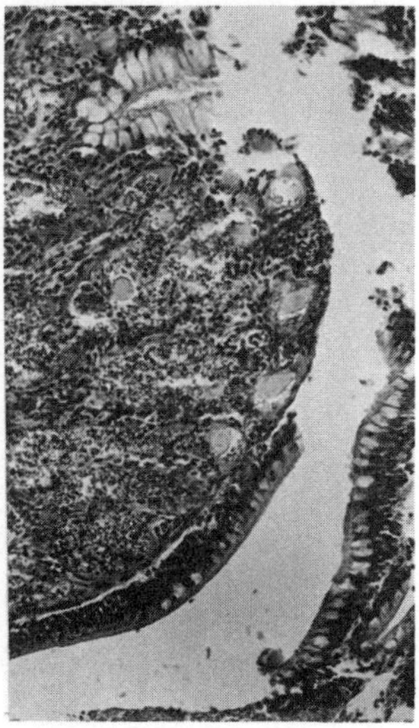

Fig. 13.7. Enlarged view of *Myxobolus cf. parvus* cyst of Fig. 6.

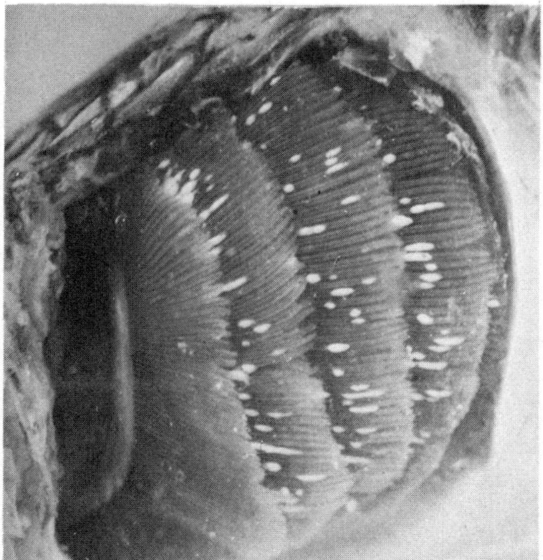

Fig. 13.8. Gill of *Mugil cephalus* infected with cysts of *Myxobolus parvus* (Sarig, 1971).

on several fishes in Mississippi, but not mullets, are the stalked species of the genus *Epistylis* Ehrenberg. Hutton (1964) reported *Epistylis* sp. from *M. cephalus* in Florida.

Ichthyophthirius multifiliis Fouquet, commonly referred to as 'ich', rates as one of the most devasting parasitic diseases of cultured fish. It infects most fish confined in freshwater ponds or aquaria, including mullet. Sarig (1971) reported it in *M. cephalus* and *M. capito* stocked in conjunction with carp, but unlike the carp, infections were mild and brief and had little effect on the hosts. *Cryptocaryon irritans* Brown, its marine counterpart, also infects *M. cephalus*. Wilkie & Gordin (1969) found the fish susceptible to the disease in sea water warmer than 15 °C. Fantham (1919) recorded mortality of maintained *M. capito* in South Africa caused by this or a related species.

Loss of fish resulting from both species of these notorious ciliates could easily discourage the hobbiest or unsuspecting fish farmer unaware of the processes taking place. In both species, a ciliate penetrates the epidermis, embeds and grows near the basement membrane of the epidermis of the body, fins and gills. After appropriate growth, it leaves the host, encysts within a gelatinous capsule and undergoes division. The resulting swarming tomites then infect a new or the same host. At least in carp infected with *I. multifiliis*, secondary changes can develop in the liver, spleen and kidneys of the host, as well as the ciliate causing direct pathological alteration of the epidermis and dermis (Hines & Spira, 1974). Hines & Spira (1973) found that some immunity persisted from previous infections.

Sporozoans

Rather than being external or among the epidermal cells, members of the next group inhabit the cells. Currently known as Apicomplexa, it includes most taxa previously regarded as Sporozoa. In this complex are the haemogregarines. They, or at least the described stages, infect the red blood cells. *Haemogregarina mugili* Carini is specific to mullets. Even though some minor differences apparently exist between the specimens reported from disjunct localities, it has been cited from *M. brasiliensis* in Brazil (Carini, 1932), from *Mugil* sp. in Guadalcanal in the South Pacific (Laird, 1958*a*), and from *M. cephalus* in Florida (Saunders, 1964). We (Becker & Overstreet, in preparation) have also encountered it in *M. cephalus* in Mississippi. *Haemogregarina bigemina* Laveran and Mesnil, a much less specific parasite for which merozoite formation occurs in white blood cells, was found by Saunders (1958) in *M. trichodon* in the Bahamas. She (1960) also recovered *Babesiosoma rubrimarensis* Saunders from *Mugil troscheli* and other fishes in the Red Sea. A life cycle of a piscine species of *Haemogregarina* has not been completed, but transmission was suggested to be conducted by leeches (Sawyer, Lawler & Overstreet, 1975; and others). In the case of the babesio-

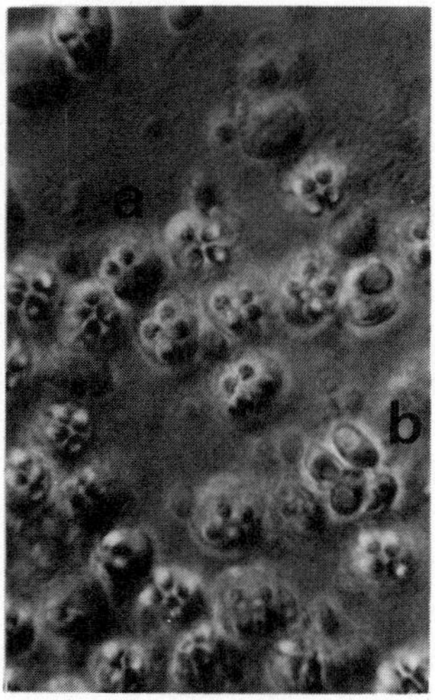

Fig. 13.9. Spores of *Kudoa* sp. (a) and *Plistophora* sp. (b) extracted from a cyst in wall of the intestine of *Crenimugil crenilabis* from the Red Sea, unfixed and unstained, phase illumination, × 1000.

somes, only a schizogonic cycle has been reported and no information exists on gametogony, on whether sporogony takes place in tissues or on whether transmission necessitates a vector.

Also in the sporozoan group are the eimeriids, some of which display destructive tendencies in various tissues of domestic birds and mammals. *Eimeria* sp. infects the grey mullet in the Caspian Sea (Mikhailov, 1958), and probably other species of these coccidians will be found prevalent around the world, once more histological studies are completed.

Microsporidans and Myxosporidans

Once considered sporozoans and then cnidosporans, the Microspora and Myxospora now constitute separate subphyla or phyla. The only microsporidan recovered to date is a species of *Pleistophora* Gurley in *Crenimugil crenilabis* from the Red Sea. These are small single-celled spores containing a single sporoplasm which extrudes through an everted polar filament into a host cell. Many of these develop within a pansporoblast, and the visible cyst comprises massive numbers of spores. Some species devastate natural populations of fish and shellfish and can do so to cultured species, given the right conditions. Pansporoblasts in *C. crenilabis* always occurred in association with the myxosporidan *Kudoa* sp. in the intestinal wall of the host (Fig. 13.9).

Myxosporidans conspicuously differ from the microsporidans and also present a much greater hazard to the health of reared mullet. From a multi-cellular origin, they possess one to several sporoplasms, one to three valves and one to six polar capsules containing extrusible filaments. They are not intracellular, and the spores' ultrastructure exhibits numerous mitochondria, structures not found in the smaller microsporidans. Spores are approximately the size of a human red blood cell and usually exist in such high numbers that a chalky-coloured cyst of spores can be seen with the unaided eye. In most of those species infecting mullet, spores have two polar filaments and two valves. Following separation of the valves' sutures, the sporoplasm exists and undergoes division, either in a bladder or among tissue. Experimental infections of most species are lacking.

To date, species of six genera have been recovered from mullets. Genera are separated on the basis of number and position of polar capsules and on location of the suture separating the valves. Identification of species, much harder to differentiate than genera, relies primarily on measurements of spores, polar capsules and polar filaments; on external sculpturing and relief on the valves; and on hosts. Unfortunately, some of these characters vary according to biological factors, as well as with method of preservation. In view of the difficulties in recognizing species, evaluations of identifications based on the literature are tenuous.

Species of *Myxidium* Butschli, *Zschokkella* Auerbach and *Ceratomyxa* Thelohan occur in the gall bladder or, more rarely, in the urinary bladder or ureters of mullets and are called coelozoic forms. Microscopic examination reveals that inside the trophic stage of most species there are two spores formed within a pansporoblastic membrane. Infections rarely harm the hosts. Known species include *Myxidium incurvatum* Thelohan in *Mugil cephalus* from North Carolina (Davis, 1917). *Zschokkella nova* Klokačewa infects both *M. cephalus* and *M. auratus* from the Black Sea basin and elsewhere in the Soviet Union. Both myxosporidans infect numerous marine and fresh-water hosts. Another species, *Z. dogieli* Pogorel'tseva, also infects mugilids in the Black Sea. We found species of *Zschokkella* and *Ceratomyxa* commonly in *Crenimugil crenilabis* from the Gulf of Elat (Figs. 13.10–13.14). *Zschokkella* sp. also occurs in *M. capito* in Bardawil Lagoon in northwestern Sinai.

Species of *Myxobolus* Butschli, *Myxosoma* Thelohan and *Kudoa* Meglitsch, on the other hand, occur in various tissues and are called histozoic forms. Cysts containing enormous numbers of spores may reach a size well over a millimeter in diameter. The specific site within the host as referenced in the literature cannot be used as an identifying character, since several species, not necessarily at the same time, infect the same tissues.

Infections with species of *Myxobolus* are the most often myxosporidans cited from mullet. Differentiation of these from those of *Myxosoma* has been based on the staining with Lugol's solution of an iodinophilous 'vacuole' in the sporoplasm in *Myxobolus*. The results of this procedure are not conclusive in every species (Walliker, 1968), including some of our material; and whether

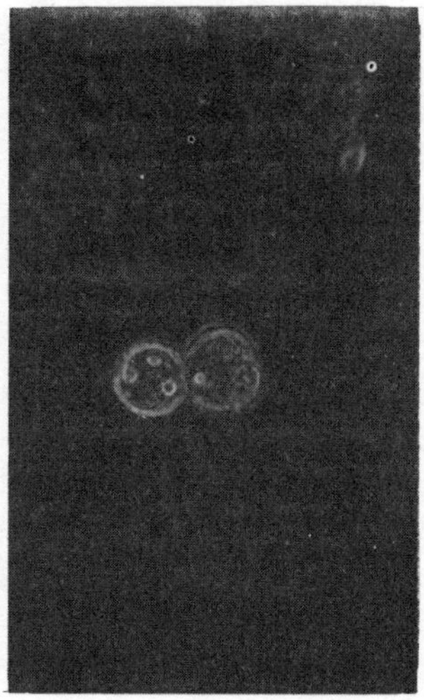

Fig. 13.10. *Zschokkella* sp. from gall bladder of *Liza carinata*, Red Sea, live specimens, dark field-phase illumination, ×450.

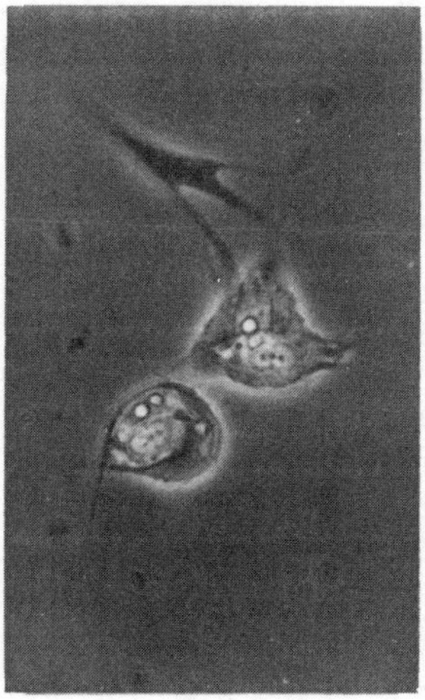

Fig. 13.11. Trophozoites of *Ceratomyxa* sp. from gall bladder of *Liza carinata*, Red Sea, live, phase illumination, ×450.

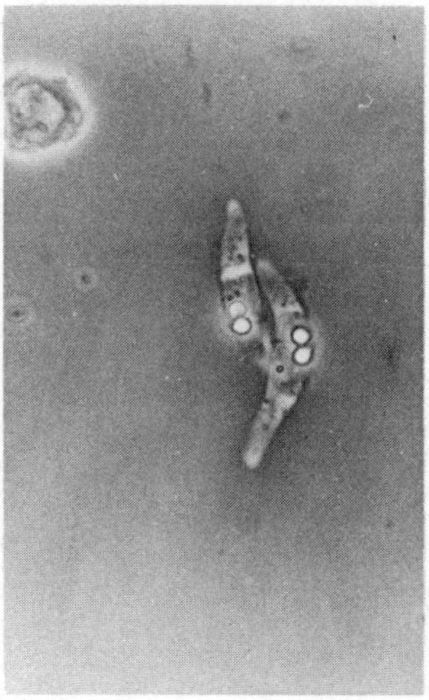

Fig. 13.12. Spores of *Ceratomyxa* sp. from gall bladder of *Liza carinata*, Red Sea, live, phase illumination, ×450.

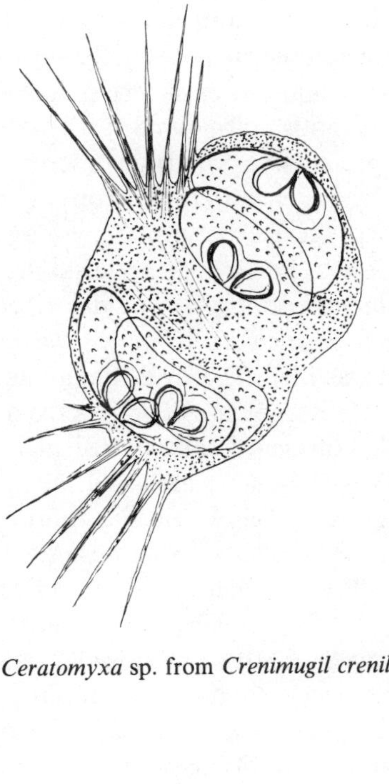

Fig. 13.13. Trophozoite of *Ceratomyxa* sp. from *Crenimugil crenilabis*, Red Sea.

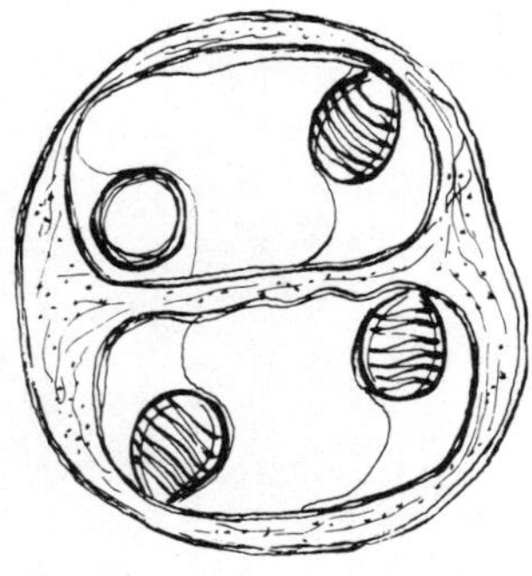

Fig. 13.14. Trophozoite of *Zschokkella* sp. from *Crenimugil crenilabis*, Red Sea.

the two genera should be synonymized has not been settled among all parasitologists. In the Sea of Japan and the Amur Basin, *Myxobolus achmerovi* Shulman occurs in the gills, fins and mesentery of *Mugil cephalus* and *M. soiuy*. In China and probably Russia, *M. cephalus* and *M. soiuy* harbour *Myxobolus parvus* Shulman in the branchial region and *M. cheni* Shulman in the musculature. *Myxobolus mulleri* Butschli and *M. exiguus* Thelohan infect mullets in the Black Sea and several areas in Russia. The former is known from *Mugil cephalus*, *M. auratus*, and *M. saliens*, and the latter is known from the above and also from *M. chelo* and *M. ramada* (= *M. capito*); both infect numerous sites in the host. According to Shulman (1966), *Myxobolus parvus* and *M. cheni* occur only in mullet, whereas the others infect a variety of both marine and fresh water fishes.

More recently, two species have both been described as *Myxobolus mugilis*. The species described by Parenzan (1967) from *Mugil chelo* in Greece is a junior homonym of the species described by Pogorel'tseva (1946) from *M. auratus* in the Black Sea. Parenzan's material needs to be renamed after it is critically differentiated from previously-described species. He additionally described *Myxobolus branchialis* from *M. chelo* in Greece.

Material from Israel was identified by Shulman as *Myxobolus parvus* (see Lahav & Sarig, 1967). Measurements of spores from mullet in the Mediterranean and the Red Sea, however, fit into two distinct size-groups (Fig. 13.5). These sizes do not vary with host or area. All the species of *Myxobolus*, as well as those of *Myxosoma* to follow, need a thorough review.

Myxosoma branchialis (Markewitsch) has been reported from the branchial region of *M. cephalus* in Russia, whereas *M. intestinalis* Narasimhamurti was described from the intestinal epithelium of *Mugil waigensis* in India; and *Myxosoma cephalus* Iversen, Chitty & VanMeter occurred in the meninges, gill arches and filaments, buccal cavity, jawbone, crop, oesophagus, intestine, liver and mesentery of *Mugil cephalus* in South Florida. Spores of a different size commonly occur in hosts in Mississippi and elsewhere along the Gulf of

Mexico coast. A distinct positive reaction with the Lugol's solution is usually lacking in the spores. More than one species of this complex infect American mullet.

Cysts of one or more species of *Kudoa* also infect *Mugil cephalus* in Mississippi and mullets, predominantly *Crenimugil crenilabis* in the Red Sea. Infections occur in the musculature and along the alimentary tract. Material is presently under study by the authors.

Host-parasite relationships of Myxosporidans

All myxosporidans are potentially harmful to their hosts. As mentioned earlier, coelozoic forms in the gall bladder rarely become pathogenic. In extemely heavy infections, the gall bladder, associated ducts and bile can be altered (Meglitsch, 1960). Such conditions, however, have not been observed in mullets, even when crowded in culture.

In contrast, histozoic forms often more seriously affect their hosts. An epizootic of *Myxobolus exiguus* led to a mass mortality of *Mugil cephalus* and *M. auratus* from Kerch and Toman Sounds in the northern Black Sea. According to Shulman (1957), the gill filaments, in some instances, became packed with cysts. When mature cysts ruptured, the adjacent tissues haemorrhaged; infected moribund fish could be recognized by a constant trickle of blood from underneath their opercula. Cysts occurred elsewhere in these fish without harm. Death apparently resulted from asphyxia or loss of blood. Approximately 500–600 kg of fish were reported as lining each kilometre of shoreline. Environmental factors in the enclosed bays of the Crimean region could have been involved in the die-off, since during the kill dissolved oxygen reached low concentrations.

In southern Florida, a heavy mortality of *Mugil cephalus* occurred in 1964. Iversen, Chitty & VanMeter (1971) thought it could have been caused by *Myxosoma cephalus*, which they described from material obtained from the brain-cavity and elsewhere. Another type of mortality occurred along the Texas coast; moribund mullet exhibited signs typical of whirling disease caused by *Myxosoma cerebralis* in salmonids. A myxosporidan was present in these fish, but as we mentioned earlier, Lewis also recovered a bacterium from the meninges.

Infections with myxosporidans differ according to site, length and geographical locality of the host, as well as yearly. Branchial and visceral infections of *Myxobolus* sp. are prevalent in *Mugil cephalus* and *M. capito*, ranging from 35 mm to 150 mm FL, in the coastal streams of the Mediterranean coast of Israel (Lahav & Sarig, 1967; Lahav, unpublished data). In ponds of the coastal region, infection occurs only sporadically, and only in those ponds receiving surface waters from the streams. In contaminated ponds, however, infections occasionally assume epizootic proportions (Sarig,

1971) (Fig. 13.8). Mortalities in ponds have never been attributed to an infection. In the hypersaline Bardawil Lagoon, visceral, but not branchial, infections prevailed during 1968 and 1969 (Lahav, personal communication), but were encountered only sporadically in later surveys of 1973 and 1974 (Paperna & Lahav, 1975). In the Gulf of Elat, infections of gills and viscera were prevalent in 50–90 mm long specimens of *Crenimugil crenilabis* and *M. subviridis*, but only sporadic in older fish (Figs. 13.6, 13.7).

Copepoda

Copepod crustacean parasites receive considerable attention because they are usually conspicuous, externally on the body surface of fins or on the gill filaments, gill rakers, mouth or associated areas. Most copepods live free, but many do infest mullets and other fishes. Parasitic species belong primarily to two major groups. The parasitic members of the Cyclopoida consist of many free-living males and, apparently, larval stages. Distinguishing parasitic females from female-members of free-living genera depends primarily on the presence of appendages modified for attachment. Ergasilids use powerful claws modified from second antennae and bomolochids use their maxillipeds. Comprising most of the remaining copepods on mullet, the members of the Caligoida, an entirely parasitic group, will be discussed later.

Ergasilids

Several species of the genus *Ergasilus* Nordmann infest mullet. They are small and barely visible to the naked eye, clinging to the gill filaments with their characteristic second antennae. Some infest a wide range of fresh-water to marine hosts, and some, because of overlapping characters, are difficult to separate. Consequently, many taxonomic problems arise. The status of North American species is the best established (Roberts, 1969, 1970; Johnson & Rogers, 1972, 1973), but still those infesting mullets require more investigation. One of the problems is the inability to rear species throughout their entire life cycle (Wilson, 1911; Gnadberg, 1949; Paperna & Zwerner, in preparation).

In the United States, *Mugil cephalus*, and in some cases *M. curema*, host *Ergasilus lizae* Krøyer, *E. versicolor* Wilson, and two other forms, according to Johnson & Rogers (1973). Additionally, a few specimens of *E. funduli* Krøyer in areas dense with cyprinodontid fishes often infest young or, occasionally, adult mullet. Johnson & Rogers (1973) suggested *E. mugilis* Vogt and *E. nanus* van Beneden, both reported from American mullets by other workers and neither being the unidentified forms mentioned above, may be synonyms of *E. lizae*. *Ergasilus versicolor* extends well inland from the coastal areas, where *E. lizae* is restricted to the coast; both infest a wide range of hosts. Considering *E. lizae* a cosmopolitan species, Roberts (1970) discussed specimens off *M. peruanus* from Chile and off *M. trichodon* from Puerto Rico. It also occurs in Israel, but was reported as *E.* cf. *nanus* of Paperna (1964*b*), *E. fryeri* Paperna and *E. seiboldi* of Lahav & Sarig (1967). In Israel, it commonly infested *M. cephalus*, *M. capito*, *M. auratus* and some non-mugilid fishes in coastal streams and fishponds (Paperna, 1964*a*). Both it and *E. mugilis* have been reported also on *M. cephalus* from both coasts of Mexico (Causey, 1960). Of all the ergasilids from mullet in the United States, *E. lizae* is the only one reported as far north as Chesapeake Bay (Zwerner & Lawler, 1972).

Also reported from the Americas are *E. longimanus* Krøyer on *M. cephalus* from Florida and *Mugil* sp. from Brazil (see citation of Skinner, 1974), and *E. cyanopictus* de Paiva Carvalho on *M. cephalus* from Brazil. *Ergasilus nanus* has been reported from *M. auratus*, *M. cephalus*, *M. capito*, *M. chelo*, *M. curema*, *M. saliens* and other non-mugilid fishes from the United States, eastern North Atlantic, Black Sea and the Sea of Azov.

Other described species include *E. shehyangensis* Wang from a mullet in Kiangsu Province of China and *Nipergasilus bora* (Yamaguti) which infests *M. auratus*, *M. cephalus*, *M. seheli* and *Crenimugil crenilabis*. After being first reported from the saline Lake Hamma in Japan (Yamaguti, 1938), *N. bora* has been seen in the Gulf of Elat and the saline mouth of the Yarkon river in Israel. Its presence in the Mediterranean could have resulted from Red Sea mullet migrating through the Suez Canal. Other species of ergasilids probably also infest mullet.

Host–parasite relationships

Ergasilids attached to gill filaments produce these small foci of eroded host tissue. Apparently, feeding involves external digestion (Halisch, 1940; Kabata, 1970); parasites produce digestive secretions which partially dissolve tissue, allowing easier ingestion. As the number of attached parasites increases, the destruction of respiratory epithelium progresses. Erosion can extend beyond the epithelial lining, resulting in obstructed branchial blood vessels. Irritation often results in responsive hyperplasia of epithelium, which, as infestation intensifies, extends over considerable areas (Kabata, 1970, Paperna & Zwerner, in preparation). Both processes reduce the respiratory function of the gills. Heavy infections are likely to occur in crowded fishponds. Establishment of ergasilids in fishponds, howevèr, depends largely on the ability of larvae to survive and develop to infective stages within ponds.

In low-saline ponds (600 to 2000 ppm Cl^{-}/l) along Israel's Mediterranean coast, epizootics of *Ergasilus lizae* during the winters of 1966 through 1968 resulted in considerable losses. In one pond, ten to seventy individuals infested each of ov^r 90% of the 100 g specimens of *Mugil cephalus* (Lahav & Sarig,

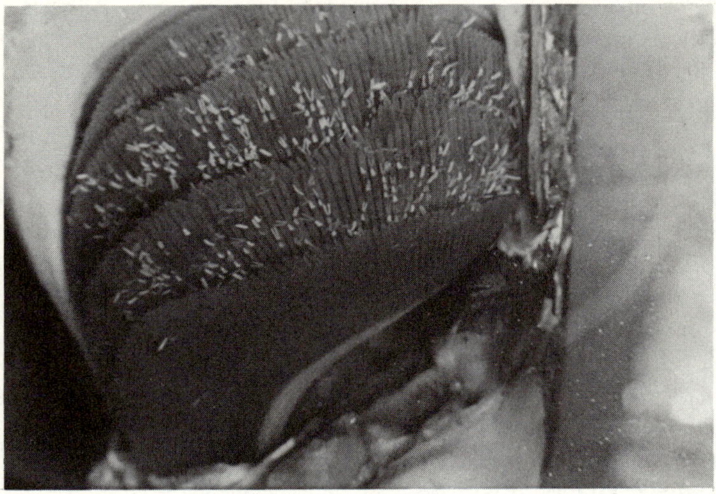

Fig. 13.15. Hyperinfection with *Ergasilus lizae* on gills of *Mugil cephalus* from brackish-water pond on the Mediterranean coast of Israel (by courtesy of Mr S. Sarig, Laboratory of Fish disease, Nir David, Israel).

1967; Sarig, 1971). Records provided by M. Lahav of the Fish Disease Laboratory at Nir David, Israel, revealed that in other ponds, 70–80 g fish each harboured 100–200 of the copepods, and 250 g fish possessed 1500–2000 individuals (Fig. 13.15). During critical periods, many fish were found dead each day and those still living were so weak that they died when netted. These emaciated mullet had recessed abdomens. By June or August, after 4–5 months of maintenance, the number of harvested fish was about 50% of those stocked, rather than the expected 90% (Yashouv, 1972).

Even though *M. capito*, *Tilapia aurea* and carp accompanied *M. cephalus*, only *M. cephalus* was involved in mortalities. Individual *M. capito* had about nine copepods compared with 100–120 on the emaciated *M. cephalus*. In one pond, 50% of the carp were infested with between ten and thirty copepods, but copepods did not infest tilapia. In a Jordan Valley pond, however, *E. lizae* heavily infested *T. aurea*, causing hyperplasia at the sites of attachment.

Infestations of mullet in Israel have only occurred in ponds of the coastal plain which feed and are fed by coastal streams. Fry and fingerlings 15–30 mm long from the mouths of streams do not display infestation when stocked. Infestation of wild mullet occurs after the fish continue up the streams and mix with infested mullet that measure 60 mm or more. *Mugil capito*, *M. auratus*, eels, *Tilapia zillii* and *T. galilaea* occurring in the streams and possessing light infestations of *E. lizae* probably serve as reservoir hosts, releasing larvae which are subsequently introduced into the ponds through the water-supply pipes or drains.

Ergasilids also heavily infest *M. cephalus* and *M. curema* in ponds at the Rockefeller Refuge near Grand Chenier, Louisiana. Neither these mullet from the Louisiana Fish and Wildlife ponds nor the ones infested (somethimes with several hundred specimens) from the nearby Price Lake and associated areas appeared emaciated. Even though *E. lizae* is potentially dangerous to fish in ponds in the United States, the individuals involved in an epizootic on centrarchid fishes in a brackish lake in Alabama reported by Kelly & Allison (1962) as involving only *E. lizae* consisted mostly of *E. arthrosis* Roberts with a limited number of *E. liza* (see Johnson & Rogers, 1973).

Other cyclopoids

Other cylocopoids also infest mullets. *Bomolochus concinnus* Wilson, also known as *B. mugilis* Pearse and *B. nitidus* Wilson, commonly occurs on *Mugil cephalus* in the southeastern United States. Skinner (1974) reported that the gills of twenty of eighty-three examined fish from Biscayne Bay, Florida, had two to twenty-five specimens per fish. Causey (1960) reported it from the Pacific coast of Mexico. *Bomolochus teres* Wilson and *B. exilipes* Wilson infested *M. cephalus* in Texas (Pearse, 1952; Causey, 1953) and *B. decapteri* Yamaguti infested *Mugil* sp. from Hainan Island, China (Shen, 1957).

The anchor worm, *Lernaea cyprinacea* Linnaeus, embeds into its hosts, often near the tail or fins, and makes the fish unsightly as a consumer product. In Israel, it affects carp detrimentally. Mullet appear refractory to heavy infestations and often remain uninfested in ponds where carp are heavily infested (Yashouv, 1959). Since this parasite prevails elsewhere in the world, most fresh-water facilities are potentially vulnerable to its epizootics.

Caligids

Caligoid copepods belonging to at least five families are capable of damaging mullets by scraping the fish's gill filaments or integument with their mandibles. Most common of these are members of the Caligidae. Like some ergasilids, several caligids have a wide geographic range and pose taxonomic problems. The general life cycle has been worked out (Wilson, 1905; Gurney, 1934) and even some larval stages of mullet-species have been described (Russel, 1925; Hewitt, 1971). After a short free-living larval stage, a chalimus stage attaches, usually to a gill filament, where it progresses through three or four moults to the adult stage. Most adults infest the buccal cavity or integument, whereas larvae attach, first by the second antennae and then by a chitinous secretion called a frontal filament, to the gills or, in heavy infestations, elsewhere on the body.

Caligus pageti Russel infested *Mugil capito* in hypersaline ponds in the Nile Delta (Russel, 1925) and *M. auratus* in Algeria (Argilas, 1930; Brian, 1931).

Fig. 13.16. *Caligus minimus* var. *mugilis* ♀, from mouth of *Mugil chelo*, Mediterranean coast of Israel, embedded in Polyvinyl – lactophenol, stained in cotton blue, × 10.

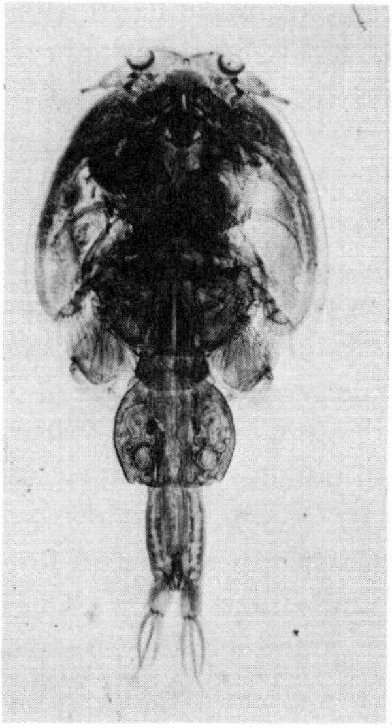

Fig. 13.17. *Pseudocaligus apodus* ♂, from skin of *Mugil cephalus* sea-water ponds, Israel, Polyvinyl – lactophenol embedding, × 10.

Brian (1935) also described *C. minimus* var. *mugilis* from the northwestern Mediterranean. In Israel, several species, some from sea-water ponds, are currently being studied by the senior author (Fig. 13.16). *Caligus platytarsis* Bassett-Smith and *C. bombayensis* Rangnekar, possibly synonymous species, infest mullet in India, whereas *C. epidermicus* Hewitt can be found on *M. cephalus*, *Aldrichetta forsteri*, *Liza argentea* and *Myxus elongans* from the hypersaline waters of Mitchell River in Victoria, Australia (Hewitt, 1971) and on *Valamugil sehli* and *Crenimugil crenilabis* in the Gulf of Elat. *C. communis* Shen has been reported on *Mugil cephalus* and *C. orientalis* Gussev on mullet in China, and *C. punctatus* Shiilno has been reported on *Liza menada* in Japan. Several other species have also been listed from the mouth or on the integument of *Mugil cephalus* in North America, including *Caligus schistonyx* Wilson, *C. rufimaculatus* Wilson, *C. praetextus* Bere, *C. bonito* Wilson, *C. irritans* Heller, *C. mutabilis* Wilson and *C. curtus* Muller. The last species was also reported from *M. chelo* in Northern Ireland as *C. mulleri* (see Parker, Kabata, Margolis & Dean, 1968).

Pseudocaligus apodus Brian, also reported as *Pseudolepophtirius mediterraneus* Paperna, infests several mullets over a wide geographical range. It occurs on *Mugil auratus*, *M. capito*, *M. cephalus*, *M. saliens*, *Crenimugil crenilabis*, *Liza carinata* and *Valamugil seheli* and is known from the Mauritanian shore of the Atlantic Ocean, the eastern Mediterranean Sea, and the Gulfs of Elat and Suez in the Red Sea. It also has established itself in sea-water ponds holding *M. cephalus* in Israel (Paperna & Lahav, 1974) (Fig. 13.17).

Caligids generally stay restricted to marine and hypersaline habitats. Only rarely do they infest mullet in brackish water. Some American species, however, may be more able to tolerate conditions of low salinity than those from other areas.

Other caligoids

The caligoids mentioned below are more specialized than members of the Caligidae which appear much like free-living forms and can move about on their hosts. Most of them stay attached to their mugilid host with little or no change in position. An exception is the pandarid *Perissopus communis* Rathbun. Since this copepod from a mullet in Texas normally infests sharks, Causey (1953) assumed that it was probably an accidental infestation.

Those remaining copepods reported from mullets do not appear like copepods, to the observer only familiar with free-living species. Two are gill infesting anthostomids: *Lernanthropus mugili* Shishido (syn *L. shishidoi* Shiino) parasitizes *Mugil cephalus* from Japan and India, and *M. soiuy* from the Pacific Soviet Union; whereas *L. mugilis* Brian infests *M. auratus* in the western Mediterranean. Of the lernaeocerids, *Lernaeenicus longiventris*

Parasites and diseases

Wilson embeds near the tail and on the fins of *Mugil cephalus* in the Atlantic and Gulf coasts of the United States, and *L. neglectus* Richiardi embeds in *M. cephalus*, *M. saliens*, *M. capito*, *M. chelo* and *M. auratus* in the western Mediterranean and the Adriatic Sea. In the Gulf of Elat, a species of *Lernaeenicus* Lesueur, possibly *L. cerberus* Leigh-Sharpe, occasionally invades mullet fry which share the same habitats as infested atherinids (Paperna, unpublished). *Mugicola bulbosus* Tripathi burrowed into the tissues of *Mugil tade* and *M. parsia* in the River Malta and a fish-farm in India. *Naobranchia lizae* Krøyer, a naobranchiid, attaches to the gills of *Mugil cephalus* and *M. curema*. It has been reported from the Atlantic and Gulf of Mexico coasts of the United States and the Pacific coast of Mexico. Lerneopodids occur in or near the protected base of pectoral fins or occasionally on the gills. *Brachiella oblonga* Valle infests *M. cephalus* along the Atlantic coast of the United States and *M. cephalus* and *M. saliens* from the Adriatic Sea. *Clavellopsis robusta* Wilson, *Alella longimana* (Bere) and *Clavella inversa* Wilson all infest *M. cephalus* from the Gulf of Mexico and Atlantic coasts of the United States.

Probably, many other specific and nonspecific copepods infest mullets. In nonparasitological works, workers occasionally mention that 'copepods' were on mullets. One such example was by Kesteven (1942) reporting unidentified specimens from the fins of *Mugil dobula* in New South Wales, Australia.

Host–parasite relationships

Reports of lesions on the body and mouth resulting from caligid infections often appear in current literature on fish diseases. Such involvement occurs both in fish from natural habitats and those confined in culture (Kubota & Takakuwa, 1963; Kabata, 1970). In mullets, caligid epizootics resulting in injuries, stress and death have taken place in cultured individuals from Egypt and Israel and in a natural population from a river in Victoria, Australia.

During epizootics, salinity values often rose higher than normal. Russel (1925) reported an epizootic of *Caligus pageti* on *Mugil capito* in an experimental pond on the shores of Lake Maryut in the Nile Delta. Being fed from a salt spring, the pond had salinity values up to 45‰. Fish covered by sores became emaciated but no mortalities were reported. Hewitt (1971) reported high intensities of *Caligus epidermicus* on the skins of several mullets and a bream. A further epizootic developed along the lower reaches of the Mitchell River in Australia between January and March during a severe drought. The water's temperature rose from 14 °C to 21.8 °C and its salinities from 4.5 to 28‰. Hewitt reported infested fish being in poor condition, but presented no further details.

Epizootics have also occurred in small, 500 m^2 ponds containing sea water at the Fish and Fishbreeding Research Station, Dor on the Mediterranean coast of Israel. These ponds were stocked with *M. cephalus* and *M. auratus*. Following a fish-kill during the winter of 1972 which apparently affected only *Mugil cephalus*, parasitological examination of the dead fish revealed numerous adult males and females, as well as chalimi, of *Pseudocaligus apodus* and a single chalimus of *Caligus* sp. (Paperna & Lahav, 1974). Dying fish had large exposed integumental lesions. The following winter, another die-off occurred in the same pond, once again affecting only *M. cephalus*. This time however *C. pageti* occurred on the moribund fish. Individuals of *M. auratus*, the only remaining species, had no ectoparasites when checked in November 1974.

Lesions resulting from attached copepods and other ectoparasites invite secondary infections of bacteria and fungi, which can result in death. In fact, Kubota & Takakuwa (1963) reported a secondary infection associated with *Vibrio* sp. and *Pseudomonas* sp. in caged marine fish subsequent to infestations with caligids and a species of *Benedenia*.

Branchiura

Species of *Argulus* Müller and related genera occasionally get grouped with copepods, but these entirely parasitic crustaceans differ in several respects. What interests people rearing fish most about members of the genus *Argulus*, commonly called 'fish-lice', is the way they feed and lay eggs. In order to feed, they protrude an organ referred to as a pre-oral sting, or stylet, through a tubular mouth and pierce the host. Rapid thrusting of this organ accompanied with the release of toxic secretions can result in severe pathological changes in the host. Rather than carrying eggs in egg-sacs as do copepods, female argulids leave their host and lay eggs in mucus strips on rocks, weeds, submerged objects and the sides of aquaria. Confined fish are easily infested by the larval stage. Argulids use no intermediate hosts.

These parasites on mullet, easy to see darting about on the fish, can be recognized by the two separated compound eyes and two stalked suckers. The suckers are modified from the second maxillae. *Argulus mugilis* has been reported from a mullet in Kiangsu Province in China (Wang, 1964), and *A. funduli* Krøyer and *A. chesapeakensis* Cressey from *Mugil cephalus* along the southeastern Atlantic coast of the United States (Rawson, 1973*b*; Skinner, 1974). *Argulus flavescens* Wilson and *A. floridensis* Meehan infest mullet along the Gulf Coast of the United States (Cressey, 1972). We found in Mississippi (Overstreet, 1974) what was apparently a rather host-specific species (Fig. 13.18) that we sent to Roger Cressey of the Smithsonian Institution. He considered it a new species.

Since species of *Argulus* can easily decimate stocks in confined areas, they should be considered a threat to mullet in aquaculture. Ljaiman (as cited by Sarig & Lahav, 1959) reported that argulids kill fry and fingerlings by the

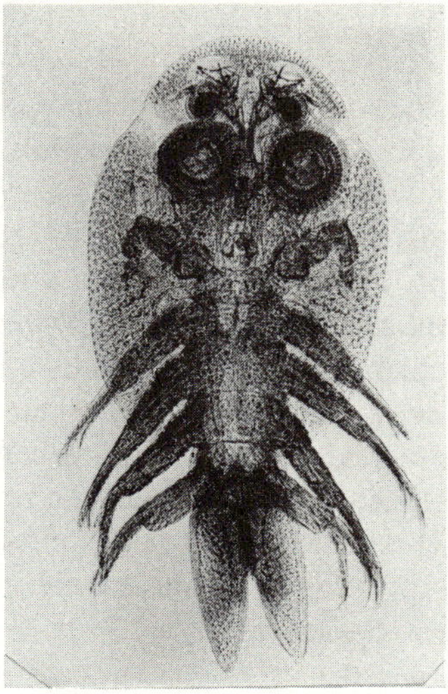

Fig. 13.18. Ventral view of fixed and cleared *Argulus* sp. from *Mugil cephalus* collected in the Mississippi.

thousands. In Israel during the summer, marketable carp are stored for up to several weeks in 0.5–2.4 ha ponds at densities up to 10000 fish/ha. Infestations of argulids and *Lernaea cyrpinacea* cause ugly wounds, loss in weight and an unmarketable product. Such infestations are now controlled with periodic treatments with pesticides (Sarig, 1968).

Isopoda

Isopods parasitize both fresh-water and marine fishes. Most parasitic species infecting fishes differ only slightly from free-living ones. They have prehensile anterior appendages, modified mouthparts and a somewhat altered alimentary tract. Gnathiids, However, differ considerably, as we discuss later.

According to the literature, several species of isopods infest mullets. Those that do present serious threats to the host's health, especially to fish caged in their natural habitat. A lack of knowledge concerning isopod life histories, however, prohibits predicting the effect of these external parasites, especially cymothoids, on reared mullet. Apparently, specific fish act as intermediate hosts for larval cymothoids (Heard & Overstreet, unpublished observations), and if known, these fish, usually different species than final hosts, could be eliminated from ponds.

Before discussing cymothoid isopods, we should mention that Pearse (1952) reported the sphaeromid *Ancinus depressus* (Say) from *Mugil cephalus* in Texas. The presence of this normally free-living isopod on the mullet was probably an accidental infestation (Heard, personal communication).

Cymothoids

Cymothoids comprise about half the species of isopods infesting fish. Several of those infesting estuarine fishes in Mississippi can cause enormous lesions on the body, destroy most of the gill filaments, or, as speculated by Lawler, Howse & Cook (1974), transmit viral diseases or irritate the tissue allowing a virus to enter. Secondary infections often accompany infestations, and one isopod seems to remain on its host for a considerable period, in some cases for most of both their lives. There are several species infesting mullet. Richard Heard (personal communication) found *Nerocila lanceolata* (Say) attached to the tail and base of the fins of large numbers of mullet in Georgia. *Nerocila acuminata* Schioedte and Meinert, either a synonym of or a species closely related to *N. lanceolata*, infests *M. cephalus* in Texas. In the western Mediterranean (France, Italy, Tunis), *Mugil cephalus*, *M. auratus*, *M. capito* and *M. chelo* harbour *N. orbignyi* (Guerin-Meneville) (see Trilles, 1965 and other papers by the same author). Three species of *Indusa* Schioedte and Meinert have also been reported from the mouths or gill chambers of mullets. *Indusa carinata* Richardson infests *Mugil hospes* from the west coast of Panama and *I. malayi* infests *M. ophueseni* from India, as does *I. ophueseni* Pillai. *Meinertia gilberti* (Richardson) attacks *M. hospes* from Mazatlan, Mexico. One remaining cymothoid from mullet is *Aegathoa oculata* (Say). The genus is a collective group-name for larval stages of some species. Apparently, several species of cymothoids have been referred to as *A. oculata*.

Cymothoids attract attention because some are protandrous hermaphrodites. After the molt from juvenile into a male, the isopod will eventually become a female.

Gnathiids

Other parasitic isopods that deserve much attention are larval species of gnathiids. These often neglected isopods are best known from the work of Monod (1926). Members of this family are unique because they possess five rather than seven thoracic appendages (peraeopods). Only the pranizae, the larval stage, parasitize the gills and body of fish. Because the larvae appear so different from the free-living adults, they were described as members of the genus *Praniza*, from which the general name for the larva was derived. Those pranizae parasitizing elasmobranchs seem to be more specific than those infesting teleosts. Moreover, the presence of numerous larvae in the benthos with adults suggests to some workers that some species act as non-obligate parasites. The males, appearing much like ant- or termite-workers, possesss large mandibles. These structures, however, probably serve as secondary sexual characters.

Pranizae of *Parangnathia formica* Monod and an unidentified species infest mullet and other hosts (Monod, 1926). Gnathiids, because of their

Fig. 13.19. *Mugil subviridis* infested with *Gnathia* pranizae, after being left overnight in a floating cage in shallow water (Red Sea).

indiscriminate selectivity of hosts and characteristic blood-feeding, pose a risk to confined marine fishes. Preliminary studies in progress in the Gulf of Elat (Paperna, in preparation) show *Crenimugil crenilabis* and other fishes confined in cages, as well as those caught in nets, can be immediately attacked by a large number of pranizae (Fig. 13.19). Within 2–4 hours, the larva's modified midgut becomes engorged with host-fluid and the isopod leaves its victim. Pranizae attacked fish only at night, but some engorged individuals remained until the following morning. Victims exhibited numerous integumentary wounds and severe anemia. Some died while still infested, and others succumbed the following day.

Gnathiid larvae appear to be attracted to confined or otherwise stressed fish. Once fish became accustomed to their cages, infestations subsided. Those fish most seriously afflicted had been in floating cages anchored near shore in water of 1–2 m depth. Few larvae infested fish in floating cages anchored about 100 m offshore at depths of 6–8 m. Young free-living pre-parasitic and infective larvae occurred in greater numbers in inshore rather than offshore plankton samples.

Preliminary observations on the life cycle of the Red Sea species showed that the praniza feeds three separate times on fish before reaching maturity. Between meals, it moults and increases in size. Detached pranizae deprived of one or more meals could not moult into a male or female. At 24 °C, larvae moulted into adults 6–8 days following their last meal. Eggs developed within a brood pouch and larvae emerged from a slit at the anterior portion of the female's pouch 22–24 days after she matured. This was 30 days following the last meal for the praniza. Females died shortly after producing young. A full-sized female produced ninety larvae, and these free-living larvae became infective to fish immediately after their release from their parent.

Monogenea

As ectoparasites on the gills or general body surface, monogeneans attach to mullet by means of hooks, anchors, suckers, clamps or combinations of these structures on their posteriorly located haptor. Development of these parasites is direct. The larva, called an oncomiracidium, hatches from an egg, which for most species is filamented, which may infest a new host or remain on the original host. In the gyrodactylids, embryos are produced within the parent and such an embryo may produce another embryo within it. Because of this direct cycle, the parasites have easy access to hosts in culture situations and heavy infestations can be harmful.

Species belonging to several families infest mullets: Gyrodactylidae, Dactylogyridae, Capsalidae and Microcotylidae. Also, a udonellid infests caligid copepods on mullet and occasionally the mullet itself. All but the microcotylids are monopisthocotyleans and feed primarily on mucus and epithelial tissues. Microcotylidae is a polyopisthocotylean family and many member species typically feed on the host's blood.

Monopisthocotyleans

Gyrodactylids may infest the gills, body surface and fins of their hosts, and mullets harbour several species. *Gyrodactylus zhukovi* Mo-en and *G. mugil* Zhukov infest *Mugil soiuy* in the Sea of Japan; *G. mugelus* Rawson infests *M. cephalus* in Georgia (Rawson, 1973*a*), as does *Polyclithrum mugilini* Rogers which occurs in association with it. A new species placed in a new genus infests *M. cephalus* in Florida and is described in an, as yet, unpublished thesis (Skinner, 1974).

Also small and difficult to identify without careful examination, but distinct from those species listed above because they possess eggs rather than embryos and because they have a few differing morphological features, are the dactylogyrids. *Ancyrocephalus vanbenedenii* (Parona & Perugia) apparently covers the widest host and geographical range of any dactylogyrid in mullets;

it infests *Mugil auratus*, *M. capito*, *M. chelo*, *M. cephalus*, *M. labeo*, *M. saliens*, and *Liza carinata* and ranges from the Gulf of Mexico and the Atlantic coasts of the United States and Europe, through the Mediterranean area and the Black Sea, to the Indo-Pacific to northern Red Sea region, and the Pacific region of the Soviet Union (Gussev, 1955; Paperna, 1964*a*, and unpublished). Populations of *A. vanbenedenii* display morphological variability that may be attributable to host and locality differences. As a result, we believe some populations of *A. vanbenedenii* have been viewed as distinct species. We suspect two of these are *Haliotrema mugilinus* (Hargis) from the Gulf of Mexico and *Ancyrocephalus fluviatilis* Bychowsky from *Mugil abu zarundiji* in Iran. Other dactylogyrids associated with mullets include *Ancyrocephalus alatus* Chauhan from *M. parsia* in India; *Haliotrema mugilis* (Tripathi) from *M. subviridis* in India and from the same species, as well as *Valamugil seheli* and *Crenimugil crenilabis*, in the northern Red Sea; and *Ergenstrema mugilis* Paperna from *M. capito* in the eastern Mediterranean Sea and the Adriatic Sea. Some reports have not identified species (Vlasenko, 1931; Ergens, 1960), and possibly additional species still exist.

Generic definitions of *Ancyrocephalus* Creplin, *Haliotrema* Johnston & Tiegs and other ancyrocephalines still are disputed by workers of this group. Gussev (1963) suggested to retain, provisionally, members of all closely related groups under the inclusive name '*Ancyrocephalus* s.l.' until a better system can be established. *Haliotrema mugilis* is closely related to *A. vanbenedenii*!

Undonella caligorum Johnston demands special attention because it infests caligid copepods which infest *Mugil cephalus* and numerous other fish. Skinner (1974) found that of eighty-three mullet in Biscayne Bay, Florida, twenty-three were infested with *Caligus rufimaculatus* Wilson or *Caligus* sp. and of these half had the monogenean. In many cases she observed eggs, juveniles and adults on the same copepod. When the worms were present, approximately 80% of the copepods were infested per host and each usually harboured many worms.

Capsalids that infest mullets include *Benedenia monticellii* (Parona & Perugia) on *Mugil auratus*, *M. cephalus* and *M. capito* from the western and eastern Mediterranean Sea, *Neobenedenia pacifica* Bravo-Hollis on *M. cephalus* from Baja California and mullets from the Red Sea (Fig. 13.20) and Mexico.

Polyopisthocotyleans

The microcotylids present taxonomic problems because characteristics of several species overlap. They have been differentiated primarily by the pattern of the genital armature, the number of haptoral clamps and the number of testes. Most species infesting mullets belong to either *Metamicrocotyla*

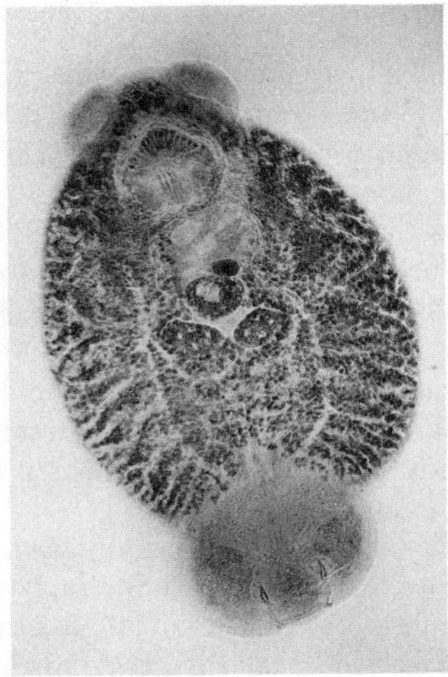

Fig. 13.20. *Benedenia* sp. from *Liza carinata*, Red Sea, whole mount stained in Gower's Carmin, × 10.

Yamaguti or *Microcotyle* van Beneden and Hesse. All infest the gills of their hosts. The species of *Metamicrocotyla* include *M. inoblita* Buhrnheim from *Mugil platanus* in Brazil, *M. macracantha* (Alexander) from *Mugil cephalus* in Baja California, Mexico, and the southeastern United States, *Metamicrocotyla cephalus* (Azim) from *M. cephalus* from Egypt and France, *M. filiformis* Yamaguti and *M. bora* Yamaguti from *M. cephalus* in the Celebes, *M. manaarensis* Unnithan from *Mugil troschelii* in India and *M. mugilis* Yamaguti from *M. cephalus* in Hawaii. Fewer species of the related *Microcotyle* have been described: *M. pseudomugilis* Hargis on *M. cephalus* from the southeastern United States, *M. mugilis* Vogt on *Mugil auratus*, *M. cephalus*, and *M. saliens* from Japan, the Mediterranean Sea, and the Black Sea, and an unidentified species on *Mugil oligolepis* from New Hebrides (Laird, 1958*b*). We are aware of what may be new species of both the above genera from the eastern Mediterranean Sea and the Gulf of Elat. *Bivagina sillaginae* (Woodcock) infested the gills of *Mugil macrolapis* in India and *Diplasiocotyle johnstoni* Sandars infested *Aldrichetta forsteri* from New Zealand and South and Western Australia.

Host–parasite relationships

Monogeneans represent a real threat for mullet in confined areas. The combination of being crowded, being stressed and allowing easy access for infestation provides the potential for death or severe disease in stocks. As mentioned earlier, the monopisthocotylids primarily feed on mucus and epithelial tisue, whereas the microcotylids feed on blood. Most examples of mortalities in confined areas concern monopisthocotylids, but microcotylids have been implicated in deaths of fishes such as those from New York Aquaria (Nigrelli, 1943), *Seriola quinqueradiata* (yellowtail) cultured in Japan (Kubota & Takakuwa, 1963) and *Siganus luridus* from holding tanks of a pilot culturing project in Elat. Infestations on *S. luridus* resulted in severe anaemia. Hematocrit value for host blood decreased to 15% of its normal level in the examined fish (Paperna, unpublished data).

Of all the monogeneans found on mullets, only *Benedenia monticelli* on the gills of mullets from the Gulfs of Elat and Suez appeared pathogenic to its hosts. In the northern gulfs of the Red Sea, it infests *Liza carinata*, *Crenimugil crenilabis*, *Mugil subviridis*, and *Valamugil seheli*. Individuals reach 4 mm long by 1.5 mm wide and normally attach to the oral mucosa. Both juvenile and adult fish harbour worms, usually fewer than six, but twelve have been recovered from a 144 mm long fish. We have not seen mullet smaller than 60 mm infested. In fish with typical infestations, the lips and oral mucosa display signs of irritation with extensive submucosal hemorrhaging, particularly on the roof of the mouth between the palate and lips. Mullet from their natural habitat subsequently stocked in open 0.75 m^3 and 1.0 m^3 culturing tanks began to die after two weeks in captivity. Experimentally infested mullet died six to eight weeks following initial exposure to the worms. In heavily infested fish, worms spread from the mouth to the entire body. In particular, they concentrated in the intermandibular groove and at the base of the ventral and dorsal fins. Lesions occurred at all these sites. In tanks, mortalities ceased two months following exposure, and examined fish were free of the parasites by the third month. The loss of worms could be a 'self cure' process, probably involving acquired resistance with or without an immune response. A similar elimination of a capsalid was described for *Neobenedenia melleni* (MacCallum) (Nigrelli & Breder, 1934; Nigrelli, 1937).

Death of fish resulting from infestations of *B. monticelli* also occur naturally in the Gulf of Suez. In April 1974 and again in February 1975, numerous emaciated moribund hosts were discovered in El Bilaim Lagoon on the Gulf's eastern shore. They exhibited signs similar to reared hosts with extensive inflammation in the mouth and about the lips. Many worms covered the body and oral region, again especially in the intermandibular groove. Worms frequently occurred under the eyelids. In many fish, one or both eyes were severely irritated, often with an opaque cornea (Fig. 13.21). Pathological

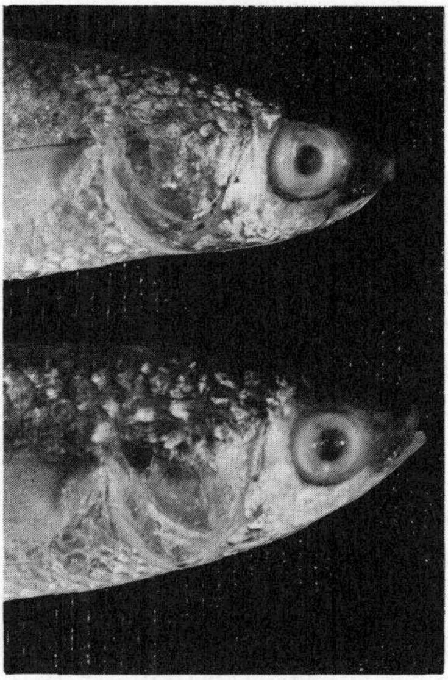

Fig. 13.21. Changes in the eye cornea of *Liza carinata*, heavily infected with *Benedenia monticelli*, fish collected from El Bilaim lagoon, Gulf of Suez during *Benedenia* epizootic.

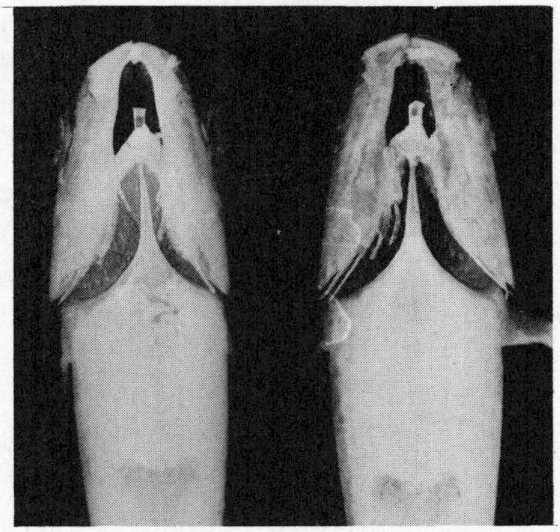

Fig. 13.22. *Mugil subviridis* dead from Benedeniasis, laboratory infection, note the extensive erosion in the gular and opercular regions.

alternations also included extensive necrosis and erosion of the oral region. In fact, many individuals had the underlying oral bones completely exposed and the gular membrane perforated Fig. 13.22. Many were so lethargic that they could be caught by hand. The role that environmental conditions played concerning disease of the fish remained uncertain. El Bilaim supports a massive mullet population as well as being heavily polluted with oil escaping from underwater oil-wells.

Other capsalid species also harm their captive hosts as exemplified by *Benedenia seriolae* (Yamaguti) in the yellowtail from Japan and *Neobenedenia melleni* in many fishes from the New York Aquarium (Nigrelli, 1943).

Digenea

Of all the helminths infecting mullets, digenetic trematodes or flukes, generally number the most, both in species and individuals. These endoparasitic flatworms undergo an alternation of generations and hosts. Digenean eggs, usually deposited within the faeces of a vertebrate host, contain a larva known as a miracidium which infects a mollusc, and its germinal cells undergo division, ultimately forming up to hundreds of thousands of short-lived, free-living cercariae. After emerging from the mollusc, these larvae, usually specific to particular molluscs, encyst externally or penetrate or become eaten by the second intermediate host, which can include a wide range of substrata, invertebrates or vertebrates, depending on the species. When a proper definitive host feeds on this metacercarial stage, the larva, either free in the intermediate host or encysted, develops to maturity in a wide variety of sites. Hosts and sites are usually characteristic for each digenean. Mullet act as intermediate hosts for numerous flukes and as final hosts for many more. These small, for the most part barely visible to the naked eye, unsegmented hermaphroditic worms can be recognized because they possess one or usually two muscular suckers and an intestine. Some contain one or two eggs, and other species contain thousands.

Because of the large number of digeneans infecting mullets, we will separate adult from larval worms and present them in tabular form. A spacial limitation prohibits listing the references, which number over 200. Such a large number of species necessarily introduces problems with synonyms, which we have tried to reduce to a minimum.

Adults

Being conservative with the number of families, we list (Table 13.1) flukes from eleven different ones. Most of these species consist of haploporids, haplosplanchnids and hemiurids. Members of the first two probably are transmitted mostly by metacercariae encysted in detritus, vegetation and debris, whereas some hemiurids use small crustaceans such as copepods as intermediate hosts. Cercariae of *Cardicola mugilis* and the two trans-versotrematids penetrate mullet directly, in contrast to the cercaria of *Pauci-vitellosus fragilis* on which the mullet feed directly. Acquisition of the remaining, results from mullets feeding primarily on small intermediate hosts. *Bucephalopsis southwelli* is acquired presumably when mullet feed on fish, a rare dietary item for most mullets.

Table 13.1. *Adult digeneans and their sites in mullets of the world*

Parasite	Host	Site	Locality
Bivesiculidae Yamaguti, 1934			
Paucivitellosus fragilis Coil, Reid & Kuntz, 1965	*Chelon troscheli*, *Mugil cephalus*, *Crenimugil crenilabis*, *Liza argentea*	Intestine	Taiwan, Queensland (Australia)
Transversotrematidae Yamaguti, 1954			
Prototransversotrema steeri Angel, 1969	*Aldrichetta forsteri*	Under scales	South Australia
Transversotrema laruei Velasquez, 1958	*Mugil* sp.	Under scales	Philippines
Aporocotylidae Odhner, 1912			
Cardicola mugilis Yamaguti,1970	*Mugil cephalus*	Heart and blood vessels	Hawaii
Fellodistomidae Nicoll, 1913			
Tergestia agnostomi Manter, 1954	*Aldrichetta forsteri*	Intestine	New Zealand
Bucephalidae Poche, 1907			
Bucephalopsis southwelli Nagaty, 1937	*Mugil* sp.		Red Sea
Hemiuridae Lühe, 1901			
Aphanurus microrchis Chauhan, 1945	*Mugil parsia*	Alimentary canal	West coast of India
Aphanurus stossichi (Monticelli, 1891) Looss, 1907	*Mugil* sp.		Black Sea
Bunocotyle mugilis Yamaguti, 1970	*Mugil cephalus*	Stomach	Hawaii
Dichadena galeata (Looss, 1970) Skrjabin & Guschanskaja, 1954	*Mugil auratus*	Intestine	Egypt, Black Sea
Hemiurus appendiculatus (Rudolphi, 1802) Looss, 1899	*Mugil cephalus*, *M. auratus*, *M. capito*	Stomach	Black Sea, Triest
Hysterolecitha elongata Manter, 1931	*Mugil cephalus*, *M. platanus*	Stomach, intestine	North Carolina, Mississippi (USA), Brazil
Lecithaster helodes Overstreet, 1973	*Mugil cephalus*, *M. curema*	Intestine, pyloric caeca	Mississippi and Florida (USA)
Lecithaster mugilis Yamaguti, 1970	*Mugil cephalus*	Intestine	Hawaii Island (China)

Table 13.1. (*cont.*)

Parasite	Host	Site	Locality
Leicithochirium polynemi Chauhan, 1945	*Mugil parsia*	Intestine	West coast of India
Lecithocladium glandulum Chauhan, 1945	*Mugil speigleri*	Intestine	West coast of India
Saturnius segmentatus Manter, 1969	*Mugil cephalus*	Stomach	Queensland (Australia)
Didymozoidae Poche, 1907			
Epithelionematobothrium sp. Skinner, 1974	*Mugil cephalus*	Body cavity	Florida (USA)
Haplosplanchnidae Poche, 1926			
Haplosplanchus bivitellosus Zhukov, 1971	*Mugil cephalus*, *M. soiuy*	Intestine	Posyet Bay and Liao Ho River (China)
Haplosplanchnus mugilis Nahhas & Cable, 1964	*Mugil cephalus*, *M. curema*	Intestine	Curaçao, Florida (USA)
Haplosplanchnus pachysoma (Eysenhardt, 1829) Looss, 1902	*Mugil cephalus*, *M. auratus*, *M. chelo*, *M. capito*, *M. labeo*, *Crenimugil crenilabis*	Intestine	Egypt, Mediterranean Sea, Black Sea, Red Sea
Haplosplanchnus purii Srivastava, 1939	*Mugil waigiensis*	Intestine	Bay of Bengal, Arabian Sea
Haplosplanchnus sp. Manter, 1947	*Mugil curema*	Intestine	Florida (USA)
Hymenocotta manteri Overstreet, 1969	*Mugil cephalus*	Intestine	Georgia to Louisiana (USA)
Hymenocotta mulli Manter, 1961	Mullet, *Mugil cephalus*	Intestine	Fiji Islands, Australia
Laruea caudata Srivastava, 1939	*Mugil waigiensis*	Intestine	Bay of Bengal, Arabian Sea
Schikhobalotrema elongatum Nahhas & Cable, 1964	*Mugil cephalus*, *M. curema*	Intestine, pyloric caeca	Curaçao, Jamaica, Florida (USA)
Schikhobalotrema sparisomae	*Mugil cephalus*	Intestine	Black Sea
Schikhobalotrema sp. Skinner, 1974	*Mugil cephalus*	Intestine	Florida (USA)
Schikhobalotrema sp.	*Mugil cephalus*	Intestine	Florida (USA)

(including Waretrematidae)

Chalcinotrema mugilicola (Shireman, 1964) Overstreet, 1971	*Mugil cephalus*	Intestine	Louisiana (USA)
Chalcinotrema simonei Travassos, Freitas & Bührnheim, 1965	*Mugil platanus*	Intestine	Brazil (South America)
Dicrogaster contracta Looss, 1902	*Mugil chelo*, *M. capito*, *M. cephalus*, *M. labrosus*, *M. saliens*	Intestine	Mediterranean Sea
Dicrogaster fastigata Thatcher & Sparks, 1958	*Mugil cephalus*	Intestine, pyloric caeca	Georgia to Louisiana (USA)
Dicrogaster perpusilla Looss, 1902	*Mugil chelo*	Intestine	Mediterranean Sea
Haploporus benedeni (Stossich, 1887) Looss, 1902	*Mugil chelo*, *M. cephalus*, *M. capito*, *M. saliens*, *M. auratus*, *M. labrosus*	Intestine, pyloric caeca	Mediterranean Sea, Britain, Israel
Haploporus lateralis Looss, 1902	*Mugil auratus*, *M. chelo*, *M. capito*	Intestine	Mediterranean Sea
Haploporus longicollum Wlassenko, 1931	*Mugil cephalus*, *M. auratus*, *M. saliens*	Intestine	Black Sea, Caspian Sea
Haplotrema flecterotestis Zhukov, 1971	*Mugil soiuy*	Intestine	Liao Ho River (China)
Isorchis manteri Martin, 1973	*Mugil cephalus*	Intestine	Queensland (Australia)
Lecithobotrys putrescens Looss, 1902	*Mugil cauratus*, *M. cephalus*, *M. capito*	Intestine	Mediterranean Sea
Lecithobotrys sprenti Martin, 1973	*Lisa argentea*, *Mugil cephalus*	Intestine	Queensland (Australia)
Lecithobotrys vitellosus Sharma & Gupta, 1970	*Mugil parsia*	Intestine	India
Paralecithobotrys brisbanensis Martin, 1974	*Mugil cephalus*	Intestine	Queensland (Australia)
Parasaccocoelium mugili Zhukov, 1971	*Mugil soiuy*	Intestine	Japan and China
Paraunisaccoides lobolecithus Martin, 1973	*Mugil cephalus*	Intestine	Queensland (Australia)

Table 13.1. (*cont.*)

Parasite	Host	Site	Locality
Saccocoelioides beauforti (Hunter & Thomas, 1961) Overstreet, 1971	*Mugil cephalus*	Intestine, pyloric caeca	North Carolina to Louisiana (USA)
Saccocoelioides pearsoni Martin, 1973	*Mugil cephalus*, *Trachystoma petardi*	Intestine	Queensland (Australia)
Saccocoelium obesum Looss, 1902	*Mugil cephalus*, *M. capito*, *M. auratus*, *M. chelo*, *M. saliens*, *M. labrosus*	Intestine, caeca	Mediterranean Sea, Red Sea, Black Sea, Caspian Sea, Egypt, Britain
Saccocoelium tensum Looss, 1902	*Mugil chelo*, *M. cephalus*, *M. auratus*, *M. capito*, *M. saliens*, *M. labrosus*	Intestine	Black Sea, Mediterranean Sea, Caspian Sea
Skrjabinolecithum spasskii Belous, 1954	*Mugil soiuy*	Intestine	Russia
Spiritestis arabii Nagaty, 1948	*Mugil* sp.	Intestine	Red Sea
Waretrema piscicola Srivastava, 1939	*Mugil waigiensis*	Intestine	Arabian Sea
Unisaccoides vitellosus Martin, 1973	*Mugil cephalus*, *Lisa argentea*	Intestine	Queensland (Australia)
Unisaccus brisbanensis Martin, 1973	*Mugil cephalus*	Intestine	Queensland (Australia)
Unisaccus spinosus Martin, 1973	*Lisa argentea*	Intestine	Queensland (Australia)
Monorchiidae Odhner, 1911			
Hurleytrema sp.	*Mugil curema*	Intestine	Gulf of Mexico
Lasiotocus glebulentus Overstreet, 1971	*Mugil cephalus*	Intestine	Mississippi to Florida (USA)
Lasiotocus mugilis Overstreet, 1969	*Mugil cephalus*	Intestine	Florida and Georgia (USA)
Paraproctotrematoides mugil Nagaty & Abel-Aal, 1972	*Mugil* sp.		Red Sea
Opecoelidae Ozaki, 1925			
Podocotyle reflexa (Creplin, 1825) Odhner, 1905	*Liza haematochila*		Sea of Japan

Numerous additional digeneans infect mullets. Even though a few species, such as two described by Looss (1902), could be synonyms of others as suggested by Dawes (1946), numerous other species remain undescribed. We are involved personally in describing several additional species. As can be observed from Table 1, a large number of the species have been described only recently (Manter, 1969; Overstreet, 1969, 1971, 1973; Sharma & Gupta, 1970; Yamaguti, 1970; Zhukov, 1971; Martin, 1973*a*, 1973*b*, 1973*c*, 1973*d*, 1974).

Even with only limited data on geographical distribution, we expect most of the digeneans would infect a wide range of mullets, but not most other fishes, should they come in contact. Mullets have been exemplified as 'ecological bridges' spreading haploporids from marine to fresh-water hosts (Manter, 1957). Although admittedly important, we agree with Cable & Isseroff (1969) that the euryhalinity of the fish provides a less decisive factor than the adaptation of marine molluscs to fresh-water habitats.

Adult trematodes probably minimally harm mullets in natural habitats. In cases where mullets are trapped with little water and high densities of snails, the hazard increases. By experimentally infecting juvenile *Mugil cephalus* with large numbers of *Dicrogaster fastigata*, we could kill the hosts (Heard & Overstreet, in preparation) and such infections could occur in ponds under improper management. Other haploporids probably would produce similar results.

Few life cycles of species from the Haploporidae, that infect mullet and related families, have been reported. Fares & Maillard (1974) infected mullet in the western Mediterranean with *Saccocoelium obesum* from a rissoid snail and with *S. tensum* and *Haploporus benedeni* from hydrobids. *Saccocoeliodes pearsoni* from the hydrobid *Posticobia brazieri* infected mullet in fresh-water areas of Australia (Martin, 1973*b*). Heard & Overstreet (in preparation) have determined some life cycles of mullet trematodes utilizing hydrobids in Mississippi. Snails from the northern Black Sea have also been reported to harbour parasites which probably infect mullet (Dolgikh, 1968). Since snails for these digeneans often have geographic limitations determined by narrow salinity concentrations, the presence in a mullet of specific trematodes can sometimes indicate where the individual fish had been.

Metacercariae

Digeneans from nine families utilize mullets as intermediate hosts to infect birds, mammals, and to a lesser extent, fishes. Table 13.2 lists reported species, most of which belong in the family Heterophyidae. Once again we reinforce our belief that many more exist. Pearson (1964) synonymized several species of haplorchines not listed in our table. More important, the piscine intermediate host of some heterophyid and related cryptogonimid species is less specific than the snail host, making mullets possible hosts for many species. The

Table 13.2. *Digenean metacercariae from mullets of the world*

Larva	Host	Locality
Strigeidae Railliet, 1919		
Cardiocephalus medioconiger Dubois & Perez-Vigueras, 1949	*Mugil cephalus*	North Carolina (USA)
Cyathocotylidae Poche, 1926		
Mesostephanus appendiculatoides (Price, 1934) Lutz, 1935	*Mugil cephalus*, *M. curema*, *M. trichodon*	Florida (USA)
Prohemistomum vivax (Sonsion, 1893)	*Mugil capito*	Egypt
Diplostomidae Poirier, 1886		
Bolbophorus confusus (Krause, 1914) Dubois, 1935	*Mugil saliens*	
Bucephalidae Poche, 1907		
Rhipidoctyle lepisostei Hopkins, 1954	*Mugil cephalus*, *M. curema*	Louisiana (USA)
Didymozoidae Poche, 1907		
Didymozoid larva	*Mugil cephalus*	Mississippi (USA), Israel
Echinostomatidae Poche, 1926		
Echinochasmus liliputanus (Looss, 1896) Odhner, 1910	*Mugil capito*	Egypt
Echinochasmus perfoliatus (Ratz, 1908) Gedoelst, 1911	*Mugil capito*	Egypt
Acanthocolpidae Luhe, 1909		
Stephanochasmus sp.	*Mugil cephalus*	Florida (USA)
Heterophyidae (Leiper, 1909) Odhner, 1914		
Adleriella minutissima (Witenberg, 1929) Witenberg, 1930	*Mugil cephalus*, *M. capito*	Israel
Ascocotyle coleostoma (Looss, 1896) Looss, 1899	*Mugil capito*, *M. saliens*	Caspian Sea, Black Sea
Centrocestus asadai Mishima, 1959	*Mugil cephalus*	Japan
Centrocestus caninus Leiper, 1913	*Mugil saliens*, *M. cephalus*	Hainan Island (China), Hawaii, Israel
Cryptocotyle concava (Creplin, 1825) Luhe, 1899	*Mugil capito*	
[illegible]	*Mugil cephalus*, *M. capito*	Israel

Galactosomum sanaensis Kobayashi, 1941	*Mugil affinis*	Hainan Island (China)
Galactosomum timondavidi Pearson & Prévot, 1971	*Mugil auratus*	France
Haplorchis parataichui Pearson, 1964	*Pseudomugil signifer*	Brisbane (Australia)
Haplorchis pumilio (Looss, 1896) Looss, 1899	*Mugil affinis*, *M. capito*, *M. cephalus*	Tunisia, Israel, Taiwan
Haplorchis taichui (Nishigori, 1924) Witenberg, 1930	*Mugil* sp., *M. affinis*, *M. cephalus*	Taiwan, Hainan Island (China), Hawaii, Japan
Haplorchis yokogawai (Katsuta, 1932) Chen, 1936	*Mugil dussumieri*, *M. affinis*, *M. cephalus*	Taiwan, Philippines, Hainan Island (China), South Sea Islands, Japan
Heterophyes aequalis Looss, 1902	*Mugil auratus*, *M. capito*, *M. cephalus*, *M. saliens*	Mediterranean Sea, Israel, Egypt
Heterophyes dispar Looss, 1902	*Mugil* sp., *M. auratus*, *M. capito*, *M. cephalus*, *M. saliens*	Egypt, Israel, Mediterranean Sea
Heterophyes heterophyes (Siebold, 1853) Stiles & Hassall, 1900	*Mugil auratus*, *M. capito*, *M. cephalus*, *M. saliens*, *M. curema*, *M. japonicus*, *Lisa* sp.	Mediterranean Sea and adjacent areas, Japan
Heterophyes nocens Onji & Nishio, 1916	*Mugil cephalus*, *M. japonicus*, *Lisa haematochila*	Japan
Heterophyes onchii Koga, 1922		Japan
Heterophyopsis continua (Onji & Nishio, 1916) Price, 1940	*Mugil* sp., *M. affinis*, *M. cephalus*	Japan
Heterophyopsis expectans (Africa & Garcia, 1935) Tubangui & Africa, 1938	*Mugil* sp., *M. dussumieri*	Philippine Islands
Metagonimus minutus Katsuta, 1932	Mullet, *Mugil cephalus*	Taiwan, Japan
Metagonimus takahashi Suzuki, in *Takahashi*, 1929	*Mugil cephalus*	Japan
Metagonimus yokogawai (Katsurada, 1912) Katsurada, 1913	*Mugil cephalus*	Japan
Phagicola italicus (Alessandrini, 1906) Price, 1932	*Mugil* sp., *M. capito*	Tunisia, Israel, Egypt
Phagicola longicollis Kuntz & Chandler, 1956	*Mugil cephalus*	Egypt

Table 13.2. (*cont.*)

Larva	Host	Locality
Phagicola longus (Ransom, 1920) Price, 1932	*Mugil cephalus*, *M. curema*, *M. trichodon*	Southeastern United States, Egypt, Israel
Phagicola sinoecum (Ciurea, 1933) Sogandares-Bernal & Lumsden, 1963	*Mugil cephalus*, *M. capito*	Black Sea
Phagicola witenbergi (Ciurea, 1933) Price, 1932	*Mugil cephalus*, *M. capito*	Rumania, Israel
Procerovum calderoni (Africa & Garcia, 1935) Price, 1940	*Mugil dussumieri*	Philippines
Procerovum varium Onji & Nishio, 1916	*Mugil affinis*, *M. cephalus*	Hainan, South Sea Islands, Japan
Pygidiopsis genata Looss, 1907	*Mugil* sp., *M. capito*, *M. dussumieri*	Philippines, Egypt
Pygidiopsis marvillai Refuerzo & Garcia, 1937	*Mugil* sp., *M. dussumieri*	Philippines
Pygidiopsis summa Onji & Nishio, 1916	*Mugil* sp., *M. cephalus*	Japan
Stellantchasmus falcatus Onji & Nishio, 1915	*Mugil* sp., *M. cephalus*, *M. dussumieri*, *M. capito*	Japan, Philippines, Hawaii, Egypt, Israel, Australia
Stictodora adulescentia (Onji & Nishio, 1916) Yamaguti, 1958	*Mugil cephalus*	Japan
Stictodora fuscata (Onji & Nishio, 1916) Yamaguti, 1958	*Mugil cephalus*	Japan
Stictodora guerreroi Garcia & Refuerzo, 1936	*Mugil* sp., *M. dussumieri*	Philippines
Stictodora manilensis Africa & Garcia, 1935	*Mugil* sp., *M. dussumieri*, *M. cephalus*	Philippines, Hong Kong
Stictodora perpendiculum (Onji & Nishio, 1916) Yamaguti, 1958	*Mugil* sp., *M. cephalus*, grey mullet	Japan
Stictodora pusilla (Onji & Nishio, 1916) Yamaguti, 1958	*Mugil cephalus*	Japan
Stictodora sawakinensis Looss, 1899	*Mugil* sp., *M. capito*, *M. cephalus*	Egypt, Israel, Russia, Philippines
Sticodora sp. Hutton & Sogandares-Bernal, 1960	*Mugil trichodon*	Florida (USA)
Cryptogonimidae (Ward, 1917) Ciurea, 1933		
Metadena oviformis (Kobayashi, 1915) Overstreet, 1971	*Mugil cephalus*	Japan

reverse, however, does occur for some heterophyids. Since metacercariae of different species appear similar, experimental feeding to uninfected birds or mammals must confirm many identifications. The identities of many larval cysts have never been established. Some, such as *Metacercaria nasegawai*, have been named (Fan, 1970).

As with most helminths, infections with digeneans debilitate young fish much more readily than their adult counterparts. Reports of the pathological changes caused by larval trematodes in mullet, however, are minimal. Lee & Cheng (1970) described minimal necrotic degeneration of myofibres adjacent to a metacercaria of *Stellantchasmus falcatus* in *Mugil cephalus*. Large fat cells occupy the spaces vacated by the degenerated cells. Ponds with snails infected with any mullet-trematode could affect the enclosed mullet, especially the larval or young individuals.

Heterophyid infections

Because of the significance of heterophyids to public health, we will expand on some members of that family. Extremely prevalent in the eastern Mediterranean waters, heterophyid metacercariae turned up in the muscles of over 90% of the mullet which reached local markets in Israel (Paperna, 1975) and Egypt (Wells & Randall, 1956). Nearly every mullet in the brackish Lake Manzala is infected (see citations by Hamed & Elias, 1970). They were also common in mullet from the markets of Ankara (Mimioglu & Sayin, 1957), Tunis (Balozet & Callot, 1939), and elsewhere where mullet are sold, as implied by Table 13.2. Since final hosts constitute piscivorous birds and mammals, humans, dogs and cats can be hosts. Some heterophyids infect but a few definitive hosts, whereas others have been reported from a wide range of hosts.

Heterophyids constitute a public health problem wherever people eat raw, salted, or otherwise improperly prepared mullet or other infected fishes. *Heterophyes heterophyes*, the predominant species encysting in eastern Mediterranean mullets, has been reported frequently to infect humans in Egypt, where it is endemic in communities of fishermen situated along lakes of the Nile Delta. From one particular village near Lake Manzala where up to 90% of the school-children and 22% of the adults were infected, Kahlil (1933, 1937) considered the large number of children and adults who operated fishing boats maintained the infections. Similarly, a majority of dogs and cats in the area also possessed *H. heterophyes*, in addition to other heterophyids which do not infect man (Kuntz & Chandler, 1956). Witenberg (1929) reported a few humans and 70% of the dogs and 80% of the cats to be infected in Israel. *Heterophyes nocens*, possibly a synonym of *H. heterophyes*, infects people in the Far East. In Japan, Korea and China, prevalence of human infections with *H. nocens* was as high as 18% in some communities where people consumed raw mullet (Onji, 1915; Asada, 1928). At least fourteen species of heterophyids, not all acquired from mullet, have been reported from man (Faust *et al.*, 1970). Humans infected with *Stellantchasmus falcatus* have been associated with the consumption of mullets in Hawaii and are also implicated in Japan. Other references cited by Faust, Russell & Jung (1970) reported *Diorchitrema pseudocirratum* in native Hawaiians who ate raw mullet. In the Phillipines, six different heterophyids were recovered from human hosts and metacercariae of at least four of them occur in mullet.

Symptoms of infection with *H. heterophyes* include abdominal discomfort, nausea, headache, vomiting, diarrhoea, and, in severe cases, dysentery (Khalil, 1937). Similar symptoms also characterize infections with other heterophyids. Such infections can result in complications. Africa & co-workers (Africa, Garcia & De Lyon, 1935, 1937) working with a few different heterophyids and a microphallid, found that in some cases, eggs from the intestine have infiltrated into various organs and resulted in lesions in the heart, liver, lungs and central nervous system. Kean & Breslau (1964) reported 14.6% of the fatal heart cases in the Philippines resulted from heterophyid myocarditis. Adult *H. heterophyes* have also been reported being encapsulated in the human brain (Deschiens, Collomb & Demarchi, 1958).

To present an example of the biology of heterophyid, we will discuss *Heterophyes heterophyes* from the Middle East in some detail. Other species, of course, differ in some of the details, but they all utilize snails and fishes as intermediate hosts. Aspects of some of these species also will be presented.

Khalil (1937) presented a full account of the life history of *H. heterophyes* in Egypt. According to Martin (1959), who also described the life cycle, *H. heterophyes* in Egypt may represent two species or subspecies. *Pirenella conica* (Fig. 13.23) feeds on the infective eggs and serves as the molluscan host which sheds pleurolophocercous cercariae which then penetrate and infect the mullet. Several nonmugilid fishes also become infected. The same snail can host other heterophyids, and the brackish-water snail *Cerithidia cingula alata* acts as the snail-host for *H. nocens* in Japan (Asada, 1928). *Pirenella conica* thrives in hypo- and hyper-saline shallow-water lagoons; it is found in large numbers in lakes of the Nile Delta and in the Bardawil Lagoon of the northern Sinai coast. Additionally, it occurs in seasonal or perennial salt-water swamps. Marine culturing ponds dug along the Mediterranean coasts became heavily populated with it within one to two years after their construction. In ecosystems with extreme biotic conditions where the water's salinity can fluctuate between 5 and 20‰ or 40 and 100‰ and the temperature range between 5 °C in the winter to over 38 °C in the summer, *P. conica* often establishes dense populations and dominates the habitat, particularly near the water's edge. In Lake Burullus in the Nile Delta, up to 98% of the snails harboured heterophyids (Kuntz, 1957).

To give an idea of cercarial production, *P. conica* infected with *H.*

Fig. 13.23. *Pirenella conica*, snail intermediate host of heterophyiids in the Near East, × 10.

heterophyes yielded 488 cercariae per day, all infective to mullet (Khalil, 1939). Most of these are shed during midday, when the temperature is high, and remained active for 24 to 48 h. Some specimens of *P. conica* from the Little Bitter Lake infected with one or more species of *Heterophyes* continued to release cercariae, but in gradually decreasing numbers, for approximately one year (Martin, 1959). Just like the snail and fish hosts, the cercaria showed great tolerance to changes in salinity. It was reported to have survived increasing salinities from 22 to 60‰, and not only up to 44‰ as reported by Khalil (1939).

The brackish and hypersaline lagoons along the southeastern Mediterranean coast where *P. conica* thrives, provide important nursery and feeding grounds for mullets and other euryhaline fishes. Commercial fishermen thus utilize several of these lagoons heavily. Additionally, the habitats support aquatic birds, particularly gulls and terns, some of which harbour adult heterophyids.

In commercial-sized mullets from eastern Mediterranean waters where they nearly all act as host to heterophyids, the load of metacercariae in the muscles may be great. Witenberg (1929) reported finding over 1000 cysts in 1 g of the muscle of *Mugil cephalus*. From *M. capito* in the Bardawil Lagoon, Paperna (1975) counted as many as 6000 cysts/g of muscle tissue, and 1136 cysts/g from *M. cephalus* taken at the same locality. Paperna estimated the total number of metacercariae in the above-examined 255 mm long *M. capito* at 582000 (Fig. 13.24).

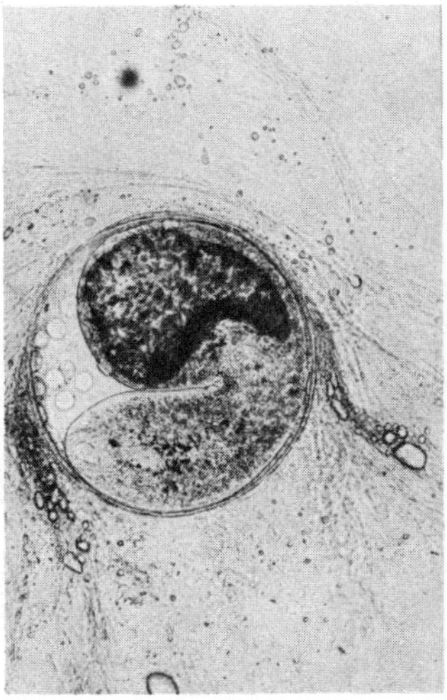

Fig. 13.24. Metacercaria of *Heterophyes heterophyes* in muscles of *Mugil capito* from Bardawil lagoon, × 100, live specimen.

Such a concentration of metacercariae results in extemely heavy infections in definitive hosts. Witenberg (1929) counted 13000 specimens of *H. heterophyes* from one dog. A massive infestation is not only a public health menace, but it might also be detrimental to the fish. Although we lacked conclusive data concerning cause, emaciated fish from the Bardawil Lagoon have been found heavily infected with metacercariae. In one such catch, eleven emaciated *M. capito* taken in January 1974 had 1730 to 6000 metacercariae/g of tissue. Young fish, because of their small size, are apparently more vulnerable to harm by heavy infections.

The burden of infection in the fish examined varied among species of mullet as well as between individuals. It could be due to differences in susceptibility or habitat preference. In the Bardawil Lagoon where incidence of infection for all mullets was 100%, *M. capito* carried the heaviest infection load, followed by *M. cephalus*, *M. auratus*, *M. saliens* and *M. chelo* (see Paperna, 1975). Heterophyids, including three species of *Heterophyes*, also occurred in other fishes, but not in the numbers encountered in mullets (Table 13.3).

In the Bardawil Lagoon, where active transmission of heterophyids occurred, metacercariae infected mullet as small as 21–30 mm. After reaching 70 mm in length, all mullet were infected. Young fish hosted *Stictodora sawakiensis* and *S. aequalis*, in addition to *H. heterophyes*. In the commercial-sized fish, however, *H. heterophyes* predominated (Table 13.4). In habitats such as those along the eastern Mediterranean coast, where *P. conica* was rare and occurred only in a few isolated localities, heterophyid infections in

Table 13.3. *Infections by heterophyid metacercariae in fishes from Bardawil Lagoon (Paperna, unpublished) and lakes of the Nile Delta (after Wells & Randall, 1956)*

Species	Bardawil Lagoon			Nile Delta Lakes	
	No. of fish examined	Prevalence (%)	Maximal no./g of tissue	No. of fish examined	Prevalence (%)
Mugil capito	37	100	6000	67	91
Mugil cephalus	88	99	1136	114	100
Mugil auratus	38	74	362	105	100
Mugil chelo	7	100	55	—	—
Mugil saliens	6	83	69	—	—
Dicentrarchus labrax	42	43	58	—	—
Dicentrarchus punctatus	22	86	88	—	—
Sparus auratus	54	37	3	—	—
Argyrosoma regius	20	5	7	44	45
Solea vulgaris	10	0	0	84	55
Tilapia nilotica	—	—	—	108	69
Tilapia zillii	—	—	—	58	67

Table 13.4. *Percentage of different heterophyids recovered from dogs fed mullet (a) and from digested mullet (b)*

Species and length of fish (mm)	Locality	Percentage				
		Heterophyes heterophyes	*Heterophyes dispar*	*Heterophyes aequalis*	*Stictodora sawakiensis*	*Stellantchasmus falcatus*
Mugil cephalus	Bardawil					
300 (a)		100	0	0	0	0
260–320 (b)		100	0	0	0	0
Mugil capito	Bardawil					
170–190 (b)		77	0	18	5	0
Mugil auratus	Bardawil					
170–190 (b)		100	0	0	0	0
Mugil sp.	Bardawil					
20–40 (a)		37	0	15	48	0
50–60 (a)		22	0	57	21	0
Mugil capito	Coastal					
225 (a)	Israeli	100	0	0	0	0
90–150 (a)	stream	0	0	0	0	100
Mugil spp.	Israeli[a] markets	94	5	1	0	0

[a] Based on values obtained by Witenberg (1929).

young-of-the-year mullet in coastal streams were low, and, if present, they consisted mostly of *Stellantchasmus falcatus*. *Heterophyes heterophyes* was present in larger fish from these habitats, but the hosts apparently acquired the infections elsewhere.

Composition of the different species of heterophyids present in an area depends on prevalent definitive hosts, as well as molluscan hosts. In the lakes of the Nile Delta, humans, dogs and cats comprise the main reservoirs, whereas in the Bardawil, located in the sparsely populated sand desert of northern Sinai, predominant reservoirs appear to be wild mammals and aquatic birds. Witenberg (1929) reported nine heterophyids, of which *H. heterophyes* was the most common, from dogs, cats and rabbits that had been fed infected mullets from local markets in Israel.

Safeguards for human consumption

Various areas of the world pose different problems concerning human heterophyid infections. The heterophyids differ, as do the habitats and dietary customs of people eating mullet and other fishes from those areas. With changes in methods of preparing mullets, the prevalence of human infections could change. In areas such as the southeastern United States where most fish is cooked, few human infections of *Phagicola longus* occur, but the consumption of cold-smoked, salted or raw mullet could easily modify the public health picture. *Phaicola longus* is of interest to us because host-specificity for it differs from that for *H. heterophyes*. At least five species from three different genera of snails host *P. longus* (Richard Heard, personal communication), but only mullets apparently act as piscine hosts. Adults occur in both birds and mammals. To give an idea of intensity of infections, we cite the work by Courtney & Forrester (1974) which reported an average of 11 849 worms in each of fourteen brown pelicans examined from Louisiana.

Eating infected mullet does not necessarily result in human infections. Several methods exist for safeguarding the product. Heterophyid metacercariae in fish musculature survived between nine and thirteen days when refrigerated at 4°–6 °C or iced at 2°–4 °C (Khalil, 1937; Hamed & Elias, 1970; Paperna, unpublished). Freezing the fish at −10° or −20 °C for 30 h was not sufficient to kill all the parasites (Hamed & Elias, 1970), but after 33 h no metacercariae were viable. We found deep-freezing at −18 °C for 24 h killed all metacercariae. According to Khalil (1937), heating the infected meat to 50 °C for a few minutes will also kill the worms. Hamed & Elias (1970), however, found worms alive after 180 min at 50 °C and after 10 min at 100 °C. Metacercariae in fish embedded in powdered salt did not survive longer than 7 days. Checking fully-processed, salted, infected mullet sold from El Arish markets in Sinai, we found no viable metacercariae. Hamed & Elias (1971) reported that the antibiotics oxytetracycline and chlortetracycline, currently used to prolong storage-life of fresh fish, were ineffective in killing metacercariae. Concentrations of the drugs necessary to kill the worms far surpassed those limits acceptable for human consumption.

In some areas of the Far East, people commonly eat raw fish, and consequently, infections are prevalent. In the Near East, few people consume raw fish, and those who become infected probably do so mostly from consuming insufficiently cooked or salted fish. Restricting the sale of inadequately-processed fish could substantially reduce risk of infection.

In some communities, special traditions of cookery will certainly result in human infections. During their extended fishing trips, Egyptian fishermen grill mullet over wooden sticks, removing a blackened fish to eat before it becomes burned. Using similar methods, Hamed & Elias (1970) found the temperature of inner portions of the irregularly processed fish only reached 54°–61 °C and 20–30% of the fish contained live metacercariae. In Israel, prior to 1948, heterophyasis appeared sporadically among Jewish women. Ben Ari (1957) reported sixty-two cases from the years 1934–39 (0.71% of the Jewish women) and eighty cases from 1940–47 (0.19%). Since 1948, no cases have been reported. Infection in those women appeared to result from their tasting minced raw fish during preparation of fish-cakes (Witenberg, personal communication). After 1948, carp gradually replaced mullets as the fish of choice for such dishes.

Cestoda

Mullets, though usually not infected with adult tapeworms, often act as intermediate and paratenic hosts of several plerocercoid larvae. Most of these can be recognized by their characteristic scolex, which is a symmetrical terminal structure armed with attachment organs. The only known adult cestode from mullet is the unsegmented amphilinid *Gyrometra mugilitaenia* Parveen described from the body cavity of *Mugil* sp. from a fish market at Karachi, Pakistan, by Parveen (1971).

When most people think of cestodes, they picture the hermaphroditic ribbon-like adults with a long chain of segmented proglottids. Plerocercoids are seldom segmented, but they do contain a scolex which may be similar to the adult, like in trypanorhynchs, or different, like in tetraphyllideans. Like the adult, they possess calcareous corpuscles and lack an alimentary tract. They are presumably acquired when the mullet feed on infected crustaceans.

With the possible exception of an unidentified pseudophyllidean reported from *Agonostomus monticola* in Cuba (Moravec & Baruš, 1971), all plerocercoids in mullet mature in elasmobranch fishes. Euzet (1962) tried unsuccessfully to infect young *Mugil auratus* with procercoids of the pseudophyllidean *Fistulicola plicata* (Rudolphi) from copepods of the genus *Acartia*.

The most common cestodes in mullet are larval tetraphyllideans, for which

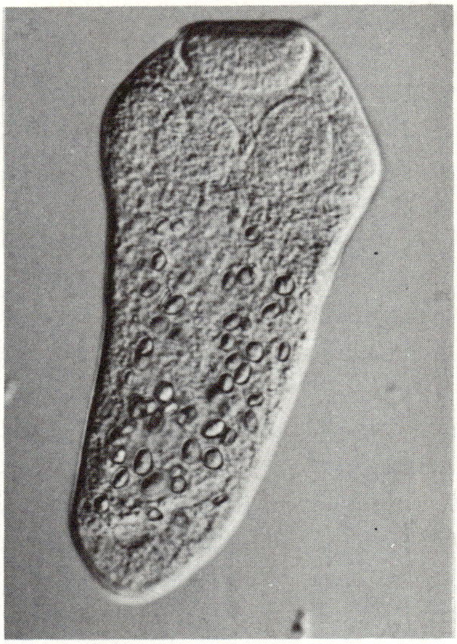

Fig. 13.25. Living *Scolex polymorphus* (Cestoda, Tetraphyllidea) from the intestine of juvenile *Mugil cephalus*, in Mississippi, × 100, phase illumination.

the identifications are presently undetermined since the scolex, the prime identifying character, usually consisting of four borthridia, is altered considerably in morphology with maturity. Some collective group-names include *Scolex pleuronectis* and *S. polymorphus*. Varieties of these designate a large number of species. *Scolex pleuronectis* has been reported from the intestine and pyloric caeca of *Mugil auratus*, *M. capito*, *M. cephalus*, *M. chelo*, *M. labeo* and *M. saliens* from Israel and the Black Sea (Reshetnikova, 1955; Paperna, 1964*a*). A different species also designated *S. polymorphus* occurs, at least, in the cystic duct of *M. cephalus* from Florida and Mississippi (Fig. 13.25).

An unarmed *S. polymorphus* from the cystic duct of a sea trout was fed to the sand shark by Curtis (1911) and apparently developed into the well-armed *Phoreiobothrium trilocutatum* Linton.

A species similar to those found in the Middle East also commonly infects juvenile *M. cephalus* and *M. curema* in Mississippi. This worm, classified under the group-name *Scolex polymorphus*, appears to be highly pathogenic to young mullet when present in large numbers. We observed as many as 4500 individuals packed within the intestine of a small mullet. Observations concerning infections and the effect on the host are covered later in a section on host-related aspects. Because the larvae most certainly differs from most other similar appearing species, we present average measurements of twenty specimens, with ranges given in parentheses. The tetraphyllidean is 229 μm (150–287) long by 101 μm (85–120) wide and has unloculated bothridia 38 μm

(32–45) long by 34 μm (30–38) wide. The apical sucker is 30 μm (25–32) long by 39 μm (35–47) wide.

As already indicated, there are probably several species present in mullet. Another minute tetraphyllidean, *Echeneibothrium tobijei* Yamaguti, also infects *M. capito* in Israel and, as more mullet, especially juveniles which feed on small crustaceans, become examined, similar larvae will be reported. An unidentified *Rhinebothrium* sp. with bi-lobed bothridia encysts in the mesentery of *M. cephalus* in Mississippi.

Easier to identify because they possess four tentacles with characteristic hooks and hook-patterns, but less common in mullets, are trypanorhynchs. An unidentified species was reported in the intestine of *M. curema* in the Gulf of Mexico (Nikolaeva & Prukhin, 1968), and from the viscera of juvenile *M. capito* in Bardawil Lagoon in Sinai. *Lacistorhynchus tenuis* (van Beneden) and *Otobothrium mugilis* Hiscock occur in the body cavities of *Aldrichetta forsteri* in New Zealand and *M. cephalus* in Australia, respectively. '*Tetrarhynchus*' *mugilauratus* Wagener infects *M. auratus*.

Infections are not necessarily widespread. Hiscock reported that specimens of *O. mugilis* were observed in 75% of the mullet from the estuarine Brisbane River between early 1935 until he published his paper, but none was seen previous to that date nor in commercial mullet caught from other localities.

Nematoda

Mullets host relatively few roundworms and those worms that do infect mullets are usually larvae. Infection for some occurs when they feed on intermediate hosts and for others when they feed on eggs or larvae. Primarily elongate spindle-shaped worms, both large and small individuals possess a relatively tough cuticle underlayed by bands of longitudinal muscles. These muscles and the absence of circular muscles cause the worms to exhibit a characteristic wave-like movement. For most species it is difficult to distinguish males from females in the larval stage, which most often occurs in cysts or free in mesentery, visceral organs or musculature.

Of the nematodes infecting mullet, some present potential human health hazards and others may discourage people from eating mullet because of the aesthetically displeasing appearance of the product. One good example of the latter is the larval *Contracaecum robustum* Chandler which infects *Mugil cephalus* and *M. curema* in the southeastern United States. The life history and the adult form of this worm, presumably from a bird, remain uncertain, but not the unwillingness of the consumer to purchase fish infected with the 2–3 cm long reddish to cream-coloured worm in the liver, kidneys and adjacent tissues. Coiled, curved, or straight, this relatively-large larva can replace much host tissue. It was reported from Texas (Chandler, 1935) and the Adriatic Sea (Nikolaeva & Naidenova, 1964). The larva also occurs in

M. cephalus and *M. curema* from Louisiana, Mississippi, Alabama and Florida; we noted it most heavily from near Grand Chenier, Louisiana, where the infection may have affected the hosts' health. Infections in small fish probably result in many dead fish or make hosts especially vulnerable to prey.

Several unidentified species of larval *Contracaecum* Raillet & Henry, some of which can or have been assigned with the genus *Thynnascaris* Dollfus, a group whose species infect fishes as adults rather than birds or mammals, have also been reported from mullets. These include some from Israel (Paperna, 1964*a*), Florida (Skinner, 1974) and New Zealand (Brunsdon, 1956); certainly others exist elsewhere. Brunsdon (1956) also reported a related larval *Anisakis* sp. from *Aldrichetta forsteri*.

Larval ascaridoids, which comprise all the above-mentioned nematodes, constitute a potential human health hazard if the infected fish is inadequately prepared. Some species cause lesions in man and produce symptoms similar to those caused by an ulcer or cancer of the alimentary tract. Numerous reviews on the 'anisakiasis' problem have been written, as exemplified by the one of Oshima (1972). Norris & Overstreet (in press) provided a recent discussion on larval *Thynnascaris* species with their implications to public health.

Cuccullanus mugili Belous occurred as an adult in the intestine of *Mugil soiuy* in the Primorsk region of USSR (Belous, 1965), and one of eighty-three *M. cephalus* from near Miami, Florida, harboured *C. dodsworthi* Barreto (Skinner, 1974). Another species, in addition to one of the related *Cucullanellus* Tornquist, infected *M. cephalus* in New Zealand (Brunsdon, 1956). The deterimental effects on the hosts of these nematodes, as well as those to be discussed below, are not established; but they are probably minimal, at least for moderately heavy infections in natural populations of adult hosts. Other nematodes reported from mullets include *Philometra biglobocerca* Belous in the body cavity of *M. soiuy* from USSR and immature specimens of *Philometra* sp. in the intestine of *M. cephalus*, *M. labeo* and *M. saliens* from Israel. Infecting mullets also are spiruroids such as *Spinitectus agonostomi* Moravec and Barus in Cuba, *Heliconema* sp. of Bilqees & Khanum in Pakistan, and *Hedruris spinigera* Baylis in New Zealand, from *Agonostomus monticola*, *M. hamiltoni* and *Aldrichetta forsteri*, respectively.

Rawson (1973*b*) reported light infections of *Capillaria* sp. from *M. cephalus* in Georgia (USA). Possibly of significance are the human infections in the Philippines with *C. philippinensis* Chitwood, Valesquez & Salazar. Human deaths were attributed to infection with the nematode, and most infected individuals had a history of eating raw fish and shrimps.

Acanthocephala

Several acanthocephalans, also known as spiny-headed worms, infect the intestines of mullets. Members of this phylum superficially appear like nematodes, but closer examination of any reveals a protrusible spiny proboscis and no intestine. They have simple life-cycle. Once the embryo within a diatom-like shell reaches a certain point of development, it can be released from the female worm and continues development only when eaten by a specific arthropod. There the acanthor stage metamorphoses, soon to become a juvenile which is infective to the fish. Occasionally, paratenic hosts are involved in cycles.

With two exceptions, species reported from mullet belong to the family Neoechinorhynchidae. Those exceptions are the diplosentid *Pararhadinorhynchus mugilis* Johnston & Edmonds in *Mugil cephalus* from Australia and the gigantorhynchid *Mediorhynchus sharmai* Gupta & Lata from India. Species of the latter archiacanthocephalan genus typically infest terrestrial birds. An unreported exception, much smaller in size, infects young mullet from Bardawil Lagoon in Sinai.

The neoechinorhynchid species invite studies on the zoogeography and speciation of the worms and on the phylogeny of mullets, using the acanthocephalans as indicators, since the parasites presumably evolved at a slower rate than their hosts. From India and Pakistan are reported *Neoechinorhynchus bangoni* Tripathi (in *Mugil tade*), *N. chilkaensis* Podder (in *M. cephalus*), *N. elongatus* Tripathi (in *M. subviridis* and *M. dussumieri*) and *N. karachiensis* Bilqees (in *M. spegieleri*), as well as *Gracilisentis mugilus* Gupta & Lata (in *Mugil* sp.) and *Dispiron mugili* Bilqees (in *M. buchanani*). From nearby Yugoslavia and adjacent areas *N. agile* (Rudolphi) (in *M. cephalus* and *Mugil* spp.) is reported. The same species also occurs in *M. cephalus* and *M. capito* along the Mediterranean coast of Sinai and Israel, as well as the Gulfs of Elat and Suez in the Red Sea. In the Red Sea it infects *M. subviridis*, *Crenimugil crenilabis* while another species *N. coiliae* appears in *Liza carinata*. In the more distant Americas occur *Floridosentis mugilis* (Machado Filho) (in *Mugil* sp. and *M. platanus* from Brazil), *F. pacifica* Bravo-Hollis (in *M. curema* from the Pacific coast of Mexico) and *F. elongatus* Ward (in *M. cephalus* and *M. curema* from the Atlantic coast of Florida to Texas, from Puerto Rico and from Baja California, Mexico). Unidentified species have been reported (Travassos, Kohn & Da Silva Motta, 1963) and probably other species exist. In general, none of the above acanthocephalans causes harm to mullet in their natural environment.

An unidentified juvenile of the neoechinorhynchid genus *Paulisentis* infected the intestine of *Agonostomus monticola* in Cuba (Moravec & Baruš, 1971). The mullet could be an accidental or a paratenic host for that worm. Most known life cycles of neoechinorhynchids incorporate ostracods as intermediate hosts, with snails and crayfish presumably occasionally acting as paratenic hosts (Merritt & Pratt, 1964). The exception is *Paulisentis fractus* Van Cleave & Bangham, which uses a copepod (Cable & Dill, 1967). Life cycles for those neoechinorhynchids infecting mullet have not been determined, but considering the consistency in acanthocephalan life cycles, mullets probably acquire the worms from feeding on ostracods, copepods and snails.

Hirudinea

The only annelid group infesting mullets are leeches. One could argue that leeches act as micropredators rather than parasites, but some leeches exhibit considerable host-specificity. One leech, *Myzobdella lugubris* Leidy, may affect a mullet detrimentally when heavy infestations of it occur. From estuarine and fresh-water habitats in Mississippi, *Mugil cephalus*, *Paralichthys lethostigma* and *Fundulus grandis*, in addition to the blue crab, *Callinectes sapidus*, act as primary hosts for the species, whereas *Ictalurus catus* replaces or serves in conjunction with the mullet as a primary host in some other areas (Sawyer, Lawler & Overstreet, 1975). *Mugil curema* and other fishes also host *M. lugubris* and we also note that both Rawson (1973*b*) and Skinner (1974) reported *Piscicola* sp. from *M. cephalus* along the Atlantic coast of the United States. In both reports, the leech could be *M. lugubris*. In some cases in Mississippi, we noted lesions on the fish associated with attached leeches. Excessive blood-loss probably occurs only with intense and prolonged infestations. Small hosts, of course, could be more easily harmed and secondary infections can follow any infestation.

Leeches ingest blood from their mullet hosts, and, as discussed by Overstreet (1974), Sawyer *et al.* (1975) and others, they probably act as vectors for the protozoan parasites inhabiting blood of mullet and other fishes.

Glochidia

Glochidia attach to the gills, fins or other sites of many fresh-water fishes. They are larval stages of fresh-water bivalves of the Unionidae. Unless closely examined to determine the presence of a bivalve shell and hooks or small spines, an individual could be confused with a larval trematode. The larva released from the parent clam becomes incorporated by host-tissue of a fish or amphibian and, if the host is an acceptable one, the glochidia will develop and metamorphose into a young clam. Some can severely harm their hosts.

Even though the record by Mikailov (1958) of a glochidia on grey mullet in the Caspian Sea is the only citation for mullet, these fish are potential hosts whenever they occupy fresh-water habitats.

Mullet as toxicants to man

Disease-causing agents other than parasites can be obtained by and be harmful to man when he eats mullet. Probably the most serious of the diseases is ciguatera poisoning (Halstead, 1967). Ingestion of a wide range of fishes, including *Chelon engeli*, *C. vaigiensis*, *Crenimugil crenilabis* and *Mugil cephalus*, whether cooked or not, can result in the disease. Neurological and gastrointestinal symptoms of a wide and varied nature characterize the poisoning. Although the origin of the toxin or toxins remains unknown, implicated have been noxious substances from benthic algae acquired during feeding activities of herbivorous fishes which can then be concentrated in predators after feeding on those fishes, apparently without harming them. Humans also concentrate the toxin, without symptoms, but when a threshold value is surpassed, symptoms are dramatic and they may last for weeks or months. Death occurs in about 7% of those poisoned.

Considering the amount of mullet eaten throughout the world, this group of fishes is rarely implicated in the disease. In fact, sometimes a species of fish from one area will be toxic whereas individuals from a nearby area are not. Banner & Helfrich (cited by Halstead, 1967) reported such a case for mullet in the Marquesas Islands.

More specifically associated with mullet is 'hallucinatory mullet (or fish) poisoning'. It occurs in mullet and several other fishes in the tropical Pacific and Indian Oceans and affects primarily the central nervous system of man. Symptoms such as loss of equilibrium, hallucinations, mental depression, sensation of a tight constriction around the chest and conviction of approaching death or other fantasy can develop within a few minutes to two hours after ingestion of the mullet and persist for a day or longer. Deaths have not been reported. Other than the fact that cooking does not destroy the toxin, little is known about it. Helfrich & Banner (1960) first reported the toxin from mullet in Hawaii and it, or a similar substance, is now known to occur in a few other fishes.

Hyuga fever, also known as Kagami fever, has for its aetiologic agent *Rickettsia sennetsu*. Kitao, Farrell & Fukuda (1973) report that this disease may occur in man after ingestion of raw *Mugil cephalus*. They have collected experimental data implying the aetiologic agent is transmitted by endoparasites in the mullet. The disease needs additional study.

Taints

The above toxins and infective agents consist of those toxic to humans, but not the mullet. Hydrocarbons which taint the tissue of mullet can be toxic both to man and fish. Man is primarily concerned because the taste of the fish-product makes it unpalatable or, at least, distasteful. The taint can be acquired from hydrocarbons synthesized by aquatic animals or derived from petroleum pollutants. Connell (1974) identified the taint as a mixture of hydrocarbons, closely resembling commercial kerosene, in *Mugil cephalus* from the Brisbane River in Australia. His evidence suggested the fish acquired the taint by consuming sediments or by absorption through the gills. He also cited articles on other tainted mullets and other fishes.

Consumers of mullet from the Mediterranean coast of Israel, especially from near large cities and ports, commonly complain of off-flavour and taints. The mullet congregate in heavily polluted areas where they feed both on sewage and on matter saturated with petroleum products. Presumably, the acquired pathogenic bacteria, toxic organic substances and heavy metals can all be transmitted to man when he eats the contaminated fish.

Unpublished studies conducted at the Gulf Coast Research Laboratory by Julia Lytle (personal communication) indicated that, in addition to altering the flavour of the mullet's flesh, some hydrocarbons kill the fish. She was able, even with low concentrations, to cause mortalities in *M. cephalus* with Arabian crude oil and with specific hydrocarbons. She exposed, seasonally, *M. cephalus* between 1 and 30 cm in total length to various concentrations of crude oils. Lethal concentrations of the oils for 50% of the fish after 96 h varied with the different crude oils used, but not with length of fish or season. The high-sulphur Arabian crude oils containing high levels of aromatic hydrocarbons were approximately three times more toxic than the high-paraffin, low-sulphur, crude oils from the Gulf of Mexico. Mullet, however, survived much higher concentrations of the oils than an oyster and two species of commercial penaeid shrimps. In fact, *M. cephalus* withstood concentrations twenty times higher than shrimp.

Exposure to relatively low levels of oil altered the feeding behaviour of the mullet. The fish responded to food slower and ate less than non-exposed fish, resulting in loss of weight and slower growth rates. At higher levels of exposure, fish lost equilibrium and would spiral to the water's surface before falling to the bottom of their tank. This behaviour was usually repeated often. Erratic movements also characterized fish presented lethal concentrations of the oils.

When mullet had injuries or appeared unhealthy, they died quickly when in contact with the oils. Consequently, Lytle did not utilize such fish in her experiments, but the increased susceptibility of the fish most certainly relates to potential problems of environmental stress.

Other diseases

Many nonparasitic diseases also affect mullets. Several of these we discuss briefly in this or other sections. Most of these do have an influence on the effect of parasitic infections. When a host is malnourished or otherwise

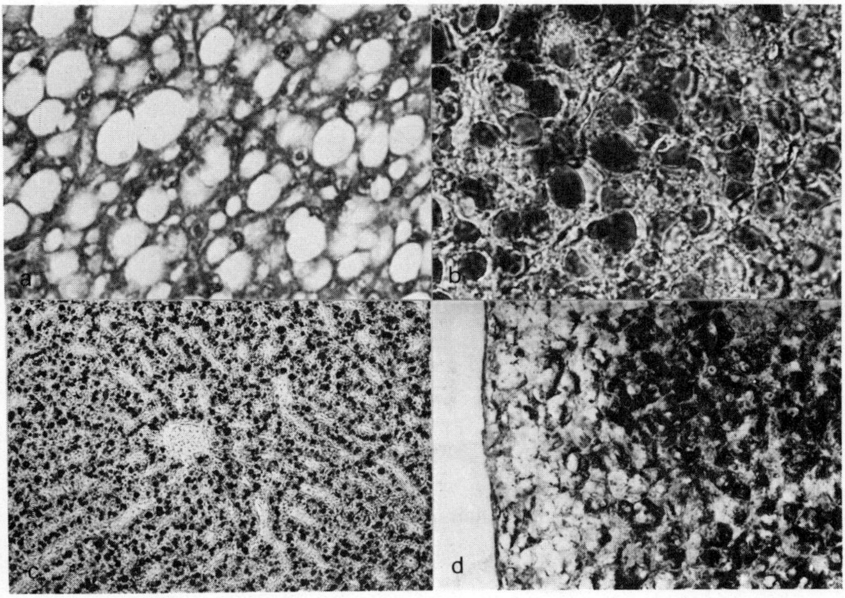

Fig. 13.26. Sections of atypical livers of *Mugil cephalus* from Mississippi. (*a*) Extensive vacuolation in hepatocytes, Masson Trichrome stain, ×450. (*b*) Same tissue, frozen sections stained in Sudan Black B, demonstrating lipid deposition, ×450. (*c*) Fatty tissue in liver of *M. cephalus*, frozen sections stained with Sudan Black B. Note dual plated muralia radiating from central vein, ×450. (*d*) Sections stained with periodic acid Schiff technique, demonstrating extensive glycogen deposites, ×450.

stressed, its resistance to some parasites is lowered and infections can kill the host.

Proper diet, an important aspect when rearing fish, influences growth, resistance and abnormalities. In addition to an abnormal external appearance, the appearance of the liver often reflects the general health of the fish. Overstreet (1974) discussed some liver conditions of *Mugil cephalus* from their natural environment (Fig. 13.26), including those fish involved in a mass-mortality and others. Ghittino (1969–70) reported 'liver lipoid degeneration (LLD) in fingerlings of *M. chelo* fed dry crumbles. A book edited by Halver (1972) covered many aspects of fish nutrition.

Abnormal-appearing mullet also occasionally occur in nature, but more often occur under artificial rearing-conditions. Such abnormalities can result from deficiencies in diet, environmental influences, including pollution, or genetic problems (Fig. 13.27).

Mugil cephalus in Biscayne Bay, Florida and the northern Gulf of Mexico display tumours (Sindermann, 1972; Lightner, 1974; Edwards & Overstreet, 1976). Edwards & Overstreet suggested that the relatively recent occurrence

Fig. 13.27. Malformed adult *Mugil cephalus* from the natural environment in Mississippi. Cause of the configuration unknown; genetic, nutritional or environmental?

of all these tumours could have resulted from increased pollution. Many features of the tumours that they described in the subcutaneous tissues of *M. cephalus* from Mississippi, suggested that the neoplasms were malignant, although no metastases were observed.

Mass mortality due to environmental crises

Mass mortalities involving mullets may be specific to mullet, include mullet or affect almost every organism in the surrounding area. They may result from a purely naturally-occurring event, a man-made condition or a combination of the two. We will discuss many of these situations, primarily focusing on *Mugil cephalus* because it is cosmopolitan in tropical and subtropical areas, because it can tolerate wider ranges in water conditions than most species and because more data concern it. Similar mortalities can effect other mullets, so we will also consider some of the differences between species.

Natural mortalities are caused by many factors other than parasites and commonly occur in specific localities. The most common causes include extremes of temperature and salinity of the water, toxic substances and decreased oxygen concentrations. Drastic changes in the water's temperature, sometimes associated with salt-concentration, probably account for the largest number of naturally-occurring fatalities. Overstreet (1974) reported a massive kill consisting almost entirely of *M. cephalus* in tidally-influenced bayous of the Mississippi following a period of freezing temperatures. Other individuals of *M. cephalus* where the salinity was greater than 6‰ survived, as did other species of fish in the area of the mortality. He attributed the deaths to the fish being incompletely acclimated to the rapid drop in temperature, combined with additional conditions of stress. In that paper he discussed similar kills and reviewed osmoregulation, acclimation, nutrition, pesticides and parasites as they related to the mortality. Roberts (1974) also presented a review on the general problems involved with acclimation. Another mullet-kill is to be reported by Moore (in press), who concluded that the lethal water temperature for *M. cephalus* near Redfish Bay, Texas, seemed to be about 4.5 °C. In laboratory experiments, Moore (1974) found that *M. cephalus* was not affected detrimentally at a temperature of 9 °C. He also observed individuals swimming in Texas localities where the temperature reached as high as 40 °C, but nearly all of them died when netted from water of 37 °C. Specimens of *M. cephalus* survived this same stress, but died under laboratory conditions in water less than 14 °C during all seasons except winter, when they became moribund at 12 °C. This tolerance is corroborated by field observations made in Alabama during January 1973, during the same period as the kills reported by Overstreet and by Moore, Swingle & Bland (1974), then found over a thousand dead adult *M. curema* in East and West Fowl Rivers, but no other dying species.

When temperatures begin to decrease rapidly during cool months or increase rapidly during warm months, *M. cephalus* and other fish in enclosed areas usually migrate to less vulnerable localities.

The reader must realize that *M. cephalus*, as well as other mullets, vary in their ability to tolerate both temperature and salinity, as well as to acclimate with changing conditions, according to geographical locality, age of fish and other factors. In Hawaii, where adult *M. cephalus* occur along open coasts and juveniles occur most commonly in estuaries, Sylvester (1974) studied the juveniles' ability to acclimate to different temperatures. They tolerated temperatures ranging from 14 to 29 °C in sea water. It took, for fish maintained at 25.5 °C, about seven days to acclimate to 27 or 29 °C and up to eleven days to acclimate to 23 or 15 °C. In the studies using low temperatures, the fish showed decreased resistance when the exposure fluctuated rather than maintaining a constant value. Experiments in Israel using fry of *M. cephalus* and *M. capito* showed that these fish suffered more from sudden changes in temperature than in salt concentration. Within 15 min, however, the fry acclimated to a 10 °C gradient (Mires, 1970).

Mugil cephalus readily adapts to most changes in salinity of the water. Parker, Klussman, Holcomb & McNeil (1971) documented this ability in fish from a marsh in Texas. In April 1968, 10 cm of rain fell in that marsh within 5 h and caused the salinity to drop from 14 to 1‰. Fishes could leave the area for one of a higher salinity, and apparently most species except *M. cephalus* and a few atheriniforms did just that, or they died.

Similarly, mullet tolerate hypersaline conditions. In Bardawil Lagoon in northern Sinai where mullets are the dominant fishes, salinity of the water ranges between 40 and 70‰. Levels during the summers of 1969 and 1970 reached 100‰ (Ben Tuvia & Herman, 1972), and at that level, mullets, especially *Mugil cephalus*, became stressed. In some isolated sectors of the lagoon, mullet became progressively emaciated with loosened scales and often developed opaque eye-lenses. No aetiological agent other than high values for salinity and temperature of the water, and relatively low oxygen-concentrations could be associated with the pathological changes. In fact, the prevalence of infections with disease-causing organisms was less than in fish during moderate conditions. According to fishermen, mass mortalities occurred during the high salinities, but no one critically examined moribund fish.

Naturally-produced toxic materials can cause mortalities. A good example of such mortality is that caused by 'red tide'. Along the Gulf of Mexico and Atlantic coasts of North America, deaths resulting from a red tide are often attributed to a lowered concentration of dissolved oxygen caused by the deteriorating dinoflagellates or a combination of dinoflagellates and bacteria. Unialgal and bacteria-free cultures of *Gymnodinium breve* Davis, and cultures of *Gonyaulax monilata* Howell with bacteria, both produced one or more substances which by themselves were lethal in relatively low concentrations to *M. cephalus* (Ray & Wilson, 1957; Gates & Wilson, 1960). Lysis of the dinoflagellates apparently releases an increase of toxic material resulting in quicker death of mullet. Gates & Wilson (1960) also reported signs of mullet exposed to *G. monilata* as frenzied activity, followed by loss of equilibrium with slow opercular movements. After lying on the substratum or floating for some time, the fish displayed a violent burst of activity followed by death with mouths and opercula open. Quick & Henderson (1974) described the behaviour of *M. cephalus* in areas with low to moderate concentrations (greater than 250000 cells/l) of *G. breve* as being similar to that resulting from severe anoxia. Gulping fish surfaced, then sank slowly, often losing their equilibrium. After first exhibiting this repetitious behaviour, they would die within 15 to 30 min. Typically 10% of a school displayed such behaviour at any one time, and about 70% of those sank upon death. Other species of fish such as *Elops saurus* and *Anchoa mitchelli* observed by Quick & Henderson revealed distinctly different signs of distress, but similar to those shown by mullet exposed to *G. monilata.* They moved rapidly and erratically, quivering, jumping and then dying suddenly. Such behaviour, suggestive of being caused by a neurotoxin, has also been observed in mullet exposed to *G. breve* by Starr (1958). Thus, mullet apparently react differently under different conditions of exposure.

Quick & Henderson also critically examined the blood of a few mullet recently killed by the 'red tide' and found overt cyanemia in 16/19 fish, increased viscosity in 11/14, debris usually consisting of various shaped dark particles accompanied by lipid-like droplets in 9/12 and greatly increased clotting time (over 60 sec) in 14/15. Abnormal percentages of various blood cells occurred in the fish, as well as enlarged spleens.

Mass mortalities have also followed blooms of the toxin-producing phytoflagellate *Prymnesium parvum* Carter in Israeli fishponds. The toxin affected all cultured fishes including *Mugil cephalus* and *M. capito*. As a brackish-water species, the phytoflagellate can bloom in ponds where the salinity of the water exceeds 0.4‰. Sarig (1971) reviewed the extensive research conducted in Israel since 1950 on this organism, its toxin, and measures for its control. Before affected fish concentrate in the shallow parts of ponds and start leaping from the water and dying, the presence of toxin can be determined. Pond water is routinely tested using a 'cofactor' which, under specific conditions, enhances the toxic effect of *P. parvum*, causing death in gambusia when the toxin's concentration in the ponds is well below that necessary to kill fish. Depending on the temperature and pH of the pond, a proper concentration of ammonium sulphate can then be administered to it in order to lyse the alga.

Fish-kills also follow blooms of blue-green algal species of *Microcystis*, *Anabaena*, *Oscillatoria* and *Spirulina* in fishponds in tropical and subtropical countries. In Israel, such outbreaks commonly fill all cultured fishes. Mortality results from oxygen depletion following decomposition of algal masses, from release to toxic substances or from a combination of the two. Sarig (1971) also reviewed the biological implications of such blooms and methods to control them.

Hydrogen sulphide, released from the sediments following high winds, can cause fish-kills. This type of climatic factor may have resulted in many dead mullet in the Ashepoo River of South Carolina during Hurricane Gracie (Burbanck, 1961).

Death in the majority of mass-mortalities in estuarine areas results from either oxygen-depletion or a combination of that with some other factor. Most of these circumstances arise from human intervention, but the following example illustrates a natural phenomenon. Along specific areas of Gulfport, Mississippi and the eastern shore of Mobile Bay, Alabama, occur what is known to local residents as a 'jubilee'. Infrequently during the summer in Mobile Bay, a series of events apparently associated with the early hours, easterly wind, rising tide and meeting of two water masses result in stunned fishes along the beaches (Loesch, 1960). In addition to the deeper water mass having a lower concentration of dissolved oxygen than the surface water, it also has higher concentrations of salt and carbon dioxide. Local people harvest the immobilized fish for eating, before they swim away. Few fish die.

Low levels of oxygen more often result from an introduction of either organic material that necessitates oxygen for degradation or inorganic material that becomes incorporated into an algal bloom which dies and subsequently depletes oxygen or that utilizes more oxygen than it produces. Wastes accumulate from both domestic or industrial pollution. *Mugil cephalus* readily makes physiological and behavioural adjustments to hypoxic environments (Cech & Wohlschlag, 1973), but, because of its abundance can nevertheless contribute the largest number of individuals in a kill caused by oxygen depletion (Grantham, see Christmas, 1973) and does constitute, along with menhaden, the most abundant species reported from most kills of undetermined origin in Mississippi (Christmas, 1973).

Man's activities also modify watersheds so that runoff, with its associated silt, pesticides and abundance of nutrients, increases. Such material often reduces areas suitable for mullet, but it may also kill or stress mullet. Pesticides rate as the most easily tested of the above pollutants. In addition to being introduced to estuarine and marine areas by runoff, pesticides are also sprayed from aircraft and accidentally or intentionally discarded from, or in, the original containers after some or most has been utilized.

Pesticides concentrate in various tissues in mullet, primarily those containing lipids, and their levels are rapidly building up in mullet throughout the world. So much variation in the ability to concentrate and tolerate various toxic materials exists among organs, individuals and populations that amounts recorded from mullets are mostly meaningless. Overstreet (1974) reviewed several cases from the southeastern United States where mullet were analyzed for pesticides or killed by them. Since *M. cephalus* often constitutes the most prevalent fish in a system, individuals often become involved in fish-kills. Even in Sussex, where grey mullet were introduced to scavenge refuse from yachts in a harbour, large numbers were killed when the organophosphate 'Phorate' was aerially spread nearby (Anon., 1970).

Nonlethal concentrations of pesticides in mullet also cause problems. Man can concentrate them in his tissues by eating the mullet, also the mullet, sometimes with pathologically altered organs, can become more susceptible to specific diseases and to other toxic materials. If a fish has a high level of a pesticide in its fat, even though low levels exist in other tissues, starvation can cause rapid release of that toxic substance from fat into its blood with resulting death.

As pointed out earlier, mullet can tolerate considerable pollution. In the Middle East they are among the few species that survive in the polluted and eutrophic estuaries and coastal streams receiving urban effluents. Moreover, mullet naturally gather at outflows of fresh water or low saline water. When this water contains excessive amounts of toxic material, mullet, because they are the dominant fish, become conspicuous victims of the pollution. Even though they have high tolerance limits for many toxic substances, they cannot endure abrupt high concentrations of metallic salts, detergents, hydrocarbons or organic solvents. Mullet in inland Israeli waters are particularly affected. Periodic kills occur in coastal streams flowing to the Mediterranean Sea and less extensive kills occur in the sea adjacent to industrial areas. The pollution could deplete mullet from the eastern Mediterranean area. Those who collect fingerlings to stock ponds complain that their catches from the mouths of the coastal Israeli streams have significantly declined in recent years.

Epizootics of bacterial or other infectious diseases, as discussed earlier, often result from polluted conditions or fish being otherwise stressed. Even though little research has been conducted on such cases, Snieszko (1974) presented a review on the subject.

Environmental aspects of parasitic infections

Most species of mullet are euryhaline fishes and therefore are capable of adapting to a wide range of salinities. Given time to acclimate, some species survive in fresh water and some species tolerate up to 100‰ salt. A reader might question whether parasites infecting mullet can tolerate the same salinity ranges as their hosts. Experimental data on these parasites remain sparse; most available information reflects extrapolation of field notes taken from various habitats. One must recognize these data as biased, since presence or absence of parasites depends on a number of ecological, as well as biological, parameters. Moreover, experimental studies have demonstrated that eggs and larvae of parasites may tolerate salinity differently than the adults (Paperna, 1964*a*). Application of data is especially questionable for endoparasites because, as such, organisms are buffered from the external environment by their host. This buffer may not always be sufficient as suggested by Overstreet (1968) who reported an inverse relationship between monthly mean numbers of the trematode *Sterrhurus musculus* Looss in the stomach of *Synodus foetens* and monthly mean salinity values.

The locality of collections could differ from that where an infection originated by many kilometres and by opposed ecological conditions. For those parasites utilizing complex life-cycles, the salinity optima and tolerances evolved for molluscan and other intermediate hosts in the case of most trematodes, and for crustacean intermediate hosts in the case of most cestodes, acanthocephalans and nematodes, bear more significance than the tolerance established for the parasite. Salinity can also regulate growth of the intermediate host and the host's vulnerability to predators. Occasionally the incidence of a parasite will fluctuate because the majority of the intermediate host population gets eaten by predators not involved in the parasitic life cycle.

Examination of conditions tolerated by mullet parasites from the eastern Mediterranean, the northern Red Sea (Paperna, 1964*a*; Paperna & Lahav, 1971), and the Black Sea (Reshetnikova, 1955) reveal few species that occur in both low and high saline environments. The only ectoparasite is the monogenean *Ancyrocephalus vanbenedenii*, whereas endoparasites include the adult trematode *Haploporus benedenii*, the metacercaria of *Heterophyes heterophyes* and the myxosporidan *Myxobolus* sp. Most parasites infecting mullet occur within somewhat restricted aquatic conditions.

Experimental evidence shows that some cercariae and their hosts tolerate wide ranges of salinity. *Pirenella conica*, the snail host for *Heterophyes heterophyes*, is even more euryhaline than are mullets. It survives oligohaline water and a saturated salt solution. The free-living cercaria can withstand, at least 22–60‰. Kuntz (1957) reported cercaria of *H. aequalis* to survive within the range 6–19‰, but not in fresh water; circumstantial evidence suggests that this larva also survives hypersaline conditions (Paperna & Lahav, 1971). Cercariae of some other heterophyids in Israel are more stenohaline. Lee & Cheng (1970) show graphically that about 20% of the cercariae of *Stellantchasmus falcatus* tested in Hawaii survive 20 h in 4.5 and 9.0‰, but none lasted over 10 h in 1.2 or 18.0‰.

Ergasilid copepods and other ectoparasites also vary in their abilities to tolerate salt. *Ergasilus lizae* in the eastern Mediterranean region and in the United States occurs in low-to-medium saline habitats. But, the related *E. versicolor*, in addition to being a coastal American species, also infests fresh-water fishes far inland (Johnson & Rogers, 1973). Reshetnikova (1955) reported the absence of *E. nanus* in water of 40‰ in the Black Sea. Many caligids, however, seem to thrive in high saline water and only rarely occur in mesohaline habitats. Also influenced by salinity are monogeneans. According to Rawson (1973*b*), the intensity of *Polyclithrum mugilis* related inversely to salinity in year-old mullet. Malmberg (1970) studied several species of *Gyrodactylus* Nordmann and only few, apparently, could adapt to a wide range of salinity concentrations. The leech *Myzobdella lugubris* has been collected from salinities ranging from 0 to 26‰, but usually from water less than 15‰ (Sawyer *et al.*, 1975). Sawyer and his co-workers described experiments which showed that the maximal salinity tolerated under abrupt changes at about 24 °C was near 15‰. Cocoons hatched in salinities of 13‰ and less, but not at 28‰.

The absence in some areas of a larval tetraphyllidean infecting mullet from open seas probably relates more to the absence of suitable elasmobranch hosts than to salinity conditions for the parasite or intermediate host.

We have been stressing salinity as if it were the single most important influencing variable for parasites. The relationship between salinity and temperature often outweighs the importance of salinity alone. For lack of good parasitic examples, we cite information concerning the brown shrimp, *Penaeus aztecus*. Eighty per cent of tested postlarval shrimp survived direct transfer from 25.5‰ to salinities ranging between 8.5 and 51.0‰, but sudden fluctuations in temperature detrimentally influenced their ability to tolerate such stresses. The combination of salinity and temperature influenced the shrimps' growth, metabolism, and regulation of blood-ions (Venkataramiah, Lakshmi & Gunter, 1974).

The relationship between temperature and epizootics must be stressed. As so succinctly stated by Hoffman & Meyer (1974), 'Both fish and parasites respond to their optimum temperature with rapid growth. Epizootics are likely when the optimum temperature for the parasite is unfavourable for the host'.

Temperature of water also strongly dominates seasonal variation of parasites. Rawson (1973*b*) related water temperature and other environmental parameters to seasonal variation for several parasites of *Mugil cephalus* in Georgia. He found infections with the digeneans *Dicrogaster fastigata*, *Saccocoeliodies beauforti* and *Lasiotocus mugilis* began to increase considerably in both young and old mullet in November, just prior to termination of his sampling. Apparently more infective larvae of those species occur during colder months. The incidence and intensity of infection with *Hymenocotta manteri* was greatest in May, the first month of sampling. The intensity of the monogenean *Ancyrocephalus vanbenedenii* reached its highest level in June when temperatures were between 23 and 36 °C. This monogenean on year-old mullet reached a much greater peak in late fall when the mullet were ripe. The highest incidence of mullet infested with argulids was also reported from spawning congregations.

In addition to the variation in the incidence or intensity of infections, worms can also vary their position with temperature. Overstreet (1968) showed statistically that the percentage of a cestode population located near the anus of its host was related directly to water temperature.

Host-related aspects of parasitic infections

Host length

The parasitic fauna of mullet undergoes a dynamic change in relationship to many different variables. One variable often considered is age of host. That change, however, is not so much one caused by a change in the host's age, but rather one caused by the fish's altered behaviour and is most related to environmental parameters. Changes have been demonstrated for *Mugil auratus*, *M. cephalus* and *M. saliens* in the Black Sea (Reshetnikova, 1955) and for the above three species, as well as *M. capito*, *M. chelo* and *M. labeo* in the eastern Mediterranean (Paperna, 1964*a*, and unpublished data). Similar acquisition and loss of parasites also characterize young *M. cephalus* in Mississippi and will be discussed below.

Changes in the parasitofauna correspond with new diets and new habitats. The most profound change occurs when young fish discontinue feeding on

Table 13.5. *Prevalence (per cent) of sampled mullet in different length-groups from Israeli coastal streams (CS) and littoral Mediterranean Israeli coasts (M) infected by specific parasites based on data of Paperna (1964) and Paperna & Lahav[a] (1971)*

	Fork length (mm)								
	15–30		31–50		51–90		91–150		150–250
Parasite	CS	M	CS	M	CS	M	CS[a]	M	CS[a]
Scolex pleuronectis	57	39	20	52	3	4	0	10	0
Larval didymozoid	52	3	18	15	0	0	0	0	0
Larval nematodes	2	4	12	20	0	11	0	20	0
Ancyrocephalus vanbenedenii	0	0	7	5	17	4	8	8	40
Heterophyids	0	0	0	0	6	0	23	0	67
Haploporus spp.	0	0	10	0	32	0	56	0	27
Saccocoelium & *Haplosplanchnus* spp.	0	0	1	5	4	29	0	0	92
Ergasilus lizae	0	0	0	0	12	0	0	0	0
No. fish examined	81	29	118	43	89	22	15	15	17

plankton in offshore waters. The fish migrate to coastal areas and gradually enter inland waters. During this period, they gradually lose larval parasites: tetraphyllidean cestodes, nematodes, and didymozoid and hemiurid trematode larvae. These parasites are replaced by different ones originating from the new habitat. Good examples include myxosporidans, trichodinid ciliates, haploporid and haplosplanchnid trematodes, larval heterophyid trematodes, and neoechinorhynchid acanthocephalans. Once established in coastal and inland areas and associated with older mullet, younger mullet gradually accumulate various monogeneans and crustaceans.

In Israel, the alterations in the parasitic composition begin in mullet of 15–20 mm TL, their size upon arriving in coastal waters. The change is mostly completed by a month or two when the fish reach 50–70 mm in length (Tables 13.5, 13.6). Those sizes especially characterize mullet entering estuaries and inland water. Mullets remaining in coastal marine areas, such as *Mugil labeo* and occasionally *M. auratus* and *M. chelo*, may retain parasites of planktonic origin because the fish apparently continue to feed on plankton in inshore areas. Such infections, however, usually decline because the fish alters its diet.

Once mullet leave the open sea, they acquire new infections and these infections may vary seasonally and geographically. Many factors such as salinity, temperature and oxygen concentration of the water, as well as available nutrients, period of light or dark, season and physical condition of

Table 13.6. *Infections of* Scolex pleuronectis *and heterophyid metacercariae as related to length of mullet in Bardawil Lagoon*

Fork-length of host (mm)	No. fish examined	*Scolex pleuronectis*			Heterophyid metacercariae		
		Prevalence (%)	Mean no. per infected fish	Maximal infection	Prevalence (%)	Mean no. per infected fish	Maximal infection
21–30	22	82	7.4	33	9	1	1
31–40	29	24	3.7	8	45	2	2
41–50	14	35	7.0	11	50	2	7
51–60	21	14	8.7	18	52	4	15
61–70	11	9	4.0	4	45	3	5
71–90	32	0	0	0	100	9	10

the host influence the infections. Mullet obviously must frequent areas where parasites occur, to become infected. They are infected with few parasites utilizing intermediate hosts capable of extensive migrations.

Components of the fauna will vary with the nature of a habitat. Enclosed or otherwise confined areas often sustain heavy infections of specific parasites, such as trichodinids, haploporids and monogeneans, in contrast to less extensive infections of those parasites in more open areas. Other parasites thrive in open areas. In the eastern Mediterranean region, metacercariae of *Heterophyes* spp. heavily infect mullet which spend time in the large lagoons of northern Sinai and the Nile Delta where *Pirenella conica*, the snail-host, abounds (Paperna & Lahav, 1971; Paperna, 1975). In Bardawil Lagoon in northern Sinai, heterophyid infections appear in mullet 20–30 mm TL, and they reach a rate of 100% in fish 70 mm long (Table 13.6). In other areas lacking the snail, infected fish only comprise those longer than 70 mm and they had immigrated from another area; only about half the fish longer than 100 mm show infections. *Ergasilus lizae* and *Ergenstrema mugilis* occur only in inland and low-salinity water (Paperna, 1964*a*). On the other hand, *Pseudocaligus apodus* and *Caligus pageti* most often occur on mullet from the sea or hypersaline lagoons (Paperna & Lahav, 1974). Mullets from Bardawil Lagoon, but not the open sea, frequently act as host *Pseudocaligus apodus*, *Gyrodactylus lagoonaris* and *Neoechinorhynchus agile*.

Relationships between host-size and infections vary geographically. Skinner (1974) presented data on infections in eighty-three specimens between 10 and 40 cm in standard length from Biscayne Bay, Florida, according to length of host. She found a tetraphyllidean larva in the gall bladder and cystic duct, but not in small fish. Only fish over 20 cm long possessed the worm and infection was most prevalent in those 30–35 cm long. In general, she found parasites with planktonic intermediate hosts in young mullet.

Skinner found few fish less than 20 cm long infected with *Myxosoma*

Table 13.7. *Parasites in juvenile* Mugil cephalus *between 8 and 28 January 1975 at Gulf Coast Research Laboratory, Ocean Springs, Mississippi*

Date	8 Jan[a]	22 Jan	27 Jan	28 Jan	31 Jan
Length of fish SL (mm)	18–21	18–25	20–25	19–23	25–31
No. examined	10	45	40	35	8
	Per cent infected/approximate average number per host				
Scolex polymorphus	100/150	100/600	100/90	100/80	100/140
Didymozoid larvae	30/1	0	0	0	0
Saccocoeliodes beauforti	0	100/11	18/8	0	0
Dicrogaster fastigata	0	11/7	88/14	49/10	38/18
Trichodina sp.	0	100/100	100/100	100/100	75/30
Other parasities	0	7/1	0	0	13/1

[a] Specimens collected from a pier in Davis Bayou; the specimens collected on other days were from an adjacent protected small harbour.

cephalus. Both the incidence and intensity of infection with that myxosporidan remained high in all groups of longer fish. *Trichodina* sp. did not infest fish shorter than 20 cm, but there were heavy infestations in fish between 20 and 35 cm long. Infestations dropped off in fish over 35 cm long. She suggested that mullet over 35 cm long could have developed a resistance to such parasites. While such an hypothesis is tenable for some monogeneans and protozoans, we tend to think that environmental conditions, diet and habitat-preference more significantly influence most parasites.

We examined young specimens of *Mugil cephalus* that had recently migrated from the high salinity water of the Gulf of Mexico to Davis Bayou in Ocean Springs, Mississippi, immediately upon collecting them (Table 13.7). The salinity values for water from the bayou ranged between 5.5 and 15.0‰ over the 3 week period of collections. Those first examined all had *Scolex polymorphus* in the intestine and three of them each had one didymozoid larva in their musculature. Fish in the next collection which were from a more protected nearby area did not have the didymozoid, but they did have other parasites. Two years earlier in the February of 1973, 67% of thirty-three fish harboured an average of 2.0 didymozoids. In December 1972, two mullet had eighteen and twenty-two specimens of the larva. Based on examination of mullet maintained for a few weeks in tanks, the didymozoid larvae did not persist in the mullet for long.

Just as others have reported, we found a larval tetraphyllidean commonly infecting young mullet. Of the forty-five fish examined on January 22, thirteen each had over 600 specimens and three each had about 4500. None of these mullet had obvious food in their alimentary tracts. We observed earlier that other mullet which did not feed well and which died in aquaria often had a similar heavy burden of the cestode larvae. Presumably, in their natural habitat, heavily-infected mullet fare less well than their less-infected schoolmates and are easy prey for numerous fishes.

Assuming the fish collected on January 8 and 22 represent the same initial population, several parasites were acquired within the period of two weeks. Two haploporids were obtained during the mullet's benthic feeding. Externally the fish acquired light infestations of *Trichodina* sp. on the gills and skin. One fish had *Scyphidia* sp. and another had a microcotylid monogenean. Once established on an individual, the infestation of such parasites can be easily transmitted to others in the school. Four fish examined from a 4‰ tidal pool several kilometres away had between fifteen and thirty specimens of *Scolex polymorphus*, but no *Trichodina* sp. Also, two of them harboured over twenty-five specimens of *Saccocoeliodes beauforti*, one had a didymozoid larva, and one had a bucephalid metacercaria.

Rawson (1973*b*) studied variations in some parasites of *Mugil cephalus* between May and November of 1970 and March and June of 1971 in Georgia. Information was available for the cyclopoid copepods *Ergasilus lizae*, *E. versicolor*, *E. funduli*, *Bomolochus concinnus* and several other parasites. The intensity of *E. lizae* showed a positive relationship with length of mullet. On young mullet, *E. versicolor* increased in both incidence and intensity of infestation, from June to December, when sampling stopped. Its incidence in large fish dropped from 100% to about 35% during the same period, while the number per fish fluctuated. *Ergasilus funduli* infested 100% of the small mullet in July and most of them in August, but otherwise was relatively uncommon. This infestation probably resulted from young mullet being closely associated with cyprinodontid fishes. *Bomolochus consinnus* was usually rare, but had infested 35% of the mullet samples on 17 August 1970, with an average of 4.1 individuals.

Location on host

Gill parasites, according to Rawson (1973*b*), also distributed themselves differently according to species. The monogenean *Ancyrocephalus vanbenedenii* was most common on the most anterior branchial arch of year-old mullet, and its frequency progressively decreased to about one-third the amount on the fourth arch. *Ergasilus versicolor*, the caligoid *Branchiella oblonga* and to a lesser degree *E. funduli* expressed this same pattern. More individuals of *E. lizae* occurred on the third arch than any other.

Ancyrocephalus vanbenedenii, which was found to infest most fish throughout the study, occurred in conspicuously higher numbers in June and November, and usually occupied the dorsal proximal region of the gill, whereas the copepods primarily resided elsewhere. *Ergasilus lizae* was most frequently found on the ventral proximal region, whereas the others were on

Table 13.8. *Numbers of known parasites from mullets*

Eastern Mediterranean (based on Witenberg, 1929; Paperna, 1964, 1975; Paperna & Lahav, 1971); northern Red Sea (Paperna & Overstreet, unpublished); Black Sea (based on Reshetnikova, 1955; Lom, 1962); Mississippi (Overstreet, unpublished); and additional reports of parasites from the southeastern United States from numerous sources

Parasites	Eastern Mediterranean	Northern Red Sea	Black Sea	Mississippi	Additional reports from southeastern US
Blood protozoa	0	1	0	2	0
Gill protozoa	1	2	3	4	2
Microsporida	0	1	0	0	0
Myxosporida	2	4	2	3	0
Monogenea	6	5	2	3	6
Digenea – adults	6	7	7	11	5
Metacercariae	12	2	2	6	3
Cestoda – larvae	2	0	1	3	0
Nematoda	4	0	1	3	4
Acanthocephalan	2	1	1	1	0
Copepoda	4	4	1	6	11
Branchiura	0	0	0	1	4
Isopoda	0	2	0	1	1
Hirudinea	0	0	0	1	1
Total	39	29	20	45	37

the middle region. *Ergasilus versicolor* and *E. funduli* were distally located on the filaments, and *B. oblonga* nearer the middle region. According to Rawson, none of the gill-parasites, including the microcotylids, caused exclusion of others.

Several other examples, however, have been reported citing specific parasites occurring together more or less frequently than would be expected statistically (e.g., Overstreet, 1968). Without naming species, Scott (1972) reported an enhancement of an infection with a copepod by one of the monogeneans infesting *Mugil cephalus* in northwestern Florida. *Ergasilus lizae* often occur in conjunction with *Myxobolus* sp. on the gills of mullet in Israel. Whether this pathogenic combination is incidental or not has not been determined (Sarig, 1971).

Regional variations in the parasitofauna

Considering the present state of knowledge on the parasites of mullet, we find any reliable comparative evaluation of regional faunas impossible. The significance of intraspecific variations in characteristics in some parasites from geographically isolated areas needs further evaluation. Species-complexes from different areas should be examined by specialists.

Mullets harbour a highly diverse parasitic fauna, which includes representatives of all major parasitic groups. To date, considering just protozoans, helminths and crustaceans, thirty-nine species have been reported from mullets in the eastern Mediterranean, twenty-nine species from the northern Red Sea where studies are continuing and twenty species from the Black Sea. We already know of forty-five species, not all reported, from *Mugil cephalus* in Mississippi, and an additional thirty-seven species have been reported from the southeastern United States, but not seen by us (Table 13.8). If all are valid, probably more than eighty-two species infect *M. cephalus* in this one relatively small area in the United States. Some barriers probably exist. In Georgia, twenty-two species were reported (Rawson, 1973*b*), whereas of the thirty-five reported from Biscayne Bay, Florida, only thirteen were shared in both areas (Skinner, 1974).

Regional differences in parasites may develop through isolation of the hosts, especially noticeable in the Atlantic and Indo-Pacific regions. Considerable speciation of parasites infecting *Mugil cephalus*, for example, can be inferred from the list of haploporid and haplosplanchnid digeneans. Mugilids constitute the hosts for many of the species in those families, especially the haploporids. Consequently, we see that the host seemed to have less influence on radiation of the species than did the snail hosts, geographic separation and ecological conditions.

Larval tetraphyllideans and didymozoids, common in juvenile mullet from the Mediterranean Sea and Gulf of Mexico, are absent or rare in the northern Red Sea. Heterophyid metacercariae dominate the parasitofauna of mullets in most areas, but they rarely occur in the Black sea and Red Sea. If not for *Phagicola longus*, heterophyids would be rare in the southeastern US. The leech *Myzobdella lugubris* and cymothoid isopods commonly infest some populations of mullet in the United States, but are absent in the Red Sea, where gnathniid isopods prevail. The cases listed above result mostly from differences in habitats. For the most part, faunas from different regions are similar.

Similarities for some infections may have more in common than we realize. Trichodinids and myxosporidans need to be studied on a world-wide basis to determine the actual number of species involved. Some parasites which are transmitted by birds, such as *Heterophyes heterophyes* and *Phagicola longus*, are known to be widely distributed.

Interoceanic canals may contribute to geographic dispersal of some parasites. Migration of mullets through the Suez Canal (Ben Tuvia, 1966; Steinitz & Ben Tuvia, 1972) permits dispersal of parasites between the previously separated northern Red Sea and eastern Mediterranean. Several parasites common to both seas include *Myxobolus* sp., *Zschokkella* sp., *Ancyrocephalus vanbenedenii*, *Benedenia monticelli*, *Saccocoelium obesum*, *Neoechinorhynchus agile*, *Pseudocaligus apodus* and *Nipergasilus bora*. Some are cosmopolitan. Circumstantial evidence, extrapolated from distribution data, however, suggests that *Nipergasilus bora* recently migrated into the Mediterranean Sea and that *Pseudocaligus apodus* became established recently in the Red Sea (Paperna, in preparation).

Control measures

Only a limited amount of experience has been accumulated on methods to control parasites and diseases of mullet in aquaculture. In comparison with the culture of trout, salmon or carp, that for mullet is still restricted to a few regions, and, with the exception of Israel, it is usually on a small scale. Obviously, general sanitation, good water quality and preventive measures already practised in most fish farms may be applied directly, or with necessary adjustments, to mullet culture. Such practices have been widely discussed in the literature on aquaculture and treatments for diseases (Schaperclaus, 1954; Amlacher, 1961; Sarig, 1971; Klontz, 1973; Hoffman & Meyer, 1974; Sindermann, 1974; and others).

In ponds with fresh and low saline water, mullet, particularly when in polyculture with carp and cichlids, acquire fresh-water diseases such as ichthyophthiriasis, saprolegniasis and infections with the anchor worm, *Lernaea cyprinacea*, and other parasites. In sea water or brackish water, mullet are protected from most fresh water diseases, but vulnerable to a multitude of others.

Chemicals may be the method of choice to treat an outbreak. If the disease can be controlled without chemicals, for example by prophylactic or environmental measures, the system may profit. Chemicals evoke trouble; they can harm the mullet directly, they can harm people that eat or raise the mullet and they can affect the water quality. In large-scale culturing operations, however, drastic means such as chemicals often are employed to provide maximal yields.

Agencies in some countries, such as the Food and Drug Administration, Environmental Protection Agency, and similar state agencies in the United States, restrict the use of most drugs and chemicals for use with fish to be consumed by humans. At present in the US, only salt, glacial acetic acid and sulphamerazine can be used legally to treat mullet reared to be eaten. Whether used legally or illegally, chemicals should be used with proper care. Malachite green may cause cancer. Potassium permanganate is a strong oxidizing agent which reacts violently when in contact with several reducing agents. Therefore, the chemical should not even be stored together with organic materials or other reducing substances. If the dust is applied to a pond, we recommend that a cotton mask, safety glasses and gloves should be worn by the person handling it. However, in practice, such precautions are difficult to enforce. Hence, the danger involved with use of chemical agents is considerable and increases with the utilization of unskilled employees.

A few insecticides have been used effectively to treat mullet diseases in Israel (reported in detail by Sarig, 1971). Successful control of the copepod *Ergasilus lizae* was achieved by spraying ponds with enough Bromex-50 (naled–50% active 1,2-dibromo-2, 2-dichloroethyl dimethyl phosphate) to produce a concentration of 0.13–0.15 ppm. Also effective against the copepod were Dylox (Dipterex, 50% W/V 0,0-dimethyl 2,2,2-trichloro-1-hydroxyethyl phosphonate) at a concentration of 0.15 ppm and Malathion (50% W/V 0,0-dimethyl S-1,2,biscarbethoxyl ethyl phosphorothioate) at a concentration of 0.20 ppm. The use of these pesticides, however, was applicable only to brackish-water ponds, since in fresh-water with chlorinity below 400 ppm/l, the lethal dose to the copepod approached that lethal to the fish. Experiments showed the lethal concentration of Bromex to kill all specimens of *Mugil cephalus* in 96 h (LC_{100}) was 0.55 ppm at 2000 ppm Cl^- and 0.10 ppm at 40 ppm Cl^-. The LC_{50} for *Mugil subviridis* after 6 h in sea water was 0.50 ppm Bromex. The sensitivity of carp and cichlids to Bromex was not modified by salinity.

Netting and handling processes during stocking, transferring from pond to pond and periodic weighing usually results in loss in the fish of some of its scales and some other minor injuries. The phycomycete fungus *Saprolegnia* sp. secondarily invades most of those injuries and can ultimately kill the fish. Therefore, a prophylactic treatment is suggested. In Israel, spraying with 0.15 ppm of the dye malachite green, zinc-free oxalate immediately after stocking and three additional doses at 3-day intervals completely rid the fungus from the fish (Sarig, 1971). Salt, of course, also eliminates the disease.

A combination of malachite green and formalin has proven successful for the treatment of *Ichthyophthirius multifiliis* in fresh-water ponds (Leteux & Meyer, 1972); it also is effectively for treatment of several external parasites and their free-living stages in marine and fresh-water systems. In the case of 'ich', however, if all fish can be removed from a system, the inefective agent will be eliminated after a week without any treatment. Resistant stages, such as occur for *Oodinium*, apparently do not occur.

As discussed above for Bromex, the activity of other chemicals also varies with salinity. Birdsong & Avault (1971) reported that the effectiveness of copper sulphate was less in sea water than fresh water. Another danger from the use of copper sulphate, besides human consumption of tainted products, is that it precipitates out of solution, but returns to solution under acidic conditions. A sudden relatively high level would then mysteriously kill all mullet in the pond.

Formalin can be used in fresh or salt water. If used in a pond, some algae will die, subsequently lowering the level of dissolved oxygen. Dipping fish in a bath of 100–250 ppm of the 40% formaldehyde solution effectively rids them of most monogeneans tested in the United States, Israel and elsewhere. The treatment, however, usually damages or kills the epithelial tissues of the gills and other surfaces, inviting secondary infection. In Israel, *Mugil cephalus* was reported to be less sensitive to formalin than carp and tilapia which were killed when exposed to 200 ppm for 10 h (Lahav & Sarig, 1972). Williams (1972) reported that prolonged treatment with 15 ppm was ineffective in eliminating *Trichodina* and monogeneans on *Mugil cephalus*. If an estuarine or marine fish can withstand a fresh-water bath, that treatment will often rid the host of marine monogeneans and protozoans. *Benedenia monticelli* was eliminated from infested *Liza carinata* after 1 h in baths of 10 and 70‰ sea water.

In Alabama, fingerling *Mugil cephalus* treated for 3 h with 100 ppm formalin, some of which were treated additionally for 24 h with 0.25 ppm Dylox, were rid of *Trichodina* sp. (Williams *et al.*, 1974). After the fish were introduced into cages in the bay at Dauphin Island, however, Williams and his co-workers found that the mullet had acquired *Scyphidia* sp. and re-established infestations of *Trichodina* sp. The incidence of harvested fish infested with copepods and monogeneans was less than for those fish originally stocked and for those noncultured fish simultaneously sampled from the environment.

References to treatments for diseases cover twenty-two pages of a bibliography by Hoffman & Meyer (1974) and are admittedly incomplete. Their book would be a good starting point for someone interested in what treatments have been tested, by whom, at what dose and with what results. The treatments, as pointed out by the authors, vary with many factors and anyone considering using a technique should first test it on a small number of fish before applying it to a large population.

Acknowledgments

Many people provided us with useful information and are cited in the text and we thank them. Theresa Ann St. Andrie helped compile many of the references and Adrian R. Lawler read the manuscript.

References

Africa, C. M., Garcia, E. Y. & De Leon, W. (1935). Intestinal heterophyidiasis with cardiac involvement. *Philipp. J. Public Health* **2**, 1–22.

Africa, C. M., De Leon, W. & Garcia, E. Y. (1937). Heterophyidiasis. VI. Two more cases of heart failure associated with the presence of eggs in sclerosed veins. *J. Philipp. Isl. Med. Assoc.* **17**, 605–9.

Almeida, L. J., Da Silva, E. J. & Freitas, Y. M. (1968). Microorganisms from some tropical fish diseases. *J. Fish. Res. Board. Can.* **25** (1), 197–201.

Amlacher, E. (1961). *Textbook of Fish Diseases*. TFH Publications, Neptune, New Jersey (English translation).

Anon. (1970). Fish kill in Chichester Harbour. *Mar. Poll. Bull.* **1** (8), 117–8.

Argilas, A. (1930). Un copépode parasite de *Mugil auratus* Risso nouveau pour l'Algerie: *Caligus pageti* Russel. *Bull. Stn Agric. Pêche Castiglione* (2), 93–106.

Asada, J. (1928). Studies on human trematodes of the genus *Heterophyes* occurring in Japan. I. Determination of the first intermediate host of *Heterophyes heterophyes*. *Tokyo med. News* 2564, 6–12 (Japanese).

Balozet, L. & Callot, J. (1939). Trématodes de Tunisie. III. Superfamille Heterophyoidea. *Arch. Inst. Pasteur Tunis* **28** (1), 34–63.

Belous, E. V. (1965). Nematodes of freshwater fish from the Primorsk region. In *Parasitic Worms of Domestic and Wild Animals*, pp. 48–65. Dalnevostochnii Gosudarstvennii Universitet, Vladivostok (in Russian).

Ben-Ari, J. (1957). Helminthic infections among the Jewish population of Jerusalem. *Harefuah* **52** (7), 166–71 (in Hebrew).

Ben-Tuvia, A. (1966). Red Sea fishes recently found in the Mediterranean. *Copeia* (2), 254–275.

Ben-Tuvia, A. & Herman, Z. (1972). Biology of the fishes of Bardawil Lagoon. *Daig U'Midgeh* **7** (4), 38–53 (in Hebrew, English summary).

Birdsong, C. L. & Avault, J. W., Jr. (1971). Toxicity of certain chemicals to juvenile pompano. *Prog. Fish. Cult.* **32** (2), 76–80.

Brian, A. (1931). Sur la synonymie du *Caligus argilasi* Brian, 1931 avec *Caligus pageti* Russell, 1925. *Bull. Soc. Hist. Nat. Afr.*, N (22), 1–157.

Brian, A. (1935). *Caligus* parassiti dei pesci del Mediterraneo (Copepodi). *Ann. Museo Civico Storia Nat. Giacoma Doria, Genoa* **57**, 152–211.

Brown, E. M. (1934). On *Oodinium ocellatum* Brown, a parasitic dinoflagellate causing epidemic diseases in marine fish. *Proc. Zool. Soc. Lond.* 583–607.

Brunsdon, R. V. (1956). Studies on nematode parasites of New Zealand fishes. Thesis, Victoria Univ., Wellington Library.

Bullock, G. L. (1971). Identification of fish pathogenic bacteria. In *Diseases of Fishes*, ed. S. F. Snieszko & H. R. Axelrod, pp. 1–45. TFH Publications, Jersey City, New Jersey.

Bullock, G. L., Conroy, D. A. & Snieszko, S. F. (1971). Bacterial diseases of fishes. in *Diseases of Fishes*, ed. S. F. Snieszko & H. R. Axelrod. TFH Publications, Jersey City, New Jersey.

Burbanck, W. D. (1961). The disappearance of *Cyathura polita* from the Ashepoo River, South Carolina, after hurricane 'Gracie' in 1959. *Ecol.* **42** (3), 606–7.

Cable, R. M. & Dill, W. T. (1967). The morphology and life history of *Paulisentis fractus* Van Cleave and Bangham, 1949 (Acanthocephala, Neoechinorhynchidae). *J. Parasitol.* **53** (4), 810–17.

Cable, R. M. & Isseroff, H. (1969). A protandrous haploporid cercaria, probably the larva of *Saccocoelioides sogandaresi* Lumsden, 1963. *Proc. Helminthol. Soc. Wash.* **36** (1), 131–5.

Carini, A. (1932). Sobre uma hemogregarina de um peixe do mar do Brasil. *Reun. Soc. argent. Patol. reg.* (2), 920–1.

Causey, D. (1953). Parasitic Copepoda of Texas coastal fishes. *Publ. Inst. Mar. Sci. Univ. Tex.* **3**, 7–16.

Causey, D. (1960). Parasitic Copepoda from Mexican coastal fishes. *Bull. Mar. Sci. Gulf Caribb.* **10** (3), 323–37.

Cech, J. J., Jr. & Wohlschlag, D. E. (1973). Respiratory responses of the striped mullet, *Mugil cephalus* L. to hypoxic conditions. *J. Fish. Biol.* **5** (4), 421–8.

Chandler, A. C. (1935). Parasites of fishes in Galveston Bay. *Proc. US Natl. Mus.* **83** (2977), 123–57.

Christmas, J. Y. (1973). *Co-operative Gulf of Mexico Estuarine Inventory and Study, Mississippi. Phase I. Area Description.* Pp. 1–71, Gulf Coast Research Laboratory, Ocean Springs, Mississippi.

Connell, D. W. (1974). A kerosene-like taint in the sea mullet, *Mugil cephalus* (L.). I. Composition and environmental occurrence of the tainting substance. *Aust. J. Mar. Freshwater Res.* **25** (1), 7–24.

Cook, D. W. & Lofton, S. R. (1975). Pathogenicity studies with a *Streptococcus* sp. isolated from fishes in an Alabama-Florida fish-kill. *Trans. Am. Fish. Soc.* **104** (2), 286–8.

Courtney, C. H. & Forrester, D. J. (1974). Helminth parasites of the brown pelican in Florida and Louisiana. *Proc. Helminthol. Soc. Wash.* **41** (1), 89–93.

Cressey, R. F. (1972). *The genus Argulus (Crustacea: Branchiura) of the United States.* Water Pollution Control Research Series, Biota of Freshwater Ecosystems Identification Manual no. 2.

Curtis, W. C. (1911). The life history of the *Scolex polymorphus* of the Woodshole region. *J. Morphol.* **22** (3), 819–53.

Davis, H. S. (1917). The Myxosporidia of the Beaufort region, a systematic and biologic study. *Bull. Bur. Fish., Washington* **35**, 199–244.

Dawes, B. (1946). *The Trematoda with Special Reference to British and other European Forms.* Cambridge University Press.

Deschiens, R., Collomb, H. & Demarchi, J. (1958). Distomastose cerebrale a *Heterophyes heterophyes.* In *Abstracts of the Sixth International Congress of Tropical Medicine and Malaria.*

Dolgikh, A. V. (1968). Helminthofauna of molluscs of the northern part of the Black Sea. *Biol. Sea, Kiev* **14**, 114–26 (in Russian).

Edwards, R. H. & Overstreet, R. M. (1976). Mesenchymal tumours of some estuarine fishes of the northern Gulf of Mexico. I. Subcutaneous tumours, probably fibrosarcomas, in the striped mullet, *Mugil cephalus. Bull. Mar. Sci.* **26** (1), 33–40.

Ergens, R. (1960). Neve bisher beschriebene Dactylogyrous-Arten (Monogenoidea) aus der Tschechoslowakei. *Helminthol.* **2** (1), 3–8.

Euzet, L. (1962). A propos de *Fistulicola plicata* (Rudolphi, 1819) (Cestoda, Pseudophyllidea). *C. r. Congr. Socs. Sav. Montpellier* 631–3.

Fan, P. C. (1970). The metacercarial infection in fresh-water fishes in Taiwan Islands. *Southeast Asian J. Trop. Med. Public Health* **1** (1), 155–6.

Fantham, H. B. (1919). Some parasitic protozoa found in South African fishes and amphibians. *S. Afr. J. Sci.* **15** (5), 337–8 (Abstract).

Fares, A. & Maillard, C. (1974). Recherches sur quelques Haploporidae (Trematoda) parasites des muges de Méditerranée occidentale: systématique et cycles évolutifs. *Z. Parasitenkd.* **45** (1), 11–43.

Faust, E. C., Russell, P. F. & Jung, R. C. (1970). *Craig and Faust's Clinical Parasitology*, 8th edn. Lea & Febiger, Philadelphia.

Gates, J. A. & Wilson, W. B. (1960). The toxicity of *Gonyaulax monilata* Howell to *Mugil cephalus. Limmol. Oceanogr.* **5** (2), 171–4.

Ghittino, P. (1969–70). *Piscicdtura et Ittiopatologia*, vol. 2, *Ittiopatologia.* Rivista Zootechnicia, Milano.

Gnadeberg, W. (1949). Beitrage zur Biologie und Entwicklung des *Ergasilus sieboldi* v. Nordmann (Copepoda Parasitica). *Z. Parasitenkd*, **14** (1–2), 103–80.

Gurney, R. (1934). The development of certain parasitic Copepoda of the families Caligidae and Clarellidae. *Proc. Zool. Soc. Lond.* 177–217.

Gussev, A. V. (1955). *Ancyrocephalus* (s.l.) *vanbenedeni* (Par. et Per.) (Monogenoidea) and its geographical distribution. *Zool. Zh.* **34** (2), 291–294 (in Russian, English translation, VIMS Transl. Ser. 11).

Gussev, A. V. (1963). New species of Monogenoidea from fishes of Ceylon. *Bull. Fish. Res. Stn Ceylon* **16** (1), 53–93.

Halisch, W. (1940). Anatomie und Biologie von *Ergasilus minor* Halisch. *Z. Parasitenkd.* **11**, 284–330.

Halstead, B. W. (1967). *Poisonous and Venomous Marine Animals of the World.* 2. *Vertebrates.* US Govt Printing Office, Washington, DC.

Halver, J. E. (ed.) (1972). *Fish Nutrition.* Academic Press, New York.

Hamed, M. G. E. & Elias, A. N. (1970). Effect of food-processing methods upon survival of the trematode *Heterophyes* sp. in flesh of mullet caught from brackish Egyptian waters. *J. Food Sci.* **35** (4), 386–8.

Hamed, M. G. E. & Elias, A. N. (1871). Effect of oxytetracycline and chlortetracycline on survival of the trematode *Heterophyes* sp. in the flesh of mullet caught in brackish Egyptian waters. *J. Food Sci.* **36** (1), 67–9.

Helfrich, P. & Banner, A. H. (1960). Hallucinatory mullet poisoning. *J. trop. med. Hyg.* **63** (4), 86–9.

Hewitt, G. C. (1971). Two species of *Caligus* (Copepoda, Caligidae) from Australian waters, with a description of some developmental stages. *Pacif. Sci.* **25** (2), 145–64.

Hines, R. S. & Spira, D. T. (1973). Acquired immunity of the mirror carp (*Cyprinus carpio* L.) to ichthyophthiriasis. *Refuah vet.* **30** (1), 17–9.

Hines, R. S. & Spira, D. T. (1974). Ichthyophthiriasis in the mirror carp *Cyprinus carpio* (L.). III. Pathology. *J. Fish Biol.* **6**, 189–96.

Hiscock, I. D. (1954). A new species of *Otobothrium* (Cestoda, Trypanorhyncha) from Australian fishes. *Parasitology* **44** (1–2), 65–70.

Hoffman, G. L. & Meyer, F. P. (1974). *Parasites of Freshwater Fishes: A Review of Their Control and Treatment.* T.F.H. Publications, Hong Kong.

Hutton, R. F. (1964). A second list of parasites from marine and coastal animals of Florida. *Trans. Am. Microsc. Soc.* **83** (4), 439–47.

Iversen, E. S., Chitty, N. & VanMeter, N. (1971). Some myxosporida from marine fishes in south Florida. *J. Protozool.* **18** (1), 82–6.

Janssen, W. A. (1970). Fish as potential vectors of human bacterial diseases. In *A symposium on diseases of fishes and shellfishes.* Ed. S. F. Snieszko. Am. Fish. Soc. Spec. Publ. **5**, 284–290.

Johnson, S. K. & Rogers, W. A. (1972). *Ergasilus clupeidarum* sp.n. (Copepoda, Cyclopoida) from clupeid fishes from the southeastern US with a synopsis of the North American *Ergasilus* species with a two-jointed first endopod. *J. Parasitol.* **58** (2), 385–92.

Johnson, S. K. & Rogers, W. A. (1973). Distributions of the genus *Ergasilus* in several Gulf of Mexico drainage basins. *Bull. Alabama Agric. Exp. Stn* 445, 1–74.

Kabata, Z. (1970). Crustacea as enemies of fishes. In *Diseases of Fishes*, vol. 1, ed. S. F. Snieszko & H. R. Axelrod. T.F.H. Publications, Jersey City, NJ.

Kean, B. H. & Breslau, R. C. (1964). Cardiac heterophyidiasis. In *Parasites of the Human Heart*, pp. 95–103, Grune & Stratton, New York.

Kelly, H. D. & Allison, R. (1962). Observations on the infestation of a fresh water fish population by a marine copepod (*Ergasilus lizae* Kroyer 1863). In *Proceedings of the Sixteenth Conference of the South East Game Fish Commission*, vol. 16, 236–39.

Kesteven, G. L. (1942). Studies in the biology of Australian mullet. I. Account of the fishery and preliminary statement of the biology of *Mugil dobula* Gunther. *Bull. Counc. sci. ind. Res.* Melbourne, **157**, 1–147.

Khalil, M. (1933). The life history of the human trematode parasite *Heterophyes heterophyes* in Egypt. Lancet, **225** (5740), 537.

Khalil, M. (1937). The life history of the human trematode parasite *Heterophyes heterophyes. C. R. Congress. Int. Zool., Lisbon* **3**, 1889–99.

Khalil, M. (1939). On the bionomics of the free living phase of *Cercaria Heterophyes*. *S. Yoshida Jub. Vol.* 161–6.

Kitao, T., Farrell, R. K. & Fukuda, T. (1973). Differentiation of salmon poisoning disease and Elokomin fluke fever: fluorescent antibody studies with *Rickettsia sennetsu*. *Am. J. Vet. Res.* **34** (7), 927–8.

Klontz, G. W. (1973). *Syllabus of fish Health Management. I. Fish Culture Methods. II. Fish Disease Diagnosis*. Texas A & M University – Sea Grant Program, TAMU-SG-74-401.

Kubota, S. S. & Takakuwa, M. (1963). Studies on the diseases of marine cultured fishes. I. General description and preliminary discussion of fish diseases in Mie Prefecture. *J. Fac. Fish. Pref. Univ. Mie-Tsu*, **6**, 107–24. (English translation, Fish. Res. Board. Can., Biol. Stn, Nanaimo, BC, 1966).

Kuntz, R. E. (1957). Experimental studies on the biology of *Heterophyes aequalis* Looss, 1902 in Egypt. *Proc. Helminthol. Soc. Wash.* **24** (2), 110–16.

Kuntz, R. E. & Chandler, A. C. (1956). Studies on Egyptian trematodes with special reference to the heterophyiids of mammals. I. Adult flukes with descriptions of *Phagicola longicollis* n. sp. *Cynodiplostomum namrui* n. sp. and a *Stephanoprora* from cats. *J. Parasitol.* **42**, 445–59.

Lahav, M. & Sarig, S. (1967). *Ergasilus sieboldi* Nordman infestation of grey mullet in Israel fish ponds. *Bamidgeh* **19** (4), 69–80.

Lahav, M. & Sarig, S. (1972). Control of unicellular parasites using formalin. *Bamidgeh* **24** (1), 3–11.

Laird, M. (1958*a*). Parasites of South Pacific fishes. I. Introduction, and Haematozoa. *Can. J. Zool.* **36**, 153-65.

Laird, M. (1958*b*). Parasites of South Pacific fishes. III. Trematodes from the Solomons and New Hebrides, with a description of *Daitreosoma parva* n. sp. (Monogenea) from Guadalcanal. *Can. J. Zool.* **36** (3), 441–6.

Lawler, A., Howse, H. D. & Cook, D. W. (1974). Silver perch, *Bairdiella chrysura*: New host for lymphocystis. *Copeia*, (1), 266–9.

Lee, F. O. & Cheng, T. C. (1970). The histochemistry of *Stellantchasmus falcatus* Onji and Nishio, 1915 (Trematoda: Heterophyidae) metacercarial cyst in the mullet *Mugil cephalus* L. and histopathological alterations in the host. *J. Fish Biol.* **2** (3), 235–43.

Leteux, F. & Meyer, F. P. (1972). Mixtures of malachte green and formalin for controlling *Ichthyophthirius* and other protozoan parasites of fish. *Prog. Fish Cult.* **34** (1), 21–6.

Lewis, D. H. (1973). *Predominant aerobic bacteria of fish and shellfish*. Texas agric. mech. TAMU-SG-73-401.

Lewis, D. H., Grumbles, L. C., McConnell, S. & Flowers, A. I. (1970). *Pasteurella*-like bacteria from an epizootic in menhaden and mullet in Galveston Bay. *J. Wildl. Dis.* **6** (3), 160–2.

Lightner, D. V. (1974). Case reports of ossifying fibromata in the striped mullet. *J. Wildl. Dis.* **10** (4), 317–20.

Loesch, H. (1960). Sporadic mass shoreward migrations of demersal fish and crustaceans in Mobile Bay, Alabama. *Ecol.* **41** (2), 292–8.

Lom, J. (1962). Trichodinid ciliates from fishes of the Rumanian Black Sea coast. *Parasitol.* **52** (1–2), 49–61.

Lom, J. & Lawler, A. R. (1973). An ultrastructural study on the mode of attachment in dinoflagellates invading gills of Cyprinodontidae. *Protistol.* **9** (2), 293–309.

Looss, A. (1902). Eine Revision der Fasciolidengattung *Heterophyes* Cobb. (Notizen zur Helminthologie Egyptens. 5). *Zentbl. Bakt.* **32** (12), 886–91.

Malmberg, G. (1970). The excretory systems and the marginal hooks as a basis for the systematics of *Gyrodactylus* (Trematoda, Monogenea). *Arkh. Zool.* **23** (1), 1–235.

Manter, H. W. (1957). Host specificity and other host relationships among the digenetic trematodes of marine fishes. In *Premier Symposium sur la Spécificité Parasitaire des Parasites de Vértebrés*, pp. 185–98. Institute of Zoology, University of Neuchatel.

Manter, H. W. (1969). Three new genera of hemiurid trematodes from Australian fishes. *J. Fish. Res. Board. Can.* **26** (4), 787–92.

Martin, W. E. (1959). Egyptian heterophyid trematodes. *Trans. Am. microsc. Soc.* **78** (2), 172–81.

Martin, W. E. (1973*a*). A new subfamily, two new genera, and three new species of haploporid trematodes. *Proc. Helminthol. Soc. Wash.* **40** (1), 112–17.

Martin, W. E. (1973*b*). Life history of *Saccocoelioides pearsoni* n. sp. and the description of *Lecithobotrys sprenti* n. sp. (Trematoda: Haploporidae). *Trans. Am. microsc. Soc.* **92** (1), 80–95.

Martin, W. E. (1973*c*). *Isorchis manteri* sp. n. from Australian mullet and a key to haploporid trematode genera with two testes. *Proc. Helminthol. Soc. Wash.* **40** (2), 237–40.

Martin, W. E. (1973*d*). A new genus and species of haploporid trematode (Haploporidae: Trematoda) from Australian mullet. *Bull. South. California Acad. Sci.* **72** (3), 166–8.

Martin, W. E. (1974). *Paralecithobotrys brisbanensis* sp. n. from Australian mullet (Trematoda: Haploporidae). *Proc. Helminthol. Soc. Wash.* **41** (1), 16–18.

Meglitsch, P. A. (1960). Some coelozoic Myxosporidia from New Zealand fishes. I. General and family Ceratomyxidae. *Trans. R. Soc. NZ* **88** (2), 265–365.

Merritt, S. V. & Pratt, I. (1964). The life history of *Neoechinorhynchus rutili* and its development in the intermediate host (Acanthocephala: Neoechinorhynchidae). *J. Parasitol* **50** (3), 394–400.

Meyer, F. P. (1970). Seasonal fluctuations in the incidence of disease on fish farms. In *A Symposium on Diseases of Fishes and Shellfishes*, ed. S. F. Snieszko, pp. 21–29. Am. Fish. Soc. Spec. Publ 5.

Mikailov, T. K. (1958). Parasitofauna of *Mugil saliens* Risso of the Caspian Sea. *Zool. Zh.* **37** (3), 373–8 (in Russian).

Mimioglu, M. & Sayin, F. (1957). Ankara köpeklerinde tesbit edilen ilk *Heterophyes heterophyes* (v. Siebold, 1852) stiles re Hassal 1900. *Vak'asi Vet. Fac. Dergisi Ankara Univ.* **4**, 1–5 (in Turkish, German summary).

Mires, D. (1970). Preliminary observations on the effects of salinity and temperature of water changes on *Mugil capito* fry. *Bamidgeh* **22** (1), 19–24.

Monod, T. (1926). Les Gnathiidae, essai monographique (morphologie, biologie, systématique). *Mém. Soc. Sci. Nat. Phys. Maroc* **13**, 1–667.

Moore, R. H. (1974). General ecology, distribution and relative abundance of *Mugil cephalus* and *Mugil curema* on the south Texas coast. *Contrib. Mar. Sci.* **18**, 241–55.

Moore, R. H. (1976). Observations on fishes killed by cold at Port Aransas, Texas, 11–12 January 1973. *The Southwestern Naturalist* **20** (4), 461–66.

Moravec, F. & Baruš, V. (1971). Studies on parasitic worms from Cuban fishes. *Věst. čsl. Spol. Zool.* **35** (1), 56–74.

Nigrelli, R. F. (1937). Further studies on the susceptibility and acquired immunity of marine fishes to *Epibdella melleni*, a monogenetic trematoda. *Zoologica, NY* **22** (2), 185–92.

Nigrelli, R. F. (1943). Causes of diseases and death of fishes in captivity. *Zoologica, NY* **28** (22), 203–16.

Nigrelli, R. F. & Breder, C. M., Jr. (1934). The susceptibility and immunity of certain marine fishes to *Epibdella melleni*, a monogenetic trematode. *J. Parasitol.* **20** (5), 259–69.

Nikolaeva, V. M. & Naidenova, N. N. (1964). Nematodes of pelagic and benthopelagic fish of seas of the Mediterranean basin. *Tr. Sevastopol'. Biol. Stn.* **17**, 125–58 (in Russian).

Nikolaeva, V. M. & Prukhin, A. M. (1968). Study of helminths of fish in the Gulf of Mexico. Studies of Central American Seas (Yankovskaya). *Kiev* (*Nankova Dumka*) **2**, 126–49.

Norris, D. E. & Overstreet, R. M. (1976). The public health implications of *Thynnascaris* nematodes from shellfish. *J. Milk Food Technol.* **39** (1), 47–54.

Onji, Y. (1915). On a species of Heterophyes whose encysted cercaria lives in *Mugil japonicus*. *Tokyo Med. News*, **1918**, 875–83 (in Japanese).

Oshima, T. (1972). *Anisakis* and anisakiasis in Japan and adjacent area. *Prog. Med. Parasitol. Jap.* **4**, 301–93.

Overstreet, R. M. (1968). Parasites of the inshore lizardfish, *Synodus foetens*, from south Florida, including a description of a new genus of Cestoda. *Bull. Mar. Sci.* **18** (2), 444–70.

Overstreet, R. M. (1969). Digenetic trematodes of marine teleost fishes from Biscayne Bay, Florida. *Tulane Stud. Zool.* **15** (4), 119–75.

Overstreet, R. M. (1971). Some adult digenetic trematodes in striped mullet from the northern Gulf of Mexico. *J. Parasitol.* **57** (5), 967–74.

Overstreet, R. M. (1973). Some species of *Lecithaster* Lühe, 1901 (Digenea: Hemiuridae) and related genera from fishes in the northern Gulf of Mexico. *Trans. Am. Microsc. Soc.* **92** (2), 231–40.

Overstreet, R. M. (1974). An estuarine low-temperature fish-kill in Mississippi, with remarks on restricted necropsies. *Gulf Res. Rep.* **4** (3), 328–50.

Paperna, I. (1964*a*). The metazoan parasite fauna of Israel inland water fishes. *Bamidgeh* **16** (1–2), 3–66.

Paperna, I. (1964*b*). Parastic Crustacea (Copepoda and Branchiura) from inland water fishes of Israel. *Israel J. Zool.* **13**, 58–68.

Paperna, I. (1975). Parasites and diseases of the grey mullet (Mugilidae) with special reference to the seas of the Near East. *Aquaculture* **5** (1), 65–80.

Paperna, I. & Lahav, M. (1971). New records and further data on fish parasites in Israel. *Bamidgeh* **23** (2), 43–52.

Paperna, I. & Lahav, M. (1974). Mortality among grey mullets in a seawater pond due to caligiid parasitic copepod epizootic. *Bamidgeh* **26** (1), 12–15.

Paperna, I. & Lahav, M. (1975). Parasites of fish of the hypersaline Bardawil Lagoon, North Sinai. *Rapp. Comm. Mer. Mediterr.* **23** (3), 127–8.

Parenzan, P. (1967). *Myxobolus mugilis* e *Myxobolus branchialis*, due nuovi missosporidi parassiti di *Mugil chelo* dello Jonio. *Biol. Soc. Nat. Napoli* **75**, 3–8.

Parker, J. C., Klussman, W. G., Holcomb, H. W., Jr. & McNeill, J. C. (1971). Distribution of aquatic macrofauna in a marsh on West Galveston Bay, Texas and possible effects thereon resulting from impoundments for shrimp culture. In *Proceedings of the Second Annual Workshop, Wildlife and Mariculture Society*, Galveston, Texas Jan. 28–9 1971, 97–117.

Parker, R. R., Kabata, Z., Margolis, L. & Dean, M. D. (1968). A review and description of *Caligus curtus* Müller, 1785 (Caligidae: Copepoda), type species of its genus. *J. Fish. Res. Board. Can.* **25** (9), 1923–69.

Parveen, I. (1971). *Gyrometra mugilitaenia* sp. nov. (Cestode: sub-family Gyromatrinae Yamaguti 1954) from the body cavity of a *Mugil* sp. *Rec. Zool. Surv. Pak*, **2** (1), 145–53.

Pearse, A. S. (1952). Parasitic Crustacea from the Texas coast. *Publ. Inst. Mar. Sci. Univ. Tex.* **2** (2), 5–42.

Pearse, L. (1972). A note on a marine trichodinid ciliate parasitic on the skin of captive flatfish. *Aquaculture* **1**, 261–6.

Pearson, J. C. (1964). A revision of the subfamily Haplorchinae Looss, 1899 (Trematoda: Heterophyidae). I. The Haplorchis group. *Parasitol.* **54** (4), 601–76.

Petter, A. J. (1969). Enquête sur les nématodes des sardines pêchées dans la région nantaise. Rapport possible avec les granulomes éosinophiles observés chez l'homme dans la région. *Ann. Parasitol. Hum. Comp.* **44** (1), 25–35.

Plump, J. A., Schachte, J. H., Gaines, J. L., Peltier, W. & Carroll, B. (1974). *Streptococcus* sp. from marine fishes along the Alabama and northwest Florida coast of the Gulf of Mexico. *Trans. Am. Fish. Soc.* **103** (2), 358–61.

Pogorel'tseva, T. P. (1964). Material on the study of protozoan parasites of fish of the Black Sea. *Tr. Ukrainsk. Respub. Nauch. Obsh. Parazit*, **3**, 16–29 (in Russian).

Purdom, C. E. & Howard, A. E. (1971). Ciliate infestations: a problem in marine fish farming. *J. Cons. Perm. Int. Explor. Mer.* **33** (3), 511–14.

Quick, J. A., Jr. & Henderson, G. E. (1974). Effects of *Gymnodinium breve* red tide on fishes and birds: a preliminary report on behaviour, anatomy, haematology, and histopathology. In *Proceedings of Gulf Coast Regional Symposium on Diseases of Aquatic Animals*, ed. R. L. Amborski, M. A. Hood & R. R. Miller, LSU-SG-74-05. Pp. 85–113.

Rawson, M. V., Jr. (1973*a*). Two new species of Gyrodactylus (Trematoda, Monogenea) from the Georgia coast. *Proc. Helminthol. Soc. Wash.* **40** (2), 180–3.

Rawson, M. V., Jr. (1973*b*). The development and seasonal abundance of the parasites of striped mullet, *Mugil cephalus* L., and mummichogs, *Fundulus heteroclitus* (L.). Dissertion Univ. Georgia.

Ray, S. M. & Wilson, W. B. (1957). Effects of unialgal and bacteria-free cultures of *Gymnodinium brevis* on fish. *Fish. Bull. Fish Wildl. Serv. US*, **57** (123), 469–96.

Reshetnikova, A. V. (1955). Parasite fauna of *Mugil cephalus* in the Black Sea. *Tr. Karadah. nauch. Stn. B*, 71–95.

Roberts, J. L. (1974). Temperature acclimation and behavioral thermoregulation in cold-blooded animals. *Fed. Proc., Fed. Am. Soc. Exp. Biol.* **33** (10), 2155–61.

Roberts, L. S. (1969). *Ergasilus arthrosis* n. sp. (Copepoda: Cyclopoida) and the taxonomic status of *Ergasilus versicolor* Wilson, 1911, *Ergasilus elegans* Wilson, 1916, and *Ergasilus celestis* Mueller, 1936, from North American Fishes. *J. Fish. Res. Board. Can.* **26**, 997–1011.

Roberts, L. S. (1970). *Ergasilus* (Copepoda: Cyclopoida) revision and key to species in North America. *Tran. Am. microsc. Soc.* **89**, 134–61.

Russel, F. S. (1925). A new species of *Caligus* from Egypt, *Caligus pageti* sp. n. *Ann. Mag. Nat. Hist.* (*S*) **9**(90) 15, 611–18.

Sarig, S. (1968). Possibilities of prophylaxis and control of ectoparasites under conditions of intensive warm-water pondfish culture in Israel. *Bull. Off. Int. Épizoot.* **69** (9–10), 1577–90.

Sarig, S. (1971). The Prevention and Treatment of Diseases of Warmwater Fishes under Subtropical Conditions, with Special Emphasis on Intensive Fish Farming. In *Diseases of Fishes*, ed. S. F. Snieszko & H. R. Axelrod, T.F.H. Publications Inc., Ltd, Hong Kong.

Sarig, S. & Lahav, M. (1959). The treatment with lindane of carp and fish ponds infected with the fish-louse, *Argulus*. *Proc. Tech. Pap. Gen. Fish. Counc. Mediterr.* **5**, 151–6.

Saunders, D. C. (1958). The occurrence of *Haemogregarina bigemina* Laveran and Mesnil, and *H. dasyatis* n. sp. in marine fish from Bimini, Bahams, BWI. *Trans. Am. Microsc. Soc.* **77** (4), 404–12.

Saunders, D. C. (1960). A survey of the blood parasites in the fishes of the Red Sea. *Trans. Am. microsc. Soc.* **79** (3), 239–52.

Saunders, D. C. (1964). Blood parasites of marine fish of southwest Florida, including a new haemogregarine from the menhaden, *Brevoortia tyrannus* (Latrobe). *Trans. Am. microsc. Soc.* **83** (2), 218–25.

Sawyer, R. T., Lawler, A. R. & Overstreet, R. M. (1975). Marine leeches of the eastern United States and the Gulf of Mexico with a key to the species. *J. Nat. Hist.* **9** (6), 633–67.

Schäperclaus, W. (1954). *Fischkrankheiten.* Akademic Verlag, Berlin.

Scott, D. F. (1972). Host-parasite relationships: metazoan ectoparasites of grey mullet, *Mugil cephalus. ASB Bull.* **19** (2), 98.

Sharma, P. N. Gupta, A. N. (1970). *Lecithobotrys vitellosus* n. sp. (Haploporidae: Trematoda) from India. *Riv. Parassit.* **31** (3), 175–8.

Shen C.-J. (1957). Parasitic copepods from fishes of China. 1. Cyclopoida. *Acta Zool. Sinica,* **9** (4), 297–327.

Shulman, S. S. (1957). Pathogenicity of the myxosporidian *Myxobolus exiguus* and epizootics produced by it. In *Parasites and Diseases of Fish.* ed. G. K. Petrushevskii, *Izo. uses. Inst. Ozern. Rechn, Ryb. Khoz., Leningrad,* **42**, (in Russian, translated by Nat. Sci. Found. & US Dept. Interior 1960).

Shulman, S. S. (1966). *The Myxosporidian Fauna of the USSR.* Nauka, Moscow-Leningrad (in Russian).

Sindermann, C. J. (1972). Some biological indicators of marine environmental degredation. *J. Wash. Acad. Sci.* **62** (2), 184–9.

Sindermann, C. J. (ed.) (1974). *Diagnosis and Control of Mariculture Diseases in the United States.* Technical Series Report No. 2, Middle Atlantic Coastal Fisheries Center.

Skinner, R. (1974). Parasites of the striped mullet, *Mugil cephalus,* from Biscayne Bay, Florida, with descriptions of a new genus and three new species of trematodes. Thesis, Univ. Miami, Coral Gables.

Snieszko, S. F. (1974). The effects of environmental stress on outbreaks of infectious diseases of fishes. *J. Fish Biol.* **6**, 197–208.

Snieszko, S. F., Bullock, G. L., Hollis, E. & Boone, J. G. (1964). *Pasteurella* sp. from an epizootic of white perch (*Roccus americanus*) in Chesapeake Bay tidewater areas. *J. Bact.* **88** (6), 1814–15.

Starr, T. J. (1958). Notes on a toxin from *Gymnodinium breve. Tex. Rep. Biol. Med.* **16**, 500–7.

Steinitz, H. & Ben Tuvia, A. (1972). Fishes of the Suez Canal. *Israel J. Zool.* **21** (3–4), 385–9.

Swingle, H. A. & Bland, D. G. (1974). A study of the fishes of the coastal watercourses of Alabama. *Ala. Mar. Resour. Bull.* **10**, 17–102.

Sylvester, J. R. (1974). Thermal response of juvenile Hawaiian mullet *Mugil cephalus* (L.) to acclimation time and fluctuating low temperatures. *J. Fish Biol.* **6** (6), 791–6.

Travassos, L. P., Kohn, A. & Da Silva Motta, C. (1863). Excursao à Ilha de Marambaia, Estado do Rio de Janeiro. *Atas Soc. Biol. Rio de Janeiro* **7** (3), 4–9.

Trilles, J. P. (1965). Spécificité parasitaire chez les isopodes Cymothoidae méditerranéens. Note préliminaire. *Vie Milieu* **15** (1), 105–16.

Venkataramiah, A., Lakshmi, G. J. & Gunter, G. (1974). Studies on the effects of salinity and temperature on the commercial shrimp, *Penaeus aztecus* Ives, with special regard to survival limits, growth, oxygen consumption and ionic regulation. *Vicksburg, US Engineer Waterways Experiment Station Contract Report H-74-2.*

Vlasenko, P. V. (1931). Zur Helminthofauna de Schwarzmeerfische. *Trudȳ Karadah. Nauch. Sta.* **4**, 88–136 (in Russian, German summary).

Walliker, D. (1968). The nature of the iodinophilous vacuole of myxosporidian spores, and a proposal to synonymize the genus *Myxosoma* Thélohan, 1892 with the genus *Myxobolus* Bütschli, 1882. *J. Protozool.* **15**, 571–5.

Wang, K. N. (1964). Parasitic crustaceans of freshwater fishes from Kiangsu and Shanghai. *Tung Wu Hsiieh Pao,* **16** (3), 465–73 (in Chinese).

Wells, W. H. & Randall, B. H. (1956). New hosts for trematodes of the genus *Heterophyes* in Egypt. *J. Parasitol.* **42** (3), 287–92.

White, H. C. (1940). 'Sea lice' (*Lepeophtheirus*) and death of salmon. *J. Fish. Res. Board, Can.* **5**, 172–5.

Wilkie, D. W. & Gordin, H. (1969). Outbreak of cryptocaryoniasis in marine aquaria at Scripps Institution of Oceanography. *Calif. Fish Game,* **55** (3), 227–36.

Williams, E. H. (1972). Parasitic infestation of some marine fishes before and after confinement in feeding cages. *Ala. Mar. Resour. Bull.* **8**, 25–31.

Williams, E. H., Jr., Gaines, J. L., Jr. & Phelps, R. P. (1974). Parasites of pompano, *Trachinotus carolinus* (L.) and striped mullet, *Mugil cephalus* L. before and after cage culture. In *Proceedings of Gulf Coast Regional Symposium on Diseases of Aquatic Animals,* ed. R. L. Amorski, M. A. Hood & R. R. Miller, LSU. LSU-SG-74-05, 125–35.

Wilson, C. B. (1905). North American parasitic copepods belonging to the family Caligidae. I. The Caliginae. *Proc. US Natn. Mus.* **28**, 479–672.

Wilson, C. B. (1911). North American parasitic copepods belonging to the family Ergasilidae. *Proc. US Natn. Mus.* **39**, 263–400.

Witenberg, G. G. (1929). Studies on the trematode-family Heterophyidae. *Ann. trop. Med. Parasitol.* **23** (2), 131–9.

Yamaguti, S. (1938). Studies on the helminth fauna of Japan, 24. Trematodes of fishes, V. *Jap. J. Zool.* **8** (1), 15–74.

Yamaguti, S. (1970). *Digenetic Trematodes of Hawaiian Fishes.* Keigaku Publishing Co. Ltd. Tokyo.

Yashouv, A. (1959). On the biology of *Lernaea* in fish ponds. *Bamidgeh* **11** (4), 80–9.

Yashouv, A. (1972). Efficiency of mullet growth in fishponds. *Bamidgeh* **24** (1), 12–25.

Zhukov, E. V. (1971). New trematodes of marine and freshwater fish from the Japan and Yellow Sea basins. *Parazitol.* **5** (2), 155–61 (in Russian).

Zwerner, D. E. & Lawler, A. R. (1972). Some parasites of Chesapeake Bay Fauna. In *A check list of the biota of lower Chesapeake Bay. With inclusions from the upper bay and the Virginia Sea.* Compiled by M. L. Wass, *Spec. scient. Rep. Va. Inst. mar. Sci.* **65**, 78–94.

Note added in proof: This manuscript was submitted for publication in 1975. Thus, all information presented herein reflects the state of knowledge to that date. Since 1975 extensive studies have been carried out on wild and cultured mullets in Israel and the southern United States by the authors. Some of these results are already published and others are now being prepared for publication. A review of diseases of cultured mullets was submitted for distribution by the FAO cooperative programme of research on aquaculture (COPRAQ).

Species index

Species index

Species index

Subject index